URBAN STORM DRAINAGE

URBAN STORM DRAINAGE

Proceedings of the International Conference held
at the University of Southampton, April, 1978

Edited by P. R. Helliwell
Department of Civil Engineering,
University of Southampton

PENTECH PRESS
London : Plymouth

First published, 1978
by PENTECH PRESS LIMITED
Estover Road, Plymouth, Devon

ISBN 0 7273 2102 1
© The several contributors named
   in the list of contents, 1978

2nd impression, 1979

British Library Cataloguing in Publication Data

Urban storm drainage.
   1. Storm sewers — Congresses
   I. Helliwell, P R
   628'.24   TD665

   ISBN 0 7273 2102 1

Printed in Great Britain by
Billing & Sons Limited,
Guildford, London and Worcester.

PREFACE

Planners and designers of storm water systems show a continuing preference for the established methods, and make only limited use of recent research results. The shortcomings of some design procedures are well-known. The investment in new storm sewer systems is very large, so that the penalty for inaccurate design is also large.

Research into various topics bearing more or less directly on urban hydrology is continuing in many countries. There is a considerable British programme of investigation, a major portion of which is under the sponsorship of the Department of the Environment.

The papers deal with recent applicable research, including water quality considerations, costs and benefits and meteorological considerations, as well as papers on simulation models and design methods for urban storm drainage systems. Papers came mainly from authors in the U.K., North America and Europe, particularly the Scandinavian countries. Six papers received too late for printing in the main body of the book are included in an appendix.

This publication will be of interest to researchers and to practising civil engineers, in that it presents the latest views and results in this important and topical field.

I wish to acknowledge the help and advice received from the members of the Department of the Environment/National Water Council Working Party on the Hydraulic Design of Storm Sewers, and for the interest shown by the International Water Research Association, particularly the group in the U.S.A.

P. R. Helliwell

CONFERENCE OPENING

THE LORD NUGENT OF GUILDFORD

This Conference has attracted much interest and drawn dele-
gates not only from our country, but from all over the world.
I extend a very warm welcome to visitors from overseas, many
of whom have travelled long distances to attend.

The papers which are before the Conference are impressive. I
would like to congratulate Dr. Helliwell and Southampton
University on organising this Conference. We believe that
Britain may have the best, and the strongest system of water
management to be found anywhere in the world. We have by the
1973 Act combined the whole hydrological cycle under one
single management structure. There are now 10 Regional
Authorities which cover England and Wales, which have
responsibility for every aspect of water management, rivers,
supply, sewage disposal and treatment, fisheries and
recreation. At national level, the National Water Council
co-ordinates the activities of the 10 Regional Authorities.
This system has been in operation for 4 years and is leading
to a more efficient management of water generally.

The Planner is outside everybody's control, possibly his own
as well! I did wonder, Dr. Helliwell, why you didn't invite
any of them here. Perhaps you did, and they didn't dare
come to meet all you engineers. But, nevertheless, there is
no doubt about it that the problems you experts contend with
in your respective spheres, cities and countries were very
largely the faults in town planning, in the past. If only
our forebears had had the sense not to build towns and cities
in flood plains and not to build houses in depressions which
naturally flood, drainage would be much easier. The most
common problem that we face is universal, and is the urban
area problem in big cities. In this country, flooding occurs
most years in the built up area of cities which have been
extended decade by decade after the original major sewage
system of the city was built in the last century. As further
additions are made, following further development of the city,
connections into the original system are made and inevitably
surcharging and flooding occurs with every minor storm. This
is one of the major problems to be considered by the Conference.

There are occasionally dramatic storms in Britain. 10 years ago a storm produced 180 mm. of rain in 24 hours. This was quite a record in that particular area in the south east of England, and caused serious flooding. The tributaries of the River Thames were overtopped and some 10,000 houses on the outskirts of London were flooded as a result of it. I had direct responsibility, as the Chairman of the Thames Conservancy, for river management, and I felt that the degree of loss and discomfort suffered by the residents in the area was such, that a flood relief scheme was needed to provide a new drainage channel which would give protection against a repeat. In countries susceptible to tropical cyclones, four times as much rain can occur. Indeed, I saw that even last month a cyclone in Australia produced 900 mm. in 4 days.

The solution of urban drainage problems lies in bringing together the expert studies of a wide range of engineers and scientists - such as you have assembled in the papers presented to the Conference.

Meteorologists must work out predictions of the return period, the intensity and duration of storms. Secondly, the engineers and hydrologists must estimate the capacity of the existing urban drainage system, plus surface flows when the sewers are surcharged, plus surface water depths in the low-lying areas. Possible raw sewage discharges to streams and ditches and the pollution intensity must also be found. The structural engineers must work out the structural damage to buildings when flooding takes place. Economists and engineers must work out the cost of increasing the capacity of the system to protect the local community from a recurrence of their loss and discomfort. And then finally, the Government of the country concerned must take the decision as to whether it is justified. This great range of highly expert technical work which you perform provides the facts on which these vital decisions must be taken. Resources in every country are always limited and, therefore, every scheme, however desirable, cannot be done everywhere. So it is essential that it should be possible to identify the right priorities.

The common theme lies in perfecting the methods of predicting the incidence of storms, the methods of protecting communities, especially cities, from the resulting damage, and the methods of estimating the cost of these complex operations. On these highly expert measurements and calculations depends the decision whether a particular flood relief scheme shall be executed, and that, if it is, it shall be adequate. I have very much pleasure in declaring this Conference open.

CONTENTS

AN ENGINEERING EVALUATION OF STORM PROFILES FOR THE DESIGN
OF URBAN DRAINAGE SYSTEMS IN THE U.K.

A.J. Price   R.A. Howard
Partner   Section Engineer
John Taylor & Sons   John Taylor & Sons

INTRODUCTION

When a British Engineer is confronted by the design of any,
but the simplest,urban storm water U.K. drainage system, he
will invariably use the TRRL hydrograph method to compute
the design flows and pipe diameters within the system. In
addition to the basic catchment input data of impermeable area,
pipe length, gradient and roughness this method requires
the profile of a suitable design storm.

Prior to 1975, unless the engineer devised his own profiles
from local experience, his choice would have been limited to
those profiles produced for the 1963 edition of Road Note 35
(RN35(1))which had return period as the sole variable.
However, with the publication in 1975 of the NERC Flood
Studies Report (FSR), profiles became available which varied
not only with return period, but with duration, location,
catchment area and peakedness, thus presenting the engineer
with an infinite number of storm profiles from which to
select his design storm. In 1976, this situation was eased
by the publication of the 2nd edition of Road Note 35 (RN35(2))
which offered a simplified set of profiles that varied only
with return period and location.

When the FSR profiles first became available an extensive
investigation into the elimination of the flooding problems
of Torquay was being carried out with the TRRL hydrograph
method, using the RN35(1) profiles for the design storms.
The design of the relief works for a large and complex
catchment such as Torquay could be expected to benefit from
the improved accuracy of flow prediction available by using
the FSR profiles as the design storms. Therefore, these
were immediately applied to the computer network currently
in use.

1

However, this first series of runs showed no marked differences in the calculated flows between the FSR and the RN35(1) results and in most cases the resultant differences of pipe diameter was generally insignificant.   Thus, the RN35(1) storms were predicting with reasonable accuracy the FSR results, and furthermore, doing it with considerable economy of computer and design office time, when compared with the procedure required for using the FSR storms.

Having made the assumption, in the absence of any flow measurement data to check the accuracy of the FSR results, that the FSR storms gave the best estimate of design flows, the emphasis of the evaluation changed to determining how well the RN35(1) storms predicted the FSR design flows.

With the publication of the 2nd edition of Road Note 35 in 1976, a set of profiles became available which superseded the RN35(1) profiles, in so far as the RN35(2) storms were a summary of the FSR results.   It therefore became necessary to repeat the work, using the RN35(2) storms.

The work with RN35(1) storms is now only of historical interest, and this paper concentrates on the results obtained using the FSR and RN35(2) profiles, although naturally our work on the RN35(2) storm was influenced by the results from the RN35(1) storms.   In addition, the effect of the FSR and RN35(2) storms on two other catchments is reported, together with some work carried out calculating the volumes of balancing tanks.

ADVICE ON RAINFALL PROFILES TO THE ENGINEER

The profiles given by RN35(1) were derived mainly from the work of Bilham and generally represent, in terms of FSR profiles, 75 percentile summer storms for lowland southern England. These profiles only vary for return period and have a duration of 120 mins.

From the information given in FSR Vol II and V, profiles can be derived which vary for duration, location, peakedness, catchment area, as well as return period.   At present the Meteorological Office advise that the design storm is the 50 percentile storm of such duration that gives the peak flow in the pipeline being considered. Initially the Meteorological Office advised that the storm with this critical duration was the storm of length $3 \times (t_c - 5)$min, where t_c is time of concentration. Subsequently, this was revised, and engineers are now advised to use the storm with duration of $2 \times t_c$.

Finally, with the publication in 1976 of RN35(2) a set of profiles of 120 mins duration became available which varied for location throughout the U.K.   These profiles were deemed accurate enough to use on catchments with tc's of less than 60 mins and where peak flows in pipes only are required. For lowland southern England these profiles approximate to the 90 percentile FSR profile with a duration of 120 mins.

CATCHMENTS USED IN THE INVESTIGATION

The location of the three catchments: Torquay, Horsell and Theale, where the various profiles were tested, is shown in Fig. 1.   All these catchments are located in lowland, southern England and therefore the results should not show any significant variation between RN35(1) and RN35(2) because of location.   However, it also means that the factor for variation in location of the RN35(2) storms is not really tested and checked against the FSR storms.

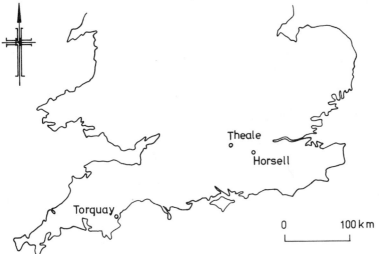

Figure 1. Location of Catchments

PROFILES USED IN INVESTIGATION

The various profiles used in the investigation are given in Table 1.

| | 50 percentile FSR storms-duration in Mins | | | | | | | | | RN35 (1) | RN35 (2) |
	15	30	36	60	90	120	240	270	480		
Torquay	*	*		*	*	*		*		*	*
Horsell	*		*	*		*				*	*
Theale	*	*		*		*			*		*

Table 1. Storm Profiles used in the investigation

The profiles used on the Torquay network are shown in Fig. 2.

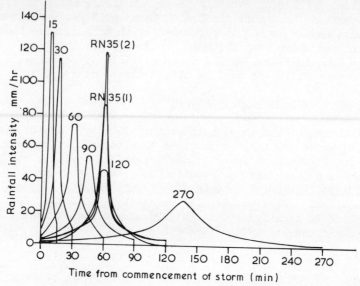

Figure 2. Profiles used at Torquay

TORQUAY

The Torquay Catchment

The Torquay catchment comprises three catchments all of which
drain to a 2.14m dia. tunnel, constructed in the 1870's, which
outfalls into the sea.   The main features are shown in Fig.3.
Characteristically, the catchments are steep sided valleys
in which the trunk sewers run along the lines of the old streams
and virtually all the drainage in each valley, including flows
from open areas, drains to the combined sewers.   Torquay is
typical of many other British towns in that the basic drainage
system was constructed in the 19th century and since then the
system has become progressively overloaded by subsequent
development.   Over the years, whenever flooding or gross
pollution has occurred, the system has been extended or
modified until today it comprises trunk mains of dual or even
triple pipes, storm overflow chambers with overflow pipes
which rejoin the main flow further downstream, and a cross
catchment overflow.

By the early 1970's it was obvious that immediate action was
required to relieve the flooding, which was by now occurring
in many parts of the catchment, and in particular with
increasing regularity and severity in the town centre, which
was flooded three times in 1973.   Figs. 4 and 5 show the
flood of the16th July 1973 in the Town Centre. It is
estimated, using Bilhams results, that this was produced by a

storm of 3 year return period.

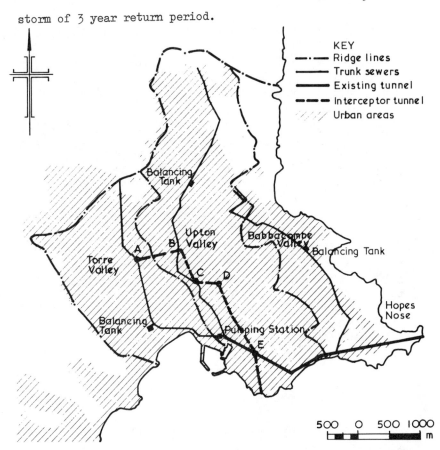

Figure 3. The Torquay Catchment

To identify the causes of flooding a detailed network comprising 154 sub-areas, with a total imp. area of 470 ha, and 325 pipe-lines was built up. After checking that results from the network correlated reasonably well with the known locations of flooding, the network was modified to allow for future development and to model our proposals for the elimination of flooding.

One of the design requirements on the catchment was the provision of a graded protection, in terms of the design storm return period, for various parts of the town. The town centre would receive 10 year protection, areas of relatively dense domestic development, already subject to flooding, 5 year protection, and the outer areas of the town 2 year protection.

To ensure that the 5 and 10 year areas were provided with their design protection, it was necessary to consider the

Figure 4.   Flooding in Torquay town centre
16th July, 1973

Figure 5. Flooding in Torquay town centre
16th July, 1973

effect of run off from the upper areas not being carried away by the sewers in those areas when using the design storm for the lower areas.   Some of this run off will pond in local low spots but most will travel overland on roads, or in stream channels, and eventually enter an area with a higher return period of design protection, where unless it is carried away by the storm sewers, would cause flooding at lower than the design return period.   Detailed consideration was therefore given to the paths such overland flow would take and these were added to the sewer network.

About 9% of the total catchment area comprised open areas, and as streams from these areas entered the sewerage system an estimate had to be made of their contribution.   This was done by giving them impermeable areas, and times of entry, using the data derived from the Flood Studies Report. To model the stream flow conditions, notional pipes were designed, which had the same characteristics as the channels, viz . length, velocity and capacity.

The effect of these flows from the open areas and the overland flow from upper areas was to produce a secondary network of channels which considerably increased the complexity of the overall network.

The Profiles used at Torquay

Table 1 lists the storms used at Torquay: RN35(1), RN35(2) and FSR 50 percentile storms of durations 15,30,60,90,120 and 270 mins.   The results given are for 5 year storms.   The 90 and 270 mins. storms were used in our early work when initially we tried to select the critical storms on the basis of 3 (tc -5)min; these storms being suited to the outfall and a major Storm Overflow Chamber on the system. The other profiles were used when it was realised that knowledge of pipe flow through the catchment was needed.   This required storms of critical durations suited to all tc's and the use of a range of storms.

Flow results at Torquay

With 325 pipes in the catchment, of which 142 were unrestricted and free to calculate flows showing the affect of the various profiles, a difficulty arises in presenting the results.   However, 62 of these pipes in the catchment are upper branches with tc's of less than 5 mins and an examination of these pipes showed that the 15 min FSR storm gave the maximum flows.   Table 2 gives the flows in pipes for two trunk sewers in the catchment; the 100 line running down Upton Valley and the 200 line down the Torre Valley and its extension into the interceptor (500 and 511) shown as ABCDE on Fig. 3.

Pipe No	tc mins	FSR storms-duration in mins.						RN35 (1)	RN35 (2)
		15	30	60	90	120	270		
100.00	5.8	528	526	438	338	313	202	477	577
100.01*	6.9	2079	2038	1619	1206	1116	712	1791	2194
100.02*	7.4	2211	2244	1968	1698	1116	1002	2083	2326
100.03*	8.2	1959	1876	1625	1450	1576	1192	1737	1957
100.04*	9.1	1777	1745	1601	1482	1393	1314	1663	1802
100.05	9.4	1440	1457	1396	1348	1446	1277	1418	1484
100.06*	10.0	1498	1513	1454	1392	1332	1302	1486	1532
100.07	10.5	3665	3983	3774	3477	1373	2892	3862	4139
100.08	11.7	5819	6405	6015	5304	3405	4043	6226	6902
100.09	11.8	5924	6671	6225	5487	5144	4163	6431	7093
100.10	12.5	5934	6751	6343	5595	5323	4248	6545	7175
100.11 $_1$	12.6	6352	7480	6996	6163	5984	4607	7195	7880
100.12 ^{1}	13.6	7290	8498	8193	7502	7695	6105	8525	9395
100.14	55.6	6044	6973	7710	7273	7153	6562	7591	8565
100.15 $_2$	56.1	6638	7171	7734	7646	7746	7256	7613	7521
100.17 ^{2}	57.3	8794	10223	10700	10423	10577	9975	10910	11080
100.18	59.6	5449	5578	5695	5634	5614	5525	5693	5801
200.00	15.4	177	229	234	208	206	152	234	257
200.01	16.1	229	281	294	267	266	196	293	323
200.02	16.9	1926	1894	1573	1305	1148	872	1728	1977
200.03	17.3	1863	1785	1473	1212	1146	835	1605	1939
200.04	18.8	2222	2390	2049	1684	1441	1132	1984	2360
200.05	19.3	2401	2373	2038	1677	1592	1127	2189	2574
200.06	21.7	3554	3594	3166	2592	2463	1703	3364	3922
200.07	22.4	3765	3922	3314	2787	2637	1803	3645	4265
200.08	22.5	3528	3631	3238	2675	2559	1778	3440	3965
500.01	30.2	4709	4758	4151	3412	3247	2336	4459	5203
511.01	85.1	10617	12812	13608	13102	13158	12230	14140	13245
511.02	89.8	9962	12234	14520	14156	14879	13507	14378	15189

Flows in l/sec
Peak flow from critical FSR storm underlined
* Overflow on pipe discharges into stream channels
1 Upton Balancing tank downstream of pipe 100.12
2 Overflow to interceptor
Table 2. Peak flows from storms tested at Torquay

The values from the 100 line have been given to show the
typical results on a complex length of sewer. Despite the
effect of overflows and a balancing tank on this sewer line,
it can be seen that the duration of FSR storm giving maximum
flow increases with the length of tc. Thus, on this
particular line the 15 min FSR gave peak flow, with one
marginal exception, up to tc's of 9.1 min, between 9.1 min
and 15 min the 30 min storm gave the peak flows and at the

long tc's in the pipes below 100.14 the 60 min. storm gave
the peak flows.

The flows from the 200 line show the results for a relatively
simple sewer line.   Again, it can be seen that, with the
exception of pipes 200.00 and 01, there is an increase in the
duration of the critical FSR storm with tc. Pipe 200.00 has
a very long tc for an upper branch, as it collects a remote
area of development through a long length of pipe, and this
may explain why the 60 min FSR storm gives the maximum flows
in these two pipes.

Table 3 gives an analysis of the 142 pipes free to enlarge
showing the critical FSR storm giving the maximum flow at
different tc's.

FSR Storm Duration	Computer print out tc mins				
	0 - 5	5 - 10	10 - 20	20 - 40	40 +
15	62	29	3	3	5
30	–	13	9	4	2
60	–	–	2	–	6
90	–	–	–	–	–
120	–	–	–	–	4
270	–	–	–	–	–

Table 3. Affect of tc on critical storm duration (Torquay)

Although this confirms the general trend for critical FSR
storm duration to increase with tc, shown on the 100 and 200
lines, it also shows that at all tc's there were anomalies
which would have made the selection of the critical FSR storm
on the basis of the computer print out value of tc invalid.

As there was no data available on measured flows in the
catchment we could not check the accuracy of the FSR results.
In checking the performance of the RN35(1) and (2) storms we
were forced to assume that the FSR storms gave the correct
design flows and measure the accuracy of the RN35 storms
against the FSR storms.

A plot of the ratio of FSR peak flows to RN35(2) peak flows
against tc is given in Fig. 6, mainly for pipes in the 200
line.   This shows that, over the whole range of tc's plotted,
the RN35(2) storm predicted flows to within 15% of the FSR
storm peak flows.   This result is an improvement over
the RN35(1) storms where, although at tc's over 30 mins,
flows were only slightly overestimated, at shorter tc's they
were underestimated by up to 20%.

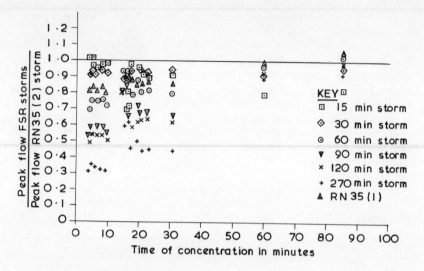

Figure 6. Variation in ratio of Peak Flow with
time of concentration (Torquay)

Results of Pipe Diameter

The computer model used for the Torquay work designed the
sewer pipe sizes in increments of 75mm, thus tending to even
out the effect of the variations in predicted flow. Table 4
gives the diameters of some pipes in the catchment using the
various storms tested.

pipe No.	tc mins	FSR storms – durations in mins						RN35 (1)	RN35 (2)
		15	30	60	90	120	270		
101.00	4.4	225	225	225	225	225	225	225	225
102.00	5.2	460	460	460	380	380	305	530	460
100.07	10.5	990	1070	1070	990	990	915	1070	1070
100.08	11.7	1220	1300	1300	1140	1140	1070	1220	1300
100.09	11.8	1300	1370	1370	1300	1300	1140	1370	1370
100.10	12.5	1300	1370	1370	1300	1300	1300	1370	1370
100.11	12.6	1370	1450	1450	1370	1370	1220	1450	1520
200.00	15.4	305	380	380	380	380	305	380	380
200.03	17.3	690	690	610	610	610	530	690	690
200.05	19.3	1070	1070	1070	990	915	840	1070	1140
200.06	21.7	1370	1370	1300	1220	1220	1070	1370	1450
200.08	22.5	690	690	610	610	610	530	690	690
500.01	30.2	1520	1600	1600	1520	1520	1370	1600	1675
515.01	57.6	1140	1300	1370	1300	1370	1300	1370	1370
511.01	85.1	2060	2210	2290	2290	2290	2210	2360	2290
511.02	89.8	1905	2060	2210	2210	2290	2130	2210	2290

diameters in mm.
Table 4. Variation in pipe diameters (Torquay)

In Table 4 the largest diameters of pipes given by the FSR
storms are underlined. It can be seen that generally at
least two of the FSR storms produce the maximum pipe size
so that the selection of the wrong critical storm would have
little effect on the design pipe diameter, provided that the
durationsof the storms were close to the critical duration.
If only the 15 min and 60 min FSR storm had been used then
out of the 142 pipes checked that were free to enlarge, then
7 would have been undersized by 75mm. If the 30 min.
and 120min FSR storm had been used, then only 5 would have
been undersized by 75 mm.

Table 4 also shows that the RN35(2) storm estimated relatively
accurately the diameters given by the FSR storms. On the whole
catchment, out of 142 pipes, where the model enabled pipe
sizes to enlarge, the effect of the RN35(2) storms, compared
with the maximum diameter given by the FSR storm, was to
reduce 8 by 75mm whilst enlarging 12 by 75mm, therefore
tending to nominally overestimate pipe diameter. Similarly,
comparing the RN35(1) results with the FSR results out of
the 142 pipes 2 were enlarged by 75mm whilst 34 were reduced
by 75mm, indicating that the RN35(1) storm tender to underestimate the pipe diameter. It is, therefore, concluded
that despite the complexity of the Torquay catchment, in
terms of estimating the pipe diameter the RN35(2) storm
represents a better and more reasonable approximation
to the alternative of using a range of FSR storms than does
the RN35(1) storm.

THE HORSELL CATCHMENT

The network at Horsell was used for the design of storm
sewers on a large new housing estate with a total catchment
area of 277 ha of which 123 ha was estimated as being
impermeable. As the design was for a completely new
separate storm drainage system the model was free to design
the diameter of every pipe.

Four 50 percentile FSR storms were run on the network
with durations of 15, 36, 60 and 120 minutes, where the 36
min storm was suited to the tc of the outfall on the basis
of 3 x (tc - 5) min. Both RN35(1) and RN35(2) storms were
also used.

Table 5 gives flows in an upper branch and the main trunk
sewer. At tc's less than 10 mins the maximum flow was
given by the 15 min FSR storm. Out of the 42 pipes with
tc's longer than 10 mins, 14 of them would receive maximum
flows with the 36 min FSR storm, and the remainder with the
15 min FSR storm. However, it can be seen that the variation
in flow between the 15 and 36 min FSR storms is small

compared with the actual flows, and it would appear that
this transition between the 15 min and 36 min FSR
storms confirms that the critical FSR storm duration
increases with tc.

Pipe No.	tc mins	FSR storms-duration in mins.				RN35 (1)	RN35 (2)
		15	36	60	120		
4.00	3.3	_140_	100	83	57	92	217
4.01	3.7	_300_	218	182	125	199	258
4.02	3.9	_448_	330	275	190	300	416
4.03	4.3	_431_	325	272	189	295	416
4.04	4.4	_451_	343	291	202	311	416
1.00	11.2	1270	_1292_	1226	939	1197	1315
1.01	11.8	1321	_1356_	1293	998	1256	1372
1.02	12.1	_1684_	1615	1527	1193	1469	1703
1.03	12.4	_1657_	1609	1524	1192	1467	1689
1.04	13.0	_2313_	2256	2102	1624	2057	2347
1.05	13.1	_2455_	2410	2261	1757	2208	2515
1.06	13.2	_2536_	2524	2369	1850	2313	2600
1.07	13.5	2634	_2655_	2495	1957	2432	3113
1.08	13.7	2680	_2709_	2552	2011	2483	3113
1.09	14.0	2718	_2773_	2609	2064	2542	3113
1.10	14.2	2739	_2809_	2650	2097	2577	3113
1.11	14.7	3022	_3060_	2871	2268	2792	3684
1.12	14.8	_3607_	3549	3276	2567	3204	3730
1.13	15.1	_3599_	3552	3303	2590	3230	3751
1.14	15.4	_3587_	3476	3324	2619	3251	3744
1.15	15.7	_3594_	3585	3361	2654	3287	3775
1.16	15.8	_3606_	3608	3393	2687	3319	3797
1.17	15.8	_4499_	4484	4174	3275	4109	4697
1.18	16.0	_4715_	4676	4342	3400	4279	4917
1.19	16.1	_4736_	4682	4356	3420	4293	4937
1.20	16.3	_4766_	4729	4433	3491	4370	4988
1.21	16.5	_5005_	4998	4689	3681	4613	5238
1.22	16.9	4995	_5010_	4699	3707	4623	5249
1.23	17.0	5006	_5035_	4746	3755	4669	5278
1.24	17.2	4949	_5026_	4746	3755	4654	5242
1.25	17.9	4982	_5163_	4938	3978	4831	5383

Flows in l/sec
Peak flows from critical FSR storm underlined
Table 5. Peak flows from storms tested at Horsell

Fig. 7 gives a plot of the ratio of FSR storm flows to
RN35(2) flows against tc for the points in Table 5.   This
shows that at tc's greater than about 12 mins the RN35(2)
storm overestimated the peak flows from the critical FSR
storms by up to 5%, and at tc's shorter than 12 mins
underestimated peak flows by up to 15%.

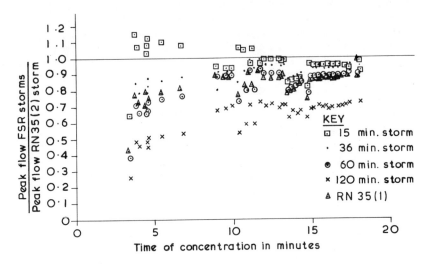

Figure 7. Variation in ratio of Peak Flow with
time of concentration (Horsell)

Table 6 gives the pipe sizes corresponding to some of the
flows given in Table 5, again with the largest diameters
from the critical FSR storm underlined.

Pipe No.	tc mins	FSR storms – durations in mins				RN35 (1)	RN35 (2)
		15	36	60	120		
4.00	3.3	305	225	225	225	225	305
4.01	3.7	460	380	380	305	380	380
4.02	3.9	530	460	460	380	460	460
4.03	4.3	530	460	460	380	460	460
15.02	5.2	460	380	380	305	380	460
1.00	11.2	840	840	760	690	760	840
1.01	11.8	840	840	840	760	840	840
1.02	12.1	915	915	840	760	840	915
1.03	12.4	915	915	840	760	840	915
1.04	13.0	1070	1070	990	915	990	1070
1.07	13.5	1070	1140	1070	990	1070	1140
1.08	13.7	1140	1140	1070	990	1070	1140
1.09	14.0	1140	1140	1070	990	1070	1140
1.11	14.7	1140	1140	1140	1070	1140	1220
1.13	15.1	1220	1220	1220	1070	1220	1220
1.15	15.7	1220	1220	1220	1070	1220	1220
1.19	16.1	1450	1450	1370	1300	1370	1450
1.20	16.3	1450	1450	1450	1300	1370	1450
1.22	16.9	1450	1450	1450	1300	1370	1450
1.24	17.2	1450	1450	1450	1300	1370	1450
1.25	17.9	1450	1450	1450	1300	1370	1450

diameters in mm.

Table 6. Variation in pipe diameters (Horsell)

Table 6 did not confirm the findings at Torquay where the
duration of the critical FSR storm appeared unimportant
when calculating pipe sizes.   If the 36 min storm had been
used instead of the 15 min storm, 77 pipes would have been
underestimated by 75mm and 3 by 150 mm , but if the 15 min.
storm had alone been used then only 1 pipe diameter would
have been underestimated by 75mm.

Comparing the performance of the RN35(2) storm in estimating
the diameters given by the critical FSR storm showed that
out of 154 pipes the use of the RN35(2) storm would
have reduced the diameters of 38 by 75mm and increased
3 by 75mm, thus tending to underestimate pipe diameter.

BALANCING TANKS

On the Theale catchment a network analysis was carried out
to facilitate the design of a storage pond at the lower end
of the drainage system, the majority of which was already
in existence.   Total catchment area was 317 ha of which
110 ha was impermeable.   As much of the system drained areas
outside the natural catchment of the balancing pond and the
model was not permitted to enlarge pipe diameters, most of
the excess flows in the contributing pipes were overflowed
and discharged off the catchment.   Consequently an analysis
of sewer flows yielded very little information.

Our early work at Torquay had shown that FSR storms of
differing durations produced a marked variation in the
volumes of the balancing tanks and that there was an FSR
storm with a critical duration which gave a maximum volume.
At Theale therefore we purposely ran a wide range of 50
percentile storms with durations of 15, 30, 60, 120, 240
and 480 mins.

Although there is no evidence that the design volume
of a balancing tank is given by the maximum volume from the
critical 50 percentile FSR storm, if the engineer wishes to
use a tank on a complex sewerage system, where the only
available method of analysis is the TRRL hydrograph method,
then he must make some decision on the design storm.   As the
50 percentile FSR storms are a median profile shape it was
assumed that these were the design storms to use and they
were applied to the Theale network.

Table 7 gives the calculated volumes of the pond for the
various FSR storms.   Run 2 was a modification of Run 1
network to reduce the peak storage volume required.

Storm duration mins	Storage Volume m^3	
	Run 1	Run 2
FSR 15	3123	–
FSR 30	7253	–
FSR 60	10883	–
FSR 120	15281	13382
FSR 240	21872	18520
FSR 480	20692	15813

Table 7. Variation in Storage Volume

The RN35(2) storm was also used on Run 2 yielding a volume of 12,700m^3, or 68% of the maximum volume given by the FSR storms.

USING THE FSR PROFILES

The FSR storms can either be derived using Volumes II and V of FSR or will be produced on request by the Met. Office using a special computer programme developed for the purpose. Initially, the FSR profiles used were supplied by the Met. Office and these were received in a numerical matrix form easily entered into the computer and involved little extra office work.   Subsequently, the profiles were produced from the maps and charts in FSR which took about 1 man hour to the numerical matrix stage.   Further profiles for the same location then took an additional $\frac{1}{2}$ man hour each.

Running a range of profiles on the network obviously increased the computer time over the time taken by the RN35 profiles by the number of profiles used. With several computer print outs to examine and compare, instead of merely the one when using the RN35 storms, the office time required for assessment of the flows considerably increased.   Even so this extra work is not really significant when compared with the total time taken in preparing the data for the network and that taken in considering the implications of the design flows and any subsequent modifications to the network that they required.

SUMMARY OF RESULTS

Flows
There was a general tendency for the duration of the critical

FSR storm to increase with tc but on the complex Torbay
network there were some significant exceptions where the
shortest FSR storm gave the maximum flow in some pipes
with the longest tc's.

The work did not confirm either of the relationships given by
the Meteorological Office, for estimating the critical duration
of the FSR storm, viz.3 x (tc - 5)min or 2 x tc mins.   The
use of the computed tc to select the critical FSR storm on the
basis of either of the recommendations would have given rise to
the underestimation of flows in certain pipes on the Torquay
system.

The RN35(2) storms estimated with reasonable accuracy at
Torquay the flows given by the critical FSR storms with a
variation of approximately ±5%. At Horsell the RN35(2) storm
overestimated flows by up to 5% at tc's longer than 12 mins,
and underestimated peak flows by up to 15% at shorter tc's.
These results were an improvement over those given by RN35(1)
storm.

Pipe Diameter

The way in which the model designed pipe diameter, in 75mm
increments so that pipes produced commercially were generally
specified, tended to iron out the effect of variations in
peak flows produced by the various storms.   At both Torquay
and Horsell if pipe sizes had been the sole interest this would
have allowed less FSR profiles to be used.   In fact on both
catchments the 15 min and 60 min FSR storms would have been
sufficient.Comparison of estimated pipe diameters computed by
the RN35(2) and FSR storms method showed that at Torquay 86%
of pipe diameters were the same and 14% changed by 75mm.
At Horsell 73% were the same and the remaining 27% changed by
75mm.   The results were an improvement over those given by
RN35(1) storm.

Balancing Tanks

At both Theale and Torquay the FSR storms gave very large
variations in storage volumes between storms of differing
duration and it was shown that there is an FSR storm of a
critical duration giving the peak volume for any given tank.

If the maximum storage volume given by the critical FSR
storms is considered to be the design storage then using
RN35(2) to calculate the storage volume would have produced a
significantly different lower volume.

CONCLUSIONS

If the flows in U.K. urban storm water drainage systems are required, then assuming that the FSR storms give the most accurate design flows, the range of FSR storms should be used as design storms, even if only flows at a given point are of interest.

If pipe diameters only are required, then, on the catchments tested, FSR storms with duration of 15 and 60 mins, would have yielded virtually the same results as using FSR storms with durations of 15,30,60, and 120 mins.

The RN35(2) storms predicted with reasonable accuracy the diameters given by the critical FSR storms but it is debatable whether this prediction was accurate enough. It may be argued that the use of a design storm of a given return period underestimates the return period of flooding in the final design and the variations in diameter given by the RN35(2) would be unimportant.   The diameters given by the RN35(2) were accurate enough for development of the alternative proposals to relieve flooding, which were being carried out at Torquay, but for a final detailed design check, if the catchment is large and complex, then the range of FSR storms should be used. It is interesting that on the complex catchment of Torquay the RN35(2) storms were more accurate than on the simple Horsell catchment and it was least accurate at short tc's. However, it can be said that the RN35(2) storm represents an improvement over the RN35(1) storm.

For the design of balancing tanks the FSR storms should be be used.   However, it should be noted that neither the TRRL method nor the FSR profiles were produced for the calculation of volumes, and although they probably represent the best design approximation at the present time, for tanks on complex systems, this is a field where further research is required.

ACKNOWLEDGEMENTS

We wish to acknowledge the Borough of Torbay and the South West Water Authority in permitting us to publish the results from work carried out in Torquay.   The photographs of flooding at Torquay are reproduced with the permission of the Editor of the Torbay Herald Express.

REFERENCES

Natural Environment Research Council (1975) Flood Studies Report.

Road Research Laboratory (1963) A guide for engineers to the design of storm sewer systems. Road Note No. 35 HMSO.

Transport and Road Research Laboratory (1976) A guide for engineers to the design of storm sewer systems. Road Note 35. Second edition HMSO.

LOCAL CONSIDERATIONS AFFECTING THE CHOICE OF DESIGN STORM
FREQUENCY

Marshall, J.K.

Mander Raikes & Marshall

INTRODUCTION

Before designing works for the alleviation of flooding, a
decision has to be made on the frequency of flood flow which
the works will contain without overtopping. In works designed
to alleviate river flooding in urban areas, there is an
increasing demand for this decision to be made by means of
cost/benefit analysis. It may be assumed that if works are
constructed to alleviate flooding, the frequency and cost of
flood damage will be reduced but will not be eliminated. By
means of hydrological, engineering and economic techniques, a
frequency of flood flow which the works should be designed to
contain, may be selected at which the total of the cost of
works and the cost of residual flood damage is a minimum.

However cost/benefit analysis of this kind is not used in the
design of urban storm sewerage in the United Kingdom. There
has not only been lack of an appropriate methodology, but
there has also been little demand upon designers to provide
it. It has not been uncommon for designers to select a design
storm frequency from one of three options (1, 2 or 5 year
return period) on no better principle than "rule-of-thumb".

It has been stated (National Water Council, 1976) that the
purpose of storm sewerage is to convey storm run-off from
developed areas so that the occurrence of flooding is reduced
to an acceptable frequency. For the purpose of this paper
and for reasons given below, an acceptable frequency of
flooding of residential or commercial property is taken to be
of the order of 35 year return period. Certainly no designer
supposes that a sewer designed for 1, 2 or 5 year return
period storm will result in significant flooding of the
development it serves at say 2, 4 and 10 year return period
respectively. The designer knows by experience that a much
larger safety factor exists, but little attempt seems to have

been made in the past to quantify it.

THE SAFETY FACTOR IN CURRENT STORM SEWERAGE DESIGN

Current methods of storm sewerage design contain a safety
factor made up of three elements : an element of margin
between designed sewer capacity and its surcharged
performance; an element of margin between the onset of surface
flooding and the onset of damage; and an element concerned
with the overland movement and destination of flood flows.

Surcharged performance
A storm sewerage system may be deemed to comprise trunk sewers
and their manholes, branch sewers throughout the reticulation
including pipes for highway gullies and those within the
curtilage of properties for roof and yard gullies, the manholes and chambers on all these branches, and all gullies.

Pipes are only commercially available in standard steps of
diameter. It is customary to provide the next larger
available diameter to that theoretically required for trunk
and major branch sewers. For the smallest pipes, it is
customary to provide a minimum diameter regardless of calculated requirements, chiefly to avoid risk of blockage. Oversize pipes provide potentially larger flow carrying capacity
although the additional margin will vary widely through the
system.

It is normal practice to design for the specified performance
without surcharge. Surcharge of one pipe length may substantially increase its discharging capacity, but the
effective total discharging capacity of the system may not be
very significantly increased if surcharging occurs throughout.

In the surcharged condition an increased volume of underground
storage will become available. Many oversize pipes may
contribute, and manholes may make a significant contribution.

Research to investigate the surcharged performance of a storm
sewerage system will no doubt continue, but it does not seem
likely that surcharging will prove to be the major element in
the overall safety factor.

The damage threshold of flood depths
On a perfectly flat developed site, flood damage caused solely
by rain falling on and immediately around the property affected would be very rare. There are exceptions such as
properties with unprotected basements, but as a rule ground
floors in England are raised approximately 200 mm above the
surrounding paved area. In England, 200 mm of rainfall does
not occur in a period of less than 48 hours with a return
period of less than 100 years. In England, a much more likely

event would be the overland movement of flood flow from a
developed area which resulted in an accumulation of flood
water to a damaging depth in some location elsewhere.

Overland flood flow

This paper discusses the overland movement of flood flow in a
residentially developed area, when storm sewers have become
fully surcharged. The paper is not presented as a basic
research contribution but to promote interest and discussion
in an area of investigation which seems to merit more attention than it has received. Unless the storm sewer designer
makes some attempt to assess the quantity and behaviour of
total run-off from rare storms, he will remain in ignorance of
the consequences of his choice of the design storm frequency
for the sewer.

AN ACCEPTABLE FREQUENCY OF FLOODING

Mention was made earlier that an acceptable frequency of
flooding of residential or commercial property is taken as 35
years for the purpose of this paper. This figure is only used
for illustration. In any particular case, some other
acceptable frequency might be selected by means of a strictly
rational cost/benefit analysis. Such an analysis should
provide valuable guidance but damage to property by flooding
carries with it ill effects which are not easily quantified.
A decision made on strictly rational economic grounds may be
acceptable to the body of rate-payers and their administrat-
ors, but individuals affected may not be so content to live
with regular flood damage even if their direct costs are re-
imbursed. Twice in a lifetime is probably as frequently as
the less tangible ill effects of flooding can be accepted.

A PRACTICAL APPLICATION IN THE CITY OF HEREFORD

The method of assessing overland flow discussed in this paper
is more easily explained by reference to a practical
application than by a theoretical exposition on its own.
Figure 1 shows diagrammatically a part of the City of Hereford
which has been used for illustration. This part of the City
is almost entirely devoted to residential development, old
and recent.

Since 1960 three major floods have occurred on the Ledbury
Road where it crosses the Yazor Brook. The first, in December
1960 was caused by an exceptionally high level of water in the
River Wye 0.6 km to the south. This event is not of interest
for the purpose of this paper. The second flood occurred
after a rainstorm of 90 minutes duration and of about 15 year
return period in September 1971, and the third flood occurred
after a rainstorm of similar duration and of about 20 year
return period in September 1976. No damage to property

occurred on either occasion but the forecourt of a petrol
filling station adjoining the Ledbury Road near the Yazor
Brook was flooded. On the occasion of the last flood, the
Ledbury Road had to be closed to traffic.

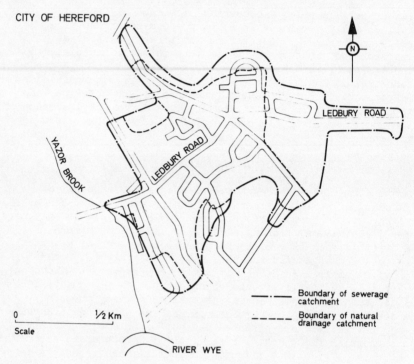

Figure 1. The area used for illustration.

The existing sewerage system

The boundary of the area served by sewers draining down
Ledbury Road to the City's trunk outfall alongside the Yazor
Brook is shown in chain-dotted line on Figure 1. The sewerage system contains a mixture of separate foul and surface
water sewers, partially separate sewers, and combined sewers.
The system has been developed piecemeal over many years and
does not conform to a single standard of design.

As part of a flood study carried out for the County Council
by the author's firm in 1975/76, data on this part of the
City's sewerage system was included for tabulation in
performance tests using the T.R.R.L. program. Many locations
of surcharging resulted, using a standard 2 year return
period storm profile, but at the time of that study the
program was not run with shorter return period storms.

It is not in fact possible to establish the capacity of the

existing sewerage system in terms of rainfall of one
particular uniform intensity and duration which just brings
the whole system to the fully surcharged condition. Using the
Lloyd-Davies method (Lloyd-Davies, 1905-6), it is calculated
that the system would be approximately at full capacity with
uniform rainfall intensity of 14 mm/hour of 30 minutes
duration, an event having 6 months return period. While
recognising the errors in the approximation, it is assumed for
the purpose of this paper that when the system is at full
capacity according to this method of calculation, it is fully
surcharged to surface.

The natural drainage catchment

Shown on Figure 1 in dotted line is the boundary of the
natural drainage catchment to Ledbury Road at the Yazor Brook.
If no provision for surface water drainage had existed, this
catchment is the area from which effective rainfall would
contribute to flooding on the Ledbury Road.

It is assumed that in rainfall of high intensity and
sufficiently long duration the existing storm sewerage system
is fully surcharged and merely provides one defined route by
which a known part of the run-off is conveyed. In order to
predict the behaviour of the remainder of the run-off,
conveniently referred to as the excess run-off, the
characteristics of the natural catchment need to be examined
in some detail.

In hydrological terms, the catchment is too complex for an
accepted method such as that provided in the Flood Studies
Report (Natural Environment Research Council, 1975) to be
used to predict the hydrograph of flood flow. Field investigation of the area suggested that the highways within the
catchment constitute an open channel reticulation, and that
this reticulation provides the dominant characteristic of the
catchment response.

Within the area shown on Figure 1, four different types of
features were identified. These are illustrated diagrammatically in Figures 2(a), 2(b), 2(c) and 2(d) respectively.

Figure 2(a) illustrates the cut-off defining the head of the
open channel, where the highway rises to a peak. Although
areas upstream are drained by the storm sewerage system, the
highway peak effectively diverts all excess run-off from the
areas upstream into gardens.

Figure 2(b) illustrates partial flow diversion at a highway
junction.

Figure 2(c) illustrates a low point to which a section of
highway falls, where a public footpath, along which the storm

sewerage system has been laid, leaves the highway at a
higher level than the lowest kerb levels of the highway. The
result is that excess run-off from the highway will pass into
gardens instead of following the footpath.

Figure 2(d) illustrates the assumption that excess run-off
from all paved areas of properties standing on higher ground
than the highway will discharge on to the highway, while the
excess run-off from all paved areas of properties standing on
lower ground than the highway will discharge on to gardens.

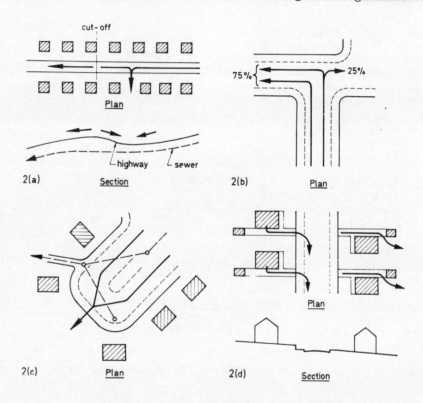

Figure 2.   Typical features defining drainage to highways and
           gardens.

Figure 3 illustrates diagrammatically the results of the
field investigation of the area, defining the extent to which
the highways (shown cross-hatched) can form an open channel
reticulation draining to the Ledbury Road at the Yazor Brook,
and defining those parts of the paved areas (shown stippled),
other than the highway itself, which will drain excess runoff to this reticulation. For the purpose of subsequent
calculations all paved areas were assumed to have an effective

impermeability of 100% and all non-paved areas an effective
impermeability of zero (Transport and Road Research
Laboratory, 1976).

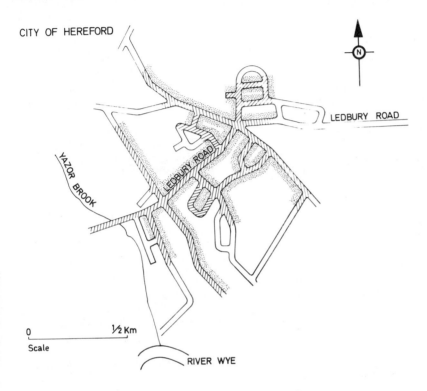

Figure 3.   Impermeable areas draining to highways.

METHOD OF CALCULATING EXCESS RUN-OFF

Modern storm sewer design practice tends to considerable
sophistication, using equations which take account of the
effects of storage. Suitable equations could no doubt be
developed for the open channel reticulation, but it seemed
pointless to explore a new concept with such sophistication in
a paper intended chiefly to promote discussion. The simple
Lloyd-Davies approach has therefore been used in the belief
that if reasonably sensible results were obtained with this
method, and the concept found any favour, more sophisticated
methods could be developed as a future task.

The relation between time of concentration and rainfall
On Figure 4 is shown the rainfall intensity/duration relationship for return periods of 6 months and 35 years respectively
for Hereford (Keers and Westcott, 1976).

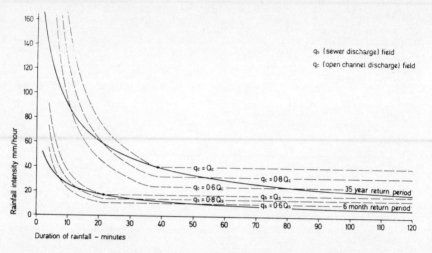

Figure 4.   Rainfall intensity/duration relationships and
             discharge fields.

It is assumed for the purpose of discussion that the time of
concentration of the storm sewerage system is 30 minutes and
that rainfall of 14 mm/hour uniform intensity and 30 minutes
duration just brings the system to full capacity at the outfall with all points upstream becoming fully surcharged to
surface. The rainfall intensity and duration required to
achieve this condition is shown as a point on the 6 month
return period rainfall curve. Around this point, a discharge
field of straight and curved lines has been constructed
representing lines of constant qs, where qs is the rate of
flow at the outfall. The line marked qs = Qs, where Qs is
the rate of flow at full capacity, is drawn at constant rainfall intensity for increasing durations of rainfall, because
the storm sewerage system is maintained at full capacity at
the outfall by rainfall of this intensity for all durations
exceeding the time of concentration.

For durations less than the time of concentration, rainfall
of higher intensity than 14 mm/hour is required to maintain
the outfall at full capacity because the area contributing to
run-off reduces with the duration of rainfall.

Below the line marked qs = Qs are shown lines for qs = 0.8 Qs
and qs = 0.6 Qs, established by similar reasoning, to
demonstrate the nature of the discharge field of qs.

The relationship between the 6 month return period rainfall
intensity/duration curve, and the discharge field of qs
demonstrates the basic Lloyd-Davies principle that qs is
maximised at Qs when the duration of rainfall is equal to the

time of concentration for the system.

This part of Figure 4 also shows that the line marked qs = Qs
defines the boundary of the field of rainfalls of all
intensities and durations above which the outfall of the
storm sewerage system will remain at full capacity.

Referring again to Figure 4, it is assumed for the purpose of
discussion that the time of concentration at the outfall of
the open channel reticulation formed by the highway is 39
minutes. This point is shown on the rainfall intensity/
duration curve for 35 year return period. The discharge
field of qc is shown, qc being the rate of flow in the open
channel at the outfall.

Within the whole field of rainfall intensities and durations
above the line marked qs = Qs the rate of flow carried by the
storm sewerage system is constant so that the basic Lloyd-
Davies principle can be applied, namely that qc is maximised
at Qc when the duration of rainfall used for calculation of
flow in the open channel reticulation is equal to the time of
concentration for that reticulation.

The highway as an open channel
In calculation, by the Lloyd-Davies method, of run-off
conveyed by an open channel instead of a pipe the designer
has to select a depth of flow instead of a pipe diameter.
The calculations done for the purpose of this paper, were
made from tables prepared by computer. Three questions were
considered in their preparation.

For simplicity, no attempt was made to allow for local
variations of surface roughness. Manning's formula was used
and the roughness coefficient n was taken as 0.016 throughout,
representing an average value.

Again for simplicity, a constant cross-section was assumed.
Several cross-sections were measured at random points within
the area. The dimensions did not differ very significantly
from those specified for a standard local distributor
(Department of the Enviroment, 1974). The dimensions of
this standard were used throughout.

Whereas sewers are invariably designed to be laid with
constant gradient between manholes, highways generally have
curved profiles in longitudinal section. Some theoretical
work on the variation in the time of flow, suggested that
sufficiently accurate results, consistent with the order of
approximation being made elsewhere, would be obtained by
using eye-fitted linear segments to curved longitudinal
profiles, within a maximum departure of 10% of the fall over
the segment.

Calculation for excess run-off

In the Lloyd-Davies calculation, an elemental catchment is
used, impermeable area and time of concentration being
cumulated to form successively larger elements, proceeding
downstream. The rainfall intensity applied to the cumulative
impermeable area is derived from the cumulative time of
concentration to obtain rainfall run-off. This procedure
establishes the maximum rate of flow which will arise at each
point in the system under the most unfavourable rainfall of
selected return period. Being a design procedure, it is not
entirely appropriate as a procedure for calculating performance, but it does maintain each stage of the calculation on
the selected return period rainfall intensity/duration relationship. It has therefore been used initially, if only to
provide a guide to the time of concentration.

The impermeable areas used have been based on the assessments
illustrated on Figure 2 and summarised diagrammatically on
Figure 3. The time of concentration has been calculated by
the usual trial and error process from the open channel
velocity. A "time of entry" allowance of 3 minutes has been
included. From the calculated total run-off, the excess runoff has been obtained by deducting a calculated rate of flow
to the storm sewer. This calculated rate of flow is not the
capacity of the sewer at the downstream end of the element
under consideration but a proportion of that capacity,
related to other inputs to the sewer, chiefly the capacity
provided for impermeable areas draining to the sewer in its
designed mode of operation but not draining excess run-off to
the highway reticulation.

RESULTS OF CALCULATIONS

Varying intensity of 35 year return period rainfall

Calculation of the excess run-off using the normal Lloyd-
Davies method of varying the rainfall intensity with increasing time of concentration through the stages of the calculation, produced a time of concentration to the outfall of 21
minutes and a maximum rate of flow of excess run-off in the
open channel at the outfall of 1.07 m^3/sec. On the usual
Lloyd-Davies assumptions, the total volume of excess run-off
according to these results is 1350 m^3.

Constant intensity rainfall, with existing sewers

By trial and error, a constant intensity of rainfall was used
in all stages of the calculation, with the object of giving a
time of concentration at the outfall equal to the duration of
35 year return period rainfall of that intensity. The time
of concentration found was 23 minutes using a rainfall
intensity of 56 mm/hour. The maximum rate of flow of excess
run-off in the open channel at the outfall was found to be
0.98 m^3/sec. The total volume of excess run-off according to

these results is 1350 m^3.

Constant intensity rainfall, with enlarged sewers

In order to determine the effect on the excess run-off of providing enlarged sewers, the whole of the existing storm sewerage system was re-designed by the normal Lloyd-Davies method, using a 2 year return period storm as the design criterion.

The calculation in the preceding section was then repeated. The time of concentration was found to be 33 minutes using a rainfall intensity of 44.7 mm/hour. The maximum rate of flow of excess run-off in the open channel at the outfall was found to be 0.22 m^3/sec. The total volume of excess run-off according to these results is 440 m^3.

Rainfall of September 1976, with existing sewers

Although the main rainfall lasted 90 minutes the highest average intensity recorded over about $\frac{1}{2}$ hour only represented an event of about 5 year return period. Constant intensity rainfall was used to simulate this event with the existing sewers. The time of concentration at the outfall of the open channel was found to be 26 minutes using a rainfall intensity of 29 mm/hour. The maximum rate of flow of excess run-off in the open channel at the outfall was found to be 0.30 m^3/sec. The total volume of excess run-off according to these results is 470 m^3.

A detailed survey around the petrol filling station shows that the calculated volume of excess run-off would flood the forecourt to a depth of 0.43 m but would hardly have necessitated closure of the road to traffic. A likely explanation is that the 90 minutes of rainfall caused the City's trunk outfall sewer alongside the Yazor Brook to become heavily surcharged, blowing off unsealed manhole covers in Ledbury Road. The trunk outfall sewer has sealed manhole covers and a time of concentration of about 72 minutes at this location.

As an example for cost/benefit analysis, the case illustrated is not a good one. Only nuisance is caused by the flooding and there is no direct damage. The City Authority have in the past compensated the owner of the petrol filling station by rate rebate. Enlargement of the local storm sewerage system to reduce excess run-off could not be considered in isolation from its effect on the trunk sewerage system in this case.

CONCLUSIONS

Subject to the validity of the concept put forward in the paper being received favourably, there is need for a more refined technique to be developed, preferably in the form of a standard computer program. If the destination and quantity of excess run-off at a location of flood damage risk can be identified, and a theoretically acceptable and practical method can be established of relating the hydrograph of excess run-off to the design storm frequency for proposed urban storm drainage, the methodology for cost/benefit analysis will then be available.

In addition to this general objective, there may be scope for other economies in the provision of urban storm drainage for new housing development. Within the constraints of good town planning, much might be done to reduce the excess run-off emerging from any new development by careful attention to the highway layout, its levels and its discontinuities as an open channel reticulation.

ACKNOWLEDGEMENTS

The author expresses his thanks to W.D. Peters FRTPI, Dip.TP (Lond), DMS, MBIM, County Planner, the County Council of Hereford and Worcester for permission to refer to a study commissioned by his Authority, to G.J. Roberts, C.Eng, MICE, FIMunE, City Engineer & Surveyor and Joint Planning Officer, the City of Hereford, for providing additional data, and to the author's colleagues in his firm who assisted in the calculations and presentation.

REFERENCES

Department of the Environment (1974). Roads in Urban Areas, Metric Supplement. H.M.S.O. London.

Keers, J.F. and Westcott, P. (1976). A Computer-based Model for Design Rainfall in the United Kingdom. Scientific Paper No.36. Meteorological Office, Bracknell.

Lloyd-Davies, D.E. (1905-6). The Elimination of Stormwater from Sewerage Systems. Proc.Inst.C.E.Lond. 164 2 : 41-67

National Water Council (1976). A Review of Progress March 1974 - June 1975. Working Party on the Hydraulic Design of Storm Sewers. National Water Council, London.

Natural Environment Research Council (1975). Flood Studies Report. Natural Environment Research Council, London.

Transport and Road Research Laboratory (1976). A Guide for Engineers to the Design of Storm Sewer Systems. H.M.S.O. London.

INVESTIGATION OF THE ACCURACY OF THE POSTULATE "TOTAL RAINFALL FREQUENCY EQUAL FLOOD PEAK FREQUENCY"

Sieker, F.

Institut für Wasserwirtschaft, Technische Universität Hannover
Callinstr. 32, 3ooo Hannover BRD

STATEMENT OF THE PROBLEM

It is usual to calculate urban drainage systems by so-called design rainfalls. These design rainfalls are characterized by:

- A specified duration which is, in general, the longest estimated time of travel in the catchment area.
- A frequency function of the total rainfall.

The design rainfalls are taken from sets of curves, which are derived from the three characteristic quantities of rainfall: duration, frequency and total rainfall, represented by measured values. Primarily, this is a graphic representation with total rainfall as the x-ordinate, frequency function F(x) as y-ordinate and the duration as a parameter for the set of curves. This method of representation results by fitting the random samples to one of the known and proven frequency distribution functions. The normal practice of this graphic representation is the duration as x-ordinate, the total rainfall as the y-ordinate and the frequencies or the time of return as the curve parameter. The second method is easily derived from the first.

In every case, a design rainfall results wherein the rainfall intensity remains a constant during the total rainfall duration. This means a marked simplification compared with actual conditions and one must question if, in given cases, too low or too high values are obtained in the calculation of the flood peaks. In this connection, the general question arises if the postulate "total rainfall frequency equal flood peak frequency" in reference to drainage

31

systems, is correct. This question will be investigated in the following discussion.

GENERAL VIEW OF THE PROCEDURE METHODS

On one hand, a frequency function of the peaks shall be calculated, derived from natural rainfall events (that is, rainfall events with intensities, which actually occured) and on the other hand, a frequency function of the peaks which are calculated from the design rainfall with a constant intensity. It is understood, of course, that both frequency functions must refer to the same catchment area. The comparison of the two functions should give an answer to the question introduced in section 1.

Unfortunately, there is almost no gauging station for sewage systems where measurements are taken for many years in catchment area of which discharge-influencing conditions have remained relatively constant. The latter is but an assumption for the stationarity of the discharge characteristics which, in such an investigation, must be taken as a basis.

A test area, in which measurements of discharge and rainfall were made simultaneously for over 2 years, was available to the author. Furthermore, a rainfall gauging station somewhat outside the area with records of 24 years was available, too. The recorded rainfall-discharge-events during the 2-year period were analyzed. Determinations were made for:

- The discharge-rainfall quotient, of the individual events, and its bivariate regression function with rainfall duration and total rainfall.

- An averaged response-function of the discharge gauging station with a unit impuls of 1 mm effective rain per 5 minutes.

From the 24-years of recorded rainfall events were determined:

- A set of curves corresponding to the details in section 1. From the set, design rainfalls with varying frequencies of exceedance of total rainfall amounts based on the given duration were derived.

- A set of data of extreme natural rainfall events (approximately 2 per year) with intensities recorded every 5 minutes.

Both the design rainfall and the natural rainfall
events were reduced, through the regression function
to effective rainfall. The superimposing of the effective rainfall with the response-function results
in:

- an extreme-value series from the flood peaks -
  calculated from the desing reinfall with constant
  intensity, and

- an extreme-value series from the flood peaks -
  calculated from the natural rainfall events with
  their sequential intensities.

Both extreme-value series were adjusted by means of
a suitable statistical distribution function. It
will be shown if, and how, the resultant distribution functions differ from one another.

INVESTIGATIONS

Test area/measurement results

Appendix 1 shows the site plan of the test area,
an approximately 54 hektar residential area north
east of Hamburg. The part of the sealed surface
amounts to 38 % of the total area. The area is
drained by a combined sewerage system. At the
drainage outlets of the area the water level fluctuations are recorded by means of a graphic recording
gauge. During, and in between, the floods, velocity
measurements are conducted with current meters, so
that a well-defind water level-discharge relation
could be established. From this, the discharge fluctuations are derived. For this relativ small catchment area, one self-recording rainfall gauge station
was deemed adequate. Local deviations from the
centrally located rainfall gauge station were disregarded.

Appendix 2 shows a typical rainfall-discharge event,
which was recorded during the 2-year measuring period.

Evaluation of rainfall-discharge events

Discharge-rainfall quotient   Although there is a
combined sewerage system, whereby a basic dry
weather discharge occurs, the superimposed floods
are relatively easy to separate through straight lines
(see appendix 2). In this way it is possible to
integrate distinct floods and corresponding distinct
rainfall events and to set each in ratio to each

other. The computed discharge-rainfall quotients
are bivariate linear regressions subject to the
criteria of rainfall duration and total rainfall.
The following equation results:

$$A/N = 0,2758 + (0,0122 \cdot N) - (0,000497 \cdot D) \qquad (1)$$
A   = discharge in millimeters
N   = total rainfall in millimeters
D   = rainfall duration in minutes

<u>Impulse-response function</u>   Next, in the search for
the most appropriate impulse-response function for
various events, different procedures for calculation of this function will be applied:

- The unit-hydrograph-procedure on the basis of an
  overdetermined set of equations

- the procedure of a linear storage model

- the procedure of a linear cascade-storage-model

- a flood-routing-procedure known as the Muskingum-
  procedure.

The reverse computation of these events with the
averaged response function shows that the procedure
of the linear storage-model is the most suitable
for this area. The lag-time was calculated as
k = 750 seconds.

Rainfall events for flood simulations

<u>Natural rainfall events</u>   The known 24-year recorded
rainfall events are devided in 5-minute intervals.
Then it is possible to plot the especially significant rainfall events' variations in intensity and
use them for the flood simulation. From every years'
observations, approximately 2 rainfall events are
selected which, based on their intensities and
variations in intensity, are expected to produce
the simulated flood peak as a years' maximum.

<u>Statistically determined rainfall events</u>   Appendix
3 shows the statistical analysis of the rainfall
events observations of the 24-year-period. The extreme I-distribution (Gumbel-distribution) serves
as a distribution function type for the frequency
function, dependent on the total rainfall with a
given duration. The frequency, which is the curve
parameter of the plot in appencix 3, is here the
reciprocal value of the periodic intervals in years.

Simulation of flood peaks

Simulation of annual extreme peaks from natural rainfall events

The selected rainfall events are reduced through the regression function to effective rainfall. The superimposing of the reduced natural rainfall to the impulse-response function and the selection in each case of the respective largest flood peak from each year results in an annual extreme-value series, which can be adjusted to a distribution function.

Simulation of flood peaks resulting from a design rainfall

From appendix 3, rainfall events of various frequencies of exceedance (return intervals) were taken. With regard to the flood simulation, since it is not certain, from the beginning, which rainfall duration results in the respectively highest flood peak. A series of rainfall events will be taken for each of the different rainfall durations. The extracted "design rainfalls" are known in regard to their total volumen but not in regard to their intensitiy-distributions. Therefore they are taken as rainfall events with constant intensities, which is the usual general practice. The reduction to effective rainfall results from the regression-function. Then the superimposing of the impulse-response function, can, also in this case, be performed.

Distribution function of the simulated flood peaks

The flood peaks, determined in the two ways, are plotted in a common coordinate-system, with the discharge Q as y-ordinate and the time of return T_R as x-ordinate (see appendix 4). The discharges, determined from the natural rainfall events, can be fitted by a distribution function, type Pearson-III (full line). The values determined from the design rainfall form a distribution function by themselves, since the corresponding rainfall events are taken from a distribution function (dotted line).

Results

Appendix 4 shows that the distribution functions, determined in two ways, intersect at about $T_R = 1$ year. For greater time intervals of return, the distribution function of the natural rainfall results in higher values, for example, with $T_R = 1o$ years, about 5o % higher. The distribution function determined from the natural rainfall events can be regarded as remaining closer to reality. Consequently, it must be stated that the principle

of design rainfalls with constant intensity during
the duration assumed - in this case of greater
time intervals of return, yields too small values,
which means, therefore, that the sewerage cross-
sections will be designed too small. To what extent
this result can be generalized, particulary with
greater catchment areas with longer times of travel
(which means longer durations of the design rainfalls), further investigations, which are being
made at the time of this report, should demonstrate.

However, one can already state that the postulate
"total rainfall frequency = flood peak frequency"
is not obviously accurate and that, with this
postulate, one remains throughout on the uncertain
side.

Recommendations

The object remains, to compute the sewerage cross-
sections according to the true frequency of the
flood peaks. In the author's opinion, there are
three possibilities to attain this object:

- Through further investigations of the foregoing
  type, a definable ratio between the total rainfall frequency and the flood peak frequency will
  be sought. This correlation probably must include the rainfall duration and the design-"time
  interval of return" as parameters.

- In general, a statistical investigation of rainfall will be eliminated. Instead of this, all
  significant natural rainfall events of the
  available records of rainfall would be used
  successively as input, which is today, through
  a computer, no longer an unsolvable operation.
  The designed flood peak will then be determined
  from the simulated flood peaks by statistical
  methods.

- An attempt to establish the design rainfall not
  only through its total rainfall (with constant
  intensity), but also through its "probable"
  intensity-distribution. The latter has to be
  established by additional investigations (Sieker,
  1977).

References

Abraham, Ch.et.al. (1976) Grundlegende Untersuchungen zur Ermittlung von Bemessungsregen für die

Stadtentwässerung in Hamburg (Basic investigations for calculation of design storms with regard to the sewerage system of Hamburg)
Mitteilungen des Instituts für Wasserwirtschaft der Technischen Universität Hannover, Heft 37.

Sieker, F. (1977) Statistical simulation models based on analysis-of-variance, paper for the 3rd International Hydrology Symposium held in Fort Collins, June 1977.

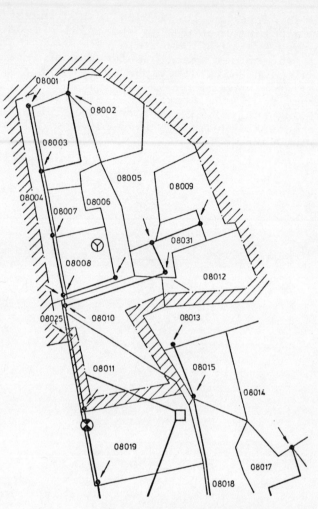

⊗ rainfall gauging station
⊗ discharge gauging station

Appendix 1 - Test area

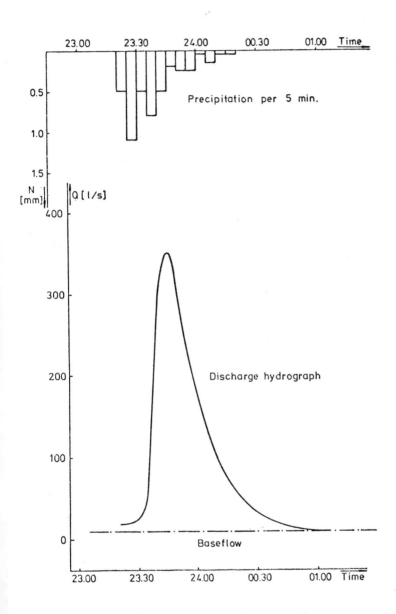

Appendix 2 - Recorded rainfall-discharge event

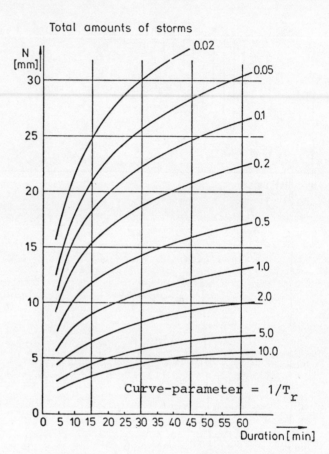

Appendix 3 - Rainfall statistics

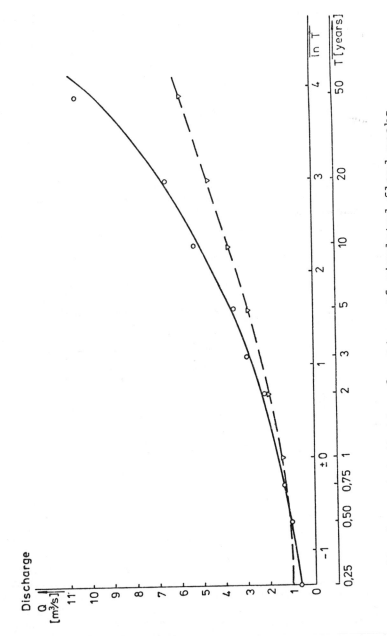

Appendix 4   —   Distribution functions of simulated flood peaks

THE SHORT-PERIOD WEATHER FORECASTING PILOT PROJECT AND ITS
RELEVANCE TO STORM SEWER OPERATION

C G Collier and K A Browning

Meteorological Office Radar Research Laboratory
Royal Signals and Radar Establishment, Malvern, UK

1   INTRODUCTION

The purpose of this paper is to describe plans, now being
implemented, for a 5-8 year Pilot Project (Browning 1977)
based at the Meteorological Office Radar Research Laboratory.
The project is aimed at improving short-period forecasts of
precipitation, ie forecasts up to several hours ahead. Such
forecasts would have many applications, but one specialist
area is in the operation of sewage treatment systems within
urban areas.

The operation of sewage systems within urban areas, and
in particular the utility of short period forecasts of
precipitation, depends very much upon the design and condition
of the system itself. The Birmingham sewage network for
example, requires a negligible amount of pumping, and has no
important cross-connections. Hence, there is no possibility
of diverting storm flows in response to rainfall forecast
information. In Manchester, most of the sewers are over a
hundred years old, and there are many storm overflow points
in the network which, although unsatisfactory in themselves,
provide a natural regulation such that only storms of long
duration cause flooding. Here short period rainfall forecasts
would be of only limited use. Where improved forecasts of
rainfall are likely to be most valuable is in the case of
sewage networks such as those in London which contain a number
of cross-connections between the sewers.

In the London network the main sewers run north-south and
the interceptors run east-west (three on the north bank of the
river Thames, and two on the south bank). These cross-
connections permit, in principle, the diversion of storm flows
from badly affected areas to others less affected. .Storm
overflow sewers, designed to carry away storm flow which
cannot be accepted by the interceptors, run parallel to the

north-south sewers, and may also offer prospects for the
diversion of storm flows. If accurate advance warning of
heavy rainfall were available, then it would be possible to
pre-release or pre-pump from storm sewers out to the river
Thames. This would not only reduce the risk of flooding,
but would also permit good effluent quality to be maintained
during the subsequent storm flow. Given sufficient warning
it might even be possible to accomplish the pumping in
periods of low tariff electricity.

Improved short period rainfall forecasts could also
assist in the planning of the disposal of residual solids on
agricultural land (since the land should not be wet when
deposition takes place), and in chemical conditioning
processes (since chemical conditioners are expensive and
their rate of use depends upon the magnitude of the flow
through the treatment works). It could also assist in the
scheduling of sewer maintenance. When work is carried out
underground in sewers, it is usual for a lookout to be posted
at the sewer entrance to warn of the occurrence of heavy
rain. Unexpected heavy rainfall can have disastrous consequences to temporary underground structures, and may lead to
blockages producing a temporary build-up of water which, when
released, may endanger life. A description of such a combination of circumstances, leading to the loss of two lives, has
been recorded by Holford (1976).

A detailed description of the present weather cannot be
provided by the routine hourly observations from the existing
network of meteorological stations in the UK. To provide the
required description of the current weather, and in particular
of cloud and rain, other observing techniques must be used.
A radar can detect rain falling at ranges up to about 200 km,
and make quantitative measurements of rainfall intensity up
to ranges of about 100 km. Cloud type and height may be
observed with equipment mounted on satellites, and this
information can be used to infer precipitation rates. Techniques based upon the use of radar and satellites, can enable
data to be collected and processed virtually in real-time.
This Pilot Project will exploit these techniques. In the first
few years of the project the emphasis will be on what is sometimes referred to as "nowcasting" (Scofield and Weiss, 1976).
Nowcasting involves the use of a good description of the
current weather to assess trends for a few hours ahead, thereby providing data on which very short-term forecasts may be
based.

2   OBJECTIVES OF THE SHORT-PERIOD WEATHER FORECASTING PILOT
     PROJECT

The Pilot Project is being set up to develop the
necessary technical capacity and meteorological experience

and understanding to improve short-period forecasts of
precipitation (the Pilot Project is also concerned with forecasting strong winds, but in this paper we limit our attention
to precipitation), and to optimise the impact of such developments on the forecasting capability of the Meteorological
Office. It is intended to be a balanced program of fundamental
and operational research. Specific objectives are:

1   To establish and operate facilities to provide
mesoscale (Meteorologists use the term mesoscale to
refer to weather systems having dimensions of kilometres
to hundreds of kilometres) observational fields of cloud
and precipitation over, initially, a limited part of the
country; and, in the light of practical experience, to
optimise the accuracy, reliability, and the clarity and
timeliness of presentation of the data.

2   To exploit these data to improve our understanding
of the structure, mechanism, evolution, and predict-
ability of precipitation.

3   To develop simple analytical procedures to optimise
the use of these data by a local Meteorological Office
for the provision of improved forecasts of precipitation,
initially over a period of a few hours, but with a view
to extending the period of improved forecasts up to 6 to
12 hours.

4   To assess from practical experience the utility of
the actual and forecast fields of precipitation to users
(such as the Water Industry for example).

5   To assess the desirability, and most cost-effective
way, of extending the mesoscale observational network
and forecasting techniques to other parts of the country.

3   THE MESOSCALE OBSERVATIONAL FACILITIES USED IN THE PILOT
      PROJECT

Several types of data will be acquired during the Pilot
Project. These include sequential upper air ascents from the
Malvern area throughout the passage of some of the precipitation systems and, in the later stages of the Project, data
from automatic weather stations sited in remote areas. However, the primary observational facilities in this Project
will be (1) a network of weather radars with overlapping
coverage, and (2) suitably processed data from both geo-
stationary and polar-orbiting satellites. Much of the initial
work in the Project will be aimed at establishing these
facilities.

3.1 The radar network

The basic radar network (Figure 1) will consist of three
Meteorological Office radars situated at Camborne (Cornwall),
Upavon (Wiltshire) and Clee Hill (Shropshire) (subject to
successful negotiations for the site rental). A fourth radar
is to be installed at Hameldon Hill (East Lancashire). This
radar is being jointly financed by the North West Water
Authority, who will co-ordinate the installation, the Meteorological Office, the Central Water Planning Unit, the Water
Research Centre, and the Ministry of Agriculture, Fisheries
and Food. It is intended to operate the first three radars
from about the end of 1978. The radar on Hameldon Hill, which
unlike the other radars will be unmanned, is planned to begin

128 x 5 Km

FIGURE 1: Approximate area within which precipitation can be
observed using radars located at (1) Camborne (Cornwall),
(2) Upavon (Wiltshire), (3) Clee Hill (Shropshire), and
(4) Hameldon Hill (East Lancashire). Areas with qualitative
coverage are hatched (broken hatching for radar No 4) and
areas with quantitative coverage are cross hatched (broken
shading for radar No 4). The square frame denotes the
approximate boundary of the television display on which data
from all of these radars will be combined.

operation by the end of 1979. All of the radars are to be operated as nearly as possible 24 hours a day.

The radars will be calibrated in real-time using a number of clusters of telemetering raingauges, each cluster comprising at least two raingauges. Each radar will be scanned in a sequence of low elevation angles repeated at intervals of about five minutes. Data will be processed digitally using a mini-computer at each radar site and then transmitted by low-cost telephone lines (see Taylor and Browning (1974)). The principal data formats will be:

(a) rectangular Cartesian matrices of rainfall averaged over 5 km squares (also a limited amount of data averaged over 2 km squares).

(b) areal rainfall totals over river or urban sub-catchments, integrated over periods specified by the users.

3.2 Satellites

During the Pilot Project satellite data giving cloud cover and cloud height (VIS and IR) will be available from polar orbiting satellites and from the European geosynchronous satellite, Meteosat. The spatial resolution of the polar orbiting satellite data is good (1-2 km), but these data will be available only every six hours, and therefore their usefulness for short-period forecasting is diminished. The Meteosat data, however, will be available every 30 minutes with a spatial resolution at the latitudes of the Pilot Project of about 5 km in a west-east direction and 10 km in a north-south direction.

These data will be received at Malvern in two forms: (a) in analogue form covering much of the North Atlantic and Europe, and (b) in radar-compatible digital format over the more limited area shown in Figure 2.

4   REAL-TIME DATA OUTPUTS IN THE PILOT PROJECT

Radar rainfall data on a 5 km grid will be transmitted every 15 minutes via Post Office telephone lines to remote users in 3-bit form (8 levels of rainfall intensity). The remote users will receive data direct from individual radars in an 84 x 84 format, which will be displayed on a colour television via a special radar data receiver. This receiver contains a store capable of holding nine pictures at any one time which may be replayed at several different speeds in order to assess the motion of precipitation areas. By using a simple audio tape recorder, the interval between the replayed pictures can be selected to suit the needs of the user. As well as the television picture, a user may also receive a simple print-out of areal rainfall totals integrated

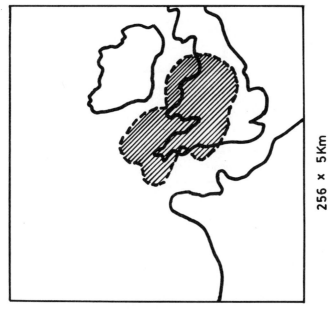

256 x 5Km

FIGURE 2: Proposed area of coverage for the television
display of satellite data in radar-compatible format. Shaded
areas indicate the coverage of radars 1, 2 and 3 in figure 1.

over any time period for about 100 river or urban subcatchments. The complete user terminal, developed by a team
at the Royal Signals and Radar Establishment is shown in
Figure 3.

        Radar rainfall data on a 5 km grid will be transmitted
every 15 min via Post Office telephone line to the Radar
Research Laboratory at Malvern in 8-bit form (208 levels of
intensity). There, the data from each site will be composited
and displayed in real-time using a 128 x 128 format covering
the area shown in Figure 1. A monochrome photograph of the
colour television showing the composite display is reproduced
in Figure 4. The radar-compatible satellite data will be
received at Malvern in a 256 x 256 format, and the facility
will be provided to superimpose or alternate the cloud and
radar-rainfall data on the same display (Figure 2) to help
clarify their spatial correspondence. In this way the radar
will serve to 'calibrate' the satellite cloud data in terms
of probable rainfall intensity where the satellite coverage
extends beyond the radar network.

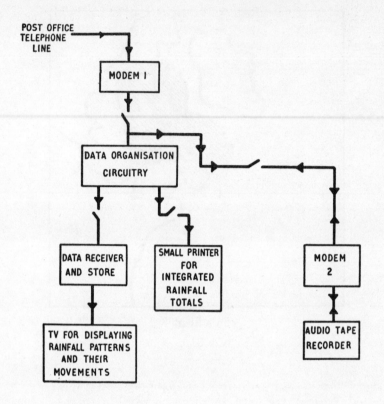

FIGURE 3:   The terminal equipment available to a user for
receiving data from an individual radar via a Post Office
telephone line. The user may have all the equipment shown,
or any part of it with the exception of modem 1 and the
data organisation circuitry which are mandatory.

5   CONCLUDING REMARKS

In order to exploit the special facilities becoming
available at Malvern, a small team of forecasters will be
established to develop short-period precipitation forecasting techniques and to implement them on an experimental
basis. They will investigate, amongst other things, both
subjective and computer-based methods of forecasting cloud
and precipitation, based on simple extrapolation. The forecasting period likely to be achieved by these techniques is
limited to less than an hour in the case of individual
shower clouds but, as shown by Wilson (1966), Tatehira et al
(1976) and Hill et al (1977), the forecast period can some-

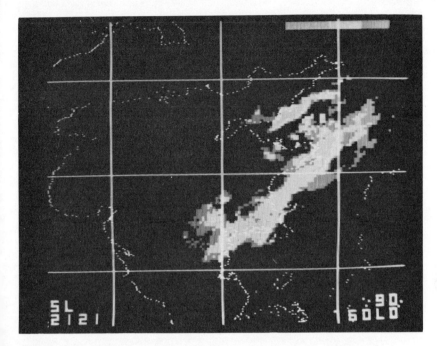

FIGURE 4: Monochrome photograph of the colour television
display showing the rainfall distribution at 0709 BST on
12 December 1975 as it was seen in real-time at Malvern using
data transmitted by telephone from radars in north Wales and
south-west Wales. Rainfall cells are 5 x 5 km and correspond
to the National Grid. Large squares are 200 x 200 km.

times be extended to several hours by restricting attention
to mesoscale features with dimensions of 20 km or more.

The Pilot Project will also provide the opportunity to
start building up a systematic body of mesoscale experience
not previously available. It can be expected that this
experience will lead to the development of physical models
and climatological rules. Enabling the broad trends and local
topographically-induced effects to be predicted on a time
scale of several hours. For longer forecast periods,
techniques of mixing the radar-cum-satellite information with
the output of a new generation of mesoscale numerical models
will need to be developed.

We believe that the development of improved techniques
for forecasting precipitation, and the rapid dissemination
of these more detailed forecasts using methods of communication based on the Teletext and Viewdata concepts, will
result in significant benefits to a variety of industries.

REFERENCES

Browning, K A (1977) Unpublished report available from the
Meteorological Office Radar Research Laboratory, Malvern.

Hill, F F, Whyte, K W and Browning, K A (1977) Met.Mag., 106,
pp 69-89.

Holford, I (1976) British Weather Disasters, publ.
David and Charles, Newton Abbott.

Scofield, R A and Weiss, C E (1976) Preprints 6th Conf. on
Weather Forecasting and Analysis, Amer. Met. Soc., Boston,
pp 67-73.

Taylor, B C and Browning, K A (1974) Weather, 29, pp 202-216.

Tatehira, R, Sato, H and Makino, Y (1976) J Met.Res.,
Tokyo, 28, pp 61-70.

Wilson, J W (1966) US Weather Bureau Contract CWB-11093,
Final Report, The Travelers Research Centre, Inc.

RECENT AND PLANNED RAINFALL STUDIES IN THE
METEOROLOGICAL OFFICE WITH AN APPLICATION TO URBAN
DRAINAGE DESIGN
C K Folland and M G Colgate,
Meteorological Office

ABSTRACT

As a result of the project described in the U.K.
Flood Studies Report (1975) improved advice is
available for the rainfall input to urban drainage
design. The availability, since 1975, of a
steadily increasing rainfall archive based on
autographic records, with a fine resolution in
time and rainfall amount, progressively allows
more complete studies of short duration rainfall.
The archive is discussed. Studies are now being
carried out (or initiated) which utilise the
archive to determine firstly the rainfall totals
to be expected in short durations for common
return periods down to at least 10 times a year,
secondly the uncertainty in estimates in common
return periods from short (real) records and
thirdly to examine the variations in return period
resulting from short period climatic fluctuations.
Recent work on the areal reduction factor, based on
studies of 2-day rainfall, is also discussed.
Proposals for further work are made.

INTRODUCTION

Application of the information contained in the
U.K. Flood Studies Report (NERC 1975) led to
improved rainfall advice to urban drainage
designers. This improvement was reflected in the
revision of a manual on urban drainage design and
led to the publication in 1976 of the currently
recommended procedures in Road Note 35 (Second
Edition). The Flood Studies Report shows how
estimates of the rainfall total for a given
duration may be made, for any return period down to

six months, anywhere in the U.K. Values of an
areal reduction factor, which allow the statistics
of point rainfall to be converted to those of areal
rainfall, were also produced for the Report,
together with a simple method for arriving at an
approximation to the variation of rainfall
intensity with time through a storm (the storm
profile). As far as the urban drainage designer
in the U.K. is concerned, some rainfall problems
still remain unsolved; those of current interest
can be broadly summarised as:-

a. What rainfall totals (for a given duration)
are to be associated with the more common rainfall
events? This information is needed for example,
for the design of balancing tanks and storm overflows.

b. What is the nature of the areal reduction
factor for these more common events and, indeed,
how is the factor to be defined for these events.

c. What is the uncertainty in estimates of point
or areal rainfall totals for more common events?
Factors to be considered here are sampling errors,
variations on different time scales, and
uncertainties in the representativeness of the
statistics for the geographical location being
studied.

d. Are the storm profile statistics in the Flood
Studies Report, which allow only the variables of
duration, peakedness and return period of the
storm profile to be considered in combination,
adequate for all purposes.

No further reference will be made to storm movement in this paper. This is often mentioned as a
further problem in urban drainage design and the
reader should refer to some recent work by
Shearman (1977) and Marshall (1977).

The urban drainage engineer is usually interested
in short rainfall durations of a few minutes to a
day or so, and in return periods from one tenth of
a year to about one hundred years. The amount of
readily accessible data on short duration rainfall
was limited when the Flood Studies Project was in
progress, but is now increasing. These newly
available data can obviously assist in an answer
to questions a-d. A description of the new data
now follows.

THE PRECISION ENCODED AND PATTERN RECOGNITION (PEPR)
MACHINE AND THE DIGITISED RAINFALL ARCHIVE

The short duration rainfall data available to the
Flood Studies team consisted mainly of tabulations
of hourly rainfall totals and tabulations of the
maximum rainfall total within a range of specified
durations for each month and each year. Neither
data are satisfactory descriptions of short
duration rainfall with a very common return period;
only a continuous data stream, preferably of
minute-by-minute values, is adequate.

In 1973 the Meteorological Office set up a
cooperative project with the Nuclear Physics
Laboratory at Oxford University for the automatic
digitisation of hyetograms (Davey et al 1974,
Wiley et al 1974). A computer-controlled, flying-
spot scanner, originally developed by the Nuclear
Physics Laboratory to digitise bubble chamber
photographs and known as a precision encoded and
pattern recognition machine (PEPR) has been
employed. A contract was placed for the digitisation of about 1¼ million daily Dines Tilting
Syphon and Natural Syphon recorder hyetograms from
locations chosen by the Meteorological Office.
These rainfall recorders are described in
Meteorological Office (1956). The Greater London
Council also became associated with the project
and about 350,000 of the 1¼ million hyetograms
being processed are from the dense autographic
raingauge network in and around London.

The procedure is briefly as follows. The original
hyetograms are scrutinised and cleaned up manually
to remove features which might confuse the
automatic scanner (a laborious task). The daily
rainfall records from an adjacent 5" manually read,
check gauge, (almost always available at
autographic stations) are punched onto cards for
each month of the autographic record, together
with information about the state of the autographic
record. This information is needed during digitisation, for instance, to inform the scanner that a
trace is broken or that the zero level of the chart
is misaligned. Occasionally charts are retraced
when the original hyetogram is considered
unsuitable for the scanner to follow: the charts
are then microfilmed. The microfilm is placed
before the scanner (at Oxford) which produces a
series of time/rainfall amount coordinates at a
fixed interval of about 2mm along the trace. The

resolution in rainfall amount and time therefore
depends on the angle of the rainfall trace to the
horizontal. With standard Met Office charts which
have a vertical depth equivalent to 5mm of rain,
an almost vertical trace would give a resolution
of about 0.2mm and perhaps no time increment,
while a horizontal trace, corresponding to no
rainfall, would have a time resolution of
17 minutes. Other comparable resolutions would
result from some of the other types of chart
digitised. A consequence of the resolution
limitations is that sharp turning points on the
trace are smoothed a little, but this effect is of
slight consequence when the data are processed to
support urban drainage design.

The rainfall trace is rescaled to agree with the
daily read check-gauge total, provided that the
apparent total rainfall measured by the autographic
gauge for the 24 hour period 0900-0900 GMT is
within 90% to 120% of the check gauge total and
also within 1.5mm of the check gauge total.
Allowance is also made for constant but non-
standard starting and finishing times of the
autographic record and for small irregular timing
errors; large timing errors are "flagged".
Adjustments are also made if the vertical pen
excursion across the chart at the occurrence of
siphoning is non-standard i.e. (usually) not
exactly equivalent to, say, 5mm of rain. Further
adjustments can also be made.

The resulting magnetic tape contains the digitised
hyetogram rainfall values at increments of
hundredths of a mm and associated (variable) times
(to the nearest minute) at which the increments
were achieved, together with the information from
the punched card and any corrective action taken
during digitisation. Other information stored
includes the starting and stopping times of each
days' trace, as calculated by the PEPR machine,
the number of siphon tips and the hourly rainfall
totals.

Computer quality control programs are then run at
the Meteorological Office to ensure that the traces
have been correctly analysed. These include
comparisons of the check gauge rainfall total with
the hyetogram totals, a check that there is an
adequate number of digitised points in a day and
that the rainfall increments between successive
digitised rainfall trace points have reasonable

values.

As a result of these and other tests, error
messages are produced, as well as messages
describing the corrective action taken at the
digitisation stage. Subsequent corrections involve
manual scrutiny and may involve the manual
digitisation of complete traces or parts of traces,
deletion of records or perhaps a reassessment of
the check gauge total. These corrections are
themselves subjected to quality control and the
process is repeated until all errors have
apparently been resolved or poor data deleted.
The records are also scrutinised for notable
rainfall totals.

The resulting archive is rather unwieldy for some
analytical purposes. Some records have been
reorganised in such a way as to eliminate the
archiving of long dry periods and also to avoid
too much fragmentation of the rainfall data into
many small events. Other forms of archiving have
also been used depending on the problem in hand.

MAGNETIC TAPE EVENT RECORDER (MTER) DATA

Only a brief description is given here and fuller
details appear in Folland, Harrold and Hooper
(1978). The Meteorological Office as a program
underway to install MTERS in gaps in both the
autographic and daily rainfall networks, though
some recorders have been placed at Meteorological
Office Stations to allow comparisons to be carried
out against both Dines Tilting Siphon and daily
raingauges. MTERS have been placed in both urban
and remote rural areas. They record rainfall on
a magnetic tape cassette in 0.2mm increments
measured by a tipping bucket raingauge, and have
a timing resolution of 1 minute. Unattended
operation over 1 month is usual and a longer
interval is feasible. It is expected that about
100 stations will be operational within two years
or so; 60 are already so. The rainfall
increments and associated times of occurrence are
transferred to a central archive in the
Meteorological Office, after translation of the
tapes. Simple forms of quality control are carried
out including corrections for timing errors. An
adjacent check gauge is not always maintained.
MTERS may become valuable as secondary, non real-
time, calibrating raingauges for precipitation
radars, as their data are more quickly retrieved

than those from Dines charts, especially if the
cassettes are changed at frequent intervals. In
the meantime MTERS will produce useful
supplementary data for urban drainage studies.

THE DISTRIBUTION OF AUTOGRAPHIC RAINFALL STATIONS

Fig. 1. shows the locations of autographic rainfall
stations believed to be operating in the U.K. in
October 1977. It must be emphasised that data are
not regularly received by the Meteorological Office
from most of the stations and at many the data
recovery will be incomplete. The data from the
magnetic tape event recorder stations identified
separately in Fig. 1. are, however, all archived
regularly and data are, on average, about 90%
complete.

The need to provide a good geographical coverage
combined with the need for some dense autographic
networks has, to a considerable extent, influenced
the selection of autographic records for
digitisation by the PEPR system. An outstanding
constraint however has been the need for good
quality and fairly lengthy records. At the
moment, digitisation of records is only well
advanced for England and Wales. It should be
pointed out that in addition to the PEPR stations,
a dense network of about 20 autographic stations
has been maintained in connection with the
Gloucester Surface Water Study and data from these
have been regularly digitised. Some of these
stations have now been converted from Dines
Tilting Siphon to Magnetic Tape Event Recorder
stations (Fig. 1.).

Fig. 2. shows the "PEPR stations" with over 10
years of data which are currently being, or are
planned to be, digitised. Data from a further
50 or so stations, mostly in the London area, with
less than 10 years of record, have also been
digitised.

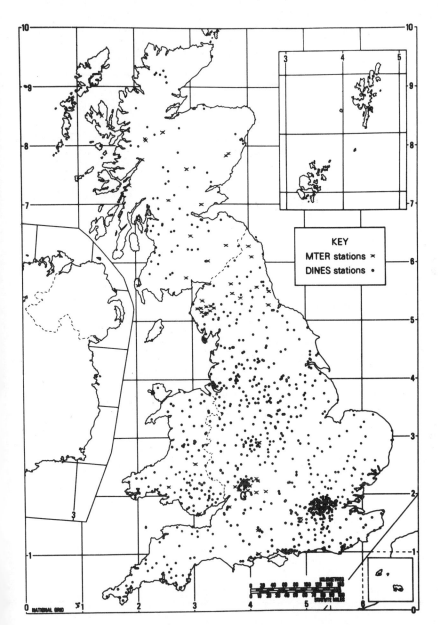

Figure 1.   Distribution of autographic raingauges
October 1977 (Gloucester area marked G)

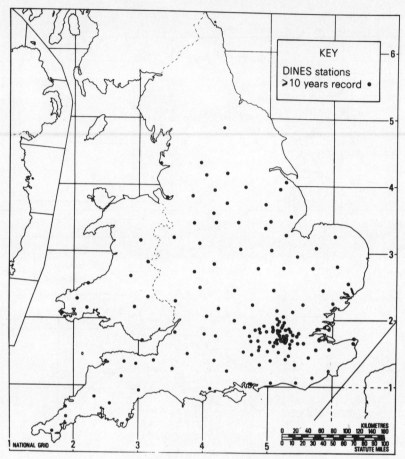

KEY

DINES stations
≥10 years record •

1 NATIONAL GRID

KILOMETRES
0   20   40   60   80   100   120   140   160
0   10   20   30   40   50   60   70   80   90   100
STATUTE MILES

Figure 2.   Distribution of PEPR stations with
over ten years of record already
or being digitised

THE APPLICATION OF PEPR AND DAILY RAINFALL DATA
TO THE ANALYSIS OF EVENTS HAVING COMMON RETURN
PERIODS

The construction of "growth' curves for common
return periods

In United Kingdom practice it is customary to use
the concept of a design storm having a (given)
return period that is associated with the total
rainfall that occurs within the duration of the
storm. This concept is being retained in the new
drainage design model being developed in this

country (Price 1978). The following description
refers to a program of work which is attempting to
extend the Flood Studies rainfall statistics to
include events with common return periods, for use,
for example, as the input to such a drainage design
model.

In the Flood Studies Report Volume II (FSR (2))
relationships ("growth" curves) are produced (for
a given duration of rainfall) between rainfall
amount and return period using specific ranges of
once in 5 year (M5) rainfall as a grouping factor.
The Report goes on to show that the growth curves
have a form that is largely independent of the
rainfall duration considered. Consequently any
growth curve may be simply and effectively
identified by points specifying firstly the M5
rainfall amount and secondly values of MT rainfall
(where T is any return period) where the latter
are obtained by multiplying the M5 value by a
factor depending on the value of T tabulated in
the Report. The Report makes it clear that
although return periods as small as twice a year
have been assessed, the reliability of the
rainfall totals corresponding to such common
return periods is lower than for some greater
return periods. FSR(2) used an annual maximum
series so that only the largest rainfall value of
a given duration in a given year was used in the
analysis. These values were converted in FSR(2)
into the equivalent partial duration series for
return periods of twice a year (2M) and once a
year (1M). The partial duration series of
maximum rainfalls is essentially a complete series
of all the values recorded above a certain
threshold, no matter when they occurred in the
years of record. The partial duration series has
been used in all the subsequent analyses in this
paper. The annual maximum series obviously
cannot be used to produce rainfall totals
corresponding to return periods of several times
a year. The partial duration series is also
thought preferable to, say, a monthly or weekly
maximum series because of seasonal effects.

A pilot study was carried out on 12 stations in
southeast and central England, each with about 20
years of record. The PEPR data were converted
into minute by minute rainfall totals. Rainfall
durations of 5, 10, 15, 20, 30, 40, 60 and 120
minutes were chosen. The total quantity of
rainfall in each of these durations was calculated

for all 20 years; a count was begun each minute
when rain was falling. A partial duration series
of maximum rainfalls was formed for each duration
(N minutes) of interest in the following way. The
maximum rainfall depth of that duration occurring
in the whole record was found and its date and time
of occurrence noted. Any further period of N
minutes which started during the N minutes period
of the maximum rainfall was rejected before
proceeding to find the next largest depth. This
was done to reduce the effects of persistence by
preventing overlapping of the extracted rainfall
totals. The process was repeated until an
average of 100 depths per year of record for a
given duration were extracted, the number in any
year varying somewhat. This process effectively
defined a fixed rainfall threshold below which no
further analysis was done. A test for
independence of the extracted data was carried out
for 5-minute rainfalls at one of the stations,
Hampton, near London, by rearranging the partial
duration series into chronological order and
calculating the correlation between successive
rainfall values in the resulting series, at
various time increments apart. Table 1
illustrates the results.

TABLE 1   Serial correlation of Hampton 5-minute
           rainfalls at various time increments
           apart (1954-1973 data)

Time Increment	0-30 min	30-60 min	60-90 min	90-180 min	3-12 h	12-24 h
Correlation	.33	.27	.36	.18	.10	.00
Number of Observations	500	443	279	337	510	256

The correlations are significant up to and
including events 3-12 hours apart. The effect of
the correlations is however, only to reduce the
variance by about 10% compared to independent
data so for the purposes of the pilot study the
data were treated as independent. However further
work is required to estimate the true equivalent
number of random 5 minute rainfalls and those of
other durations. With this reservation in mind
statistical distributions were then fitted to the
rainfall totals ranked according to their return
periods.

Estimates of the best fitting distribution for the
"growth" factor   Although some statistical models
for the partial duration series are known, it is
not clear, due to the unknown underlying
distribution, which of several likely distributions
is best (eg see Volume I of FSR (1975)). For events
corresponding to return periods of 10 times a year
and greater it is not sufficient to assess the
return period merely from the position of the
individual rainfall total in the ranked series, as
some smoothing is required. Four likely
distributions were tried and chi-square and
Kolmogorov-Smirnov tests applied to estimate the
likely best fit (Siegel 1956). Table 2 shows the
result of testing the best fit distribution using
the chi-square test for the twelve stations for
each of six durations in the range 5 mins to
120 mins. The figures in the table refer to the
number of occasions where the null hypothesis that
the distribution was a good fit was rejected at
the 5% level. For details of these distributions
see Kendall and Stuart (1969).

TABLE 2   Comparison of distributions using
           chi-square test.

Distribution	Pareto	Exponential	Pearson Type III	Log Pearson Type III
Number of rejections at 5% level	10	19	29	7

There is obviously little to choose between the
four distributions, none of which are always a
good fit. However the Pareto and Log Pearson
Type III are of comparable goodness of fit, a
conclusion supported broadly by the Kolmogorov-
Smirnov test (not shown). The Pareto distribution
was chosen for the pilot study because of its
simpler form; it can be expressed in the form of
Equations 1 and 2 as follows:-

$$\frac{x_2}{x_1} = \left(\frac{T_2}{T_1}\right)^{-\frac{1}{b-1}} \tag{1}$$

where x_1 and x_2 are rainfall totals of a given
duration and T_1 and T_2 are the corresponding
return periods.

b is a parameter given by

$$b = 1 + \cfrac{1}{\overline{\log_e x} - \log_e x_{min}} \qquad (2)$$

where $\overline{\log_e x}$ is the average logarithm of all the rainfall values and x_{min} is the threshold rainfall value.

The above solution for b was obtained by the maximum likelihood method. (Kendall and Stuart 1969). Further analysis on more stations is required to confirm the choice of the Pareto distribution.

<u>Growth factor for common return periods</u>   The relationship between the 5-year return period rainfall total (0.2M) and lesser return period totals was investigated for durations of 5, 10, 15, 20, 30, 40 and 60 minutes at S. Farnborough, Hampshire. (Note that the convention adopted in FSR(2) for the nomenclature for partial duration series return periods in the form $T^{-1}M$ and annual maximum series return periods in the form MT has been retained; thus a partial duration series return period of 5 years is denoted by 0.2M while that of one fifth of a year is denoted by 5M.) Fig. 3. shows an example of the resulting growth curve for return periods from 0.2M to 10M using the Pareto distribution where all durations have been combined into one curve using the appropriate values of 0.2M as the normalising factor. The vertical bars show the total range over which the ratio $T^{-1}M \div 0.2M$ varied.

The Flood Studies estimates for S. Farnborough are shown as open circles. The conclusion is that the use of 0.2M as a normalising factor is quite appropriate for this range of durations confirming the use of M5 (the annual maximum version of the once in five year rainfall) in FSR(2) at least at S. Farnborough. The FSR(2) values agree well with the values presented here except for the 0.5M value This is probably because FSR(2) used the M2 value, the rainfall of return period 2 years in the annual maximum series, which can be shown, given independent rainfall data, to be equivalent to the 0.69M partial duration value (Langbein 1949). The black dot (on Fig. 3) shows the Flood Studies value plotted accordingly at a return period of 1.45 years. It can also be shown that there is

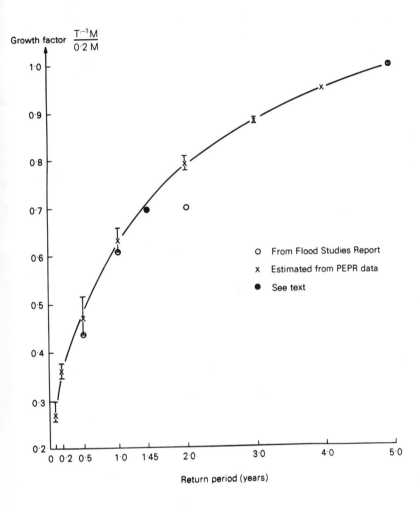

Figure 3.   Ratio of $T^{-1}M$ to 0.2M rainfall for short
             durations (5-60 minutes) at S. Farnborough

little statistical difference between 0.2M and
M5 estimates, which accounts for the good agreement
between these values for S. Farnborough, while
FSR(2) itself used partial duration series estimates
for return periods of 1 year and 6 months.

Uncertainty of return period estimates

FSR(2) states that local records should be used to
estimate short return period rainfall values. An
investigation was carried out into the
consequences of using records only five years long
to estimate return periods from 0.2M to 10M at
Hampton. Fig. 4 shows the percentage difference
from the mean of 4 successive 5 year estimates
(1954-1973) of rainfall totals associated with
given return periods and durations of 5 minutes,
30 minutes and 2 hours. The total spread of values
increases as duration decreases and is still
considerable even when the return period is much
less than the length of the data record. For
example the 1M value estimates vary by as much as
± 20% for 5 minute rainfall and ± 10% for
120 minute rainfall. This result demonstrates the
fluctuations in return period determined from data
of different periods and which may be random or
systematic in time. Thus stable values from short
period records require the pooling of data from
several adjacent and topographically similar
stations. The result also implies that a design
rainfall total, however good in the long term, may
not always be entirely appropriate to a particular
sub-set of years. This natural variability in
rainfall may set constraints to the accuracy with
which other design variables are really required
in sewer design, and is worthy of further
consideration.

Areal reduction factor appropriate to common return periods

The areal reduction factor
calculated in FSR(2) is really one corresponding
to a return period of 2-3 years. The definition
used in FSR(2) cannot be used to estimate areal
reduction factors for other return periods as it
stands. It is thought appropriate, however, to
retain the basic principle of the FSR(2)
definition, which, for a given duration essentially
forms, in a fixed catchment, the ratio of the
point values of rainfall to the corresponding areal
values, each having the same return period. Storm-
centred definitions that work from the maximum
point value near the centre of the storm are not
really appropriate to drainage design as they do
not compare point and areal values of the same
return period. This fact is the main reason why
storm-centred areal reduction factors have
different values from the factors in FSR(2),
usually being less, since on average, the areal
values in given storms have a lower, more common

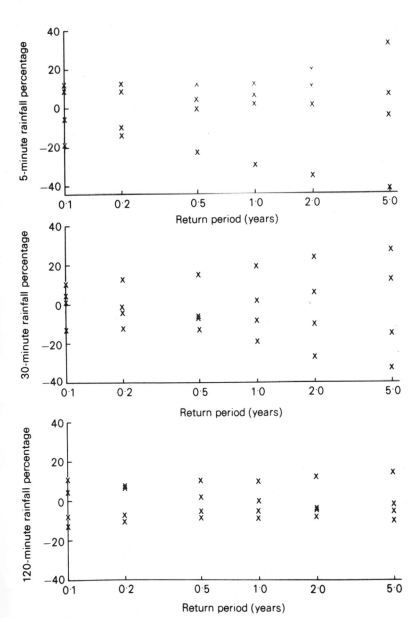

Figure 4. Rainfall totals associated with given return periods
(as determined from successive 5 year periods of
data) expressed as a percentage departure from the
mean of the 4 successive 5 year periods.

The 0.5M value on the graph (return period two
years) was obtained in a slightly different way as
it cannot be reliably obtained from five years of
data (a record length chosen only to reduce the
great computational time required to calculate the
common period rainfall totals). The 0.5M areal
reduction factor was calculated from a series of
annual maximum point and areal rainfalls in a
similar way to FSR(2) except that the median rather
than the mean value of the 47 reduction factors was
finally calculated. The median is thought more
appropriate than the mean because the distribution
of areal reduction factors is not normal. The
resulting factor has a return period rather less
than that in FSR(2) because of the use of the
median. It is difficult to estimate the value
precisely but it is quite near 0.5M and it has been
allocated this value in Fig. 5.

The crosses on Fig. 5 represent the range over
which the central 83% of the areal reduction factor
estimates occurred at each return period. Although
the values of the factor shown in Fig. 5 may not
apply to the smaller areas and durations
appropriate to drainage design it is clear that the
assumption of an areal reduction factor independent
of the return period of the event breaks down for
relatively common events. This must be so in an
area of homogeneous rainfall climate, since rainfall
averaged over the area must reduce to be the same
as rainfall at a point, when the rainfall is
totalled over all storms of all return periods. For
rare events, areal values are always less compared
to point values, because of the restricted size of
intense storm cells. Very common events, however,
making up about half of total rainfall, are
associated with areal falls larger than point falls.
This reflects the fact that the number of dry days
at a point is greater than that over an area. The
relative distributions of areal and point rainfall
are well illustrated in a recent analysis of radar
data for short duration rainfall over relatively
small areas in the USA (Frederick et al 1977).

Investigations have also been carried out into the
seasonal and geographical variation of areal
reduction factors over 1000 km^2 areas for 2-day
rainfall totals associated with the 2-year return
period event. The conclusion is that there is only
a slight seasonal and geographical variation. An
analysis of such areal reduction factors for events
with return periods greater than 2 years showed

(areal) return period than that of the maximum
(point) value near the centre.

A study was undertaken, using daily rainfall data,
to calculate the value of the areal reduction factor
for 2-day rainfall totals for return periods from
1000M to 2M. The frequency of occurrence of all
rainfall values, in increments of 1mm, was found
for both the areal values and values at each point.
Five years of data from 47 circular areas, each of
1000 km^2, from all over the UK, were used with, on
average, 20 stations in each area. The areal value
was calculated by interpolating the 2-day rainfall
totals onto a regular grid with a spacing of 5km
(Shearman and Salter 1972). The factor was then
estimated as the ratio of the areal to point
rainfall amounts having the same frequency of
occurrence or return period. Fig 5 shows the
areal reduction factor calculated in this way,
averaged over the 47 1000 km^2 areas in the UK for
return periods between 100M and 2M for 2-day
rainfalls.

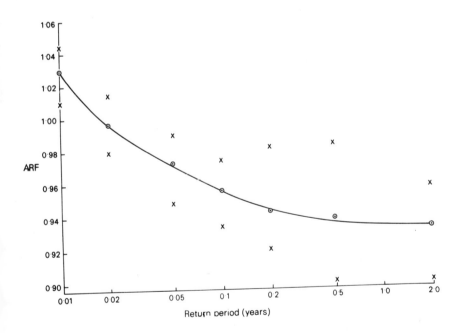

Figure 5   Variation of the areal reduction factor
              (ARF) with common return periods

little variation with return period. For further
details see Colgate and Folland (1978, unpublished).

FUTURE WORK PLANNED IN THE METEOROLOGICAL OFFICE

a. <u>Work on rainfall with a frequent return period</u>
The Meteorological Office hopes to carry out much
more extensive work on this important subject,
utilising data from rainfall stations illustrated
in Fig 2. Further work will be undertaken to
determine the best way of assembling the data and
the results will be presented in a form suitable
for engineers, much as was done in FSR(2).

b. <u>Work on variability of return period estimates</u>
This should be greatly extended and include
investigations of systematic changes in the
frequency of heavy rainfalls with length of, and
through, a record, for rainfall durations of a
few minutes to 1-2 days. Statistical techniques
such as maximum entropy power spectrum analysis
(Ulrych and Bishop (1975), and time series
filtering (Craddock 1968) can be used to study the
changes in the frequency of occurrence of heavy
rainfall above different thresholds. Other
techniques, such as Shearman's w statistic,
(Craddock 1968) can be used to draw some
conclusions about the rarer events as these events,
by definition, involve a very limited data sample.

c. <u>Areal reduction factor</u>
A determined attack on this problem for durations
and areas appropriate to urban drainage design must
await extensive radar precipitation data. However,
in the interim, it would be possible to carry out
limited work around London and Gloucester for areas
of several tens of km^2, for all short durations.
It is likely that a pilot study will be carried out
for rainfalls having frequent return periods as an
extension of (a).

d. <u>Further work on storm profiles</u>
Currently it is thought that, for the purposes of
assessing peak flow in storm sewers (Price 1978),
the symmetrical profiles produced in FSR(2) are
adequate. However it is possible that for the
assessment of flood volumes and the design of
balancing tanks etc, further investigation into the
structure and sequence of storms, particularly
those of a longer duration e.g. several hours,
would be valuable. The Meteorological Office has
currently no plans to undertake this work but it

would undoubtedly be possible, in principle, to
carry out useful work using the **PEPR** data combined
with advanced statistical techniques such as
principal component analysis. Such a technique
might be used to transform storm profiles to a set
of orthogonal functions representing different
storm patterns which might then be ranked according
to their relative importance, should distinctive
patterns exist; Butters and Vairavamoorthy (1977),
have given some classification of actual storm
profiles and have shown, for example, that the
bimodal storm profile is not uncommon.

REFERENCES

Butters, K. and Vairavamoorthy, A. (1977)
Hydrological studies in some river catchments in
Greater London. Proc.Instn.Civ.Engrs.
Part II, 63:331-361.

Colgate, M.G. and Folland, C.K. (1978)
Relationships between point and areal rainfall
for rainfalls totalled over two days. (unpublished,
copy available in National Meteorological Library)

Craddock, J.M. (1968) Statistics in the Computer
Age. English Universities Press.

Davey, P.G., Harris, J.F., Hawes, B.M., Lamb, P.R.,
Lohen, J.G. and West, N. (1974) Technical
Developments in the Oxford PEPR system. Oxford
Conference on Computer Scanning. Oxford University.

Folland, C.K., Harrold, T.W. and Hooper, A.H. (1978)
The use of recording raingauges in the Meteorological
Office (to be published).

Frederick, R.H., Myers, V.A., and Auciello, E.P.
Storm Depth - Area Relations from Digitised Radar
Returns. Water Res. Research, 13, 3:675-679.

Kendall, M.G. and Stuart, A. (1969) The Advanced
Theory of Statistics Vol. I. Charles Griffin and
Company Ltd.

Langbein, W.B. (1949) Annual floods and the
partial-duration flood series. Wash.,Trans.Amer.
Geoph.Union, 30:879-881.

Marshall, R. (1977) PhD thesis, University of
Bristol.

Meteorological Office, (1956) Handbook of
Meteorological Instruments, Part I. HMSO.

NERC (1975) UK Flood Studies Report, Vols 1-5
(Vol II: Meteorological Studies).

Price, R.K. (1978) A New Design and Simulation
Method for Storm Sewers. Int. Conf. Urban Storm
Drainage, Southampton.

Shearman, R.J. (1977) The Speed and Direction of
Movement of Storm Rainfall Patterns. (Unpublished,
copy available in National Meteorological Library).

Severn-Trent Water Authority. (1975) Gloucester
Joint Surface Water Study - progress report and
further research. (Unpublished).

Siegel, S. (1956) Non parametric statistics for
the behavioural sciences. McGraw Hill.

Transport & Road Research Laboratory (1976) A
guide for engineers to the design of storm sewer
systems. Road Note 35, Second Edition. HMSO.

Ulrych, T.J. & Bishop, T.N. (1975) Maximum Entropy
Spectral Analysis and Autoregressive Decomposition.
Rev. Geophys. and Space Phys. 13, 1:183-200.

Wiley, R.L., Hopkins, J.S. and Walker, L.R. (1974).
The automatic digitisation of hyetograms using PEPR.
Oxford Conference on Computer Scanning. Oxford
University.

ANALYSIS OF RAINFALL DATA FOR USE IN DESIGN OF STORM SEWER
SYSTEMS

Viktor Arnell

Department of Hydraulics, Chalmers University of Technology,
Fack, S-402 20 Göteborg, Sweden.

ABSTRACT

The paper describes a comparison of calculated storm water discharges with two kinds of rainfall data: design rainfalls developed from intensity-duration-frequency relationships or from
measured rainfall data and real measured time series of rainfalls or time series generated by statistical methods. These
two rainfall approaches have been compared by simulation of
the runoff by a runoff model for a 0.154 km² catchment area,
Bergsjön, in Göteborg, Sweden. Data from two years of rainfall-runoff measurements have been analysed. Different types of
design rainfalls have been derived and the 40 heaviest real
rainfalls have been selected for simulations. The statistical
analysis of the simulated peak flows shows that the real rainfalls give the best results. The conclusion is that the use of
design rainfalls give a more uneven dimensioning of storm sewer
systems. With real rainfalls it is possible to make a design
from a statistical point of view and to find out what happens
at discharges with frequencies lower than the design frequency.

INTRODUCTION

The need for precipitation data is dependent not only on the
problem studied but also on the design method used. For example
you need different kinds of precipitation data for the design
of a detention storage than for the design of a single pipe.
But you also need different rain data if you use the so-called
rational method than if you use a more detailed design method.

The rational method used for design of storm sewer systems is
defined as (Arnell & Lyngfelt, 1975 b):

$$Q(T) = \varphi \cdot i_m(T,t) \cdot A \tag{1}$$

$Q(T)$ is the calculated runoff with return period T, $i_m(T,t)$
is the average rain intensity with return period T and dura-

tion t, and A is the area of the catchment. φ is a non-dimen-sioned runoff coefficient which defines the relationship be-tween the statistical distribution functions for the peak flow Q (T) and the rain intensity I (T,t),(Schaake, Geyer & Knapp, 1967). The assumption is made that the runoff coefficient is independent of the recurrence interval, and therefore the only statistical analysis you need is the one you obtain through the intensity-duration curves. The peak flows are assumed to have the same frequency as the rainfalls. The use of intensity-duration-frequency relationships is connected with the use of the rational method and gives for this method enough statisti-cal information of the rainfall. The definition above of the runoff coefficient also implies that it is not possible to cal-culate the runoff for single real rainfall events with the ra-tional method.

During the last few years new and more detailed design methods have begun to come into use. Examples of such methods are ILLU-DAS, (Terstriep & Stall, 1974), SWMM, (Storm Water Management Model, 1971), RRL-method, (Watkins, 1962), and the CTH-model, (Arnell & Lyngfelt, 1975 a). With these methods it is possible to simulate the runoff for real rainfall events. This means that you can apply the statistical analysis to the calculated runoffs instead of to the rainfalls. Since the runoff is the interesting design parameter, this method is more attractive.

Detailed design methods require a different type of precipita-tion data from that of the rational method. Input is here a series of rain intensity values describing the variation in time of the rainfall. When designing a storm water system, you can choose between different kinds of precipitation data:
- design rainfalls developed from intensity-duration-fre-quency relationships or from measured rainfall data

- real measured time series of rainfalls or time series gene-rated by statistical methods.

A design rainfall is usually an average value of many rainfalls and is developed for a certain designing recurrence interval. The simulated design flows are assumed to have the same re-currence interval. Most of the design rainfalls are in one way or another connected with the intensity-duration curves.

Real measured time series of rainfalls can also be used. This means that you apply the statistical analysis to the simulated flows to find the design flow. Since volumes and time lapse vary considerably for different rainfalls, you do not need to make the rough simplifications and assumptions, which you must do when using design rainfalls developed from intensity-dura-tion curves, (Mc Pherson, 1977), (Johansen & Harremoës, 1975). Another advantage of real rainfalls is that you receive infor-mation about what is happening with flows larger than the de-sign flow.

Two types of precipitation data are described in the following, design rainfalls and historical rainfall data. These different rainfall data have been compared by simulation of the peak flows for a runoff area in Göteborg, Sweden.

DESIGN RAINFALLS DEVELOPED FROM INTENSITY-DURATION-FREQUENCY RELATIONSHIPS OR FROM MEASURED RAINFALL DATA

Analysis of intensity-duration-frequency curves (I-D-F-curves)

Since many of the design storms have been developed from I-D-F-curves, the following explanations may be helpful.

The I-D-F-curves are the results of a statistical analysis of single independent rainfalls. The independence is usually defined as a minimum time distance between the rainfalls. This time distance should be connected with the analysed flow problem and will therefore vary depending on, for example, if it is a pipe or a detention storage that is to be designed.

For each rainfall event you evaluate maximum rain volumes for different durations. The volumes for the different rain events with a certain duration are ranked, and the statistical distribution function is evaluated. This gives one function for each duration. For specified frequencies you draw curves showing the average rain intensity as a function of duration. This is the intensity-duration-frequency curves (Figure 1). Each curve con-

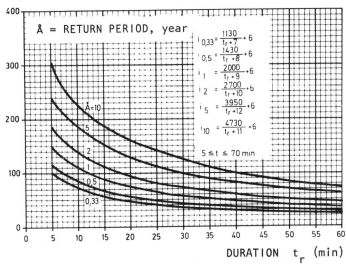

RAIN INTENSITY
l/s·ha

Å = RETURN PERIOD, year

$$i_{0,33} = \frac{1130}{t_r + 7} + 6$$

$$i_{0.5} = \frac{1430}{t_r + 8} + 6$$

$$i_1 = \frac{2000}{t_r + 9} + 6$$

$$i_2 = \frac{2700}{t_r + 10} + 6$$

$$i_5 = \frac{3950}{t_r + 12} + 6$$

$$i_{10} = \frac{4730}{t_r + 11} + 6$$

$$5 \leq t \leq 70 \, \text{min}$$

DURATION t_r (min)

Figure 1. Intensity-duration-frequency curves for Göteborg 1926 - 1971. (VAV, 1976).

tains data from several rain events since the different durations have been treated separated from each other. This means that a design storm developed from an I-D-F-curve will contain data from several real rainfalls, (Mc Pherson, 1977). The return period for the design rainfall must therefore be longer than for different parts of the I-D-F-curve.

The rain volume given by the I-D-F-curves represents only a part of the total volume of the real rainfall. The volume prior to and the one after the studied duration are not included in the analysis. Especially the rain volume prior to the studied duration influences the design of detention storages, (Mc Pherson, 1977). Table 1 shows the rain volume for different durations in comparison with the total volume for a precipitation station in Göteborg.

Table 1. The rain volumes for different durations in comparison with the total rain volume at Lundby, Göteborg, 1926-1955. Average values for rainfalls with a return period exceeding two years, (Arnell, 1974).

Duration	Rain volume corresponding to the duration	Total rain volume	Percent of the total rain volume
min	mm	mm	%
10	10,7	20,0	54
20	15,2	23,9	64
30	17,6	24,4	72
40	19,3	26,7	72
50	20,5	26,7	77
60	21,2	30,2	70
70	22,2	30,4	73

Design rainfalls developed from intensity-duration-frequency relationships

The characteristics of most of the design rainfalls developed from I-D-F-curves are such that the average intensities for different durations follow an I-D-F-curve. The easiest way of developing a design rainfall is to assume that the peak intensity is located in the middle of the rain and distribute the rest of the rain symmetrically around the peak (see Thorndal, 1971 and Figure 2).

Keifer & Chu (1957) presented a design rainfall developed from the mathematical expression for the I-D-F-curves, for example

$$i_m = \frac{a}{t+b} + c \qquad\qquad (2)$$

i_m = average rain intensity during the time t

t = duration

a,b,c = constants

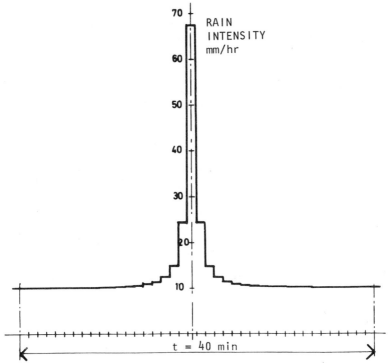

Figure 2. Design rainfall, suggested by Thorndal (1971), derived from intensity-duration-frequency relationship for Bergsjön 1973-1974. Recurrence interval 1/2 year.

From this equation it is possible to develop two expressions
describing the variation in rain intensity prior to and after
the peak intensity:

$$i = \frac{a \cdot b}{(\frac{t_f}{r} + b)^2} + c \qquad \text{(prior)} \qquad (3)$$

$$i = \frac{a \cdot b}{(\frac{t_e}{1-r} + b)^2} + c \qquad \text{(after)} \qquad (4)$$

i = instantaneous rain intensity
t_f = time counted from peak intensity towards the start of rain-
    fall
t_e = time counted from peak intensity towards the end of rain-
    fall
r = the relationship between the time prior to peak intensity
    (t_f^{max}) and the total duration (t).
$r = t_f^{max}/t$; $1-r = t_e^{max}/t$

The location of the peak intensity within the rainfall is evaluated in one of two ways. One is to study the location of the
peak intensity within the duration t for the real rainfalls. The
other way is to determine how much of the total rain volume has
been registered prior to the peak intensity. Precipitation data
for Chicago (Keifer & Chu, 1957), Cincinnati, (Preul & Papadakis,
1973), India, (Bandyopadhyay, 1972) and Czechoslovakia, (Sifalda, 1973) show that between 13/40 and 16/40 of the total rain

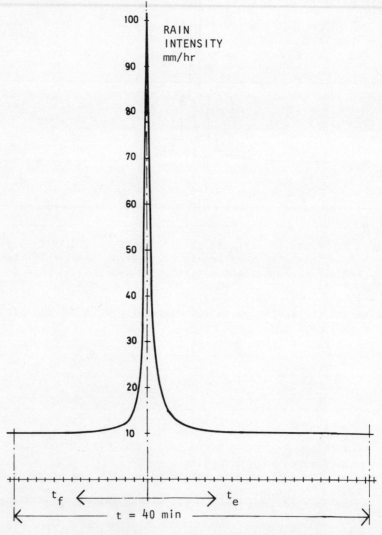

Figure 3. Design rainfall, suggested by Keifer & Chu (1957),
derived from intensity-duration-frequency relationship for
Bergsjön 1973-1974. Recurrence interval 1/2 year.

volume during a rainfall is registered prior to the peak intensity. The design rainfall is thus given an oblique distribution according to Figure 3.

Design rainfalls developed from measured rainfall data

By evaluating "typical" heavy rainfalls, you can develop design rainfalls directly from measured rainfall data. These rainfalls are rather some sort of average rainfalls than design rainfalls developed in a statistical way.

Sifalda (1973) has described a design rainfall of this type developed from data for some places in Czechoslovakia (Figure 4). The rainfall is an average rainfall for those rains, where the average rain intensity for at least one duration exceeds the I-D-F-curve with a recurrence interval of one year. The design rainfall is connected with the I-D-F-curves by part 2 for which the average intensity-duration is chosen from the curves. The average total duration of all rainfalls in the investigation was 30-35 minutes. This means that the duration of the main rainfall, part 2, on the average was only about 8 minutes. Since the rain includes the parts prior to and after the main part 2, the total volume is better described than for the design storms developed directly from the I-D-F-curves.

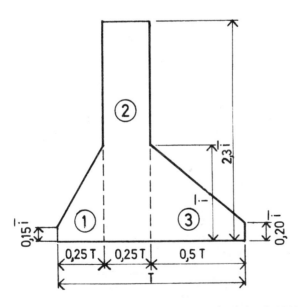

Figure 4. Design rainfall suggested by Sifalda (1973). Intensity-duration for part 2 is obtained from intensity-duration-frequency curves.

For the RRL-method in England a rainfall determined as an average of a number of heavy rainfalls is used, (Natural Environment Research Council, 1975). The rainfalls were divided into four quartiles according to the shape of the rainfalls. The shape was classified from rainfalls with pronounced peaks to more uniform rainfalls. The results are presented in tables and curves; one example is shown in Figure 5. The shapes of the curves were found to be independent of the total duration of the rainfall and the return period. Average intensity (volume) and duration for the total rainfall are given by the I-D-F-curves.

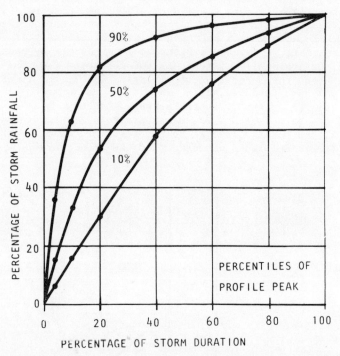

Figure 5. Cumulative percentage rainfall in England (May to October) as a function of rainfall duration. The duration, expressed as percent of the total duration, is centered around the peak intensity. The 90%-curve means that 10% of the rainfalls are more peaked than that curve. (Natural Environment Research Council, 1975).

A similar investigation has been done by Huff (1967). The resulting design storm is used in the ILLUDAS model, (Terstriep & Stall, 1974). Huff found that the peak intensity usually is located in the first quarter of the duration and therefore recommended a curve according to Figure 6. In Huff's study the rainfalls were divided into time increments of 30 minutes, and only rainfalls with a long total duration were studied. Consequently, it is difficult to judge if the result is valid for shorter durations.

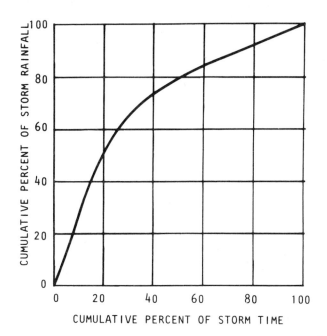

Figure 6. Design rainfall used in the ILLUDAS-model. (Terstriep & Stall, 1974).

Other design rainfalls developed directly from measured rain data have been presented by Holland (1967) and Young (1973).

USE OF REAL MEASURED TIME SERIES OF RAINFALLS

Runoff simulations for real measured rainfalls make it possible to apply the statistical analysis to the simulated flows and thereby find the design flow. With this procedure the rainfall and the runoff do not need to have the same statistical characteristics. It is more attractive to make the statistical analysis on the flow since it is the interesting design parameter.

To minimize the costs, you should select the interesting rainfalls for analysis. The number of rainfalls needed equals the number you need to evaluate the statistical distribution function for flows with interesting recurrence intervals. It should be possible to select a suitable group of rainfalls by means of some method.

Johansen & Harremoës (1975) have suggested the use of a simple runoff model to select the most interesting rainfalls. They propose that you use the time-area-method for developing a unit hydrograph for selected design points in the sewer system. Then you calculate the runoff for all rainfalls by this simple method and make the statistical analysis. The rainfalls corresponding to and close to the design frequencies are selected for more accurate simulation with a detailed design model.

Another method is based on the selection of rainfalls with certain characteristics, for example all rains with a volume exceeding some values. When designing a sewer system for peak flows, you can evaluate the times of concentration for the interesting points in the pipe system. This can be done by means of a design model and a rainfall with constant rain intensity. Knowing the times of concentration, you can, for all rainfalls, calculate maximum average rain intensities for the corresponding durations. After ranking these intensities in magnitude, you can select a group of rainfalls giving runoffs with frequencies around the design frequency. This group of rainfalls is then used in the real design.

The latter method has the advantage that you can once and for all list the rainfalls for different durations. When designing a system, you just calculate the time of concentration and choose the group of rainfalls with corresponding duration and desired frequency. The reliability of the method is probably depending on the size of the runoff area and the structure of the sewer system.

Another advantage of using historical rainfalls is that you obtain information about what is going to happen for flows larger than the design flow.

TEST OF RUNOFF SIMULATIONS WITH DIFFERENT TYPES OF PRECIPITATION DATA

Some of the rainfall approaches have been tested on the catchment called Bergsjön in Göteborg. This is a 0.154 km² large residential area with multi-family houses. The imperviousness is 38%. The structure of the storm sewer system is tree type, and the longest distance from one inlet to the outlet is 800 m. The slopes of surfaces and pipes are rather steep. Additional details of the catchment can be found in Arnell & Lyngfelt (1975 b).

For the Bergsjön area we have evaluated rainfall-runoff data for the period 1973-1974, (Arnell & Lyngfelt, 1975 b). The test described in this paper has been carried out using data on peak flows, runoff volumes, rain volumes, and average rain intensities for 1,2,3,6,9,12,15, and 20 minutes' duration. The 40 largest peak flows and average intensities have been statistically analysed and intensity-duration-frequency curves derived

(see Figure 7). The curves are described by the following
equation:

$$i_m = \frac{a}{t+b} + c \tag{5}$$

where i_m is the average rain intensity (mm/hr) for the duration
t (min) and a, b, and c are constants.

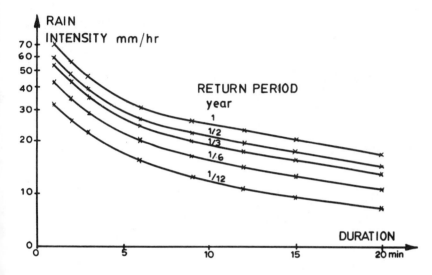

Figure 7. Intensity-duration-frequency curves for Bergsjön
1973-1974.

From the I-D-F-curves design rainfalls according to Thorndal
(1971), Keifer & Chu (1957), and Sifalda (1973) have been
evaluated for return periods of 1/12, 1/2 and 1/1 years (see
Figures 2, 3 and 4). For the Sifalda rainfall you also have
to choose duration since the central part of that rain is
obtained from the I-D-F-curves. In order to find the maximum
peak flows, we have tested the durations of 4, 6, 8, and 10
minutes for the central part giving a total rainfall duration of 16, 24, 32, and 40 minutes.

Maximum average rainfalls with a duration of 3, 4, 6, 8, and
10 minutes have also been used to simulate peak flows. All
these rainfalls have been used as input in a detailed runoff
model (see Figure 9).

The runoff model (Arnell & Lyngfelt, 1975 a), is divided into five parts: infiltration, surface depression storage,
overland flow, gutter flow, and pipe flow. The overland flow
and pipe flow are described by kinematic wave theory. The
model's capability of reproducing the statistical distribution function for the 40 largest peak flows is shown by

simulation of the runoff from the rainfalls corresponding to those peak flows (see Figure 9). For all historical storms and for all of the design storms but the maximum average rainfalls the surface depression storage is chosen to be 0.8 mm for paved areas and 0.3 mm for the roofs. For the maximum average rainfalls the storage is set at zero because these rainfalls have no rainfall prior to the main rainfall.

Time of concentration for the Bergsjön area has been determined by simulation of the runoff for a constant rain intensity of 25 mm/hr preceded by a rainfall of 3 mm/hr (see Figure 8).

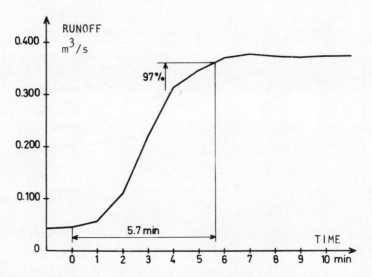

Figure 8. Runoff from Bergsjön for a constant rainfall of 25 mm/hr preceded by a rainfall of 3 mm/hr.

97% of the asymptotic flow value has been chosen as the point of time for evaluating the time of concentration (see Izzard, 1946).This time was found to be six minutes. The 40 biggest rainfalls were selected from the list of average rain intensities with six minutes' duration. These historical rainfalls are used for runoff simulations, and the result is compared with the results from the simulations for the design storms (see Figure 9).

RESULTS AND DISCUSSION

The results of all simulations are presented in Figure 9. The figure shows the real distribution function for the measured peak flows and the calculated peak flows of different kinds of rainfall data.

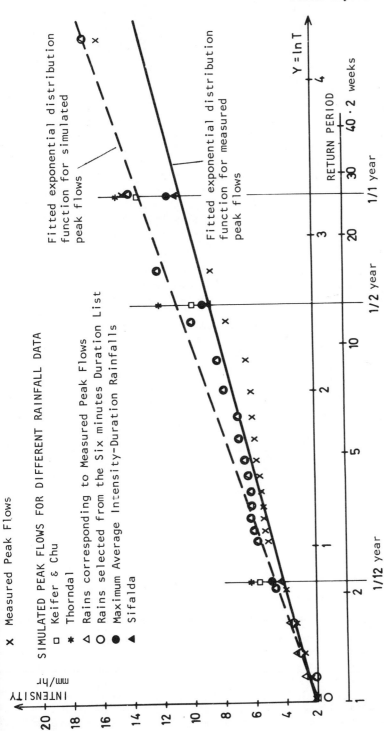

Figure 9. Results of simulation of the peak flows from Bergsjön 1973-1974 with different types of rainfall data.

The runoff model slightly overestimates the peak flows because of the difficulties in determining the areas supporting the runoff. The model is not calibrated. Input data are chosen after mapping and from the literature.

The distribution function for peak flows corresponding to the rainfalls selected from the six minutes' duration list coincide nearly exactly with the simulated distribution function for the real peak flows, except for recurrence intervals shorter than one month.

Calculated peak flows for different design rainfalls should be close to the distribution function simulated by the model (see Figure 9). Peak flows corresponding to rainfalls suggested by Thorndal and Keifer & Chu are close to or slightly larger than the distribution function, especially for the return period of 1/12 year.

Design storms presented by Sifalda and simple average-duration rainfalls give too small peak flows for return periods of 1/2 and 1/1 year. They are so close to each other because of the choice of the maximum surface depression storage. For the simple average rainfalls the storage values are set at zero. These design rainfalls are probably too great a simplification of reality. We have to remember that they were developed for use in the rational method.

CONCLUSIONS AND FUTURE WORK

This study does not indicate large differences in simulated peak flows between design storms and historical rainfall data. Simple average intensity-duration rainfalls and storms according to Sifalda may give slightly too small peak flows. However, this is a study applied to only one area and an area with a tree type pipe system. This means that the lag-time for different subareas is about the same. Marsalek (1977) compared design storms according to Keifer & Chu and historical storms. He found that the Keifer & Chu storms gave much larger peak flows. The explanation to this can probably be found in the characteristics of the runoff area.

Further studies are now being carried out on larger areas and areas with varying structure of the storm sewer system.

The use of historical storms makes it possible to carry out a good statistical analysis of simulated flows and is therefore much more attractive. The work is now mainly focused on how to analyse and use historical rainfall data when designing different storm sewer systems.

In practical engineering work the use of design storms is simple and makes the analysis cheaper. Even if the use of historical storms is more attractive from a statistical point of view, it can be worth-while trying to improve the design storms by means of a detailed runoff model and real measured rainfall data.

A long time series of rainfall data is necessary to make it possible to evaluate results and write manuals to be used in engineering design work. In collaboration with the Swedish Meteorological and Hydrological Institute, a 30 year series of measurements for a rainfall station in Göteborg is being analysed. These data will be used in future work.

REFERENCES

Arnell, V. (1974) Intensity-Duration-Frequency Relationships for Heavy Rainfalls in Göteborg during the 45 Year Period 1926-1971. Chalmers University of Technology, Urban Geohydrology Research Group, Göteborg, Report No 5. In Swedish.

Arnell, V., Lyngfelt, S. (1975 a) Rainfall-Runoff Model for Simulation of Storm Water Runoff in Urban Areas. Chalmers University of Technology, Urban Geohydrology Research Group, Göteborg, Report No 12. In Swedish.

Arnell, V., Lyngfelt, S. (1975 b) Rainfall-Runoff Measurements at Bergsjön, Göteborg 1973-1974. Chalmers University of Technology, Urban Geohydrology Research Group, Göteborg, Report No 13. In Swedish.

Bandyopadhyay, M. (1972) Synthetic Storm Pattern and Runoff for Gauhati, India. Journal of the Hydraulics Division ASCE, Vol 98, No HY5, Proc Paper 8887, pp 845-857.

Holland, D. J. (1967) The Cardington Rainfall Experiment. The Meteorological Magazine, Vol 96, No 1140.

Huff, F. A. (1967) Time Distribution of Rainfall in Heavy Storms. Water Resources Research, Vol 3, No 4, pp 1007-1019.

Izzard, C. F. (1946) Hydraulics of Runoff from Developed Surfaces. Proceedings, 26th Annual meeting, Highway Research Board, Vol 26, 1946.

Johansen, L., Harremoës, P. (1975) Selection of Rain Event for Design of Sewer Systems. Nordic Symposium on "Quantitative Urban Hydrology" in Sarpsborg, Norge, 1975. The Norwegian Committee for the International Hydrological Decade. Oslo. In Danish.

Keifer, C. J., Chu, H. H. (1957) Synthetic Storm Pattern for Drainage Design. Journals of the Hydraulics div., ASCE, Vol 83, No HY4, Aug 1957. Discussion by Mc Pherson in Vol 84, No HY1, 1958.

Marsalek, J. (1977) Runoff Control on Urbanizing Catchments. Symposium on the "Effects of Urbanization and Industrialization on the Hydrological Regime and on Water Quality" in Amsterdam, Oct 1977. IAHS-AISH Publication No 123.

Mc Pherson, M. B. (1977) The Design Storm Concept. Institute on Storm Water Detention Design, University of Wisconsin, Madison, Wisconsin.

Natural Environment Research Council, London (1975) Flood Studie Report. Vol 1 - Hydrological Studies. Vol 2 - Meteorological Stu dies.

Preul, H., Papadakis, C. N. (1973) Development of Design Storm Hyetographs for Cincinnati, Ohio. Water Resources Bulletin, Vol 9, No 2.

Schaake, J. C., Geyer, J. C., Knapp, J. W. (1967) Experimental Examination of the Rational Method. Journal of the Hydraulics Division, ASCE, Vol 93, No HY6.

Sifalda, V. (1973) Entwicklung eines Berechnungsregens für die Bemessung von Kanalnetzen. Gwf - Wasser/Abwasser 114 (1973) H9.

Storm Water Management Model (1971) Volume 1 - Final Report. Environmental Protection Agency (EPA), Water Quality Office, Washington D. C. Water Pollution Control Research Series, 11024DOCO7/71.

Terstriep, M., Stall, J. B. (1974) The Illinois Urban Drainage Area Simulator. Illinois State Water Survey. Urbana, Illionis, Bulletin 58.

Thorndal, U. (1971) Precipitation Hydrographs. Stads og havne-ingeniøren, Köpenhamn, No 7. In Danish.

VAV, Swedish Water and Sewage Works Association (1976) Manual for Design of Sewer Pipes. VAV, Stockholm, Publication P 28. In Swedish.

Watkins, L. H. (1962) The Design of Urban Sewer Systems. Dept. of Scientific and Industrial Research, London. Road Research Technical Papers No 55.

Young, C. P. (1973) Estimated Rainfall for Drainage Calculation in the United Kingdom. Transport and Road Research Laboratory TRRL, Crowthorne, England, Report LR 595.

SYNTHESIZED AND HISTORICAL STORMS FOR URBAN DRAINAGE DESIGN

J. Marsalek

Canada Centre for Inland Waters

INTRODUCTION

During the past 15 years, a number of mathematical models has
been developed for the calculation of runoff hydrographs from
urban catchments. All these models use some form of rainfall
data as one of the inputs, and the output obtained from these
models is the resulting runoff hydrograph at a selected
location. In applications of runoff models to the design of
urban drainage, the following two types of rainfall input data
are used:

      (a)   Design rainfall hyetographs
      (b)   Long historical rainfall records

      A design rainfall hyetograph completely describes
the distribution of rainfall intensity during a storm of a
known return period. Typically, such a hyetograph is derived
by synthesizing a large number of historical rainfall events
and serves as input for single-event runoff models.

      Historical rainfall records are used for continuous
runoff simulation which has so far gained a little acceptance
in the design of urban drainage. Under special circumstances,
continuous simulation can be successfully approximated by a
multi-event simulation with single-event models.

      In the following discussion, runoff peak flows
simulated for synthesized as well as historical rainfall
events are compared for a number of catchments which were
patterned after some typical urban developments in Southern
Ontario. Although the results obtained are only valid for
the conditions studied, the comparisons give a general indication of the relationship between the synthesized and
historical storms and demonstrate some shortcomings of the

approach based on the design rainfall hyetograph. The analysis is restricted to runoff peak flows on small and intermediate catchments (less than 130 ha).

SYNTHESIZED DESIGN STORMS

Traditionally, the design of urban drainage has been based on the design event concept which is well accepted by the engineering profession. The design event is characterized by a rainfall storm, either a block rainfall or, more recently, by a rainfall hyetograph. Such hyetographs are typically derived by synthesis and generalization of a large number of historical events. A probable frequency of occurrence of these design storms is estimated, and the runoff calculation proceeds under the assumption that the frequencies of occurrence of the design storm and of the calculated runoff peak are identical.

The concept of design storms and its application in urban drainage design is subject to considerable criticism. In particular, the attempts to assign mean frequencies of probable occurrence to storms of various intensities and durations are criticized, and the assumption of the identical frequencies of occurrence of the rainfall and runoff events is questioned because of the statistical non-homogeneity of rainfall and runoff data (McPherson, 1975). Although such criticism seems to be generally justified, the shortcomings of the design storm concept have never been demonstrated on historical rainfall data, or in conjunction with runoff calculations. Such an evaluation of the design storm concept was attempted in the following analysis which was limited to two typical examples of design storms, the Chicago storm and the storm proposed in the Manual of the Illinois State Water Survey (ISWS storm). The Chicago storm was selected because of its wide acceptance in the Canadian engineering practice. The ISWS storm was selected because it is closely based on the actually observed storms.

Chicago Design Storm

One of the first design storms, the Chicago storm, was recommended for the design of urban drainage more than 20 years ago (Keifer and Chu, 1957). Additional information on this storm was recently presented by Bandyopadhyay (1972), and Preul and Papadakis (1973). The Chicago storm has become fairly widespread in the North American practice, partly because the Chicago storm hyetograph can be easily derived from the existing rainfall intensity-duration-frequency curves, and partly because of the lack of other approaches. In recent years, several Canadian municipalities have adopted this type of a design storm in their design criteria for urban drainage.

In relation to the historical events, the Chicago design storm preserves the maximum volumes of water falling within the specified durations, the average amount of rainfall antecedent to the peak intensity, and the relative timing of the intensity peak.

To develop the Chicago storm hyetograph, one needs first to determine the dimensionless time of the peak intensity. This time, r, divides the hyetograph into two parts and is defined as:

$$r = t_p/T \tag{1}$$

where t_p is the time to the peak intensity measured from the beginning of the storm, and T is the total storm duration. Values of r are determined for a number of historical storms and a mean value is adopted for the design hyetograph. Both parts of the hyetograph, before and after the peak intensity, are derived from the rainfall intensity-duration-frequency curves expressed as:

$$i_{av} = \frac{a}{t_d^b + c} \tag{2}$$

where i_{av} is the average rainfall intensity over the duration t_d, and a, b, c are constants determined by fitting the above function to the observations. The total storm duration is typically selected from 1 to 6 hours; however, this duration does not affect the magnitude of the peak rainfall intensity of the storm, or the dimensionless time to peak.

The Chicago-type rainfall hyetographs of various return periods were developed for the area of interest by M.M. Dillon Ltd. (1977). These hyetographs, which are adopted in this study, were derived from a 15-year rainfall record available for the station at the Royal Botanical Gardens in Hamilton. One of these hyetographs (2-year return period) is shown in Figure 1 as an example.

Illinois State Water Survey Storm
The Illinois State Water Survey (ISWS) developed a procedure for deriving a synthesized storm for the design of urban drainage (Terstriep and Stall, 1974). In this procedure, the maximum hourly rainfall depths are derived from local data or the rainfall intensity-duration-frequency curves for various return periods. These rainfall depths are then distributed in time following the technique used by Huff (1967) to analyze heavy rainstorms in Illinois. Historical storms are first divided into a number of groups according to the relative timing of the peak intensity. For the largest group, the distributions of rainfall in time are determined, and the median distribution is adopted for the design storm.

For the rainfall record available, the maximum

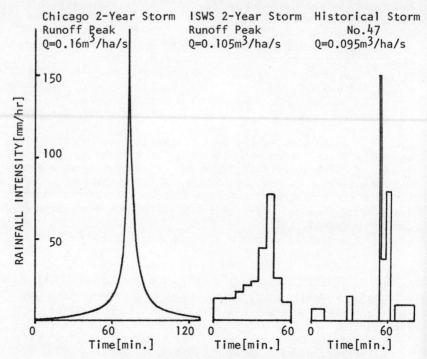

Fig.1. Synthesized and Historical Storm Hyetographs

hourly rainfall depths were taken directly from the intensity-duration curves prepared by M.M. Dillon Ltd. (1977) for the return periods of 1, 2, 5, and 10-years (see Table 1).

Table 1.   Maximum Hourly Rainfalls of Various Return
           Periods (Royal Botanical Gardens, Hamilton)

Return period [years]	1	2	5	10
Maximum hourly rainfall [mm]	22.1	26.0	33.0	38.6

To determine the temporal rainfall distribution, about 30 heavy historical storms, which are further described in the next section, were divided first into three groups according to the part of the storm in which the peak intensity burst had occurred. The majority of storms had their peak intensity occurring in the last third of the storm duration. A median rainfall distribution was determined for this group and expressed as:

$$R_{cp} = f(T_{cp}) \qquad (3)$$

where R_{cp} is the cumulative percent of rainfall and T_{cp} is the cumulative percent of storm time, and f is an empirical

function. The numerical values of this distribution, which
was adopted for the design hyetograph, are shown in Table 2.

Table 2. Median Rainfall Distribution of Predominant Storms

Cumulative % of storm time $-T_{cp}$	0	10	20	30	40	50	60	70	80	90	100
Cumulative % of rainfall $- R_{cp}$	0	5	10	15	22	30	39	56	86	96	100

An example of the ISWS design hyetograph with a 2-
year return period is shown in Figure 1.

HISTORICAL STORMS

As an alternative to the use of synthesized design storms,
several authors (Linsley and Crawford, 1974; McPherson, 1975)
proposed to transform a long historical rainfall record into
a runoff record which could directly serve for the selection
of design runoff flows. Typically, rainfall records are
transformed into runoff records by means of continuous
simulation models. Although this approach avoids the shortcomings of synthesized design storms, it has not gained much
acceptance so far. Continuous simulation may prove expensive
when sophisticated models are used, or inaccurate in the case
of simplistic models. However, certain types of urban runoff
problems, particularly those related to water quality, cannot
be effectively analyzed by any tool other than continuous
simulation.

In the design of urban drainage, most projects
deal only with runoff quantities, and then the continuous
simulation may be approximated by a series of single-event
simulations. Such simulations were performed in this study
for the selected historical storms which were likely to cause
high runoff peak flows on urban catchments. Whenever necessary, the antecedent conditions were taken into account by
adjusting the parameters of the runoff model.

Selection of Historical Events
To select the historical storms which were likely to produce
high runoff peak flows, the rainfall record was screened to
identify all the storms having either the total rainfall depth
larger than 1.25 cm or a 10-min. intensity larger than 1.5
cm/hr. In total 54 storms meeting the selection criteria were
found. Subsequently, the storms were ranked according to
their maximum 5, 10, 15, 30 and 60-min. rainfall intensities.
Because many storms were ranked among the top 20 storms in
several categories, this selection process yielded only 27
storms meeting all the selection criteria. For the purpose
of establishing the frequency of occurrence of runoff peaks
on the catchments studied, these 27 storms effectively replace

the 15-year rainfall record. The basic characteristics of the top 15 selected storms are summarized in Table 3.

Table 3. Characteristics of Top-Ranked Historical Storms

Num-ber	Storm Num-ber	Total Rainfall [mm]	Dura-tion [hr]	Antecedent Dry Weather Period [days]	5-Day Antece-dent Precipita-tion Index [mm]
1	44	37.8	0.5	8	0.5
2	2	57.7	10.3	2	11.5
3	46	31.2	1.5	2	4.3
4	10	14.2	5.4	6	10.8
5	25	44.7	4.8	3	1.5
6	36	20.8	1.0	1	7.5
7	47	15.3	1.3	1	6.3
8	20	46.5	6.5	3	4.3
9	23	22.9	0.6	1	2.2
10	6	28.7	6.3	6	0.4
11	1	30.0	9.2	3	3.5
12	8	30.7	0.7	1	10.5
13	39	17.0	4.5	3	2.4
14	54	78.5	18.4	8	0.4
15	31	27.7	2.4	0	11.8

A few observations regarding these storms are of interest. On average, the total rainfall depth was about 34 mm and the storm duration was 5 hours. Both these values are, however, affected by the definition of a storm event, i.e., the minimum inter-event time which separates the individual events.

The relationship between the antecedent dry weather period and the antecedent 5-day precipitation of these heavy storms is rather interesting. Low observed values of these parameters indicate that the catchments in the area studied are fairly dry at the beginning of heavy storms and that the effects of antecedent precipitation on runoff from design storms may be neglected. This somewhat contradicts the general criticism of design storms presented earlier.

So far, the analysis of rainfall data dealt only with statistical properties of a rainfall record. The selection of the rainfall input for the design of urban drainage is, however, also affected by other factors, such as the procedure used to calculate the runoff and the characteristics of catchments. These aspects are discussed in the next section.

RUNOFF SIMULATIONS

The analysis of rainfall data is only a preparatory step in drainage design, because eventually the designer needs to know the frequency of occurrence of runoff flows of various

magnitude. Therefore, the rainfall data described in the previous two sections were transformed into runoff flows by means
of hydrologic synthesis. Towards this end, the Storm Water
Management Model (SWMM) of U.S. Environmental Protection
Agency was used. SWMM is a single-event model which was specifically designed for simulation of urban runoff. A detailed
description of the model was presented elsewhere (U.S. Environmental Protection Agency, 1971). The values of the SWMM
hydrologic parameters were adopted from the runoff simulation
studies undertaken for a test catchment in Burlington, Ontario
(Marsalek, 1977).

        Physical catchment parameters strongly influence
runoff simulations and can to some extent influence the
selection of rainfall input. Runoff flows were, therefore,
simulated for a series of 9 hypothetical catchments of widely
varying characteristics. These catchments were patterned
after some typical urban catchments in modern residential
developments in Ontario. Three catchment sizes were used;
26 ha, 52 ha, and 130 ha. In all three cases, the drainage
density was maintained about the same. The catchment imperviousness was varied in three steps; 15%, 30%, and 45%. The
last two values are typical for modern residential areas.

        Two types of rainfall inputs were used in runoff
simulations for all the catchments. Firstly, runoff flows
were simulated for two synthesized design storms, the Chicago
and ISWS storms, of various frequencies of occurrence. The
frequencies of the runoff peaks produced by these storms were
assumed to be identical to the frequencies attributed to the
design storms.

        Secondly, runoff flows were simulated for the selec-
ted historical storms. The frequencies of occurrence of the
simulated runoff peaks had to be determined by frequency
analysis. Towards this end, the peak flows were ranked and
their recurrence intervals calculated from the Weibull
plotting-position formula (Chow, 1964) as follows:

$$T = (N + 1)/m$$

where N is the number of items, m is the order of the items
arranged in descending magnitude (thus m=1 for the largest
item), and T is the recurrence interval (T=1/P, where P is
the probability). Note that the choice of a plotting-position
formula was not very important because only the middle section
of the distribution, where all plotting-position formulas give
practically the same results, was of a particular interest.

RESULTS AND DISCUSSION

Return periods of runoff flows simulated for various historical
and design storms were plotted in Figure 2 for the smallest
catchment studied and for three values of the catchment

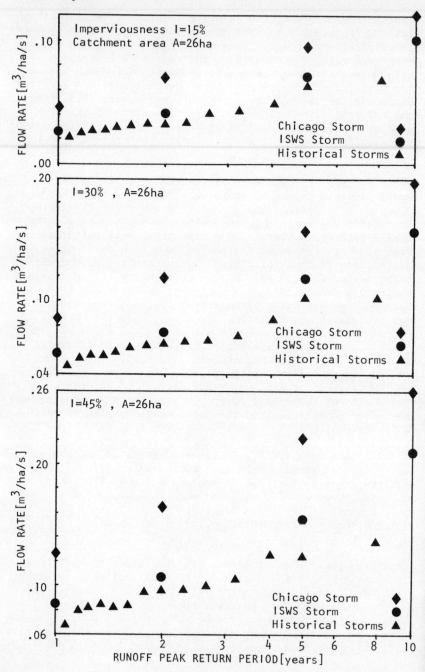

Fig.2. Return Periods of Runoff Peak Flows

imperviousness. From this graph, one can readily compare the
results obtained for the two synthesized design storms and the
historical storms.

When studying the effect of the catchment size, the
peak flows per unit area were found to be attenuated with an
increasing area. This peak attenuation was fairly consistent
and represented about a 13% reduction when comparing the
smallest (26 ha) and the largest (130 ha) catchments of otherwise identical characteristics. It is conceivable that even
larger differences could be encountered in the practice,
depending on the relation of the concentration times of the
catchments studied.

The comparison of simulated runoff peaks for historical and synthesized storms yielded interesting results.
For all the return periods, both design storms produced flows
larger than those produced by the historical storms of corresponding return periods. The overestimation was particularly
large for the Chicago storm which produced peak flows from all
the catchments about 80% larger than those produced by the
corresponding historical storms. Some explanation of this
overestimation was offered by Marsalek (1977) who pointed out
the following shortcomings of the Chicago storm:

(1)   All the maximum rainfall intensities which were observed
      for the specified durations during a number of historical
      storms, are attributed to a single design storm.

(2)   The intensity-duration-frequency curves are extrapolated
      into extremely short intervals, thus yielding peak
      rainfall intensities exceeding the 5-min. intensity
      by up to 60%.

(3)   The description of the time of the peak intensity by a
      single r-value, which is an average of all the r-values
      observed for selected storms, is questionable in view
      of the probabilistic nature of this parameter.

The ISWS storm produced better results than the
Chicago storm. The peaks simulated for the ISWS storm were
only slightly (27%) larger than those simulated for the corresponding historical storms. There is, however, some degree
of arbitrariness in the definition of this storm, particularly
in the choice of the storm duration which affects the magnitude of rainfall intensities. The ISWS storm duration of one
hour was recommended on the basis of some runoff simulations
done with the ILLUDAS model for several urban catchments
(Terstriep and Stall, 1974). The highest runoff peaks were
obtained for the one-hour storm. Similar tests were done
with the SWMM model for the rainfall data and catchments
studied here. By reducing the ISWS storm duration from one
to 0.5 hours, the runoff peaks increased by about one-third.
For the 5-hour storm duration, the simulated runoff peaks were

much smaller than those produced by the 1-hour storm. Note also that the historical storms do not lend any support to the assumed duration of the ISWS design storm of one hour. Consequently, the relatively good performance of the ISWS storm reported here may be   incidental, and the choice of durations of this storm should be further examined.

It is evident from the comparisons of runoff peaks simulated for various types of rainfall input that much more attention should be paid to the rainfall input than in the past. The synthesized design storms produced different results and these in turn differed from the results obtained for the historical storms. The uncertainty in the simulated runoff peaks which was caused by the choice of a rainfall input was larger than the uncertainty inherent to the simulation process.

In spite of some inherent shortcomings, design storms are likely to remain in use in urban drainage design, particularly when the designer deals with a conventional design for small catchments. As an alternative to the synthesized design storms, the historical storms, which produced runoff peak flows of certain   return periods, could be used. For example, the storm which produced the runoff peak with the rank of eight could be considered as a 2-year storm. For the nine catchments studied, the runoff peaks produced by storm No. 47 had an average rank of 7.9 and were ranked eighth in five out of nine cases. An average difference between the peaks produced by storm No. 47 and those having rank eight was only 1%. Consequently, for the location and types of catchments studied here, storm No. 47 could be considered as a historical design storm of a 2-year return period. The hyetograph of storm No.47 is shown in Fig.1 .

Similarly, a 5-year historical design storm would be the storm with an average rank of three. The runoff peaks produced by storm No. 46 had an average rank of 2.8 and were ranked third in seven out of nine cases. An average difference between the peaks produced by storm No. 46 and those having rank three was only 0.5%.

A 10-year historical design storm could also be derived. However, it would be desirable to work with a rainfall/runoff record longer than 15 years. For a 15-year record, the 10-year return period belongs to the tail of the distribution where the selection of a plotting formula becomes important.

Historical design storms could then be used for the design of urban drainage in a particular location and within the range of catchment parameters considered in the storm selection process. For an extreme variation in the catchment characteristics, however, the variation in the ranks of the

selected historical storms could make the use of historical
design storms impractical.

The historical storms used for runoff simulations
were selected on the basis of peak intensities for durations
of 5, 10, 15, 30 and 60 minutes. It is of interest to examine
the efficiency of this selection process. For this purpose,
the correlation between the ranks of peak intensities and runoff peaks was examined, for the individual durations, by means
of the Spearman rank correlation coefficient. When considering all 27 storms, the values of the coefficient were larger
than 0.545 which indicated a rank correlation significant at
a 0.01 level of confidence. The peak intensities appeared to
provide a good selection criterion for the identification of
important historical storms.

When attempting to directly correlate the simulated
runoff peak flows and the peak intensities of the historical
storms, the highest values of the correlation coefficient
varied from 0.629 to 0.734. This means that only 40 to 50%
of the linear variation in the runoff peaks could be explained
by the linear variation in the rainfall intensity. Evidently,
not only the storm peak intensity but also other parameters
of the rainfall distribution are important for the generation
of runoff peak flows.

Though the results presented here are only valid
for the conditions studied, the proposed methodology for the
selection of historical storms and the establishment of frequency graphs of runoff flows has a general applicability.
The graphs of runoff flow frequencies, analogous to those
shown in Figures 2 and 3, could be used for quick estimates
of runoff peaks in new areas or for checking design values.

Finally, the analysis presented did not consider the
effects of storage reservoirs in the drainage system on runoff
peaks. Such a less frequent case was analyzed previously and
it was shown that storage effectively transposes the runoff
flow frequency curve in the direction of smaller flow rates
(Marsalek, 1977). The choice of historical storms could be
significantly affected by storage.

CONCLUSIONS

The comparison of runoff peaks simulated for two types of synthesized design storms and historical storms of identical
return periods produced widely varying results. The Chicago
storm produced runoff flows 80% larger than those produced
by the historical storms of corresponding return periods.
Similarly, the use of the ISWS storm resulted in runoff flows
about 27% larger than those simulated for the corresponding
historical storms. The recommended duration of the ISWS
storm of one hour, which affected the simulated peaks

significantly, appears to be somewhat arbitrarily selected.

To establish the frequency of occurrence of runoff peak flows, continuous runoff simulation was approximated by a series of single-event simulations which were done for 27 selected historical storms. The selection of these storms, which effectively replaced a 15-year rainfall record, was based on the ranking of storms according to their peak rainfall intensities for several durations.

Runoff frequency graphs can be used to identify the historical storms which produced runoff peaks of certain frequencies of occurrence. These storms could then be used as historical design storms applicable to a particular location and types of catchments similar to those considered in the selection of historical storms.

REFERENCES

Bandyopadhyay, M. (1972) Synthetic storm pattern and runoff for Gahauti, India. J. Hydraul. Div., Proc. Amer. Soc. Civ. Engrs. 98, no. HY5, 845-857.

Chow, V.T. (1964) Handbook of Applied Hydrology: McGraw-Hill, New York, USA.

Huff, F.A. (1967) Time distribution of heavy rainfall in storms. Wat. Resour. Res. 3, no. 4, 1007-1019.

Keifer, C.J. and Chu, H.H. (1957) Synthetic storm pattern for drainage design. J. Hydraul. Div., Proc. Amer. Soc. Civ. Engrs. 83, no. HY4, 1332-1-1332-25.

Linsley, R. and Crawford, N. (1974) Continuous simulation models in urban hydrology. Geophys. Res. Lett. 1, no. 1, 59-62.

M.M. Dillon Ltd. (1977) Storm drainage criteria manual for the City of Burlington. An unpublished report, Toronto, Ontario.

Marsalek, J. (1976) Malvern urban test catchment. Canada-Ontario Agreement Research Report no. 56, Department of Environment, Ottawa, Ontario, 55 pp.

McPherson, M.B. (1975) Special characteristics of urban hydrology. In Prediction in Catchment Hydrology, pp. 239-255: Australian Academy of Science, Canberra, ACT, Australia.

Preul, H.C. and Papadakis, C.N. (1973) Development of design storm hyetographs for Cincinnati, Ohio. Wat. Resour. Bull.9, no.2, 291-300.

US Environmental Protection Agency. Water Quality Office
(1971) Storm water management model. Water Pollution Control
Research Series, Report 11024 DOC 10/71, Washington, DC, USA.

A SYSTEM FOR ANALYSIS OF PRECIPITATION FOR URBAN SEWER DESIGN

Bengt Dahlström

Swedish meteorological and hydrological institute

1. INTRODUCTION

A necessary condition for improving the design of sewerage is
an increased knowledge of the precipitation intensity climate:
one of the reasons today for the widespread use of crude
designing methods seems to be that the lack of detailed information on the characteristics of precipitation does not
justify the use of more advanced methods.

The sparsity of observational evidence on the intensity variations of precipitation is certainly the principal cause of
this deficient knowledge. In addition the fact that existing
data do not lend themselves readily to processing by a computer plays an important rôle. In section 2 below some
features of a project for increasing the information on
precipitation intensity in Sweden are described. In section 3
the use of rainfall intensity data for design purposes is
discussed.

2. TRANSFER OF HISTORICAL RECORDS OF PRECIPITATION INTENSITY TO COMPUTER MEDIA

Frequently pluviograph records are evaluated manually and in
order to reduce this laborious work only the heaviest rainfalls are selected for the study of some specific urban
drainage problem.

This practice of only partially evaluating the rainfall graphs
severely reduces the possible applications of the data. Fig 1
outlines some of the research areas where increased information
on precipitation is needed: the whole time span from one
second to ten years (or longer) is of interest as far as
planning is concerned - a fact which should be considered when
establishing a data base.

Two years ago a research project started at the Swedish meteorological and hydrological institute (SMHI) for storing and
processing precipitation intensity data. The data are based on
digitised graphs of precipitation from various instruments and
on data recorded on tape by rapid-response instruments in a
recently established network. The rapid-response equipment is
of the tipping-bucket type and time reference is given each
minut by a watch.

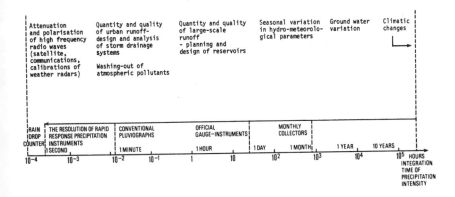

Fig. 1. Sketch of some research fields where increased information on precipitation is required. At the bottom of the
figure different types of precipitation measuring equipments
are indicated. They are arranged according to their approximate
time of sampling the data.

A system of computer programs forms the basis for the suitable
converting, checking and administration of the data. Fig 2
gives an idea of the flow of data.

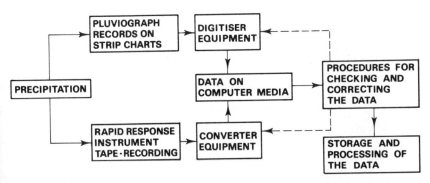

Fig 2. Flow of the collected precipitation data through the
system.

The errors in the data are often closely connected with the type of instrument. The most critical phase in the transfer of data from the graphs is the digitising procedure. To eliminate punching errors the precipitation graphs are drawn by computer and by visual inspection compared with the original ones. The plotted rainfall diagrams are also useful for certain checks when processing the data. Fig. 3 shows a computer plotted rainfall.

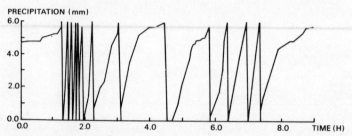

Fig. 3 Computer-plotted graph of precipitation for visual comparison with a strip chart.

Fig. 5 (section 3.1) shows how the same case appears when using an integration time of ten minutes.

When digitizing rainfall records the general rule is that the graph shall be reproduced as accurately as possible. - In many studies only precipitation amounts within fixed time intervals (e.g. 10 minutes) have been used, which blurs the characteristics of the rainfall.

3.   A SYSTEM FOR ANALYSIS OF PRECIPITATION INTENSITY FOR URBAN SEWER DESIGN

A number of mathematical models have been constructed, during the last decade, to represent the flow of water and pollutants in an urban hydrological basin. However, despite the progress of these research efforts, the power and the capability in practice of the different techniques are to a large extent still uncertain. There is at present no universal agreement on the optimum methods for designing urban sewers. The input data are in general well-defined as far as the dimensions and physical properties of the catchment area, such as the proportion of concrete surface, are concerned, whereas the suitable input of precipitation data is a controversial problem.

The problems of designing urban sewers can be attached in a diversified way if a data base containing short-term rainfall is available. Part of this diversification is a condition precedent for using the system presented here, as the input data then have to be specified.

In Fig 4 the main procedures for processing of the precipitation intensity data are indicated.

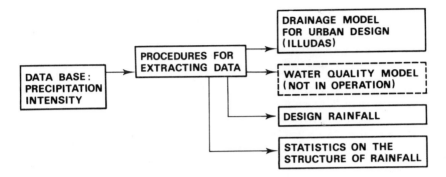

Fig 4. The system for processing of the data for design of
       urban sewers.

3.1   SIMULATION BY USE OF CONTINUOUS SERIES OF HISTORICAL
       PRECIPITATION RECORDS

The basic idea is to expose - by the computer - the urban watershed to a long sequence of real precipitation cases, for instance
the cases occurred during a decade. The precipitation is transferred by the ILLUDAS-model (Illinois Urban Drainage Area
Simulator) to a flow in the drainage system.

The input to the ILLUDAS-model is a rainfall assumed to be
uniformly distributed over the basin which is divided into sub-
basins connected with design points in the basin. Hydrographs
produced from the different sub-basins are combined and routed
downstream from one design point to the next until the outlet
is reached. The ILLUDAS-model is described by M.L. Terstriep
and J.B. Stall (1974).

Whenever overflow is indicated at any of the design points in
the sewerage, information on the duration and the quantity of
the overflow is stored. This information forms the basis for
modifying the specifications of the sewerage. The modified
sewerage is then tested on the critical storms singled out in
the previous computer run.

The fundamental characteristics of this procedure for designing
urban drainage systems seems to be that the design is based on
rainfall events which have actually occurred, rather than on
theoretical considerations.

If a continuous series of precipitation data is used,as indicated above,it is also possible to consider in detail the

processes of infiltration, evapotranspiration and interception. Such refinements, however, have not yet been incorporated in our version of the ILLUDAS-model. If the above processes are included continuously in the computations it is not necessary to process each rainfall in the complex model: in most cases a critical amount (cfr section 3.1.1 below) may be specified and only cases with a total rainfall surpassing this quantity are used as input data in a model of the type indicated above.

To give an idea of the computer capacity required a drainage area with 6 design points, and 10 subareas, was tested: An input rainfall (a hyetograph) with quantities representing 30 successive time intervals required 13.7 s CPU-time on the UNIVAC 1100/21, while a heavy storm with 327 time intervals required 23.3 s CPU-time.

The precipitation during winter has not been taken into account in our computations.

The selection of the ILLUDAS-model is not critical for the system; other fast drainage models may be applied to continuous series of rainfall. The selected model should, however, describe the essential features in the urban drainage - or flow of pollutants - as modifications then are easily introduced. Sophisticated models, based for instance on integrating the dynamic equations of flow, seem, however, to require too much computer capacity for application on continuous rainfall series. These models may - if they are found more accurate than simple models - be used as a supplementary resource for studying in detail the response of the urban hydrological basin to individual rainfall cases.

As the spatial variability of precipitation intensity is not taken into account in the system, the procedures are primarily applicable to smaller urban basins. It is believed, however, that a slightly more complicated model, making proper use of precipitation records from a limited number of stations within or adjacent to the basin, may be applied with fair results to urban basins of a somewhat larger size. The characteristics of storms are, however, still inadequately known - despite the research activities on meso-scale rainfall.

The cases where overflow has occurred may be studied in detail by using computer-plotted hyetographs where also the maximum rainfall for specified durations are shown, cf Figs 5-8. A stu of the weather situation may permit some conclusions as regards the spatial variability of rainfall.

Urban hydrological models generally use as input data the accumulated rainfall in increments of equal time intervals (usually 1, 5, 10 and 15 minutes) for the respective storm. In figures the maximum rainfall during 5 minutes (heavy line)

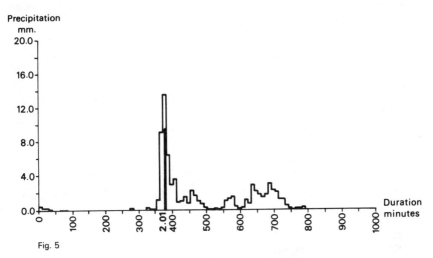

Fig. 5

Sorted with respect to : the max. rainfall
during 5 minutes (heavy pile)
Number of the rainfall case : 1
Rainfall station : KRISTINEHAMN
Period : 451001 — 541101

Start of rainfall : 510809 19.46
End of rainfall : 510810 9.48
Duration of the rainfall : 842.3 minutes
Amount of the rainfall : 78.6 mm.
Mean intensity of the rainfall : 5.6 mm./ hour
Integration time : 10.0 minutes

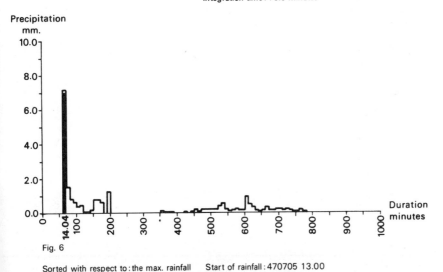

Fig. 6

Sorted with respect to : the max. rainfall
during 5 minutes (heavy pile)
Number of the rainfall case : 2
Rainfall station : KRISTINEHAMN
Period : 451001 — 541101

Start of rainfall : 470705 13.00
End of rainfall : 470706 2.13
Duration of the rainfall : 793.6 minutes
Amount of the rainfall : 23.0 mm.
Mean intensity of the rainfall : 1.7 mm./ hour
Integration time : 10.0 minutes

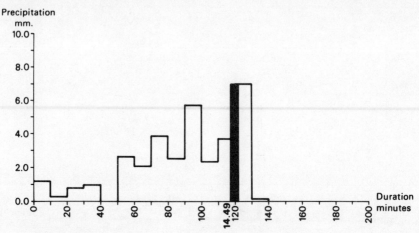

Fig. 7

Sorted with respect to : the max. rainfall
during 5 minutes (heavy pile)
Number of the rainfall case : 3
Rainfall station : KRISTINEHAMN
Period : 451001 — 541101

Start of rainfall : 490718 12.51
End of rainfall : 490718 15.05
Duration of the rainfall : 133.8 minutes
Amount of the rainfall : 33.1 mm.
Mean intensity of the rainfall : 14.9 mm. / hour
Integration time : 10.0 minutes

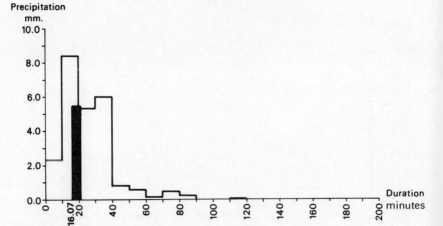

Fig. 8

Sorted with respect to : the max. rainfall
during 5 minutes (heavy pile)
Number of the rainfall case : 4
Rainfall station : KRISTINEHAMN
Period : 451001 — 541101

Start of rainfall : 480706 15.50
End of rainfall : 480706 17.48
Duration of the rainfall : 117.5 minutes
Amount of the rainfall : 24.3 mm.
Mean intensity of the rainfall : 12.4 mm. / hour
Integration time : 10.0 minutes

are plotted. It is evident that the construction of hyetographs
where the first time increment starts at the onset of rainfall
may give a significant underestimate of the maximum precipitation intensity occurred during the storm.

3.1.1 STORM SEQUENCE

Frequently several convective cells traverse the area concerned
in succession, separated by short intervals of dry weather. The
following rule is used to decide whether such a succession of
cells shall be regarded as one event or several events when
using the drainage model: A precipitation case is defined as an
event and a storm sequence as one event if the precipitation
quantity exceeds a specified value (P^*) and is delimited on
both sides by dry weather during at least a specified time
interval (D^*), cf Fig. 9 below.

By selecting a suitable value of P^* we are able to exclude
insignificant cases of rainfall. The duration D^* should equal
the estimated length of the period which allows the urban
sewer system to fully drain.

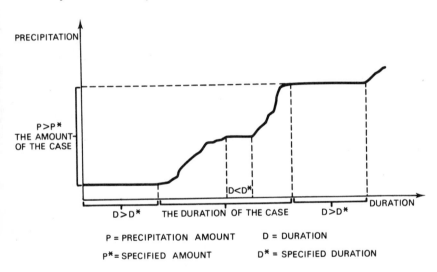

Fig. 9. Definition of a precipitation or storm sequence case.

Some information on the temporal pattern of rainfall is illustrated in Fig. 10. The statistics is based on dividing the
available period (9 years) for the station (Kristinehamn
59°18'N, 14°05'E) into dry and wet periods using $p^* = 0.1$ mm
and $D^* = 30$ minutes. The cumulative frequency of the length of
the periods is illustrated by the two curves in Fig. 6, which
for instance show that 90% of the precipitation cases and 50%
of the dry periods had a shorter duration than 4 hours. Cases
with slight drizzle are in general not represented in this
statistics.

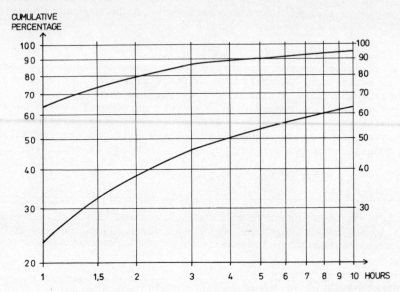

Fig 10. Cumulative distribution of the length of precipitation
events (upper curve) and dry periods (lower curve)
– Cfr the text in section 3.1.1.

3.2   DESIGN STORMS

The design of sewerage is usually based on depth-duration-
frequency statistics of precipitation or on selected individual
cases of rainfall.

Methods for obtaining design storms have been given for instanc
by Stall and Huff (1971), Sifalda (1973) and Natural Environment Research Council (1975).

Design rainfall for various return periods and durations, at
stations where continuous precipitation data are available,
may be computed by the system developed at SMHI for the analysis of rainfall intensity. In particular, this seems to be
a useful tool when designing retention basins, where short retur
periods often are of interest.

The advantage of using an individual design storm for dimensioning urban sewers is questionable as the following problems
are involved.

– the great uncertainty in the selection of the proper design
   storm – a selection indirectly requiring knowledge in
   advance of critical factors in the urban basin studied.

- the temporal pattern of a storm is not taken into account
  if a depth-duration-frequency relationship is used.

- the natural conditions concerning infiltration, soil moisture
  and evaporation are difficult to represent adequately.

- for investigation of problems related to urban water pollution individual rainfall cases are of a limited value: the
  length of the dry periods preceding intense showers is
  critical for the concentration of pollutants owing to the
  fact that sedimentation in sewers during dry periods give
  high concentrations when being washed out by intense showers.

- In Figs. 5-8 the maximum rainfall during 5 minutes is indicated by a heavy pile in the respective diagram. This small
part of the storm is in practice often the only information
used when designing sewers. It stands clear that the response
of urban basins to the cases illustrated in Figs. 6 and 7
generally will be quite different.

4.   CONCLUSIONS

A rather comprehensive system of routines is required for a
satisfactory transfer of historical records of precipitation
intensity to computer media. However, for improving the design
and planning of urban sewers - and also for other urgent
research fields - it seems of paramount importance to increase
the stocks of data on precipitation intensity available for
processing by computer.

In general continuous simulation of the rainfall - runoff
process for the design of urban basins seems to have fundamental advantages, as compared with techniques based on design
storms.

ACKNOWLEDGEMENT

This research is supported by the Swedish Building Research
Council.

REFERENCES

1.   Natural Environment Research Council (1975), Flood
      Studies Report. Vol 1 - Hydrological Studies. Vol 2 -
      Meteorological Studies. Natural Environment Research
      Council, London.

2.   Terstriep, M. L. and Stall, J. B., (1974) "The Illinois
      Urban Drainage Area Simulator, ILLUDAS", Bulletine 5-8,
      Illinois State Water Survey, Urbana, Ill.

3.   Stall, J. B., and Huff, F. A. , (1971). The structure of
      storm rainfall. Reprint 165, Illinois State Water Survey,
      Urbana, Ill.

4.   Sifalda, V., (1973). Entwicklung eines Berechnungsregens
      für die Bemessung von Kanalnetzen. Gwf - wasser/abwasser
      114 H9.

MEASUREMENT OF RUNOFF FROM AN URBAN CATCHMENT

D.A. Tupper, and D.H. Waller

Alberta Department of Environment, Edmonton, Alberta
and
Nova Scotia Technical College, Halifax, Nova Scotia

INTRODUCTION

In 1970 and 1971 quantity and quality of surface runoff and
combined sewage were measured at a catchment in Halifax, Nova
Scotia. The study may be unique in that surface runoff was
measured from two small sub-catchments within the combined
sewer area.

Water quality data obtained in this study have been discussed
by Waller (1971, 1972) and Bhatia (1973). A comparison of
combined sewage flow with flow simulated by two runoff models
was reported by Waller et al (1974).

Some features of the data collection system employed in this
study, limited by experience, time, and budgetary constraints,
were relatively crude when compared with recent recommendations for the design of urban runoff data collection systems
(Allee, [1976], Marselek [1976]). Nevertheless, the study
has yielded useful information, and experience gained from
the conduct of the study may be of value to other investigators.

As a preliminary to further application of the data obtained
in this project, the raw data, project records, and information about the catchments and the flow measurement systems
have been subjected to an intensive review. This presentation describes the catchments and the flow measurement systems, and summarizes the results of an evaluation of usable
runoff data.

COMBINED SEWER CATCHMENT

The location of the combined sewer catchment is shown by Figure 1. The principal reason for selection of this area was
the existence of a flow measurement system at its downstream

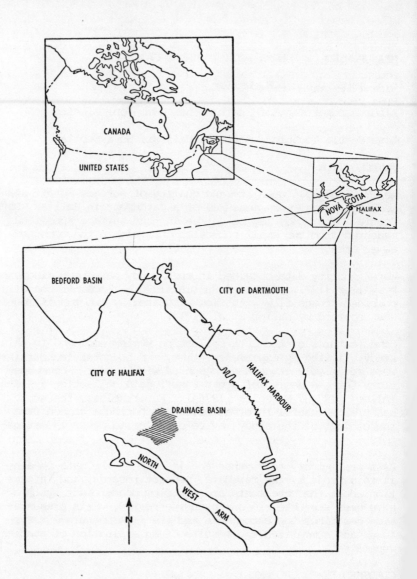

Figure 1: Location map

end.

The size of the combined sewer drainage area is 66.7 ha. This value, based on reexamination and remeasurement of the drainage boundary, is slightly smaller than the value of 67.7 ha. reported previously (Waller [1976]). A more serious problem in establishing the boundary of the area is the fact that an interconnection exists between this system and that of an adjacent drainage area, allowing an unknown proportion of the flow from an area of approximately 8 ha. to overflow during storm periods. Continuing effort is being directed to defining the configuration and behaviour of the overflow manhole.

The flow measurement system, intended for recording and control functions at the inlet to a combined sewage detention basin, consists of a critical depth meter at which flow was measured by an air bubble system. The recording system, which had been plagued with operating and maintenance problems, eventually became inoperative and was replaced for the latter part of this study with a portable pressure recorder. Recorded depths were verified by direct measurements during storm periods. Values of measured flow, which have been based on a theoretical rating curve for the meter, are being verified by field measurements.

Precipitation was measured by a tipping bucket gage located at the detention basin site. A standard rain gage at the same site, and two others at the periphery of the area, each read twice daily, provided further information on the amount and distribution of rainfall on the basin. Gage locations are shown on Figure 2.

Most of the area consists of single family residences and most buildings are at least 30 years old. The population, based on the 1971 census, was 4470 persons. A heavily travelled arterial street, bordered for part of its length by commercial buildings, bisects the area. The commercial area occupies approximately 5.5 ha., or about 8 percent of the drainage basin. All streets are surfaced with asphalt and bordered with concrete curb, gutter and sidewalks. Except for part of the commercial area all sidewalks are separated from the curbs by grass plots. The grass on these plots and on private property is well established. Roofs are required by a city by-law to be connected to the sewerage system, and an examination of buildings in the area indicates that with the exception of occasional clogged or broken gutters or downspouts, and possible spillage during periods of intense rainfall, most roof water can be expected to reach the sewerage system.

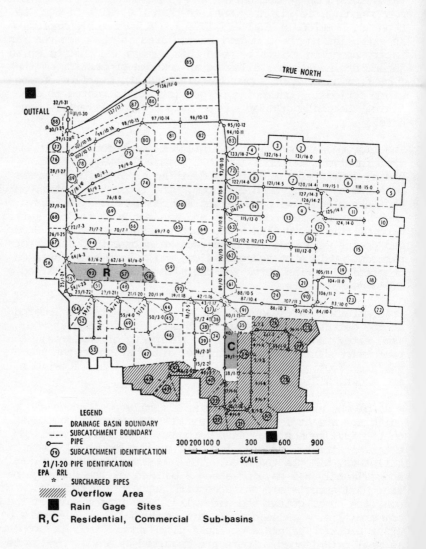

Figure 2: Pipe and subcatchment identification

SUB-CATCHMENTS

The locations of the two sub-catchments from which surface
runoff was measured are shown on Figure 2. These areas were
selected because they were considered to be reasonably representative of commercial and residential areas in the combined
sewer catchment, and because, in comparison with other areas,
the drainage boundaries appeared to be well defined. An additional requirement was that the configuration of the inlet
and catchbasin structures would permit the installation of
flow measurement and sampling equipment.

No measures were taken to change natural drainage patterns in
these drainage basins. The method by which the flow measurement systems were calibrated was intended to compensate for
the effect of flows that bypassed catchbasin inlets. Recorded
flows will be high to the extent that water enters an inlet
from an adjacent area. Visual examinations during a variety
of runoff events indicated this effect was not significant
for the commercial area. At high flows water that bypassed
an upstream inlet was observed to enter the residential catchment, thus decreasing the usefulness of data recorded at this
site.

Recording of flow rates was not the only function of equipment
and instrumentation installed at the surface runoff measurement points. The design of each installation reflected its
dual function of water quality sampling and flow measurement.
Severe design constraints were imposed by the limited space
available in the existing catchbasins.

The general arrangement of the installation at each site is
illustrated by Figure 3. Catchbasin sumps were filled with
concrete to the invert of the outlet pipe, and an aluminum
weir plate was installed across the full width of each basin.
Water depth was sensed by a bubbler system that included a
small air compressor and a pressure recorder calibrated to
register depth over the weir, both located above ground in the
sampler enclosure.

A tray (Figure 4), located under the inlet grating, housed a
float switch to activate the sampler, and served as a sample
inlet sump. The tray also served as a stilling basin, which
directed all water to the back of the weir pond with a minimum of turbulence. A removable screen above the tray permitted regular removal of debris, and of dust and dirt from the
tray, between storms.

Weirs were calibrated in-situ, using portable fibreglass lab-
oratory-calibrated weir tanks, which were carefully levelled
in the field. The tank shown in Figure 5, fitted with a 90°
V-notch, was calibrated over the range 0-0.5 c.f.s. A second

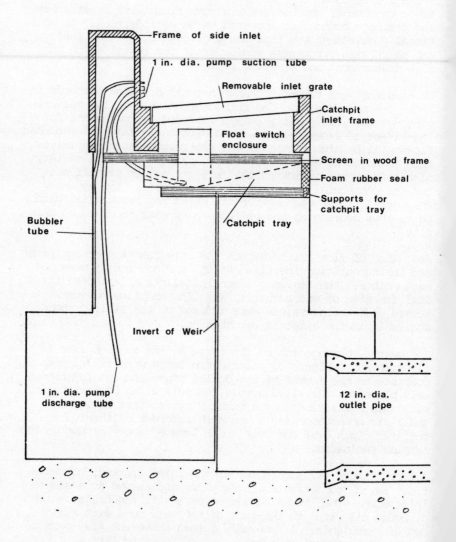

Figure 3: General arrangement of Quinpool Road catchpit

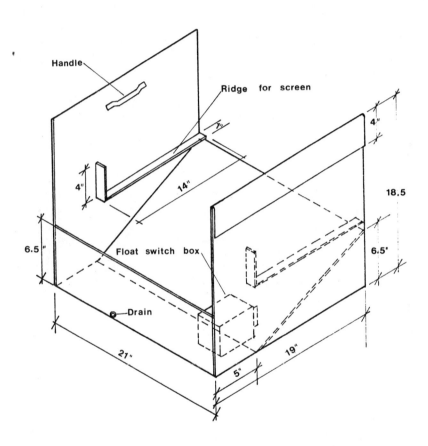

Figure 4: Cambridge Street catchpit tray

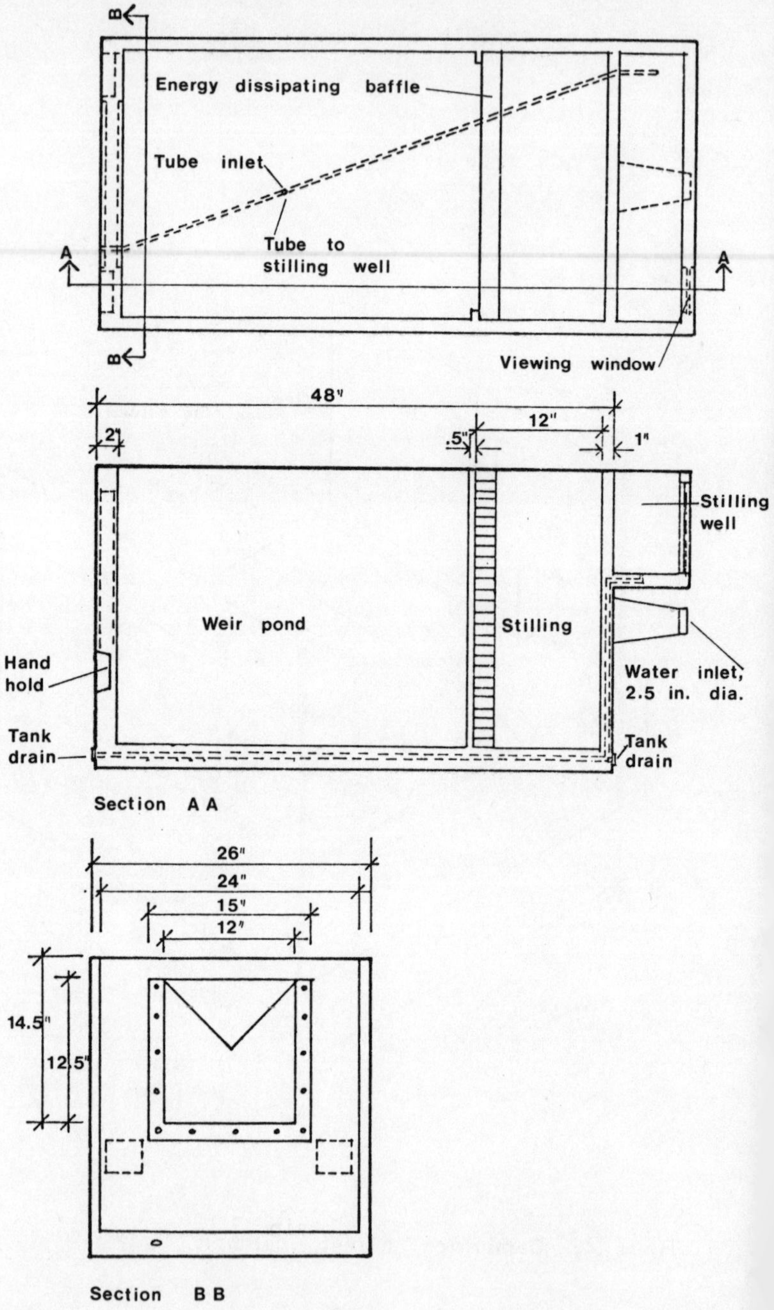

Figure 5: Plan and sections of
small (blue) portable Weir tank

tank, identical except that it was somewhat larger and was fitted with a horizontal sharpcrested weir, was calibrated up to 2.1 c.f.s. Water was supplied from nearby fire hydrants. The back of each tank was equipped with a 2½ inch fire hose fitting. The baffle separating the stilling basin from the weir pond was composed of material customarily used in air conditioning systems. No turbulence was observed in the weir ponds, even at high flow rates.

Catchbasin calibration tests were carried out between 1:00 a.m. and 5:00 a.m., when variations in water system pressure, and in hydrant flows, were minimal. This reduced, but did not remove, the necessity to adjust hydrant valve settings to maintain constant flow rates.

At the commercial site, low to moderate flows did not exceed the capacity of the catchbasin opening, but high flows caused some water to bypass the inlet. The result, shown in Figure 6, was that the measured rating curve, which corresponded closely to the theoretical catchbasin weir curve at lower flows, diverged at high flows because not all of the flow measured in the weir tank entered the catchbasin. If it can be assumed that the configuration and performance of the inlet is the same during storm periods as during the calibration tests, the measured rating curve should provide an accurate representation of flow from the catchment, despite the fact that all flow did not pass over the catchbasin weir.

At the residential site a shallow crown on the street permitted high flows to spill across the street just upstream of the catchbasin. This effect was aggravated by limited capacity of the inlet opening to accommodate peak flows. The result, shown in Figure 7, is that at high flows the measured rating curve flattens appreciably, i.e. flow increases are reflected by only minor increases in measured head at the catchbasin weirs, and measured flow records for high flows are of very doubtful value.

On several occasions leakage occurred from behind the catchbasin weirs making runoff records for some storms unreliable. The leakage was corrected after it was detected by a decline in recorded water levels between storms.

RELIABLE FLOW RECORDS

Combined sewage flow data was obtained in the period from May through November 1970 and May through July 1971. Surface runoff at the commercial site was measured from April through November in 1970 and April through July in 1971. At the residential site surface runoff measurements were obtained from August through November in 1970 and May through July in 1971.

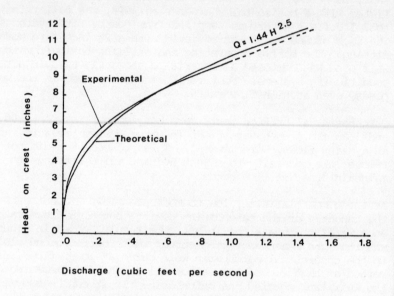

Figure 6: Discharge rating curve for
Quinpool Road catchpit

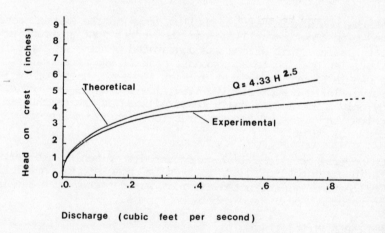

Figure 7: Discharge rating curve for
Cambridge Street catchpit

All of the flow and rainfall records, and equipment operating
and maintenance records, for these periods have been carefully
reviewed and assessed. Those records that are considered to
be reliable and reasonably complete have been identified, and
information necessary for interpretation or adjustment of
chart values has been noted.

Reasons for rejecting data included inadequate information
about time synchronization of rainfall and runoff records,
and results of mechanical, electrical, or human failure that
made records incomplete or unusable. Data that is judged to
be reliable, subject to qualifications described in previous
sections, is available at one or more of the three sites for
a total of 54 runoff events. Data was collected simultaneously at all three sites for 8 complete events. Data for the
remaining 46 events is available as follows:

Combined sewage flow only   - 2
Surface runoff from one site only   - 24
Surface runoff from both sites only   - 12
Combined sewage and surface runoff
from one site only   - 8

For some other events flow data is available that may be useful if used with due regard for its limitations.

Flow data for the 54 events is currently being tabulated and
studied. It is hoped that it will be possible to make this
data, and supporting information about the characteristics of
the catchments from which it was collected, available for
those who may be able to use it for development and calibration of urban runoff models.

PRELIMINARY RESULTS

Recorded combined sewage flow data, based on the theoretical
rating curve for the measuring flow, have been used for comparison with the results of two runoff models (Waller et al,
1976). One point that emerged from that work is that relative
areas of previous and impervious surfaces are not necessarily
a guide to the effective imperviousness of the area as a whole:
roofs and streets can be expected to contribute directly to
sewers, but most offstreet impervious surfaces drain to
grassed areas, which do not contribute to runoff except in
extreme storm events. As Table 1 indicates, the total impervious area represents 50% of the catchment. Comparison of
recorded values yielded a consistent ratio of runoff volume
to rainfall volume of 34%, except for extreme storm events
where the ratio increased somewhat as the apparent result of
contributions from pervious areas.

A similar point has been noted in preliminary analysis of data
from the surface runoff catchments. Runoff/rainfall volume
ratios of approximately 28% have been determined for the com-

mercial site. These are smaller than the ratio of impervious
surfaces shown in Table 2.

TABLE 1

Some Physical Characteristics
of Combined Sewer Drainage Basin

Percent of Total Area Occupied by:

Building Roofs (connected to sewer)	− 21.2
Garage Roofs (most drain to grass)	− 2.9
Streets	− 15.3
Other Paved Areas	− 10.5
Unpaved Areas	− 50.1

TABLE 2

Some Physical Characteristics
of Surface Runoff Subcatchments

	Quinpool Rd.	Cambridge St.
Total Area (Hectares)	0.61	0.73
Percentage of Total Area Occupied by:		
Building Roofs (connected to sewer)	27	22
Garage Roofs (most drain to grass)	−−	2
Paved Areas	33	28
Unpaved Areas	40	48

CONCLUSIONS

It is easy, in restrospect, to identify ways in which the
amount and the quality of storm runoff data produced by this
project might have been improved. Hopefully the experience
recorded here will be of some value to others who contemplate
collection of similar data, and may supplement or reinforce
the recommendations contained in the two recent documents
(Allee, [1976], Marsalek [1976]) that deal with urban runoff
data collection.

Despite the deficiencies in the data collection system, the
study described here has produced data that, used with full
understanding of its limitations, will add to the small amount
of available information about quantities of urban runoff, and
will provide a firmer basis for interpretation of the surface
runoff quality data that this study produced.

ACKNOWLEDGEMENTS

Portions of data collection programs described in this paper
were supported by Central Housing and Mortgage Corporation
and the National Research Council of Canada. Evaluation of
the data was carried out with the support of a grant provided
by Environment Canada under the Water Resources Research Support Program, Inland Waters Directorate, Environmental Management Service.

The portable weir tanks loaned for use in this study were
constructed and calibrated by Professor E.A. Guppy in the Hydraulics Laboratory at Nova Scotia Technical College. The
cooperation and assistance of the Department of Engineering
and Works, City of Halifax, and the Public Service Commission
of Halifax is gratefully acknowledged.

REFERENCES

Allee, W.M. (1976) "Guide for Collection Analysis and Use of
Urban Storm Water Data", American Society of Civil Engineers.

Bhatia, I. (1973) "Mass Discharge Rates and Variations in
Composition of Surface Runoff from Two Urban Areas", unpublished M. Eng. Thesis, Nova Scotia Technical College, Halifax.

Marsalek, J. (1976) "Instrumentation for Field Studies of Urban Runoff", Research Report No. 42, Research Program for the
Abatement of Municipal Pollution Under Provisions of the Canada - Ontario Agreement on Great Lakes Water Quality.

Waller, D.H. (1971) "Pollution Attributable to Surface Runoff
and Overflows from Combined Sewerage Systems", Central Housing
and Mortgage Corporation, Ottawa.

Waller, D.H. (1972) "Factors that Influence Variations in the Composition of Urban Surface Runoff", in Water Pollution Research in Canada.

Waller, D.H., W.A. Coulter, W.M. Carson and D.G. Bishop (1976) "Urban Drainage Model Comparison for a Catchment in Halifax, Nova Scotia", Research Report No. 43, Research Program for the Abatement of Municipal Pollution Under Provisions of the Canada - Ontario Agreement of Great Lakes Water Quality.

ELECTRONIC DATA COLLECTING SYSTEM

W. Verworn

Institut für Wasserwirtschaft, Hydrologie und land-
wirtschaftlichen Wasserbau, Technische Universität
Hannover, West Germany

1. INTRODUCTION

The simulation of the rainfall-runoff process for
urban catchments can be divided into
>   surface-runoff simulation   and
>   sewer-flow simulation

The surface-runoff calculation computes the input
hydrographs to the sewer system from the selected
design rainfall by taking into account the various
surface losses as well as the surface retention. The
sewer-flow calculation computes the peak discharge
or hydrograph at any location within the sewer system. The process is shown schematically in Figure 1.
In general there are no losses of volume within the
sewer system, therefore the total runoff volume by
given design rainfall is primarily determined by the
loss rates of the surface-runoff calculation. Especially for small catchments where the influence of
sewer retention is comparitively small the peak
runoff is substantially affected by the respective
surface losses. A wide deviation of the various loss
rates as well as a variety of types of inflow hydrographs to the sewer system can be found in the literature. J. Keser demonstrated in his paper "Compari-
tative Investigation of Computer Methods" the aggra-
vating consequences of those different loss rate
assumptions.
Corrective measures can neither consist of theoretical considerations nor of laboratory tests but only
of extensive measurements in existing urban catchments.
This paper presents an electronic data collecting
system for urban catchment by which it will hopefully be possible to determine character and size of
hydrological parameters which influence the surface

runoff.

2.   SYSTEM CONCEPTION

The data collecting system is planned to measure and
record rainfall and resulting discharge in the sewer
system. The discharge is determined implicitly via
water-level measurement by forcing a definite stage-
discharge relation at the gaging sites, in this case
by installation of weirs within the conduits.
To gain the purposes of the research project it is
not only neccessary to achieve a high accuracy of
measurement at the gaging sites but also to obtain
an exact temporal assignment of recordings of different gaging stations. Otherwise a definite fixing
of rainfall beginning and runoff beginning, for
example, would be impossible. Therefore the data
recording will not be accomplished at the gaging
sites but the data will rather be transmitted to a
central station where they are recorded synchronous-
ly or at least temporally fixed.
A galvanic connection exists between the central
station and any gaging station for data transmission
Additionally, the transducers receive the power
supply from the central station so that any gaging
station is connected with the central station via a
four-conductor cable.
The received data are recorded in a digital and
computer-compatible mode to reduce errors during
evaluation and conversion.
Figure 2 shows the schematic set-up of the system.
The data collecting system is designed for a capacity of up to five rain gages and twenty water-level
gages. The transducers convert the mass rainfall and
water level, respectively, into a d.c. voltage signal which is subsequently transmitted to the central
station. By using a d.c. voltage signal the system
becomes low-resistant and less susceptible to troubl
and maintenance and possible repair work of the tran
mission lines is more easily done. For further noise
suppression the wires of the transmission cables are
twisted 2o times per meter.
The system is operated in an "off-line" mode, e.g.
the received data are merely stored in the central
station while data conversion and evaluation will
be performed seperately on a big computer.

3.   PRECIPITATION MONITORING

For precipitation monitoring a device was developed
which was especially adapted to the requirements of
the acquisition system as well as the objects of th

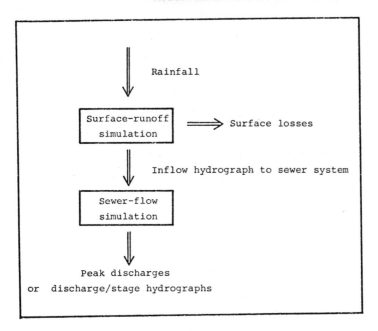

Figure 1   Rainfall-runoff process

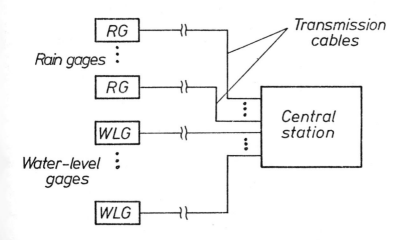

Figure 2   Schematic system set-up

research project. The rainfall is collected by an orifice of about 0.5 m in diameter and is routed into a tube. The water level within the tube is followed by a float. The float drives a single-turn low-friction potentiometer with a resistance between 0 and 5000 Ohm. Special attention was given to the reduction of friction in order to record even smallest level changes and therefore equally small amounts of precipitation.

The electronic equipment of the rain gage converts the resistance of the potentiometer into a d.c. voltage signal which is proportional to the tube level. The tube holds up to 20 mm of precipitation. The depletion of the tube is accomplished by opening a magnetic valve which is automatically activated when an upper level limit is reached. The tube as well as the feed pipes are made of lucid plastic so that any clogging or pollution is immediately evident.

Figure 3 shows the fundamental set-up of the rain gage. Long-term tests revealed that the rain gage maintains a resolution of less than 0.05 mm of precipitation even under less ideal conditions.

4.   WATER-LEVEL MONITORING

The water level in the conduits of the sewer system is sensed in a similar way as the level in the rain gages. Here, the float drives a 10-turn potentiometer with the same resistance range. Float and counterweight move in lucid-plastic pipes. In this way any interference with a correct operation of the device can be easily detected. By using low-friction potentiometers and especially designed floats it was possible to keep the diameter of the float and the protecting pipe low. Therefore the installation of the pipes do not unduly interfere with the flow in the conduits. In addition the device is mounted on one side of a manhole. At the down-stream side of the manhole a horizontal sharp-crested weir is built into the conduit to force a definite stage-discharge relation over a wide range of depth.

Figure 4 shows the set-up of the water-level gaging station.

At most gaging sites there is a small base flow so that the water level is always at the upper edge of the weir thus providing a relative value for the water-level data.

Extensive tests showed that the resolution of the water-level gage lies well under 5 mm.

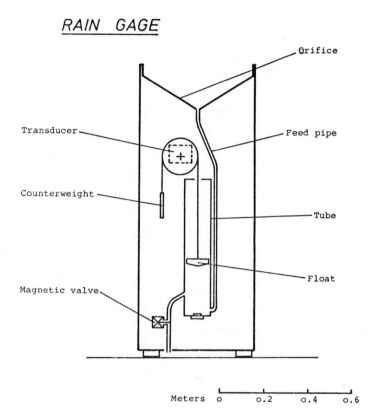

RAIN GAGE

Orifice

Transducer

Feed pipe

Counterweight

Tube

Float

Magnetic valve

Meters   o   o.2   o.4   o.6

Figure 3   Rain gage

WATER-LEVEL GAGE

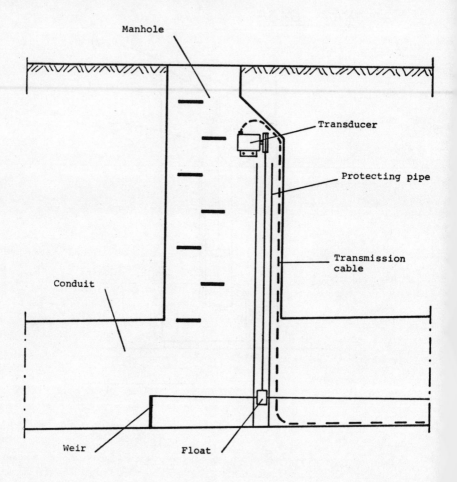

Figure 4   Water-level gage

5. CENTRAL STATION

The central station consists of a control unit, a
registration unit and a power-supply unit for the
transducers.
A paper-tape puncher serves as registration unit.
The paper tape as data-storage medium was given preference over other data-storage media, as magnetic
tape or tape cartridge, since the paper tape offers
a higher data security, also in regard to the operation of the system by less qualified personnel.
The control unit continuously scans all analog inputs in 1-minute intervals, measures the d.c. vol-
tages and displays them at a rate of one station
per second. Hence it is possible to gain an immediate survey of the performance of the system as
well as to observe the changes during rainfall
events.
The recording on paper tape takes place in adju-
stable time intervals which can be set to 5, 15 or
3o minutes. Any output comprises the actual date
and time and the number and value for all stations.
Furthermore the precipitation data are continuously
examined regarding significant changes. As soon as
the value of any rain gage alters at a rate above
a fixed limit, which means that it is raining with
a certain intensity, the control unit switches to
a data-recording in 1-minute intervals. That 1-mi-
nute cycle is kept on past the end of the rainfall
event over a period of 30 minutes in order to
record the entire runoff process in the sewer system.

6. INSTALLATION AND CALIBRATION

The electronic data collecting system is installed
in the storm drainage system of the town of Hildesheim. The selected sewer system serves a town district which is situated on the outskirts of a
range of hills and comprises delimited catchments
with greatly varying terrain characteristics such
as imperviousness, land use and land slope.
The installation of instrumentation and cables
started in spring 1976. In July, 1976, the system
went into operation with 10 gaging stations. At
present, 4 rain gages and 14 water-level gages are
attached to the system. Eighty percent of the transmission cable was layed inside the sewer system itself while the remaining twenty percent was layed
into the ground. The total cable lenght amounts to
about 35 kilometers. The distances between the cen-

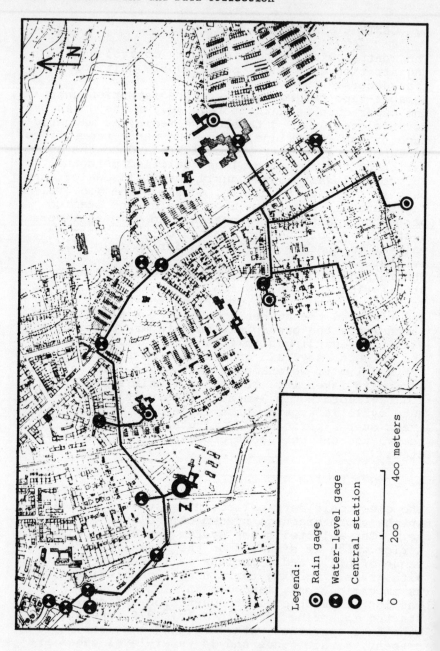

Figure 5   Sites of system components

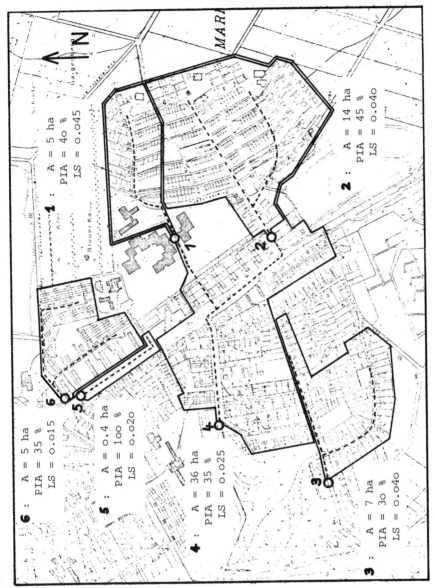

Figure 6   Selected catchment

A   : Catchment area
PIA : Percentage of impervious area
LS   : Land slope

tral station and gaging stations range from 1oo
meters up to 21oo meters.
Figure 5 gives a rough survey of the sites of the
central station, the rain gages and the water-
level gages as well as the tracing of the transmission cables.
Figure 6 shows a part of the covered district. The
catchment boundaries of the respective gaging stations and the tracing of the main sewers are outlined therein. After extensive inspection of the
territory the boundaries of the catchments were
determined as well as the percentage of impervious
area and the land use and any irregularities were
tracked. The sewer system as well was subject to a
thorough inspection. Any unintentional inter-
connections between catchments were blocked to
establish definite runoff conditions.
The different terrain characteristics can be seen
in Figure 6 where the catchment area A, belonging
to the gaging station, the percentage of impervious
area PIA and the land slope LS are specified. Note
that catchment No. 1 and No. 2 are subcatchments
of No. 4.
Gage correlations for all water-level stations were
experimentally compiled and the ranges of validity
were investigated.

7.   DATA EVALUATION

A paper tape has a capacity of about 14.ooo
measured values and hence lasts between 2 and 1o
days depending on the rainfall activity and the
selected time interval for data recording. The
paper tape is subsequently fed into the computer
of the Technical University of Hannover and stored
on magnetic tape. That is the starting point for
any further data processing.
The data which are recorded in voltage units are
intensively checked, converted to physical unities
and rainfall events are printed and plotted. At
present computer programs for the evaluation of the
extensive data material are under development.
Figure 7 shows a computer plot of a rainfall event
which is immediately drawn from the original data
using the values of rain gage RS2 and water-level
gage P3.

8.   FIRST RESULTS

The data which are recorded up until now were
checked, printed and plotted according to the above

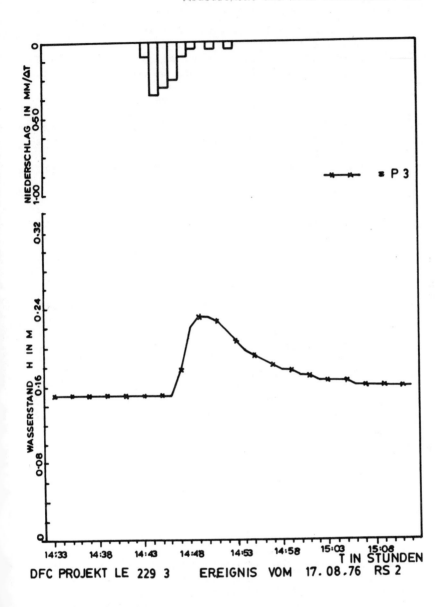

DFC PROJEKT LE 229 3   EREIGNIS VOM 17. 08.76   RS 2

Fig. 7 Recorded rainfall – runoff event

discription. Although no complete data analysis and evaluation has taken place yet, it is already possible to state some trends and qualitative findings. The lag time (the time interval between the beginning of the rainfall and the runoff) amounts to three minutes at the most. This value is considerably opposing the values which can be found in literature where generally two or three times as much is assumed. The recorded lag times which are much·shorter than usual can be explained by the heterogeneous character of the surface which exists in all urban catchments. There are alsways areas which have no suppression storage and little surface-wetting loss thus producing runoff almost immediately. A certain limiting value (the sum of suppression storage and interception and surface-wetting loss) that must be exceeded by the rainfall before the impervious area starts to produce runoff does not exist. That limiting value which is incorporated as an initial condition in many simulation models does not correspond with reality. Besides interception only the surface-wetting can be regarded as an initial surface loss which is furthermore surface specific, which means that it is dependant on the kind of surface, such as roof or pavement, for example.

The depression storage is not an initial loss but rather a time varying loss. It is surface specific, too, and the loss rate depends on the rainfall intensity. Rainfall of low intensity may produce runoff long before the depression storage is completely covered.

9.   SUMMARY

An electronic data collecting system for rainfall and runoff measurements in urban catchments has been presented which yields synchronous rainfall and water-level data of high accuracy. Based on these data the hydrological parameters that influence the surface runoff in rainfall-runoff models are to be determined in character and size. The so far recorded data make it possible to come to the following qualitative conclusions. The lag time is substantially shorter than commonly anticipated. The surface-wetting is the only initial loss for impervious areas while the depression storage is varying with time.

It is planned to continue the measurements until the end of 1979. The high standard of the so far collected data gives rise to the hope that at the

end of the project the hydrological parameters and
their state of dependence will be fixed with little
allowable variation.

10. ACKNOWLEDGEMENTS

The research project "Entwicklung und Installation
eines elektronischen Datenerfassungssystems in Ka-
nalisationen für die Erarbeitung gesicherter hydro-
logischer Daten und die Quantifizierung der hydrolo-
gischen Parameter" (Development and installation of
an electronic data collecting system for urban
sewer systems to acquire assured hydrological data
and to quantify the hydrological parameters) is
financed by Deutsche Forschungsgemeinschaft, a
federal institution.
Particular thanks are due to the municipal authorities of the town of Hildesheim for their great
personal and mechanical engagement without which
the realization of the project would have been
impossible.

URBAN RAINFALL-RUNOFF-QUALITY DATA FOR HYDROLOGIC MODEL
TESTING AND URBAN RUNOFF CHARACTERIZATION

Wayne C. Huber and James P. Heaney

Associate Professors, Department of Environmental Engineering
Sciences, University of Florida, Gainesville, Florida 32611

INTRODUCTION AND OBJECTIVES

In an effort to provide useful planning tools for abatement
of quantity and quality problems caused by urban stormwater
runoff, many mathematical models have been developed to simulate the various components of urban hydrological processes.
These models range from very simple, to very sophisticated,
yet all share a common need--adequate data for development,
calibration and verification. Specifically, these data consist of detailed measurements of rainfall, runoff and quality
parameters taken at frequent intervals during storms, such
that the full dynamic and spatially variable nature of the
urban runoff may be studied. Since most recent urban hydrologic models define the complete hydrograph or pollutograph
during a storm event, measurement of only, say, peak flows or
average concentrations is inadequate for calibration of these
models. Such models are being used in ever increasing applications and the need for relevant data has intensified.

Another important requirement arises from the need to characterize urban runoff in a variety of ways. Examples of such
needs are:
    1) determination of rainfall and runoff volumes, inten-
sities, peaks, durations, interevent times and associated
statistics;
    2) identification of quality parameters found in urban
runoff;
    3) determination of ranges, arithmetic and flow-weighted
means, medians, variances, and other statistics of quality
parameters;
    4) computation of total mass emissions of quality
parameters;
    5) computation of quality "loadings" such as pounds per
acre, pounds per curb-mile, pounds per inch of rainfall,
pounds per day, etc. and combinations, and;

138

6) evaluation of causative relationships among rainfall,
runoff, quality, demographic and abatement factors.

Several of the above needs require collection of both runoff
and quality data; e.g., calculation of total pollutant loads,
flow-weighted averages, etc. requires simultaneous measurement of flows and concentrations. Thus, concentration data
by themselves are insufficient for many required analyses.
Characterization results may then be used to synthesize data
at unmonitored locations.

Data collected for characterization purposes are not always
compatible with modeling needs since infrequent sampling times
and/or omission of key parameters are likely. However, data
suitable for model usage are usually also well suited for
characterization purposes provided enough of a sample exists.
It is desirable that characterization data be representative
of an entire year or season and thus result from samplings of
many storms since one group of data may be used for model calibration while the remaining group may be used for verification.

The project described in this paper, sponsored by the U.S.
Environmental Protection Agency, (EPA), has obtained data,
collected by others, to fulfill modeling needs as first
priority with attention also to characterization needs. As
described subsequently, there have been a surprisingly large
number of studies devoted to collection of data useful for
modeling, although collection of good quality data is more
difficult and lags the quantity data by a considerable degree.
The overall objective of this research, then, has been to
find these data and publish them.

Specific objectives are, broadly:
    1) identify sources of data;
    2) establish criteria for collection of data;
    3) acquire available data;
    4) construct initial data base;
    5) define how continuing maintenance of the data base
is to be accomplished; and
    6) define how data dissemination should be done.

These items have been addressed individually and collectively
in a published report (Huber and Heaney, 1977). Viable, current data sources are described within the report, and actual
data from these sources have been placed for easy access on
magnetic tapes. The data will also be available on the EPA
STORET (STOrage and RETrieval) data management system in the
near future. This paper can only summarize the detailed
descriptions of current and potential data sources and the
extensive reference list contained in the report.

POTENTIAL DATA SOURCES

Introduction
It has been observed during the course of the study that vast
amounts of rainfall-runoff-quality data already exist, and
even more are currently being collected. Of course, only a
minority of these data are suitable for purposes such as
modeling, although a larger fraction may be useful from the
characterization viewpoint. An even smaller fraction is
actually accessible in a well documented, tabulated fashion.
Finally, many sources, especially university studies, are
only discovered by accident; no clearinghouse for such studies
exists. Still, many data sources were uncovered during the
course of the study, and new ones continue to arrive as the
project continues. Only the ones considered most promising
from a modeling viewpoint were included in the first release
of the data base; they are described subsequently. Others
are currently being added and will be documented in addenda
to the first report (Huber and Heaney, 1977).

A considerable aid in seeking data may be found in summaries
of data published by several agencies. These are reviewed
briefly below.

American Society of Civil Engineers
The ASCE Urban Water Resources Research Council has conducted
relevant studies of urban hydrology since 1967. Among the
most widely used rainfall-runoff data are those collected at
the Northwood catchment in Baltimore (Tucker, 1968a) and the
Oakdale catchment in Chicago (Tucker, 1968b) and published
under ASCE auspices. The activities of the Council and
references to published studies are summarized by McPherson
and Mangan (1975). Recent work sponsored by the U.S. National
Science Foundation (NSF) has produced summaries of available
urban hydrologic data and modeling activities in the U.S.
(McPherson, 1975), Australia (Aitken, 1976), Canada (Marsalek,
1976), the United Kingdom (Lowing, 1976), West Germany (Massing
1976), Sweden (Lindh, 1976), France (Desbordes and Normand,
1976), Norway (Selthun, 1976), The Netherlands (Zuidema,
1977) and Poland (Blaszczyk, 1977). McPherson's (1975) report
contains a summary of U.S. and other catchments that have
actually been used for testing of several current urban hydrologic models.

Illinois State Water Survey
The Illinois State Water Survey evaluated the capabilities of
the British Road Research Laboratory (RRL) model for use in
urban drainage design (Stall and Terstriep, 1972; Terstriep
and Stall, 1969). This study included testing on ten U.S.
catchments. The Survey later extended the capabilities of the
RRL model to create the Illinois Urban Drainage Area Simulator
(ILLUDAS) model (Terstriep and Stall, 1974). For this study,

ILLUDAS was tested on rainfall-runoff data from 23 different
catchments, all of which are described by Terstriep and Stall.
The 23 include nine from the RRL study, and their report provides very useful capsulized information about each catchment.

U.S. Environmental Protection Agency
Under the EPA and its predecessor agencies, many urban runoff
studies have been conducted involving extensive sampling programs. Although better documented than most studies, many of
the earlier reports contain samples of only a few storms at
several sites or rely upon composited samples, thus making
them unsuitable for modeling applications. Such reports may
still contain useful characterization data, however, and
several are utilized for this purpose by Heaney et al. (1977).

The number of potentially useful EPA-sponsored studies is too
large to list each individually. Also, the number is increasing because of EPA Section 201 Construction Grant (for sewage
treatment plants) and Section 208 Areawide Waste Management
Grant (for integrated wet and dry weather control) studies
currently in progress under the 1972 Amendments to the
Federal Water Pollution Control Act. However, references to
several are given by Huber and Heaney (1977).

U.S. Geological Survey
The USGS has collected many of the data currently available
for urban basins, and their urban hydrology programs are
continuing. The Huber and Heaney (1977) report contains results of a very detailed rainfall-runoff-quality sampling
program conducted by the USGS at multiple sites in Broward
County, Florida, near the city of Fort Lauderdale (Mattraw
and Sherwood, 1977).

Other U.S. Agencies
Other federal agencies that have sponsored or engaged in data
collection activities include the Office of Water Resources
Technology (OWRT), the Agricultural Research Service (ARS),
the National Weather Service (NWS) and its parent agency, the
National Oceanic and Atmospheric Administration (NOAA), and
the Corps of Engineers. However, relatively few of these
studies include water quality data (in addition to rainfall-
runoff data) for urban areas.

Data Sources in Other Countries
Programs in urban hydrology in several countries have been
summarized by the ASCE as discussed previously. Several
Canadian studies are referenced in the Huber and Heaney (1977)
report, and data from Windsor, Ontario are included in the
data base. A summary of current activities related to urban
runoff in the Great Lakes region is available (Urban Drainage
Subcommittee, 1976).

As additional sources to the ASCE report on Australia (Aitken, 1976), Heeps and Mein (1974) describe rainfall-runoff monitoring in Canberra and Melbourne, and Cordery (1977) describes quality measurements in Sydney. Reports on urban runoff measurements in Paris (Coyne and Bellier, 1974) and Munich (Brunner, 1975) have also been published. Additional references to monitored West German catchments may be found in other model studies (Klym et al., 1972; Geiger, 1975). Lindh (1976) discusses data for the Bergsjon catchment near Gothenberg, Sweden. Rainfall-runoff data for this catchment may be found in reports published by Arnell and Lyngfelt (1975a, 1975b). Finally, Colyer and Pethick (1976) include references to British and other data sources in their summary of storm drainage design methods.

DATA BASE SOURCES

Urban rainfall-runoff-quality data are documented in the Huber and Heaney (1977) report for 24 catchments in eight cities. Rainfall-runoff data are included for an additional 17 catchments in 13 cities. These locations are summarized in Table 1.

Sources included in the report were chosen primarily on the basis of known high quality of the data, availability and documentation. The first consideration was checked primarily by familiarization with the sampling program, careful review of the documentation and personal conversations with the responsible personnel. The latter two considerations were the keys to actually obtaining, reducing (in some cases), key punching, etc. the data for inclusion on the magnetic tape. Since the University of Florida (UF) is distant from most of the sources, the only way in which these operations could be accomplished was to have good documentation provided in some form. In all cases, data values were inspected visually for reasonableness. Where data were key punched at UF, spot checks were made against the source listing.

At least four types of information are potentially available for each location utilized as a data source:
  1) physical, demographic, etc. descriptions of the sites, plus maps, parameters and sampling methods;
  2) published reports and other written documentation;
  3) the rainfall-runoff-quality data themselves; and
  4) associated modeling data, e.g., maps, plans, photos, etc.

The report contains item 1 in write-ups for each location. A standardized tabular format is used for all sites. Item 2 is handled through a list of references for each location. Item 3 is handled separately, wherein all data have been coded and placed on a magnetic tape. They are also in the

Table 1. Summary of Data – 1977

Location	Catchment	Area		Drainage[a] System	No. Storms With	
		ac	(ha)		Quantity	Quality
Broward County, Florida	Residential	47.5	(19.2)	S	35	35
	Commercial	39.0	(15.8)	S	14	14
	Transportation	28.4	(11.5)	S	14	4
San Francisco, California	Baker St.	168	(68)	C	4	4
	Mariposa St.	223	(90)	C	4	4
	Brotherhood Way	180	(73)	C	4	4
	Vincente St., N	16	(6.5)	S	1	1
	Vincente St., S	21	(8.5)	S	1	1
	Selby St.	3400	(1380)	C	8	8
	Laguna St.	375	(152)	C	2	2
Racine, Wisconsin	Site 1	829	(336)	C	9	9
Lincoln, Nebraska	39 & Holdrege	79	(32)	S	20	20
	63 & Holdrege	85	(39)	S	15	15
	78 & A	357	(145)	S	14	14
Windsor, Ontario	Labadie Rd.	29.5	(11.9)	S	22	22
Lancaster, Pennsylvania	Stevens Ave.	134	(59.2)	C	7	7
Seattle, Washington	View Ridge 1	630	(255)	S	30	30
	View Ridge 2	105	(43)	S	5	5
	South Seattle	27.5	(11.1)	S	31	31
	Southcenter	24	(9.8)	S	30	30
	Lake Hills	150	(61)	S	7	7
	Highlands	85	(34)	S	4	4
	Cent. Bus. Dist.	27.8	(11.3)	C	5	5
Durham, North Carolina	Third Fork	1069	(433)	S	19	4

Table 1. (Concluded)

Location	Catchment	Area ac	Area (ha)	Drainage System[a]	No. Storms With Quantity	No. Storms With Quality
Baltimore, Maryland	Northwood	47.4	(19.2)	S	14	–
	Gray Haven	23.3	(9.4)	S	29	–
Chicago, Illinois	Oakdale	12.9	(5.2)	C	21	–
Champaign-Urbana, Illinois	Boneyard Creek	2290	(927)	S	28	–
Bucyrus, Ohio	Sewer Dist. 8	179	(72.5)	C	10	–
Falls Church, Virginia	Tripps Run	332	(130)	S	10	–
Winston-Salem, North Carolina	Tar Branch	384	(155)	S	17	–
Jackson, Mississippi	Crane Creek	285	(115)	S	17	–
Wichita, Kansas	Dry Creek	1883	(762)	S	8	–
Westbury, New York	Woodoak Dr.	14.7	(6.0)	S	10	–
Philadelphia, Pennsylvania	Wingohocking	5326	(2156)	C	16	–
Los Angeles, California	Echo Park	252	(102)	S	18	–
Portland, Oregon	Eastmoreland	75	(30)	C	24	–
Houston, Texas	Hunting Bayou (Cavalcade St.)	768	(311)	S	8	–
	Hunting Bayou (Falls St.)	2509	(1016)	S	11	–
	Bering Ditch	1894	(767)	S	10	–
	Berry Creek	3110	(1259)	S	10	–

[a]C = Combined sewer; S = Storm sewer and/or open channels.

process of being placed in the EPA STORET (STOrage and RETrieval) data base management system. UF has been able to obtain a limited amount of data needed for model input, item 4. These will be available for short-term loan. The remainder of such data will have to be obtained from contacts with individuals at each location.

All key punched data samples are identified as to site location, date, time of day and parameter codes. The latter are taken from the extensive list utilized by the STORET system and represent a relatively unambiguous description. In the course of preparing the data, there was often considerable confusion as to the exact water quality parameter being reported--sampling method, type of sample (e.g., total or dissolved, fixed or volatile), laboratory procedure and units (e.g., mg/l as P vs. mg/l as PO4). Proper documentation of any study is essential.

SUMMARY AND FUTURE WORK

As indicated previously, many data sources already extant may be suitable for inclusion in the data base. In addition, there are presently under way approximately 150 EPA Section 208 Areawide Waste Management Studies, many of which are collecting storm event data of the type required. As such sources are developed, periodic addenda to the report will be issued. These will consist primarily of documentation for new sources and placement of the data on the magnetic tape with the previous sources and in the EPA STORET system. Updating of the tape for previous sources will also include addition of new storm events to those already included on the tape. Any changes in catchment parameters (e.g., imperviousness, population) will also be noted.

Present work at UF includes elementary statistical analyses of the data. These include computation of ranges, means, medians, variances, etc. of the data with allowance for flow weighting, and some computations to develop mass loadings.

Rainfall, runoff and quality data are needed for model development, urban runoff characterization, data synthesis and other purposes. Hence, potential data sources should be cultivated and added to the present data base. The University of Florida (in care of the authors of this paper) and EPA actively solicit all such data.

ACKNOWLEDGMENTS

This research has been supported by the U.S. Environmental Protection Agency under Contract 68-02-0496, The continuing support of EPA and of the many contributors of data is gratefully acknowledged.

REFERENCES

Aitken, A.P. (1976) Urban Hydrological Modeling and Catchment
Research in Australia, ASCE Urban Water Resources Research
Program, Tech. Memo No. IHP-2, NTIS-PB 260 686.

Arnell, V. and Lyngfelt, S. (1975a) Berakningsmodell for
simulering av Dagrattenflode inom Bebyggda Ormaden, (Computational Model for Simulating Daily Floods in Urban Areas), in
Swedish, Chalmers University of Technology, Dept. of Hydraulics Report No. 12, Goteborg, Sweden.

Arnell, V. and Lyngfelt, S. (1975b) Nederbords-Avrinningsmat-
ningar 1 Bergsjon Goteborg 1973-1974, (Downstream Runoff
Measurements of Runoff at Lake Bergsjon, Goteborg 1973-1974),
in Swedish, Chalmers University of Technology, Dept. of
Hydraulics, Report No. 13, Goteborg, Sweden.

Blaszczyk, P. (1977) Urban Runoff Research in Poland, ASCE
Urban Water Resources Research Program, Tech. Memo No. IHP-11,
NTIS-PB 267 871.

Brunner, P.G. (1975) The Pollution of Storm Water Runoff in
Separate Systems: Studies with Special Reference to Precipi-
tation Conditions in the Lower Alp Region. Water Resources
and Sanitary Engineering Dept. Report No. 9, Munich Technical
University. (English translation available from EPA
Cincinnati NERC Library).

Colyer, P.J. and Pethick, R.W. (1976) Storm Drainage Design
Methods, a Literature Review. Report No. INT 154, Hydraulics
Research Station, Wallingford, Oxfordshire, England.

Cordery, I. (1977) Quality Characteristics of Urban Storm
Water in Sydney, Australia. Water Resources Research, 13,
1:197-202.

Coyne and Bellier. (1974) Measurements and Evaluation of
Pollutant Loads from a Combined Sewer Overflow. (in English),
Coyne and Bellier, Consulting Engineers, 5, rue d'Heliopolis,
75017 Paris.

Desbordes, M. and Normand, D. (1976) Urban Hydrological
Modeling and Catchment Research in France. ASCE Urban
Water Resources Research Program, Tech. Memo No. IHP-8,
NTIS-PB 267 524.

Geiger, W.F. (1975) Urban Runoff Pollution Derived from
Long-Time Simulation. Proc. of National Symposium on Urban
Hydrology, Hydraulics and Sediment Control, University of
Kentucky, Lexington, pp. 259-270.

Heaney, J.P., Huber, W.C., Medina, M.A., Nix, S.J., Murphy, M.P. and Hasan, S. (1977) Nationwide evaluation of Combined Sewer Overflow and Stormwater Discharges: Vol. II. Cost Assessment and Impacts. EPA Report EPA 600/2-77-064b, Cincinnati, Ohio.

Heeps, D.P. and Mein, J.P. (1974) Independent Comparison of Three Urban Runoff Models. J. Hydr. Div., Proc. ASCE, 100, HY7:995-1009.

Huber, W.C. and Heaney, J.P. (1977) Urban Rainfall-Runoff-Quality Data Base. EPA Report EPA-600/8-77-009, Cincinnati, Ohio.

Klym, H., Koniger, W., Mevius, F. and Vogel, G. (1972) Dorsch Consult, Urban Hydrological Processes, Computer Simulation. Presented at the Seminar on Computer Methods in Hydraulics, Swiss Federal Institute, Zurich.

Lindh, G. (1976) Urban Hydrological Modeling and Catchment Research in Sweden. ASCE Urban Water Resources Research Program, Tech. Memo No. IHP-7, NTIS-PB 267 523.

Lowing, M.J. (1976) Urban Hydrological Modeling and Catchment Research in the United Kingdom. ASCE Urban Water Resources Research Program, Tech. Memo No. IHP-4, NTIS-PB 262 069.

Marsalek, J. (1976) Urban Hydrological Modeling and Catchment Research in Canada. ASCE Urban Water Resources Research Program, Tech. Memo No. IHP-3, NTIS-PB 262 068.

Massing, H. (1976) Urban Hydrology Studies and Mathematical Modeling in the Federal Republic of Germany. ASCE Urban Water Resources Research Program, Tech. Memo. No. IHP-6, NTIS-PB

Mattraw, H.C., Jr. and Sherwood, C.B. (1977) Quality of Storm-Water from a Residential Area, Broward County, Florida. J. Research U.S. Geological Survey 5, (in press).

McPherson, M.B. (1975) Urban Hydrological Modeling and Catchment Research in the U.S.A. ASCE Urban Water Resources Research Program, Tech. Memo No. IHP-1, NTIS-PB 260 685.

McPherson, M.B. and Mangan, G.F., Jr. (1975) ASCE Urban Water Resources Research Program. J. Hyd. Div. Proc. ASCE 101, HY7:347-355.

Selthun, N.R. (1976) Urban Hydrological Modeling and Catchment Research in Norway. ASCE Urban Water Resources Research Program, Tech. Memo No. IHP-9, NTIS-PB 267 365.

Stall, J.B. and Terstriep, M.L.   (1972)   Storm Sewer Design--
An Evaluation of the RRL Method.   EPA Report EPA-R2-72-068,
NTIS-PB 214 134, Cincinnati, Ohio.

Terstriep, M.L. and Stall, J.B.   (1969)   Urban Runoff by Road
Research Laboratory Method.   J. Hyd. Div., Proc. ASCE 95, HY6:
1809-1834.

Terstriep, M.L. and Stall, J.B.   (1974)   The Illinois Urban
Drainage Simulator, ILLUDAS.   State of Illinois, Illinois
State Water Survey, Bulletin 58, Urbana.

Tucker, L.S.   (1968a)   Northwood Gaging Installation, Balti-
more--Instrumentation and Data.   ASCE Urban Water Resources
Research Program, Tech. Memo No. 1, NTIS-PB 182 786.

Tucker, L.S.   (1968b)   Oakdale Gaging Installation, Chicago--
Instrumentation and Data.   ASCE Urban Water Resources Research
Program, Tech. Memo No. 2, NTIS-PB 182 787.

Urban Drainage Subcommittee.   (1965)   Report of the Urban
Drainage Subcommittee Program, Urban Drainage Subcommittee,
Canada-Ontario Agreement, Describing Projects Conducted
Between 1972 and 1976.   Training and Technology Transfer
Division (Water), Environmental Protection Service, Ottawa,
Ontario K1A 0H3.

Zuidema, F.C.   (1977)   Urban Hydrological Modeling and Catchment Research in the Netherlands.   ASCE Urban Water Resources
Research Program, Tech. Memo No. IHP-10, NTIS-PB 267 587.

THE FINNISH URBAN STORMWATER PROJECT

Matti Melanen

Helsinki University of Technology

INTRODUCTION

Basic hydrological features of Finland
The highest annual mean air temperature in the
period 1931-1960, 5.5 degrees centigrade, has been
measured on the southern coast of Finland (Central
Statistical Office of Finland 1977)(Fig.1). The
warmest month is July and the coldest are January
and February. The annual precipitation in Finland
varies between 400...650 millimeters. The highest
precipitation values occur on the southern coast
and the lowest in nothern Lapland and on the coast
of the Gulf of Bothnia. About 200...250 millimeters
of the annual precipitation comes down as snow,
30...40 % of the total in southern and central Finland and 40...50 % in Lapland. The total area of
the Finnish lakes is approximately 31,500 square
kilometers, which equals 9.4 % of the total area of
Finland (National Board of Waters 1976 a). The mean
runoff of the Finnish surface water resources is
roughly 3,000 cubic meters per second and that of
the groundwater resources approximately 46 cubic
meters per second.

Finnish sewer systems
At the end of 1975 the number of public sewage works
in Finland was 627 (National Board of Waters 1976 b).
2.95 million people (63 %) were being served by
these systems. The daily consumption of water
through water supply systems was over one million
cubic meters and the daily per capita consumption 328
liters per person. In 1975 the investment in water
supply plants and sewer systems ammounted to 780
million marks (approx. US $ 210 million) (equivalent
to 950 million marks in 1977) of which 465 million

marks was used for sewer systems. The per capita investment for sewer systems was 98 marks. Of the funds for sewer systems, 57 % covered the costs of the sewers, 36 % was used for waste water treatment plants, and the remaining 7 % was spent on pumping stations. By the end of the year the total length of sewers was 16,500 kilometers, or 5.6 meters per person; one fifth of the sewers were combined ones. The operating costs of the sewer systems for 1975 were approx. 117 million marks.

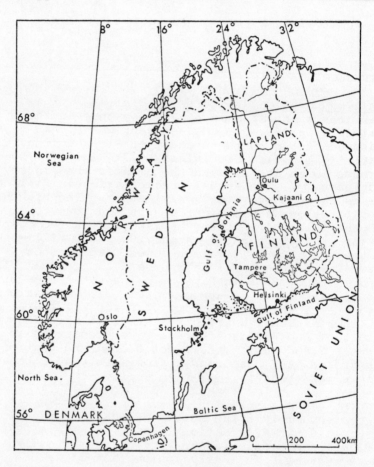

Figure 1. Location of urban experiment areas in the Finnish Urban Stormwater Project.

Need for urban runoff research in Finland

The sewer systems being built in Finland at the present are separate ones. At the same time it has become necessary to improve the efficiency of exis-

ting combined sewer systems.   Until the 1960's this
improvement was understood internationally to be
separation of combined systems.   The knowledge about
urban stormwaters has though in some cases proved
this alternative inefficient.   Also, in city areas
the separation is extremely expensive.   It has been
estimated that the complete separation of the Finnish
combined sewer areas (about 26,000 hectares) for ones
with separate sewer systems would cost about 4,500
million marks and would take 50...100 years (Yletyinen et al. 1976).   However, it is hard to predict
the effectiveness of separation.   Due to the high
costs involved the choice between combined and
separate sewer systems has become an important
question in the Finnish sewage works policy.   There
had been several years of discussion over the need
for an urban storm water project in which reliable
data on storm water quantity and quality would be
collected to give basis for the selection of the
new sewer systems and renewal of the existing ones.
The planning of this project commenced in April
1976 and the project, called here the Finnish Urban
Stormwater Project, started at the beginning of
1977, sponsored by the Maj and Tor Nessling Foundation, the National Board of Waters and four cities
(Kajosaari et al. 1976).

THE FINNISH URBAN STORMWATER PROJECT

The Project is being carried out by Helsinki University of Technology, Tampere University of Technology
and Oulu University over the 1977-1979 period.   In
1977 seven urban experimental areas and one non-
urban reference area have been included.

Objectives
The objectives of the Project are: first, to identify
and quantify the sources and transport of pollutants
to urban stormwaters and to provide information on
the processes involved in the cycle of urban pollutants; second, to analyze stormwater quantity and
quality variations and factors affecting them in
different circumstances; third, to obtain information on pollution loads caused by stormwaters in
different circumstances when compared with pollution
due to wastewaters, combined sewage and runoff from
non-urban areas and; fourth, to obtain basic knowledge for the development and improvement of technology
for reducing pollution caused by sewage works.   The
experiments are linked to the hydrological cycle
in such a way that the data achieved can also be
utilized for analyzing rainfall-runoff relationships

and for verifying existing and new mathematical
models which describe the hydrological cycle or
parts of it.

Experiment areas

Urban field experiments are being carried out in the
cities of Helsinki, Tampere, Oulu and Kajaani (Fig. 1),
each urban area representing different climatic conditions, land use type and other characteristics.
The investigation involves five residential areas of
0.15 to 0.33 square kilometers and 2,500 to 25,000
persons per square kilometer, an industrial area
with mixed industry of 0.15 square kilometers and a
traffic area of 0.05 square kilometers and 45,000
motor vehicles/day. The non-urban field area of the
Project is situated in the Siuntionjoki region in
southern Finland.

Field experiment program

In each urban experimental area the following monitoring and sampling program is being accomplished
during 1977:

1 Air temperature and humidity observation - The
   instruments used are continuously-registering
   Soviet-made thermographs and hygrographs.
2 Rainfall intensity registration - Rainfall intensity is registered continuously in the period
   May-October with German- and Soviet-made pluviographs (types Lambrecht and P-2). In each area
   there is one rain gauge.
3 Total dustfall sampling - In each area the monthly dustfall is obtained from total precipitation
   samples using two Norwegian - made NILU cylinders
   during April, August, September and October. The
   monthly samples are homogenized and the following
   laboratory analyses are performed: total dry
   solids, TOC, total nitrogen, total phosphorus, pH
   conductivity, calcium, chlorides, sulphates,
   lead, vanadium and titanium. In addition, in
   each area the total amount of dustfall is collected with a direction oriented sampler.
4 Rainwater quality observation - During storm
   events rainwater samples are collected in each
   area with a NILU cylinder. The following analyses are carried out: TOC, total nitrogen, total
   phosphorus, pH, conductivity, chlorides, sulphates, lead and vanadium.

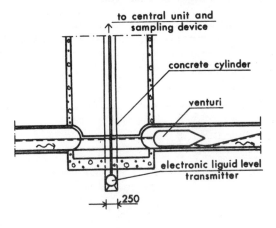

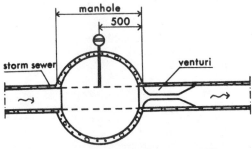

Figure 2. Principle of storm sewer flow measurement
in the Finnish Urban Stormwater Project.

5 Stormwater flow registration and sampling - In
   each area stormwater flow is registered continuously during the period May-October with a
   venturi installed in the outlet pipe of the storm
   sewer network (Fig. 2). The flow level before
   the venturi is measured with a Finnish-made
   electronic liquid level transmitter and the level
   is converted to flow with a Finnish-made FLO 110
   central unit (Fig. 3). With this device combination at least all the stormwater flows caused
   by storms with a two-year occurence interval,
   duration of ten minutes and intensity of 120
   liters per second and hectare can be registered.
   An automatic Finnish-made sampling device, SAM
   120 (Fig. 3), is controlled by the central unit
   to take either time-discrete or flow-proportional
   storm water samples. During 1977 flow-proportional composite samples are being collected from 30
   storm events in the period May-October. The
   following laboratory analyses are made on stormwater samples: total dry solids, volatile dry
   solids, suspended solids, volatile suspended

solids, settleable solids, 7-day BOD, TOC, COD
(as dichromate value), total nitrogen, total
phosphorus, pH, conductivity, chlorides, sulphates, lead, vanadium, streptococcus faecalis
and oil.

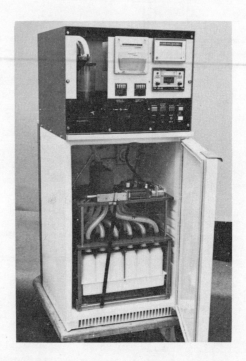

Figure 3.   Central unit and stormwater sampling device used in the Finnish Stormwater Project.

6   Groundwater level observation - In each area
    (the traffic area excluded) the groundwater level
    is observed twice monthly in 3 to 5 groundwater
    pipes.

In the non-urban field area in Siuntionjoki the
following experiment program is being carried out
simultaneously in 1977: air temperature and humidity
observation, rainfall intensity registration, runoff
registration and sampling and groundwater level observation.

Preliminary results for 1977
In 1977 approximately 7500 laboratory analyses will
have been performed in the Project.   The total cost
of the Project for 1977 will be roughly 800,000 mar
Some preliminary results of the Project from the 19
experiment period are presented in Tables 1, 2 and

In Table 1 the average monthly precipitation values
of the experiment areas are given.

Table 1. Monthly precipitation in millimeters (mm/mo) over the 1977 experiment
period in the Finnish Stormwater Project

Month	City			
	Helsinki	Tampere	Oulu	Kajaani
May	19.1	61.0	47.9	47.8
June	48.3	41.6	31.9	29.3
July	106.2	120.1	70.1	166.7
August	42.8	22.4	44.7	39.1
September	86.0	67.7	48.9	88.0
October	89.7	60.2	25.7	40.4

Table 2 shows the urban runoff quality variations
recorded in May-October and in Table 3 observed
values of monthly atmospheric dustfall fallout are
given.

Table 2. Urban runoff quality variations (min-max) over the 1977 experiment period in the Finnish Stormwater
Project

Component	Type of experiment area		
	Residential areas	Industrial area	Traffic area
Total dry solids, mg/l	83-740	2022	300-1140
Volatile dry solids, mg/l	4-157	572	70-170
Suspended solids, mg/l	37-700	1150-1460	110-1300
Volatile suspended solids, mg/l	17-190	250-276	42-220
Settleable solids (2 hours), ml/l	0.4-3.5	4.0	-
BOD_7, mg O_2/l	0-73	60-92	6-26
COD_{cr}, mg O_2/l	30-240	170-180	160-190
Total phosphorus, mg P/l	0.09-2.40	0.43-0.74	0.15-0.84
Total nitrogen, mg N/l	0.40-4.66	1.6-1.8	0.72-3.6
pH,-	4.99-7.26	7.25-7.61	6.12-8.05
Conductivity, m S/m	1.2-10.8	8.4-9.2	5.4-27.2
Chlorides, mg Cl/l	0.3-9.5	4.3-9.2	2.1-20.0
Sulphates, mg SO_4/l	1.0-21.5	8.2-12.0	6.0-40.0
Lead, µg Pb/l	16-770	365-380	415
Vanadium, µg V/l	5-64	37-50	30
Oil, mg/l	0.4-1.6	-	1.9-4.2

Table 3. Mean monthly atmospheric dustfall over the 1977 experiment period in the Finnish Urban Stormwater Project.

Table 3a

Component	Type of experiment area Residential areas		
	I	II	III
Total dry solids, g/100 m^2/mo	73.8-673.1	97.4-150.2	119.2-278.9
TOC, g C/100 m^2/mo	4.9-20.9	3.4-5.5	-
Total phosphorus, g P/100 m^2/mo	0.1-0.6	0.2-0.4	0.1-0.2
Total nitrogen, g N/100 m^2/mo	4.8-41.6	3.7-4.6	3.7-4.0
Calcium, g Ca/100 m^2/mo	1.9-13.4	2.5-5.5	2.4-3.6
Chlorides, g Cl/100 m^2/mo	2.1-6.7	0.7-3.2	3.2-3.6
Sulphates, g SO$_4$/100 m^2/mo	26.9-104.7	24.0-36.9	21.0-25.9
Lead, mg Pb/100 m^2/mo	30-750	50-620	-
Vanadium, mg V/100 m^2/mo	<20-320	<10-50	-
- - - - - - - - - - - -			
pH variations,-	3.56-6.94	4.13-7.43	4.49-6.77
Conductivity variations, m S/m	4.4-45.6	3.8-13.7	2.7-3.8

Table 3b

Component	Type of experiment area Industrial area		
	I	II	III
Total dry solids, g/100 m^2/mo	228.0	-	-
TOC, g C/100 m^2/mo	6.4	1.8-3.2	-
Total phosphorus, g P/100 m^2/mo	0.2	-	-
Total nitrogen, g N/100 m^2/mo	5.8	-	-
Calcium, g Ca/100 m^2/mo	4.8	2.6-4.2	-
Chlorides, g Cl/100 m^2/mo	1.9	0.4-0.8	-
Sulphates, g SO$_4$/100 m^2/mo	38.5	19.3-19.9	-
Lead, mg Pb/100 m^2/mo	90	80-90	-
Vanadium, mg V/100 m^2/mo	120	20	-
- - - - - - - - - - - - - -			
pH variations,-	3.72-4.15	3.83-4.15	-
Conductivity variations, m S/m	8.2-12.0	9.6-12.0	-

Table 3c

Component	Type of experiment area Traffic area		
	I	II	III
Total dry solids, g/100 m^2/mo	-	-	143.7
TOC, g C/100 m^2/mo	-	-	-
Total phosphorus, g P/100 m^2/mo	-	0.2	0.2
Total nitrogen, g N/100 m^2/mo	-	3.7-4.4	3.8
Calcium, g Ca/100 m^2/mo	-	5.1-6.2	4.7
Chlorides, g Cl/100 m^2/mo	-	1.5-1.6	4.0
Sulphates, g SO$_4$/100 m^2/mo	-	37.4-45.2	37.4
Lead, mg Pb/100 m^2/mo	-	210-290	-
Vanadium, mg V/100 m^2/mo	-	30-50	-
- - - - - - - - - - - - -			
pH variations,-	3.84-6.33	4.37-6.00	4.44
Conductivity variations, m S/m	11.8-11.9	7.8-10.0	5.2

I = April, II = August, III = September

Extension of the program during 1978-1979

During 1978-1979 the field program has been planned
to be extended as follows: for some urban field
areas non-urban areas will be defined in which ana-
logical experiments will be accomplished, the
results of which will lead to further conclusions
on the effects of urbanization; the dustfall sampling will be completed to give more reliable data
on the effects of dustfall on stormwater (at the same
time emission inventories will be carried out in the
areas); stormwater sampling will be modified so that
time-discrete samples will also be collected; finally,
in some urban field areas wastewater quantity and quality observation will be achieved to give data for
loading comparisons.

ACKNOWLEDGEMENTS

The author wishes to thank the Board and the Sponsors
of the Finnish Urban Stormwater Project for permission to publish this report and all the persons involved with the Project.

REFERENCES

Central Statistical Office of Finland (1977) Environmental Statistics 1974, Helsinki. (In Finnish,
Swedish and English)

Kajosaari, E. & Airaksinen, J. & Hooli, J. & Laikari,
H & Tiainen, V.-M. & Viitasaari, M. & Melanen, M.
(1976) Plan for the Finnish Urban Stormwater
Project 1977-1979, Helsinki. (In Finnish)

National Board of Waters (1976 a) Application of
Water Pollution Control Principles, Helsinki. (In
Finnish and Swedish with English preface)

National Board of Waters (1976 b) Water Supply and
Sewer Systems 31.12.1975, Helsinki. (In Finnish and
Swedish with English summary)

Yletyinen, P. & Ylä-Soininmäki, E (1976) Urban
Storm Water Drainage Systems. Nordic Hydrological
Conference 76, Reykjavik. (In English)

CHARACTERISTICS OF THE ABOVE-GROUND RUNOFF IN SEWERED CATCHMENTS

Jan Falk and Janusz Niemczynowicz
Lund Institute of Technology / University of Lund

ABSTRACT

Nine small asphalt catchments have been instrumented, the catchments having a size of only some few hundred squaremeters. Surface characteristics such as depression storage and response time are discussed. A single nonlinear model with time lag is used. The model is shown to fit observations; only surface slope is used to estimate model parameters.

INTRODUCTION

Runoff from small impermeable surfaces is part of the urban hydrological research carried out in our Department of Water Resources Engineering. Together with the surface runoff from pervious areas this portion constitutes the above-ground runoff or the hydrological component of urban runoff. In an urban catchment there is also a hydraulic component, namely the conduction of water in pipes below the ground. Of the above-ground runoff by far the most important contribution is that from the impermeable surfaces. The hydrological response of impermeable surfaces is largely independent of geological and climatic variables. This means that data collected should be equally pertinent everywhere.

By measuring several paved areas it should be possible to relate the physiografical factors governing the runoff process to the parameters in a mathematical model. This paper is our first attempt in this direction.

Great accuracy in measurements is important for making the data useful. In practice our field data have proven useful when the accuracy permits a complete water budget on each time step of one minute. Instruments for measuring rainfall and runoff have been developed and tested in our laboratory. Performance and reliability were tested on the adjacent impervious catchment i.e., the parking lot outside the laboratory.

INSTRUMENTATION AND DATA ACQUISITION SYSTEM

For recording rainfall intensity an instrument which operates on the "tipping bucket" principle has been constructed. In order to assure accuracy, the raingage has collecting area of 3000 cm^2. The volume of water causing tipping is 0.02 mm of rain. The opening of the raingage is situated at ground level.

Runoff is measured in gutter inlets by means of a special apparatus. The device consists of a box with a 30° Thompson weir and a vertical pipe filled with an electrolyte (0.5 % NaCl) connected to the water in the box through a rubber membrane. Two parallel platinum wires are mounted in the vertical pipe. By measuring the electrical conductivity of the wires in the electrolyte the water level can be obtained.

In order to transmit signals from the measuring units to a tape punch, a special control unit has been developed. This unit consists of a clock, a digital voltmeter and a pulse counter. The recording (tape punching) starts automatically when an initial pulse from a raingage enters the control unit. The time interval between each signal is set at one minute. That is, each minute the voltage of the runoff gages and the number of tippings for each raingage are punched. Termination of recording is determined by the elapsed time since the last pulse from the raingage was received. As all the instruments are connected to the same clock there is absolute synchronization of recording time for rainfall and runoff. All cables connecting the central unit to the gages are placed in the storm drains. A more detailed account of the instrumentation may be found elsewhere, Falk, J.; Niemczynowicz, J. (1976) and Lindh, G. (1976).

INVESTIGATED AREAS

The first area measured was a parking-lot outside the laboratory. A map of the area denoted 1.75 is presented in Figure 2.

For making a comparison between urban and rural catchment behaviour an urban catchment was chosen in the city of Lund in 1976. The catchment is dominated by six to eight story dwellings. The catchment area is 14.1 ha, of which 49 % consists of impermeable surfaces such as roads, parking-lots and roofs. In addition to measuring runoff from the whole area, the data collection system allows for four extra runoff gages. During 1976 four small impermeable surfaces were measured, during 1977 another four and in 1978 four other areas will be measured, as shown in Figure 1. By november 1977 nine small asphalt areas will have been investigated as shown in Figure 2. Each surface was surveyed very carefully around every water divide that is not clearly defined.

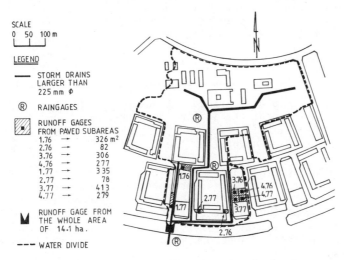

Figure 1.   Klostergården study area, Lund, Sweden

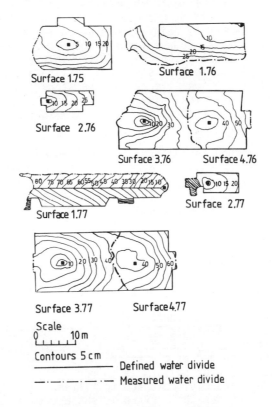

Figure 2.   Map of measured impermeable surfaces

For each surface a maximum length and a maximum slope were calculated. Weighted average values of length and slope were determined from measurements in twelve directions from the inlet.
Further details are given in Table 1.

LOSSES

Depression storage
As measurements are carried out automatically no information
is available about the initial moisture conditions at the onset of rainfall. Therefore only those rainfall events are used,
when there is no doubt that the surface is initially dry.

Experiments in the laboratory with a rainfall simulater gave
as results a storage, S_O, for wetting the surface before any
runoff occured, that was slightly higher than the depression
storage remaining on the surface after the cessation of runoff, S_D (Lindh, G., 1976). This discrepancy was thought to be
due to surface tension phenomenon because of the scale of the
model.

The field experiments showed that the behaviour of the two
storages S_O and S_D changed. The initial storage S_O calculated
as the amount of rainfall needed to generate runoff could not
be regarded as depression storage for two reasons. Some runoff from the areas closest to the inlet may occur before the
available depression storage is filled and on the other hand
the distribution in space of the storage capacity could be
such that a travel time is needed before the water enters the
inlet. The former effect giving too low and the latter too
high a depression storage.

A new method for determining the depression storage S_D was derived. By means of linear regression between total volumes
of rainfall (I_{sum}) and runoff (Q_{sum}) for all events when the
surface was   initially dry the depression storage is
determined. An example of the regression for surface 2.76 is
shown in Figure 3. The depression storage is taken as the
value of the vertical axis when runoff equals zero. Table 1
gives average values of S_O and S_D for the number of events
indicated. The loss model may be expressed

$$I_{sum} = k \cdot Q_{sum} + S_D \tag{1}$$

where k is a constant.
Depression storage, S_D is a factor whose size should primarily
depend on the geometrical quality of the surface and therefore
may be regarded as a surface characteristic. The smallest observed value of S_D is 0.13 mm for surface 1.75. This surface
is ten years old, compared to fifteen years for the other surfaces. Besides the surface is subject to a very low traffic
intensity. Surface 1.76 (with the largest value of S_D = 1.05 m

Surface identifier	1.75	1.76	2.76	3.76	4.76	1.77	2.77	3.77	4.77
No. of events used	29	22	28	28	24	24	20	24	23
Use	Park.	Park. Road	Bicy. Park	Park.	Park.	Park. Road	Bicy. Park	Park.	Park.
Traffic	Light	Heavy	No cars	Light	Light	Heavy	No cars	Light	Light
Area m^2	291	326	82	306	277	335	78	413	279
Maximum length, L_{max} m	13.5	35.0	16.1	18.4	13.2	41.0	9.3	20.7	13.4
Weighted length, L_w m	9.3	27.0	8.6	10.6	10.2	32.1	6.1	13.2	10.1
Maximum slope, s_{max} %	3.7	1.7	10.7	3.8	2.2	3.8	9.3	3.3	2.6
Weighted slope, s_w %	2.1	0.9	3.3	3.1	1.6	2.3	4.1	2.3	1.9
L Initial storage, S_O mm	0.27	1.07	0.43	0.49	0.77	0.44	0.44	0.46	0.54
O Depression storage, S_D mm	0.13	1.05	0.48	0.56	0.52	0.51	0.33	0.57	0.56
S Additional losses % of I	0.17	2.25	1.54	4.58	7.40	2.34	2.53	0.88	0.70
E									
S									
Response time, t_p min	1.3	6.4	2.7	3.2	3.8	3.5	1.9	2.7	2.5
Response time, t_{50} min	2.8	7.2	3.4	3.9	4.7	4.0	3.7	4.8	4.1
Average Manning's n	0.036	0.032	0.070	0.086	0.049	0.027	0.100	0.045	0.035
Parameter A with B = 1.5	32.0	5.7	28.0	15.0	12.0	17.0	27.0	22.0	21.7

Table 1   Surface characteristics

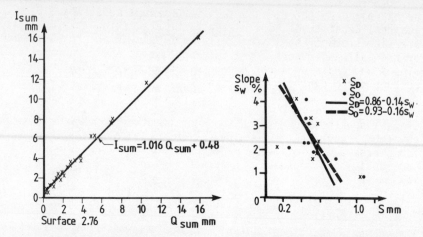

Figure 3. Linear regression between volumes of rainfall and runoff for surface 2.76

Figure 4. Storages S_D and S_o versus weighted average slope s_w for the nine investigated areas

has the most complicated geometry. Deep pools of water are built up along the gutter during rainfall and traffic intensity could be regarded as high.

The magnitude of depression storage S_D and initial storage S_o depends inversely to some extent on the weighted average slope s_w, as may be seen in Figure 4. The correlation coefficient between S_D and s_w, is 0.55 and between S_o and s_w, 0.65.

The losses due to depression storage may be related to s_w by the following equation:

$$S_D = 0.86 - 0.14 \ s_w \ \text{(mm)} \qquad (2)$$

Reasonable estimates of depression storage losses for small impervious areas having weighted average slopes in the range of from 0.9 to 4.1 % can be expected if equation 2 is used. This result differs very much from what was found for similar areas by Willeke, 1966

$$S_D = 4.1 - 1.00 \ s_w \ \text{(mm)} \qquad (3)$$

and by Viessman, 1968

$$S_D = 3.30 - 0.77 \ s_w \ \text{(mm)} \qquad (4)$$

These two equations were based on measurements from almost the same areas having slopes ranging from 0.7 to 3.4 %. For 2 % slope the three equations result in depression storages of 0.58, 2.1 and 1.76 mm respectively. The great discrepancy may

be explained by the fact that different loss models have been
adapted. Equations 3 and 4 rests on the assumption that depression storage equals the difference between the total volumes
of rainfall and runoff. More data covering a wider range of
slopes and drainage areas are needed before such equations can
be considered generally applicable.

Other losses

If no losses besides depression storage occur, constant k in
equation 1 ought to be 1.0. This is not the case. The k value
for all surfaces is slightly higher than 1.0. This indicates
some additional losses. If equation 1 is rearranged in the
following way:

$$1 - \frac{Q_{sum}}{I_{sum} - S_D} = 1 - k^{-1} = \ell \qquad (5)$$

ℓ represents additional losses.

The loss factor ℓ expressed in percent may be found in Table 1.
The range of variation is from 0.17 to 7.40 %. Again the newest
and smoothest surface 1.75 has the lowest value. The large
range in variation from one surface to another indicates that ℓ
could not be due to evaporation. Besides it is not realistic to
expect any evaporation during the rainfall or during the approximately ten minutes between the cessation of rainfall and
runoff. The only apparent explanation is infiltration through
the asphalt. This idea is supported by the fact that the best
maintained surface is subject to the lowest losses. Loss factor
ℓ may also be subject to outsplash from the catchments, but
this is also doubtful.

OTHER SURFACE CHARACTERISING PARAMETERS

Response time

Two measures of response time have been calculated. Time to
peak, t_p is taken to be the time interval between the peaks of
rainfall and runoff and lag time, t_{50} as the time interval between the time when 50 % of the effective rainfall has occured
and 50 % of the runoff has passed the gaging station. Table 1
gives average values of t_p and t_{50} for the number of events indicated. A good correlation 0.93 in all cases is found between
t_p and $S_{storage}$, t_p and S_D, t_{50} and S_o and between t_{50} and S_D. This
may be likend to a retention basin, the larger the
storage the bigger runoff lag. Catchment slope s_w shows a negative correlation (-0.64)   between the response time parameters
and catchment length a positive correlation of the same order
between t_p but somewhat smaller for t_{50}. As expected the response time is very short, amounting to a few minutes.

Manning's roughness coefficient, n

Knowing all components in the kinematic wave equation, the
Manning roughness coefficient n can be derived at equilibrium

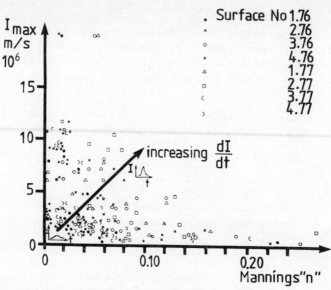

Figure 5.   Relation between Manning roughness coefficient and
rainfall intensity parameters

conditions, when the time of concentration exceeds the duration of rainfall. For further details, see for example Eagel-
son, P. (1970).

From the calculated values of n for the different surfaces it
has become evident that n has no constant value but varies
with the rainfall intensity. An explaination for this may be
that equilibrium conditions are never reached in practice because rainfall intensity is continuously varying. Raindrop
impact also changes the character of flow.

The relationship between n and maximum rainfall intensity is
shown in Figure 5. In spite of the wide scatter it may be stated that there is a correlation between n and maximum rainfall
intensity. The figure also indicates that n depends on changes
in rainfall intensity (difference between the maximum intensity and the intensity during the preceding time interval). Average values of n may be seen in Table 1.

MODELING

The relationship between runoff and surface storage forms a
base for determining what kind of model to use. In Figure 6
this relationship is plotted for surface 2.76 with all rainfall events used. Runoff is plotted against the magnitude of
storage during the preceding time, the time interval between
recordings beeing one minute. This time lag evidently gives

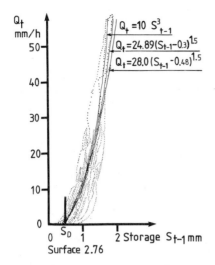

Figure 6.   Relation between storage and runoff for surface
              2.76

the smallest scatter when plotting runoff against storage. The
results indicate that the catchments act like nonlinear reservoirs. If we neglect the hysterecis effect (smaller values during the rising limb of the hydrograph than during the recession part for the same storage, the following equation is obtained:

$$0_t = A \ S_{t-1}^B \qquad (6)$$

where A and B are model parameters characteristic of a specific
catchment area

In combination with the continuity equation

$$I - Q = \frac{dS}{dt} \qquad (7)$$

equation 6 may be used for calculating runoff hydrographs from
observed hyetographs. Similar models have been used by others,
see for example Kidd (1976).

Depression storage, S_D is taken as the amount of surface storage needed before any runoff will be produced. Initial storage, S_o according to the discussion above may be a better
assumption but S_o is more difficult to measure. Also the internal correlation between S_D and S_o is 0.88, indicating they
are closely connected. Using S_D as initial loss equation 6
becomes:

$$Q_t = A (S_{t-1} - S_D)^B \qquad\qquad (8)$$

For optimizing the value of parameter A the best fit of equation 8 is obtained when B = 1.5. Both parameters are not optimized because there is a strong internal correlation between them. Trying to optimize both, B varies between 1.3 and 1.9. The only criteria used is a subjective fitting of A in equation 8 to satisfy the plotted values, see Figure 6. The figure also shows a fitting of equation 6 where no losses are assumed and B = 3.0. A can be considered a factor characterizing the physiographic conditions of the catchment in question.

Knowing the values of S_D, A and B equations 7 and 8 can be used for calculating the runoff. As the model is lagged it is possible to solve the equation on a step by step procedure. Figures 7 and 8 show typical hydrographs produced by the model. Table 1 summarises the model parameter A and Table 2, the models performance on the nine catchments.

The infiltration loss is considered to be small and therefore is neglected.

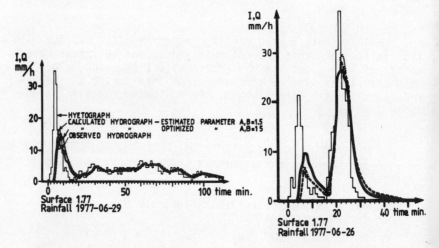

Figure 7. Model performance on surface 1.77

Figure 8. Model performance on surface 1.77

ESTIMATION OF MODEL PARAMETERS

The physiographic factor mostly correlated to model parameter A is the average weighted slope s_w, the correlation coefficient beeing 0.62. If the worst correlated surface (namely 1.75) is excluded the value increases to 0.80. The relationship between parameter A ans slope may be seen in Figure 9.

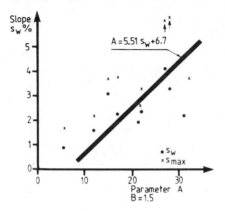

Figure 9.   Relationship between parameter A and slope

The regression equation in the figure is used as a first attempt to estimate A without any kind of calibration:

$$A = 5.51 \, s_w + 6.7 \tag{9}$$

for B = 1.5. Also needed is a value of depression storage. From Figure 4 this is chosen as follows

$$S_D = 1.0 \text{ mm} \quad \text{for} \quad s_w < 1 \, \%$$
$$S_D = 0.5 \text{ mm} \quad \text{for} \quad 1 < s_w < 3 \, \%$$
$$S_D = 0.3 \text{ mm} \quad \text{for} \quad s_w > 3 \, \%$$

From knowledge of only average weighted slope and surface area the model parameters A and S_D may be estimated and the model can be used predictively. Examples of model performance are shown in Figures 7 and 8. The figures show the fit both for calibrated and uncalibrated model runs. The performance of the model is also given in Table 2. The uncalibrated model shows almost as good agreement with observed data as the calibrated model.

Table 2. Model runs with calibrated and uncalibrated model

Rainfall event		April 23	May 1	May 13	June 1	June 26	June 29	Mean dev %
Surface 1.77								
Peak value obs mm/h		7.6	33.2	10.4	14.6	26.8	14.1	
	cal mod	10.8	31.6	8.4	16.3	27.6	15.2	+ 3.1
	uncal mod	11.9	32.7	9.0	18.0	29.4	16.2	+ 9.8
Total volume obs mm		1.0	5.1	1.7	1.8	4.8	8.5	
	cal mod	1.0	5.8	1.6	1.8	4.8	8.8	+ 3.9
	uncal mod	1.0	5.8	1.6	1.8	4.8	8.8	+ 4.0
Surface 2.77								
Peak value obs mm/h		10.2	35.0	12.9	28.2	34.1	17.7	
	cal mod	15.2	35.0	10.4	22.8	34.1	21.8	+ 0.9
	uncal mod	17.8	35.3	10.6	23.4	34.6	25.4	+ 6.5
Total volume obs mm		1.3	5.3	1.8	1.9	4.9	8.7	
	cal mod	1.0	5.8	1.6	1.8	4.8	8.8	$\pm$ 0
	uncal mod	1.2	6.0	1.8	2.0	5.0	8.9	+ 4.4
Surface 3.77								
Peak value obs mm/h		6.4	30.6	9.3	16.3	21.5	14.5	
	cal mod	11.2	32.9	8.9	18.3	29.8	15.7	+18.5
	uncal mod	11.9	32.7	9.0	18.0	29.4	16.1	+18.8
Total volume obs mm		0.7	5.0	1.7	1.6	4.6	8.7	
	cal mod	1.0	5.7	1.5	1.7	4.7	8.7	+ 4.5
	uncal mod	1.0	5.8	1.6	1.8	4.8	8.8	+ 6.3
Surface 4.77								
Peak value obs mm/h		9.3	32.4	8.0	18.8	27.6	20.6	
	cal mod	12.1	33.5	9.3	19.4	30.8	16.5	+ 4.3
	uncal mod	10.9	31.7	8.5	16.5	27.8	15.3	- 5.3
Total volume obs mm		0.8	5.1	1.6	1.6	4.6	8.7	
	cal mod	1.0	5.7	1.6	1.8	4.8	8.7	+ 5.0
	uncal mod	1.0	5.8	1.6	1.8	4.8	8.8	+ 6.5

CONCLUSIONS

Depression storage has appeared to be considerably smaller than
indicated by other authors. The Manning roughness coefficient
can not be regarded as a surface characteristic only as it
varies with rainfall intensity.

A nonlinear reservoir model with time lag gives resonable agreement between observed and calculated hydrographs. The parameter
in the model may be estimated from physiographic catchment characteristics. Such an uncalibrated model shows almost the same
result as the calibrated one. However, several more catchments
must be included in the analysis before this method can be ful
proven.

REFERENCES

Eagleson, P.S. (1970) Dynamic Hydrology. Mc Graw-Hill,
   New York.

Falk, J. and Niemczynowicz, J. (1976) Runoff from impermeable
   surfaces. Nordic Hydrological Conference 1976, Reykjavik,
   Iceland.

Kidd, C.H.R. (1976) A non-linear urban runoff model.
   Institute of Hydrology, Wallingford, UK. Report No. 31.

Lindh, G. (1976) Urban hydrological modelling and catchment
   research in Sweden. ASCE Urban Water Resources Research
   Program. Technical Memorandum No. IHP-7.

Viessman, W. (1968) Runoff estimation for very small drainage
   areas. Water Resources Research, Vol. 4, No. 1.

Willeke, G. (1966) Time in urban hydrology. ASCE Journal of
   Hydraulics Division. No HY1.

A CALIBRATED MODEL FOR THE SIMULATION OF THE INLET HYDROGRAPH FOR FULLY SEWERED CATCHMENTS

C.H.R. Kidd

Institute of Hydrology, Wallingford, U.K.

INTRODUCTION

Complementary research programmes at the Hydraulics Research Station and the Institute of Hydrology are concerned with the development of improved methods of storm sewer design. One component of any design method is a rainfall-runoff model, and the purpose of current research at the Institute is to improve the realism of such models by a separate modelling treatment of the above-and below-ground phases of the rainfall-runoff process. The above-ground phase, which is concerned with the determination of the inlet hydrograph at any junction of the sewer system, is primarily composed of hydrological phenomena, whereas the below-ground phase, which deals with pipe routing, may be seen more as a hydraulics problem. It is the first of these two phases with which the Institute of Hydrology is concerned, and this paper describes a suitable mathematical model.

Existing urban runoff models cover a considerable range in terms of complexity. On the one hand, the Rational or Lloyd-Davies (1906) Method is too simplistic in its approach, while the other extreme, typified by the Storm Water Management Model (Metcalf & Eddy et al, 1971), is generally too complicated for practical design purposes. In between these two extremes lie a number of options which attempt to achieve a compromise between sophistication and practical simplicity. Into this bracket would come the Road Research Laboratory Method (Watkins, 1962; TRRL, 1976), the theoretical basis of which has come under criticism (Newton & Painter, 1974). The model conceived at the Institute of Hydrology is of the same order of complexity, and, to this end data have been collected at the phase boundary (i.e. road gullies) to calibrate the model in terms of catchment characteristics.

The structure of the model is based on the division of

Table 1: Catchments used in regression analyses

Catchment	Sponsor	Total area (ha)	%age imperv.	Period of data	No. of events
Oxhey Road	RRL[1]	0.78	60.3	1954–59	6
WPRL, Stevenage	RRL	1.39	60.0	1955–59	26
Kidbrooke, Kent	RRL	3.42	68.2	1953–58	70
Blackpool	RRL	4.82	41.9	1953–58	46
Doncaster	RRL	5.14	30.0	1955–58	15
Leicester	RRL	59.5	36.1	1958	39
Oxhey Housing Estate	RRL	247	19.8	1953–59	19
Lordshill 1, Soton	U of S'tn	.80	41.3	1974–75	29
Lordshill 2, Soton	U of S'tn	.60	41.7	1974–75	18
St Marks Rd 1, Derby	DGWE[2]	10.4	52.9	1971–75	15
St Marks Rd 2, Derby	DGWE	8.55	51.0	1971–75	19
St Marks Rd 3, Derby	DGWE	7.23	48.7	1971–75	39
Wildridings, Bracknell	DGWE	11.60	46.2	1974–75	20
Rise Park, Nottingham	Trent Poly	62.00	31.1	1974	7

[1] Road Research Laboratory
[2] Directorate General Water Engineering, Department of the Environment.

Fig 1: School Close, Stevenage

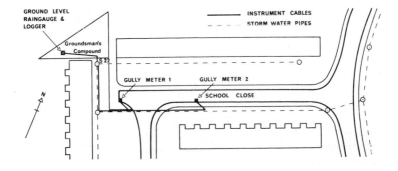

the whole process into three component parts, as follows:

i.   determination of the runoff volume for a given rainfall
      input,

ii.   temporal distribution of the net rainfall (in total
      equal to the runoff volume),

iii. surface routing of the net rainfall to the inlet of the
      sewer system.

Section 2 describes the formulation of the runoff volume
submodel (and its calibration), while Section 3 describes the
loss distribution and surface routing submodels. Section 4
describes the performance of the model on a catchment in
Oxhey, Hertfordshire.

RUNOFF VOLUME SUBMODEL

The hydrological processes which govern the runoff volume
mechanism in an urban catchment are extremely complex. Some
of these, such as infiltration and depression storage, may
yield to a physically-based modelling approach, whereas
others, such as blockages and the interaction of surface types,
do not. In general, the complexity of an urban catchment's
topography is such that a statistical approach to the problem,
as has been adopted recently for natural catchments (N.E.R.C.,
1975), is more appropriate than a physically-based approach.
Furthermore, it is considered that an approach which lumps
together the losses on a catchment-wide basis is justified,
which has allowed the use of the archive of existing U.K.
urban hydrological data currently being assembled at the
Institute of Hydrology.

    A data set derived from 368 events on 14 catchments has
been used for these analyses (Stoneham & Kidd, 1977). Brief
details of the catchments are given in Table 1. A
qualitative appraisal of the relevant processes identified 4
catchment variables (percentage impervious, soil type,
catchment slope, proportion of roofs) and 5 storm variables
(rainfall volume, duration, intensity, long-term catchment
wetness, immediate wetness condition) upon which the runoff
volume might be expected to depend.

    Analyses were conducted (Stoneham & Kidd, 1977) with a
variety of forms of the regression model using a number of
different dependent variables. The regression equation
finally adopted as most appropriate is given by:

    PRO = .924PIMP + 53.4SOIL + .0650UCWI - 33.6   ... 1
where PRO is the percentage runoff (%);
      PIMP is the percentage impervious (%),
      SOIL is the soil index (NERC, 1975)
and   UCWI is the urban catchment wetness index, given by

UCWI = 125 + 8API5 - SMD   ... 2
where API5 is the 5-day antecedent precipitation index,
and   SMD is the soil moisture deficit.

Equation 1 explains 54% of the variance in the percentage
runoff (correlation coefficient .73), and all the coefficients
of the independent variables are significant to within the
.1% level. The independent variables display some degree of
intercorrelation, and therefore the equation should be used
for prediction purposes only; and then only within the limits
of the data set (Stoneham & Kidd, 1977).

14 catchments is a smaller number than is desirable for
an analytical exercise of this type, and Equation 1 will be
updated as further data become available.

LOSS AND SURFACE ROUTING SUBMODELS

As described in the previous section, progress on runoff
volume estimation can be made by the analysis of existing
data from urban catchments. Progress on the remaining two
parts of the process necessarily depends on the collection
of rainfall-runoff data at the boundary between the above-
and below-ground phases. In 1975, the Institute of Hydrology
embarked on an extensive data collection programme following
the development of a meter for the measurement of discharge
through a road gully (Blyth & Kidd, 1977). Fifteen road
gullies have been instrumented in a total of 6 subcatchments
at Bracknell, Stevenage, Southampton and Wallingford (only
1976 data for the Bracknell, Stevenage and Southampton
subcatchments have been used in these analyses). Typical of
the type of data collection system is shown in Figure 1,
which is the experimental system (2 gully meters, .1 mm RIMCO
tipping bucket raingauge and data logger) for School Close
in the Shephall district of Stevenage. Data are recorded at
$\frac{1}{2}$-minute intervals on cassette magnetic tape, and processed
to a selected event record on the Institute of Hydrology's
UNIVAC 1108 computer.

The loss submodel
Once the volume of runoff has been determined, it is necessary
to distribute the net rainfall in time (i.e. propose a loss
model). Because the simulation performance of a model can
only be judged from data arising from both loss processes and
surface routing processes, the choice of loss model structure
is necessarily arbitrary. In this case, a two-stage model
has been adopted: (i) an allowance for depression storage
(assumed to be satisfied before overland flow occurs) and (ii)
a variable source model in which the contributing area is 100%
impervious and of such a size that the volume of runoff is
that determined by the runoff volume submodel. Thus the loss
submodel has a single parameter which is the depth of

Table 2: Depression storage estimates

Subcatchment	Location	Events	DEPSTO estimate	s.e.e.	Average slope (%)
206/1	School Close, Stevenage	6	.67	.11	0.9
206/2	School Close, Stevenage	5	.53	.10	1.7
205/1	Hyde Green, Stevenage	5	.49	.12	2.2
205/2	Hyde Green, Stevenage	5	.50	.18	2.0
204/2	Bishopdale, Bracknell	8	.45	.10	2.4
203/2	Ennerdale, Bracknell	4	.50	.29	3.1
208/1	Plover Close, Soton	11	.32	.12	3.4
208/2	Plover Close, Soton	5	.41	.29	3.0

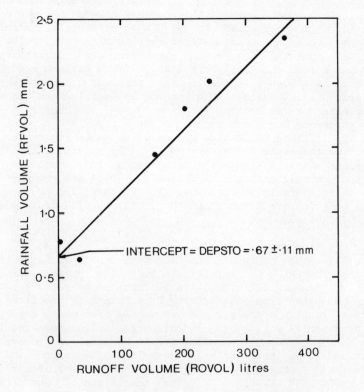

INTERCEPT = DEPSTO = ·67 ± ·11 mm

Fig. 2: Depression storage estimation for subcatchment 206/1 (School Close)

depression storage.

The depression storage cannot be measured in the field. The only published estimates (Veissman, 1966; Willeke, 1966) were based on the premise that all losses from a parking lot were attributable to depression storage. Accepting that there is some infiltration through impervious surfaces, these estimates may be high.

Data from the subcatchment areas in this study were used to provide estimates of depression storage based on the following water-balance model:

$$\text{ROVOL} = \text{PR} \times \text{PAV} \times (\text{RFVOL} - \text{DEPSTO}) \qquad \ldots 3$$

where ROVOL is the runoff volume (litres),
      RFVOL is the rainfall volume (mm),
      DEPSTO is the depression storage (mm),
      PAV is the paved area (m^2),
and   PR is the proportional runoff.

Only short (less than 15 minute duration) events of low to medium intensity on dry catchments were used, such that the contribution of pervious surface might safely be ignored. Suitable values of ROVOL & RFVOL were abstracted from the data for each subcatchment, and values of RFVOL plotted against ROVOL. A linear regression line was generated, and the intercept on the RFVOL axis taken as the estimate of DEPSTO. An example of this is shown in Figure 2 for one of the School Close subcatchments. This analysis yields an estimate of .67 mm with a standard error of the estimate of .11 mm (correlation coefficient of the regression line is .98). Table 2 lists the estimates, derived by the above method of analysis, for 8 of the subcatchments employed in this study. These estimates show a significant correlation on catchment slope, values of which are also shown in Table 2. Catchment slope has been taken along a principal represent-ative flow-path. Where more than one principal flow-path could be identified, a weighted average was calculated according to the area of which the flow-path was representa-tive. Figure 3 shows this variation of depression storage, and a regression analysis gives the following relationship:

$$\text{DEPSTO} = -.109 \text{ SLOPE} + .738 \qquad \ldots 4$$

This equation is applicable for predominantly asphalt surfaces within the limits of the data.

The surface routing submodel
Surface routing is modelled using a lumped-parameter hydrological approach. A non-linear reservoir is used, which has been shown (Kidd, 1976; Kidd & Helliwell, 1977) to

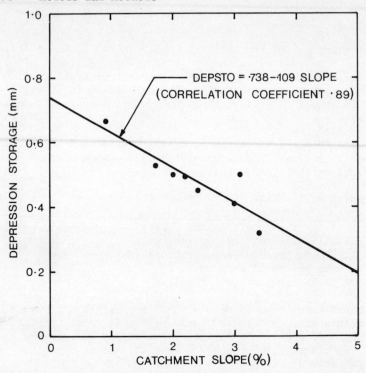

Fig. 3: Variation of depression storage with catchment slope

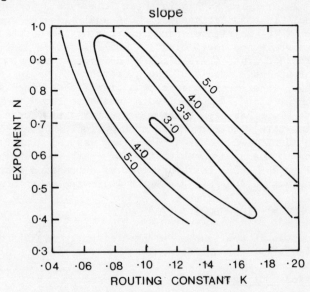

Fig. 4: Response surface for storm 206013 on catchment 206/1

Table 3:   Optimum routing constants

Subcatchment	Location	Events	K	S.D.	Paved area(m²)	Average slope(%)
206/1	School Close, Stevenage	13	.104	.016	283	.9
206/2	School Close, Stevenage	12	.168	.021	393	1.7
204/2	Bishopdale Bracknell	19	.149	.017	450	2.4
203/3	Ennerdale, Bracknell	24	.088	.022	90	3.0

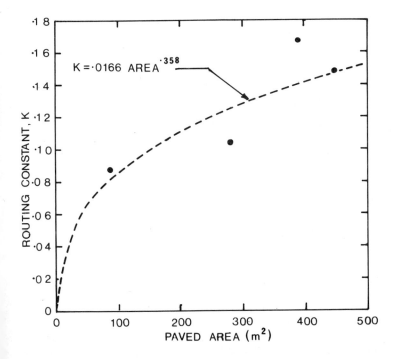

$K = \cdot 0166 \; AREA^{\cdot 358}$

Fig. 5: Variation of routing constant with paved area

provide a satisfactory synthesis of surface routing. The basic equations which describe the structure of the model are:

$$S = KQ^n \quad \text{- dynamic equation -} \qquad \ldots 5$$

$$\frac{dS}{dt} = I - Q \quad \text{- continuity equation -} \qquad \ldots 6$$

where S is the reservoir storage (mm),
        Q is the outflow (mm/hr),
        I is the reservoir inflow (mm/hr),
        t is time,
and   K and n are model parameters.

For nonzero values of input I and for general values of the exponent n, the above equations have no analytical solution. However, the two equations may be combined and then solved by a number of techniques, such as a finite-difference approach, an iterative procedure, a Runge-Kutta solution or a Varied Flow Function approach (Kidd, 1976). The relative merits of these techniques in terms of speed and accuracy are currently being investigated.

It remains to calibrate the model by determining suitable values for the model parameters and relating them to catchment characteristics. As with other models of this type (Van den Berg et al, 1977), the two model parameters are strongly interdependent, which makes efficient optimisation difficult. An example of this phenomenon is shown in Figure 4 (The objective function used is an Integral Square Error using a least-squares goodness of fit criterion). The elongated valley indicates that a satisfactory fit is possible with any number of pairs of values of n and K (within certain limits). For this reason, the value of n was fixed. If the Chezy equation is applied to a situation where the depth of flow is small in comparison with the width (such as overland or gutter flow), a value of n of .67 may be derived for equation 5. This value of n was adopted, and the subcatchment data employed to derive optimum values of the routing constant K.

For each event, objective functions based on peak estimation and on a least-squares goodness of fit criterion were used, and, where these indicated different optimum values of K, a subjective assessment of best K-value was necessary. Best K-value for each subcatchment was taken as the mean value for all events used. As yet, only four of the subcatchments have proved satisfactory for these analyses. Table 3 shows the optimum values of the routing constant derived in the manner described - the standard deviation quoted gives an indication of the spread of optimum values for a given subcatchment. Table 3 also shows values of catchment characteristics upon which the routing constant

Table 4: Data input and model parameters for Oxhey Road catchment

Parameter		Subcatchments 1, 2 and 3	Subcatchments 4 and 5	Subcatchment 6
Total area (m^2)		900	1799	1588
Paved area (m^2)		537	1074	948
PIMP (%)		59.7	59.7	59.7
SOIL		.45	.45	.45
UCWI	Event 25	93.7	93.7	93.7
PRO (%)		51.7	51.7	51.7
Contributing area (m^2)		463	926	817
UCWI	Event 43	272	272	272
PRO (%)		63.3	63.3	63.3
Contributing area (m^2)		566	1132	999
No. of gullies		2	4	6
Paved area/gully (m^2)		268.5	268.5	158
Routing constant, K		.123	.123	.102
Catchment slope (%)		.93	.93	.93
Depression storage (mm)		.64	.64	.64

might be dependent. A tentative relationship has been
derived from the few results available so far, for the
variation of the routing constant with paved area, as shown
in Figure 5 and given by:

$$K = .0166 \text{ AREA}^{.358} \qquad \qquad \dots 7$$

This equation was derived from a linear regression using
logarithmic transforms of K and AREA (correlation coefficient
.87). Because one might intuitively expect the relationship
to pass through the origin, this form of the equation is
considered superior to a linear relationship.

DEMONSTRATION OF MODEL PERFORMANCE

The calibrated model as described was tested on existing data
from Oxhey (collected by the Road Research Laboratory in the
1950s - Watkins, 1962) as a test of the model's performance.
The catchment area (see Figure 6) is a length of road in
shallow cutting, with two autographic raingauges in close
proximity and the runoff measured using a V-notch tank.
Table 4 shows the catchment data and parameters necessary for
application of the model to two recorded events, chosen to
be significantly different in storm magnitude and antecedent
catchment wetness. Catchment slope has been taken as equal
to the average pipe slope of the sewer system.

Hydrographs were routed through each pipe length by the
Kinematic Wave method (equivalent to offsetting each hydrograph ordinate by a time corresponding to steady uniform flow
at that appropriate depth). Manning's formula was used with
a value of n equal to .013 at pipe-full flow.

Results are shown in Figure 7 for the two events.

CONCLUSION

A calibrated model of the rainfall-runoff process over urban
surfaces has been developed with the aim of producing a
realistic simulation of the inlet hydrograph in sewered
catchments. Equations 1, 4 and 7 represent relationships
between model parameters and simple catchment characteristics,
and the performance of the model has been illustrated. It is
emphasised that these results are of an interim nature. It
is hoped to obtain data from further subcatchment
experiments, from laboratory catchment studies at Imperial
College, and from similar experiments in other European
countries.

This calibrated model is intended to be incorporated in
a new design method (Price & Kidd, 1978) currently under
development at the Hydraulics Research Station and the

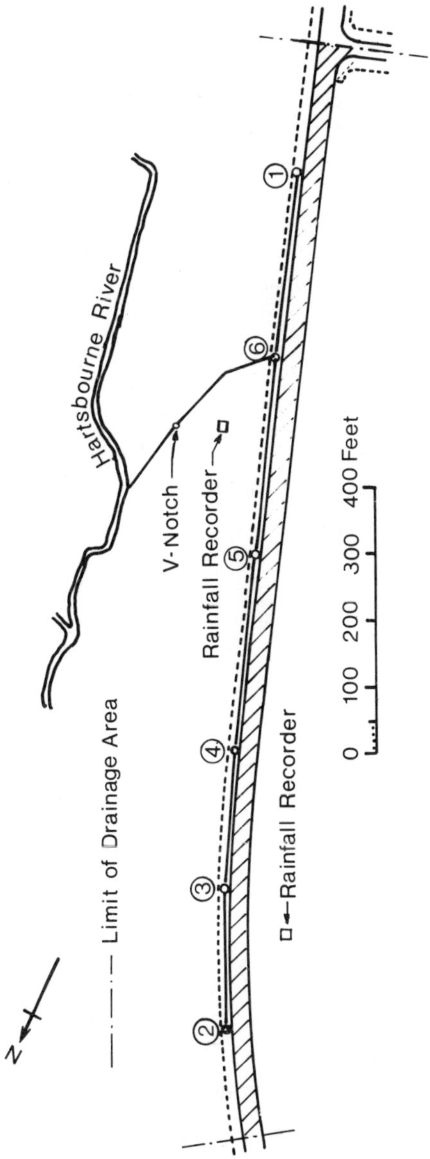

Fig. 6: Oxhey Road Catchment

N

—·—·— Limit of Drainage Area

Hartsbourne River

V-Notch

Rainfall Recorder

□ ← Rainfall Recorder

0   100   200   300   400 Feet

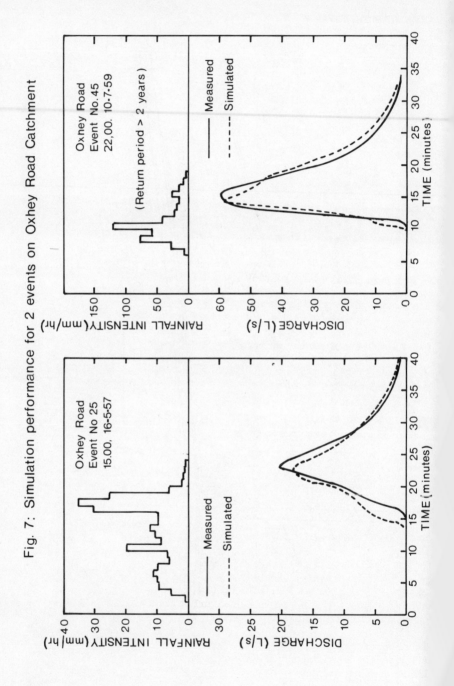

Fig. 7: Simulation performance for 2 events on Oxhey Road Catchment

Institute of Hydrology.

ACKNOWLEDGEMENTS

The author wishes to thank his colleagues, Dr M.J. Lowing for
his general supervision and helpful advice, and Mr I.W. Makin
for the collection and processing of the data. This paper
is presented with the permission of Dr J.S.G. McCulloch,
Director of the Institute of Hydrology. Work at the Institute
of Hydrology is undertaken as part of a project commissioned
by the Department of the Environment (Project No. DGR/480/38).

REFERENCES

Blyth, K. and Kidd, C.H.R. (1977) The development of a
meter for the measurement of discharge through a road gully.
The Chartered Municipal Engineer, Feb. 1977.

Kidd, C.H.R. and Helliwell, P.R. (1977) Simulation of the
inlet hydrograph for urban catchments. Jour. Hydrology,
Vol. 35, pp 159-172.

Kidd, C.H.R. (1976) A non-linear urban runoff model.
Institute of Hydrology Report No. 31.

Lloyd-Davies, D.E. (1906) The elimination of storm water
from sewerage systems. Proc. ICE, Vol. 164(2), pp 41-67.

Metcalf and Eddy Inc. et al (1971) Storm Water Management
Model, Vols. I, II, III and IV. Report Nos. 11024 DOCO7/71,
08/71, 09/71 and 10/71, Environmental Protection Agency,
Washington D.C., U.S.A.

Natural Environment Research Council (1975) Flood Studies
Report.

Newton, S.G. and Painter, R.B. (1974) A mathematical
examination of urban runoff prediction. Proc. ICE., Part 2,
Vol. 57.

Price, R.K. and Kidd, C.H.R. (1978) A design and simulation
method for storm sewers. Proc. International Conference on
Urban Storm Drainage, Southampton.

Stoneham, S.M. and Kidd, C.H.R. (1977) The prediction of
runoff volume from fully-sewered urban catchments. Institute
of Hydrology Report No. 41.

Transport and Road Research Laboratory (1976) Road Note 35.
A guide for engineers to the design of storm sewer systems.
HMSO, 2nd Edition.

Van den Berg, J.A., de Jong, J. and Schultz, E. (1977) Some qualitative and quantitative aspects of surface water in an urban area with separate storm water and waste water systems. Proc. Symposium on the effects of urbanisation and industrialisation on the hydrological regime and on water quality, Amsterdam. AHS-IAHS Publication No. 123.

Veissman, W. (1966) The hydrology of small impervious areas. Water Resources Research, Vol. 2, p 405.

Watkins, L.H. (1962) The design of urban sewer systems. HMSO, Road Research Laboratory Technical Paper No. 55.

Willeke, G.E. (1966) Time in urban hydrology. Proc. ASCE, Hyd. Div., Vol. 92 (HY1), p 13.

A LINEAR RESERVOIR MODEL FOR URBAN RUNOFF

B.B. Nussey,   E.J. Sarginson,

                        and

Sheffield City Polytechnic   Sheffield University

INTRODUCTION

The model described in the present paper is a revised version
of the two-reservoir model which has been previously described
(1, 2, 3). After deducting losses from the gross rainfall,
the nett rainfall is routed through two linear reservoirs in
series, representing overland flow and pipeflow respectively.
This results in the following differential equation:-

$$Q + (\alpha + \beta) \frac{dQ}{dt} + \alpha\beta \frac{d^2Q}{dt^2} = r(t) \qquad ---- \quad (1)$$

where Q is the discharge, $r(t)$ the nett rainfall on the
catchment, expressed as a function of time, and α and β
the constants representing overland flow and pipe flow
respectively. Previous versions of the model (2, 3) were
based on data obtained from four catchments in U.S.A. and one
in U.K., but their application to data from other catchments
led to serious errors. The revisions described in the present
paper have substantially reduced these errors.

The most important source of error was in calculating the
routing constants α and β. Originally, the pipe routing
constant β was obtained from the "tail" of the recession curve
of the discharge hydrograph after the cessation of rainfall.
It was generally found to be exponential, and it is easy to
show that this is consistent with the idea of a linear
reservoir. It is now realised that the recession curve from
the pipe system depends on overland flow as well as pipe flow,
but the lower part of the recession curve is unreliable for
the following two reasons:-

(a) It is unlikely that overland flow will long survive the
     end of a storm unless it is an exceptionally heavy one.

187

It is more likely that later than a few minutes after the
end of a storm, water remaining on the catchment surface
will either evaporate or else wet the surface without
running off.

(b)   Fig. 1 shows the relationship between storage and
discharge from a part-full pipe. When the pipe is more
than 0.3 full (i.e. the depth of flow is more than 0.3
of the diameter), the relationship is linear, and in a
previous paper (1), it was shown that the slope of this
line is equal to β. Below 0.3 full, the pipe no longer
behaves as a linear reservoir, and this has an important
effect on the lower part of the recession curve.

A better method of getting α and β from existing data is to
use the upper part of the recession curve immediately after
rainfall has ceased. Trial runs using tentative relationships
between α and β (e.g. $\alpha = \beta$, $\alpha = \beta/2$ etc.) showed that the
value of $(\alpha + \beta)$ obtained in this way was practically
independent of the ratio α/β. Several of the catchments
investigated had sufficient pipe data to enable β to be
calculated from Fig. 1, and the ratio α/β was found to vary
from 1/4 to 1/6 with a mean value 0.2. The usual practice was
to select one shown from any catchment which has a suitable
recession curve and to compute α and β from this. These
values were then applied to the other storms on the same
catchment, and this procedure worked well on the whole.

Over 37 storm events on five catchments, the new procedure
gave a mean error in peak flows of 0.6 per cent with a
standard deviation of 12.5 per cent. Fig. 2 shows good
agreement between measured and computed runoffs in respect of
one of the events.

CORRELATION OF $(\alpha + \beta)$ WITH CATCHMENT PARAMETERS

From the data available from six catchments in U.K. two in
U.S.A. and one in Canada, the value of $(\alpha + \beta)$ was found to be
correlated at a four per cent significance level with the area
and mean slope of the catchment, as follows:-

$$(\alpha + \beta) = 7.6 + .016A - 150s \qquad \text{---- (2)}$$

where A is the catchment area in hectares, and s the mean
slope, expressed as a decimal.

The adjusted correlation coefficient was 0.81 and the adjusted
standard error was 1.34. Equation (2) is a form that might
be expected. The dependence on area is comparatively small
unless the area is a very large one, because increase in area
means a larger pipe in which both the storage volume and the
discharge are increased, and the ratio is not greatly affected.

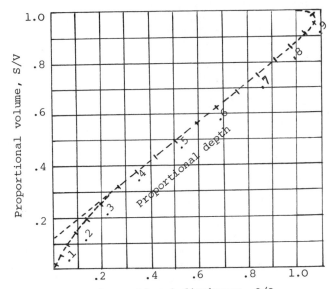

Fig.1 Properties of a part-full pipe

From the linear part of the graph, $\beta = \dfrac{0.756\,V}{Q_o}$

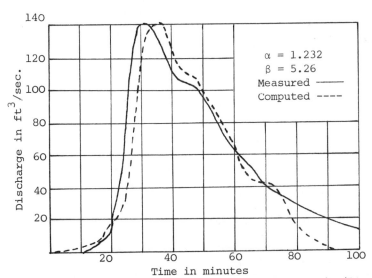

Fig.2 Discharge hydrograph from Oxhey, 23/10/54.

The effect of slope is more significant because the runoff from a steep slope is more rapid than from a flat slope, both in respect of overland flow and pipe flow. With a steep slope the storage effect both above and below ground is reduced and this reduces both α and β.

Equation (2) is useful in design, since it gives a preliminary value of $(\alpha + \beta)$, from which peak flows can be computed. Then the necessary size of pipe can be calculated, and the new value of β obtained from the pipe volume and discharge, and the calculations repeated if the new value differs materially from the old one.

TREATMENT OF LOSSES

Previously, the most accurate treatment of losses was found to be an initial deduction of rainfall to fill depression storage, followed by the deduction of infiltration loss for the remainder of the storm at a constant rate sufficient to give the correct total volume of rainfall (2). With the present method of calculating α and β, this is usually found to overestimate the peak runoff, as it gives undue emphasis to rainfall peaks, while underestimating discharges from rainfall of lighter intensity. The most effective method of heating losses in the present context is to deduct surface depression storage as before, then to reduce the rest of the rainfall hydrograph in a constant ratio to give the correct total volume, i.e. the procedure is to follow surface depression storage by proportional runoff.

There is a further point. Referring to Fig. 1, if the linear part of the graph is produced backwards, it intersects the vertical axis at 0.119V, or nearly one-eighth of the pipe volume. If pipe flow is treated as routing through a linear reservoir with its constant β equal to the slope of the linear part of Fig. 1, the volume under the actual recession curves of a storm event exceeds that predicted by the model by that amount. The resulting error in peak runoff is negligibly small if the discrepancy between the two recession curve volumes is small in comparison with the total runoff, but becomes more significant as the total volume is decreased. Thus the model is more reliable with large runoff volumes than with small ones, and this vindicates its use in design.

It was shown in a previous paper (3) that the ratio of total runoff (R) to total rainfall (I) is correlated with the catchment slope (s) and the ratio of impermeable area to total area (P) as follows:-

$$\frac{R}{I} = .0306 + .438P + 2.09s \qquad \text{----} \quad (3)$$

APPLICATION TO DESIGN

The recommended design storm is described in Road Note No. 35
(4). It is a 2 - hour storm symmetrical about the centre point
and the intensity depends on the time in minutes from the
centre point (t) and the return period in years (N). Road
Note No. 35 gives it in tabular form but for use with equation
(1) it is more conveniently expressed in functional form. The
following equation fits the tabulated figures satisfactorily:-

$$i \text{ (mm/hr)} = \frac{189.4 \ (N - 0.2)^{0.242}}{(t + 1.5)^{1.16}} \quad \text{---- (4)}$$

With a design storm of this kind, the treatment of losses is
simple. Surface depressions are filled during the long low-
intensity run-up to the storm peak, and peak discharges will
not be affected if the principle of proportional runoff is
applied to equation (4) as a whole. This is almost identical
to the procedure suggested in an earlier paper (1), in which
the design procedure was simplified by the use of a design
table which depended only on α and β. This time the design
procedure is simplified by the fact that α is a fixed
proportion (0.2) of β.

ALTERNATIVE METHOD - PIPE-BY-PIPE ROUTING

The model so far described computes the discharge from the
whole catchment treated as one unit, assuming that the taper
on the pipes justifies the idea of constant proportional depth
over the whole system at any instant. Then β can be computed
on the basis of the total storage volume of the pipes and the
discharge capacity of the outlet pipe running just full.
Alternatively, a sewer system can be designed on the basis of
a pipe-by-pipe routing procedure, so that the output from one
pipe becomes the input to the next pipe, along with the ground
runoff from the area served by that pipe. This is the
procedure adopted by the TRRL method, and the two-reservoir
model is easily adapted to the same principle. The authors
have carried out a pipe-by-pipe procedure on five sewer
networks, necessitating 195 separate pipe calculations. The
two reservoir model gave almost identical values of the peak
runoff to those obtained by the TRRL method, provided that

$$\alpha = 0.41\beta + 0.86 \quad \text{---- (5)}$$

The mean ratio in the two sets of peaks was 1.017 with a
standard deviation of 0.032. This was in nearly two hundred
separate pipe calculations where the range of flow was from
5 to 5,000litres/sec.

ACKNOWLEDGEMENT

The authors gratefully acknowledge the assistance given by Mr.C.P. Young of the Transport and Road Research Laboratory for making available the data on which the TRRW method was based.

REFERENCES

1. Sarginson, E.J. and Bourne, D.E. (1969) The Analysis of Urban Rainfall, Runoff and Discharge. J.I. Mun.E., London 96; 81-85.

2. Sarginson, E.J. Relationship of Rainfall and Runoff in Urban Areas (1973) Proc. Research Colloquium at Bristol University, CIRIA London: 901-908.

3. Sarginson, E.J. and Nussey, B.B. A Mathematical Model for Urban Runoff (1975). IAHS-AISH Publication No. 115, Bratislava; 209-213.

4. Road Note No.35, A Guide for Engineers to the Design of Storm Sewer Systems. HMSO. 1976, London.

DIRECT DESIGN OF SURFACE WATER SEWERS

Roy Hepworth

Senior Project Engineer, Ward Ashcroft and Parkman

INTRODUCTION

It would appear from the Second Progress Report[8] of
the Working Party on the Hydraulic Design of Storm
Sewers that the working party is advancing towards
a complex design procedure. There is indeed an
apparent scope for considerable refinement of
calculations in the rainfall/run off design process
but such refinement in my view does not necessarily
lead to greater accuracy in design because of
difficulty in application of a complex model.

In any case I am not convinced that excessive
refinement is required. The conclusion drawn from
my own analysis of available storm data is that
the design of storm sewers can be substantially
improved by a modicum of refinement to the existing
Lloyd Davies and Ormsby and Hart methods. This
paper describes these older methods together with
the model which has resulted from my work.

Because, in the new model, run-off is calculated
directly from the product of rainfall and area, I
have called the calculation procedure the DIRECT
method of storm sewer design. Some details of the
model have been published in the New Civil Engineer[3]
I welcome this opportunity to describe it further
and to receive comments.

FORMS OF THE DIRECT METHOD

The Direct method has two forms. The one is similar
to the Lloyd Davies method and the other is similar
to the Ormsby and Hart method. I have opted, on
this occasion, to describe the Direct method through
the medium of a description of the older methods.

THE LLOYD DAVIES METHOD

The assumptions of this method are

> 1. A short time of entry, say 2 minutes,
>   applies throughout the storm
>
> 2. Full (or half) bore velocities apply
>   throughout the storm
>
> 3. There is a linear time area diagram
>
> 4. Storm duration equals time of concentration
>   of area.

According to the theory, after the first 2 minutes
of the storm, water from the furthest part of each
minor catchment is entering its own pipe. The
flow from the most upstream increment of area
$(A_1 \times I_1)$ enters its own pipe in the third minute.
Flows similarly enter pipes all the way down the
system with the flow from the most downstream
increment of area $(A_n \times I_1)$ arriving at the design
point in the third minute.

In the fourth minute two series of events happen as
follows.

1. Water in the pipes moves downstream at full
bore velocities into the next incremental pipe
length and

2.   Increments of flow related to the intensity
in the second minute of the storm (I_2) enter the
pipes.

Thus the flow at the design point is $(A_{n-1} \times I_1)$ + $(A_n \times I_2)$, the flow in the most upstream pipe is $(A_1 \times I_2)$ and the flow in the next to most upstream pipe is $(A_1 \times I_1)$ + $(A_2 \times I_2)$.

There are similar happenings in succeeding minutes until in the $(2 + n)$th minute of the storm the flow at the design point is $(A_1 \times I_1)$ + $(A_2 \times I_2)$ +... $(A_n \times I_n)$. By definition, the time of concentration of the area is $2 + n$ minutes and therefore the storm stops. Up to this time the flow at the design point has been increasing but in the next minute the flow there is $(A_1 \times I_2)$ + $(A_2 \times I_3)$ + $(A_3 \times I_4)$ +...+ $(A_n \times I_{n+1})$ which is the same as the flow occurring the minute before. After one more minute the downstream increment of area stops contributing and the flow at the design point starts reducing. The maximum flow of $\Sigma (A_n \times I_n)$ occurs at the end of the storm during the three minute period when all the catchment is contributing at the design point.

In the Lloyd Davies method, the time area diagram is linear i.e. $A_1 = A_2 = \ldots = A_n$ and the sum of the incremental areas $\Sigma A_n = A$, the total area of the catchment. The maximum flow may therefore be written

$$Qmax = \frac{A}{n} \times \Sigma I_n \quad \text{or} \quad A \times \frac{\Sigma I_n}{n}$$

Now $\frac{\Sigma I_n}{n}$ is the average intensity which occurred over the first n minutes of the storm lasting $(2 + n)$ minutes.

Therefore Qmax = Total area x Average intensity, which is the Lloyd Davies formula.

It will be noted that if the short time of entry and full bore velocities apply throughout the storm there is nowhere else that this water, which makes up the peak flow, can be and the Lloyd Davies method is as accurate as its assumptions allow it to be.

THE ORMSBY AND HART METHOD

The assumptions of this method are

1. A short time of entry, say 2 minutes applies throughout the storm

2. Full (or half) bore velocities apply throughout the storm

3. Storm duration equals time of concentration of area.

The theory behind the method is exactly the same as that described for the Lloyd Davies method up to the point that the maximum flow is stated to be

$$\Sigma (A_n \times I_n)$$

In this case each individual incremental area is multiplied by its appropriate intensity, the products then being summed to give the maximum flow. Again, within the terms of the assumptions there is nowhere else this water can be but at the design point, and the Ormsby and Hart method is as accurate as its assumptions allow it to be.

In order to cater for an unbalanced time area diagram a storm of reduced duration, and therefore increased (average) intensity, should be applied to the various downstream parts of the area to determine whether part of the area is capable of a larger flow than the whole area. It should be noted that this exercise could also be carried out when designing to the Lloyd Davies method. The exercise is equivalent to what is referred to as the tangent method.

THE LLOYD DAVIES AND ORMSBY AND HART ASSUMPTIONS

The two methods described above are both as accurate as their assumptions allow them to be. Of the two, the Ormsby and Hart method is more accurate since it takes into account the actual distribution of the area. Notwithstanding this fact the Ormsby and Hart method (and the tangent method) have not been

in general use over the years because it was thought
that flows derived were generous. On the other hand,
the less accurate Lloyd Davies method has been in
general use.

Depending on the accuracy of the assumptions it
would appear that the above methods should continue
in use or should be disused. The assumptions are
examined below.

Time of entry
At the start of a storm, real intensities are small
and in fact it may take many minutes for a sufficient
water depth to build up on the surface so that water
can flow. To calculate the time of entry precisely
is a complicated procedure because many things have
to be taken into account. One thing is certain
however - it is generally much longer than 2 minutes.

Full or half bore velocities
Because of the generally long time of entry, water
does not start entering the pipes shortly after the
start of the storm. Records of rainfall/run-off
events demonstrate that flow at its measuring point
does not start increasing rapidly until the middle
of the storm. Clearly this is from the areas
nearest to the measuring point and it follows that
build-up of flow occurs at a similar time in pipes
of all parts of the catchment. This indicates that
during the first half of the storm, velocities can
not possibly be full or half bore value.

Linear time area diagram
The well-known "S" shape of the time area diagram
can be reasonably taken to be linear in the case of
small areas. For larger areas however it is clearly
preferable to take into account the actual shape of
the diagram. Otherwise features such as unproductive sewer lengths, or those partially unproductive,
are not correctly taken into account. The time
area diagram should not be applied at times when
velocities do not approximate to full or half bore
values.

Storm duration equal to time of concentration
This assumption, derived from the concept of full
bore velocities from an early part of the storm,
fails by virtue of the fact that full bore
velocities do not exist as assumed. To accord
fully with the theory of the Lloyd Davies method
outlined above, the storm duration should be the
time of flow of the area; this is more severe still
than the usual assumption.

Improved assumptions
The main concept of the Lloyd Davies and Ormsby
and Hart methods is "there is nowhere else where
the water can be". There is room for improvement
in the assumptions of the methods in order that the
main concept can be correctly implemented.

While the time of entry at the start of the storm
is greater than 2 minutes, later in the storm when
water has had time to build up on the surfaces the
time of entry might well be of the order of 2
minutes.

Also, later in the storm, velocities are of the
order of full or half bore value.

If the time area diagram is calculated using full
bore velocities and if these apply to the period
that a calculation covers then the use of the time
area diagram is acceptable. A linear time area
diagram is reasonable for small areas.

The assumption of length of critical storm must be
derived from factual evidence of rainfall/run-off
events.

The two forms of the Direct method are essentially
the Lloyd Davies and Ormsby and Hart methods with
improved assumptions.

SIMULATION OF RAINFALL/RUN-OFF EVENTS

Simulation procedure
Modern instrumentation has allowed simultaneous
readings of incremental rainfall and run-off. Much

valuable work has been done over the years by Watkins
and his group at TRRL[7] and more recently by other
teams.

Not a great deal of the results of these efforts
have been available to me, but from the results
which have been available I have analysed the storms
and found certain patterns.

The main idea behind my approach, like in the Lloyd
Davies and Ormsby and Hart methods, was that there
is nowhere else where the water can be.

Having established the time area diagram, I listed
the increments of area, upstream to downstream, each
with a one minute time of flow. Then, assuming a
two minute time of entry in the later part of the
storm and also full bore velocities at that time,
I multiplied the most downstream incremental area
by the intensity which occurred in the third minute
before the time of peak flow. If the assumptions
were correct, there was nowhere else this increment
of flow could be other than at the measurement
point at the time of peak flow. I then multiplied
the next area upstream by the intensity four minutes
before peak flow and added it to the first product.
I continued in this manner up the catchment until
the sum of the products was nearest to the recorded
peak flow.

It was more likely in this period, than in any other
period of the storm, that the assumptions of the
short time of entry and full bore velocities applied,
since there had been time for water to collect on
the surfaces and in the pipes. If the assumptions
did not apply during that time, they certainly did
not apply before the period.

I calculated the run-off in this manner for all the
storms which were available to me and noted the
following from the simulations.

> 1. Run-off did not start increasing rapidly
>   until about the middle of the storm.

2. In a symmetrical storm the earliest intensity to be applied was often the maximum intensity and peak run-off was at the end of the storm.

3. The number of increments of run-off calculated was approximately equal to half the length of the storm in minutes.

4. The full catchment area did not necessarily contribute to the maximum run-off. The extent of catchment which contributed had a time of flow approximately equal to half the storm duration.

5. Long storms were made up of a series of rainfall peaks, the individual peaks of which produced run-offs as if the peaks were single storms.

Results of the simulation procedure

Results of my analysis of available storms from the TRRL research have been published elsewhere[3] and compared statistically with other methods[2]. I have recently analysed some events from the Derby catchment which is being monitored as part of the H.R.S. research work.

In many cases of the recent analyses the peak run-off was not clearly at the end of the storm but the general conclusion may be drawn that the duration of rainfall which occurs before peak run-off is about equal to twice the time of flow of the contributing area. Results are given in Table 1.

It is encouraging that 8 of the 16 events analysed were consistent with the Direct method. Two could be said to be in keeping with Lloyd Davies and two had storm duration equal to about three times the time of flow of the contributing area. Results in four cases 9/2, (event/gauge) 9/4, 16/3 and 16/4 suggested that the storm was moving and not consistent over the area.

It is clear from the above that the Direct method

Event No.	Gauge No.	Time of Peak run-off after start of recorded rainfall (Mins)	Effective length of Storm (Mins)	No. of products of intensity & rainfall to give peak run-off	Is peak intensity earliest to be applied?	If not, time from peak intensity to earliest intensity applied. (Mins)	Calculated Flow (mm/hr/hectare)	Measured Flow
2	2	101	29	15	Yes	-	36.687	35.252
2	4	100	29	9	Yes	-	23.243	23.770
9	2	)						
9	4	) See comments on previous page.						
16	3	)						
16	4	)						
21	2	99	39	19	No	1	42.625	49.407
21	2	130	13	10	No	1	34.025	32.751
21	3	98	38	18	No	3	36.272	38.701
21	3	131	14	11	No	1	23.966	24.830
27	2	200	22	12	No	2	27.956	28.049
27	4	200	22	12	No	1	21.338	22.457
29	3	78	23	9	No	1	28.161	29.018
29	4	75	20	7	No	1	25.565	27.899
32	2	76	20	12	Yes	-	30.508	29.591
32	3	70	14	8	No	2	20.998	22.548

Table 1. - Results of analyses of storms from the Derby catchment

Information on storms by courtesy of Hydraulics Research Station.

does not produce total accuracy in simulation. Such accuracy cannot be expected in the complex relationship between rainfall and run-off. The interesting thing is that with a calculation system such as that employed the reasons for inaccuracies become obvious from a study of the figures. Further results will be analysed with interest as they become available.

DIRECT DESIGN OF SURFACE WATER SEWERS

I concluded from the above that Lloyd Davies was basically right for areas where a linear time area diagram could be reasonably assumed but he had got his storm duration wrong. Instead of time of concentration he should have used twice the time of flow.

I also concluded Ormsby and Hart had been correct but the second half of an incremental storm should be applied, to have the effect of a storm of duratic equal to twice the time of flow of the area and not equal to the time of concentration. Further, in the design for a particular pipe, a check should be carried out to establish whether any downstream part of the area could give a greater flow than the whole area.

These two systems, one which can be easily calculate in the office in a manner parallel with Lloyd Davie; the other preferably with the use of a computer in a manner parallel with Ormsby and Hart, form the Direct method of storm sewer design.

EVIDENCE TO SUPPORT THE DIRECT DESIGN METHOD

Perhaps the most original aspect of the Direct desi method is the thought that the duration of the critical storm is twice the time of flow of the catchment. The following evidence from the literature on the subject supports this thought.

 1. Sarginson and Bourne[5] introduced two time related parameters, α and β representing above and below ground storage. They found that t₁ solutions for particular events were most

realistic when $\alpha = \beta$. This could be interpret
-ed to mean that periods of above and below
ground attentuation were equal. This is
parallel to the Direct method in which half
the storm is required to fill above ground
storage and the second half fills below ground
storage producing peak run-off in the process.

2. Newton and Painter[4] suggested storage on
the ground was equal to storage in the pipes.
This is equivalent to saying the first and
second halves of the storm in volume are
required to fill above and below ground
storage respectively. This is again parallel
to the idea behind the Direct method.

3. The TRRL method[6,7] assesses the effect of
underground storage by calculating a primary
hydrograph of flow out of the pipe system and
then proceeds to develop a secondary hydrograph, again taking pipe parameters into
account to achieve correlation between rainfall and run-off. Two calculations are
therefore carried out to determine the effect
of underground storage with reasonable results
in simulation and this suggests that above
and below ground attentuations have a similar
effect on the outgoing flow - this again is
in keeping with the Direct method.

4. Butters and Vairavamoorthy[1] established a
formula for river catchments.
Te = (2k + 1) Tc
with k being within the limits of 0.5 and
0.75 for London catchments (Te = time of
equilibrium. Tc = time of concentration). The
authors found that their defined time of
equilibrium was applicable in a Lloyd Davies
type calculation for river catchments. It
should be noted that if the limits of k are
introduced in the equation the critical storm
duration becomes 2 to 2.5 x time of concentration which is similar to the Direct design
critical storm duration.

POINTS ABOUT DESIGN

Throughout the simulation exercises carried out,
100% and 0% run-off have been taken for paved and
unpaved areas respectively. It is reasonable that
the whole of the paved area contributes later in
the storm while any run-off from unpaved areas can
only be expected infrequently. A logical approach
in design is to continue with the above assumptions.
However, paved areas draining onto unpaved areas
should not be taken into account whereas unpaved
areas with clay subsoil draining onto paved areas or
directly connected to the pipe system should be
taken into account. All potential development in
the overall catchment should be included.

The design frequency of storm occurrence is difficult
to contemplate, so many factors being involved. The
design run-off for an overall catchment may be
equalled by a shorter storm falling on a part of the
catchment; a moving storm may have a worse effect
than a less intense stationary storm; a multi-
peaked storm may be more critical than a single
peaked storm for a large catchment. These are just
some of the possibilities which could affect the
apparent storm frequency. For Direct design purposes
a graduated scale of storm frequencies should be
applied, depending on the importance of the area.
Discussions with the Meteorological Office are
proceeding in an effort to determine suitable storms
for use with the Direct method.

In the case of a suspect existing system, a new
system should be designed. Undersized pipes should
be replaced and oversized pipes may be left working.

Flows in carrier sewers should not be decreased
except possibly in very long ones which are in any
case to be avoided wherever possible for obvious
reasons. Downstream of a carrier sewer (or
unproductive length) the time of flow in the carrier
sewer should not be ignored in subsequent
calculations.

If it is required to investigate surcharge and flooding risks, a staged calculation process is required. Firstly the system is designed with the required storm frequency and then flows are assessed in a similar calculation with a less frequent storm. Using the top water level of the receiving water as a starting point, the hydraulic levels corresponding to the less frequent storm may be then calculated back up the system. Flooding is likely to occur where the hydraulic level is above ground level. Water which is able to arrive at this point and cannot proceed downstream will leave the manhole in this vicinity. Consideration of ground levels will provide an estimate of area and depth of flooding.

CONCLUSIONS

The Direct method has been shown to be accurate in simulation for a selection of catchment areas within the limits imposed by the complexities of the natural processes involved. There is no recourse to "percentage run-off", "percentage impermeable area" and doubtful or complicated theory. The simulation procedure can be applied in design in precisely the same manner or in a simplified manual version. The two forms of the method are put forward to be used in the future as improvements to existing methods.

REFERENCES

1. Butters, K. and Vairavamoorthy, A. (June 1977) Hydrological studies on some river catchments in Greater London. Part 2 Proc. Inst. Civ. Engrs. London. 331-361.

2. Colyer, P.J. (June 1977) Performance of storm drainage simulation models. Part 2. Proc. Instn. Civ. Engrs. London 293-309.

3. Hepworth, R. (14th November 1974) A direct approach to storm sewer design - and a challenge to the TRRL method. New Civil Engineer. London.

4. Newton, S.G. and Painter, R.B. (March 1974) A mathematical examination of urban run-off prediction.

Part 2. Proc. Instn. Civ. Engrs. London. 143-157.

5. Sarginson E.J. and Bourne D.E. (March 1969) The analysis of urban rainfall run-off and discharge. Journal Inst. Mun. Engrs. London. 81-85. Also (July 1969) Correspondence. Journal Inst. Mun. Engrs London. 222-223.

6. Watkins, L.G. (1972) The Design of Urban Sewer Systems. Transport and Road Research Laboratory of the Department of Scientific and Industrial Research. London. Road Research Technical Paper No.55.

7. Watkins, L.H. (Sept 1976) The TRRL hydrograph method of urban sewer design adapted for tropical conditions. Part 2 Proc. Inst. Civ. Engrs. London 539-566. Also (June 1977) Discussion. Part 2. Proc. Inst. Civil Engrs. London 501-508.

8. Working Party on the Hydraulic Design of Storm Sewers, (October 1977) Second progress report - mid 1975 to mid 1977. Journal of the Institution of Municipal Engineers. London Volume 107 No.10.

IMPROVEMENTS IN THE USE OF TRRL HYDROGRAPH PROGRAM

C. Martin
Associate, L.G. Mouchel & Partners.

D. King
Engineer in Charge of Computing, L.G. Mouchel & Partners.

SYNOPSIS

This paper describes improvements to the basic TRRL hydrograph
method made by the authors. The improvements concern methods
of handling input data and means of producing results in a
form accessible to non-specialist engineers. The major
improvement has been to develop a method of dealing with
sewers in a surcharged condition. The analytical method has
then been extended as part of an investigation being undertaken to predict rates of flow arising from actual measured
rainfall.

Symbols

P_N Inflow to a pipe in time increment N

Q_N Outflow from a pipe in time increment N

R_N Volume of water retained in time increment N

T   Length of 1 time increment

V   Volume of pipe

A   Area of manhole

S   Difference of level between 2 ends of pipe

H_N Height of water in manhole in time increment N

A_t Area of catchment with time of flow t to the outfall

r_t Rate of rainfall at time t

1.   Introduction

1.1. Over the past fifteen years Mouchel & Partners have
been engaged on a considerable number of schemes involving
the investigation of existing surface water drainage
systems and the design of new systems. The majority of
this work has been carried out using the Transportation and
Road Research Laboratory hydrograph method. Initially the
work was carried out using a KDF9 computer operated by
International Computer Services Ltd., but of later years it
has been carried out using our own facility.

1.2. During this time we have encountered various problems
which we have endeavoured to solve. They are:-

1.2.1. Handling the input data in a way to limit the
involvement of engineers whilst providing adequate checking.

1.2.2. Dealing with sewers in a surcharged condition, the
inability to do this has been a particular difficulty.

1.2.3. Devising a means of making the output data
accessible to the client in the form of usable sewer records
capable of interpretation by non-specialist engineers.

1.3. The work we have undertaken has been based on the ICL
1900 storm analysis suite which is a well-recognised
implementation of the TRRL hydrograph method. Modern
development of mini-computers and operating systems have
made it possible to enhance the basic suite to improve its
value to the practising engineer. The main requirement
from the engineering viewpoint was to recognise manholes
rather than pipes, to reduce the results to a minimum of
essential data and to compare the one year and five year
storms on the same output. In practice, the enhancement
provided many other benefits.

2.   Overall System

2.1. When we started this work with ICSL, later changed to
BARIC, we used a standard computer orientated input data
form slightly modified for our own purposes. This was
particularly suited to their method of operation. The
preparation of the data for the computer and the checking
of that preparation was taken as the responsibility of the
Bureau, and although this often led to abortive runs it di
not involve us in any expense.

2.2. Eventually the KDF9 became obsolete and BARIC ceased
to offer this computer service. By this time we had
installed our own 'Datapoint' mini computer and we decide
it was appropriate to carry out our own processing.

This permitted an improved form of preparation of data but also meant that we had to assume responsibility for checking.

2.3.   Actual processing is done using the ICL 1900 Storm Sewer Analysis Suite but it is now the core of the larger system shown in Fig. 1. This system operates under George III, the ICL 1900 Operating System which uses a concept of a 'basic' file. This can be created by input devices such as card/paper tape readers or communications links, or by output from a program intended for punching (or paper tape) or the line printer.

3.   Data Input

3.1.   The engineer prepares a line diagram of the network and numbers the pipes in accordance with the TRRL conventions. Data for each pipe length is then inserted on to the pipe length sheet (Fig. 2) the sheets being arranged in correct order for the TRRL program logic.

3.2.   Information from the sheet is input to the mini computer via a visual display unit, the screen format following the data sheet. The VDU's flashing cursor leads the user to the right part of the data sheet for the current requirement. The mini computer prepares two disc files - 'Pipes' which contain data for the Storm Sewage Package and 'Manholes' which contains two records for each pipe length, an upstream to downstream record and a down-stream to upstream record. Gradients are automatically calculated from the given levels. Data such as levels, manhole references etc. not required for the pipe analysis go to the 'Manhole' file.

4.   Data Validation

4.1.   Two programs are available on the mini computer to assist with checking data. The first simply reads a 'Pipes' file and presents the engineer with a pictorial version (Fig. 3) of the pipe logic as it will be interpreted by the main frame program. Logic errors thus become obvious before the main run takes place.

4.2.   The second program tabulates data for each pipe, reading from both 'Pipes' and 'Manholes' and checking for consistency. It is this tabulation (Fig. 4) which the engineer checks and corrects as necessary. The files are then edited and a new check produced - a process which makes the consistency check necessary. A further point is that the main frame program relies on changes of quantities such as roughness and time of entry. The check program tabulates the current value for each pipe. Horizontal

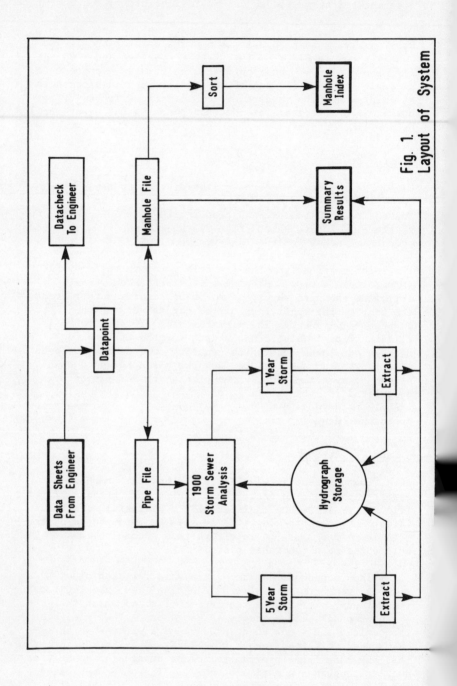

Fig. 1.
Layout of System

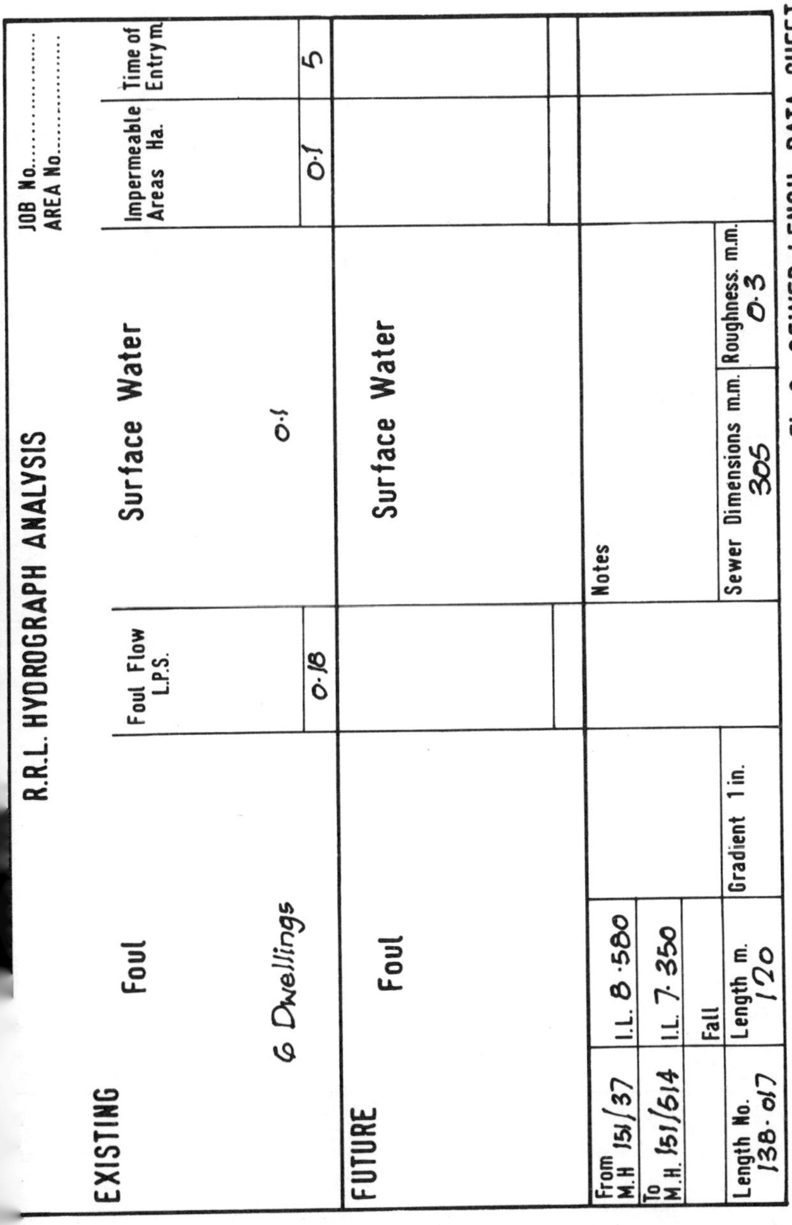

R.R.L. HYDROGRAPH ANALYSIS

JOB No. ....
AREA No. ....

EXISTING

Foul	Surface Water	Foul Flow L.P.S.	Impermeable Areas Ha.	Time of Entry m.
6 Dwellings	0.1	0.18	0.1	5

FUTURE

Foul	Surface Water

Notes

From M.H 151/37	I.L. 8.580		Sewer Dimensions m.m. 305	Roughness. m.m. 0.3
To M.H. 151/614	I.L. 7.350			
Length No. 138.017	Fall Length m. 120	Gradient 1 in.		

Fig.2. SEWER LENGH DATA SHEET

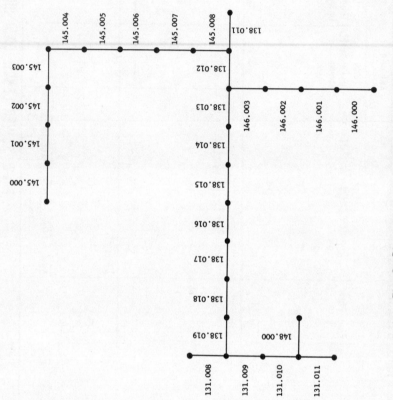

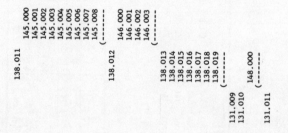

Fig. 3 Pipe logic and Diagrammatic layout

STORM SEWER DESIGN AND ANALYSIS

Pipe Number	From Manhole	From Invert	To Manhole	To Invert	Length (m)	Gradient (1 in n)	D.W. Flow (cumecs)	Existing Sewer (mm)	Roughness Coefficient	Impermeable Area (ha)	Time of Entry
145.002	146/16	29.665	146/15	28.575	75	69	.00009	229	.0003	.000	.00
145.003	146/15	28.575	146/14	26.990	90	57	.00015	229	.0003	.150	5.00
145.004	146/14	26.990	146/13	24.140	50	18	.00009	229	.0003	.080	5.00
145.005	146/13	24.140	146/12	23.254	15	17	.00000	229	.0003	.000	.00
145.006	146/12	23.254	146/10	21.815	20	14	.00030	229	.0003	.100	10.00
145.007	146/10	21.815	146/9	11.610	160	16	.00012	229	.0003	.150	5.00
145.008	146/9	11.610	146/5	11.455	15	97	.00087	305	.0003	.050	15.00
** Maximum Flow in Pipe 145.008 is .120 cumecs (Hydrograph Output HG1450081131) **											
138.012	146/5	11.455	146/4	11.335	8	67	.00000	229	.0003	.000	.00
** Maximum Flow in Pipe 138.012 is .140 cumecs (Hydrograph Output HG1380121131) **											
146.000	151/33	18.690	145/47	17.355	70	52	.00006	229	.0003	.100	5.00
146.001	145/47	17.355	146/1	17.010	60	174	.00027	229	.0003	.000	.00
146.002	146/1	17.010	146/2	16.915	10	105	.00000	229	.0003	.000	.00
146.003	146/2	16.915	146/4	11.335	65	12	.00006	152	.0003	.150	2.00
138.013	146/4	11.335	146/3	10.870	50	108	.00000	229	.0003	.000	.00
** Maximum Flow in Pipe 138.013 is .035 cumecs (Hydrograph Output HG1380131131) **											
138.014	146/3	10.870	145/48	10.080	60	76	.00000	229	.0003	.000	.00
138.015	145/48	10.080	151/36	9.455	105	168	.00000	229	.0003	.000	.00
138.016	151/36	9.455	151/37	8.580	100	114	.00000	305	.0003	.000	.00
138.017	151/37	8.580	151/514	7.350	120	98	.00018	305	.0003	.100	5.00
138.018	151/514	7.350	151/29	7.075	55	200	.00027	305	.0003	.200	5.00
138.019	151/29	7.075	151/30	6.795	20	71	.00096	305	.0003	.050	5.00
131.009	151/30	6.795	151/32	6.360	12	28	.00000	675	.0003	.000	.00
131.010	151/32	6.360	151/35	2.360	20	5	.00000	229	.0003	.000	.00

FIG. 4   INPUT DATA

lines are inserted above the first pipe of a new branch in order to clarify the logic.

5.   Hydrographs

5.1.   There is a considerable need in a practical analysis to be able to manipulate hydrographs, storing them sometimes and retrieving them later.

5.2.   The main frame system can accept a hydrograph coming from a sub-area. Such hydrographs are formed in one of two ways:-

5.2.1.   The final overflow from the end pipe of an area may be used as an input to another area. This is of value in breaking up a large network into manageable sections.

5.2.2.   Where a storm overflow is provided, the main frame program prints the hydrograph presented to the pipe and also the modified hydrograph resulting from the presence of the overflow. In fact, the difference of these two - the overflow hydrograph - is important since it represents physical input to some other part of the system.

5.3.   The results of the storm sewer analysis are scanned for hydrographs which are output as basic files, each with an unique reference consisting of the area number, the pipe number, and the storm frequency. The overflow hydrographs are computed by subtracting the two printed hydrographs and various constants for the sub-area are added. A sub-area input requires a time of concentration, an area contributing, an outfall pipe diameter and a cumulative pipe volume.

5.4.   For overflow hydrographs, the area contributing and the cumulative pipe volume are taken as zero - the other parameters are taken from the upstream pipe, as are all parameters for outflow hydrographs. These conventions preserve the hydraulic properties of the catchment regardless of the precise route of the outfall considered. These hydrographs are stored on magnetic tape and the engineer specifies on his input sheets data to determine the hydrograph to be retrieved and input at any particular point.

6.   Processing

6.1.   When a fully valid pair of files is available they are transmitted to the main frame where the pipes file is duplicated to produce one year and five year storm data, or other selected storm frequency. The hydrographs required are retrieved and edited into the relevant data files - a line of text unacceptable to the storm sewer analysis but containing instructions to the operator simplifies this.

6.2.   George III permits the results of the main run to be (a) printed on a line printer to provide a hard copy, or (b) used as input to the post processer system. Likewise the original data point disc files become pseudo card reader input, and paper tape copies are produced for long-term storage which can be edited in the event of future alterations.

7.   Surcharged Sewers

7.1.   Within the original TRRL hydrograph method two options are given for dealing with existing systems.

7.2.   In the first method, if an existing pipe is found to be overloaded then it is redesigned so that an imaginary pipe is inserted which has sufficient capacity to carry the flow. Subsequent pipes are then subjected to the full bore capacity arising from this imaginary pipe. This process is repeated down the system for each pipe which is found to be overloaded by the now enhanced flow. In analysing an existing system this also has the disadvantage of increased computer time in order to get the wrong answer!

7.3.   The second method allows for the suppression of the redesign facility. When a pipe is encountered that has insufficient capacity to carry the load, a surcharge ratio is calculated which is based on the flow reaching the pipe divided by the full bore capacity of that pipe. Since this applies to overloaded pipes, this ratio is always greater than one. It is then assumed that any flow in excess of the full bore capacity of the pipe goes to waste and is not taken into any further consideration.

7.4.   The effect of the first of these methods is to over-state the degree of overloading experienced by pipes in the lower reaches of the system since greater flows are theoretically passed to them through imaginary pipes than would be the case in practice. The effect of the second case is to understate the degree of surcharge in the lower reaches of the system since a certain proportion of the flow is run to waste and no account is taken of the increased capacity of the upstream lengths due to possible surcharge.

7.5.   The true state of the system lies somewhere between these two extremes but it is difficult to decide where, and it is unsatisfactory to explain to a client, particularly to a Committee of lay members, in the vague terms necessary due to the imprecision of the method of analysis.

7.6.   We have always found it possible to deal with the situation manually by taking account of the degree of surcharge caused by the overloading, but clearly this is impracticable for a system containing several hundred lengths of sewer.

7.7.   The method we are now developing will enable us to calculate and take account of true degrees of surcharge and to indicate the point at which the system will flood, i.e. where storm sewage will appear above ground level.

7.8.   The TRRL retention expression becomes

$$\frac{2}{T} R_N + Q_N = P_N + P_{N-1} - Q_{N-1} + 2 \frac{R_{N-1}}{T} ---(i)$$

with some rearrangement. Normally the right hand side consists of known quantities, whilst R_N and Q_N are functions of the proportional depth, and hence they can be found by numerical methods. If the upstream manhole is considered as a part of the pipe,

$$R_N = V + A. H_N -----------(ii)$$
$$Q_N \backsimeq \sqrt{\frac{S + H_N}{\gamma}} -----------(iii)$$

which γ can be shown to be a function of the pipe. Substitution into (i) with some manipulation, yields a quadratic which can be solved for H_N. A and V are rather complex numerically since they must reflect available space in pipes upstream of the manhole and any shape variation in the manhole, but this is only detail coding and the 'Manholes' file already associates a manhole description with a particular pipe. Similarly γ should reflect the Colebrook-White resistance.

7.9.   The above approach is valid provided that H_N is not sufficiently large to either overflow out of the system, or to cause interference with flow in the immediate upstream pipes. Clearly the first case causes either retention in the form of a flood, the volume and time characteristics of which can now be calculated, or an overflow hydrograph which can be reinput e.g. after running down a road.

7.10.   The second case of the previous paragraph is, at first sight, more complex. However, if it is considered as a discontinuity in the solution - i.e. the manhole concerned has a head H_N for the downstream pipe but 0 for upstream pipes, a correction q_N to Q_N to eliminate the discontinuity can be found. This correction reduces both Q_N and H_N and is a simple linear function of known variables with the exception of a term $(2 - \dfrac{q_N}{Q_N})$. This term does not pose a serious problem since clearly q_N/Q_N is much less than 1, and $\dfrac{q_{N-1}}{Q_{N-1}}$ provides a good initial approximation. It does, however, involve iteration back up the system.

7.11.   The TRRL numbering system and the data storage method on the ICL 1900 are not suitable for this back iteration. A different approach to numbering derivable from the TRRL scheme has been developed and the data storage reorganised. This has had the advantage of lifting restrictions on numbering and network size and increasing efficiency for a reduction in overall core usage.

7.12.   At the time of writing, the surcharge routine is under development and it is expected to have results when this paper is presented. The revision will alter the approach to overflow weirs although it seems unlikely that there will be a radical change numerically from the existing method of specifying a limiting flow upon which all flow is recorded as passing over the weir. There will probably be a change in timing however, and the change may be more marked if there is a surcharged pipe slightly downstream.

8.   Presentation of Results

8.1.   In order to carry out a hydrograph analysis, it is necessary to assemble considerable quantities of information concerning the sewerage system. On completion of the analysis the results arrive in computer printout form and include the presentation of a certain amount of the original data.

8.2.   The method of analysis requires the information to be presented in a particular pattern related to the shape of the sewerage system, and all of these things combined mean that the extraction of information requires a fair degree of knowledge of the analytical method, and it can sometimes be an arduous process to follow the flow pattern in order to find data concerning a particular length.

8.3.   The most convenient method of maintaining sewer records we have found is in the form of Ordnance Sheets showing the position of manholes and indicating sewer runs. We have developed an indexing system which makes reference to manholes by number and Ordnance Sheet, and prints out all the relevant information extracted from the hydrograph analysis. For each manhole it gives details of manhole number, invert level, length number, pipe size, the upstream manhole, its invert level and the drainage area number. It then gives similar information for the down-stream connections. Thus the physical properties of each manhole and its upstream and downstream connections are clearly and simply stated.

8.4.   From the area number and length number it is then possible to obtain additional information concerning the hydraulic status of the sewer.

9.   Post-Processor

9.1.   The first step in producing the results file is to extract the essential data from the main results file. This consists of peak flow, dry weather flow, pipe capacity, diameter and gradient. Hydrographs are also extracted at this stage and transferred to basic files with unique names as already described.

9.2.   The next step also depends on George III facilities. At this stage there exist a 'Manholes' file, a 'One Year Results' file, a 'Five Year Results' file, and a 'Titles' file. These are all treated as card reader files by the print program, reading 1 or 2 records from each in turn apparently from 4 card readers! Various consistency checks are performed to avoid data getting out of phase, and the results printed in similar format to the original data check. (Fig. 5). Various transformations are made, printing peak flows as a ratio of capacity, highlighting surcharging, etc. Thus the engineer ends up with two computer outputs of consistent format and paging:

9.2.1.   The input data to the system.

9.2.2.   The essential features of the results.

9.3.   Of course, the full results file is available for consultation on occasion, but there is normally no necessity to go past the condensed version.

10.   Computation Index

10.1.   The 'Manholes' file serves a second purpose besides conveying data around the main analysis. The complete set of 'Manholes' files is sorted with a standard commercial sort to produce data for an index to the complete system. Manholes are given a two part reference defining the drawing and the manhole number on

Pipe Number	From Manhole	To Manhole	Gradient 1 in	Pipe Size (mm)	Capacity litres/sec	3 DWF litres/sec	One Year Storm Flow litres/sec	One Year Flow/ Capacity	Five Year Storm Flow litres/sec	Five Year Flow/ Capacity	Surcharged
145.002	146/16	146/15	69	229	72	1	11	0.15	19	0.26	
145.003	146/15	146/14	57	229	79	1	25	0.32	45	0.57	
145.004	146/14	146/13	18	229	141	1	33	0.23	60	0.42	
145.005	146/13	146/12	17	229	145	1	33	0.23	59	0.41	
145.006	146/12	146/10	14	229	160	1	42	0.26	75	0.47	
145.007	146/10	146/9	16	229	150	1	56	0.37	100	0.66	
						Overflow Hydrograph from the following length stored					
145.008	146/9 *	146/5	97	305	128	2	61	0.47	107	0.84	s
						Overflow Hydrograph from the following length stored					
138.012	146/5 *	146/4	67	229	73	15	140	1.93	140	1.93	
146.000	151/33	145/47	52	229	83	1	12	0.13	20	0.24	
146.001	145/47	146/1	174	229	45	1	11	0.24	19	0.43	
146.002	146/1	146/2	105	229	58	1	11	0.19	19	0.33	
146.003	146/2	146/4	12	152	59	1	26	0.44	47	0.79	
						Overflow Hydrograph from the following length stored					
138.013	146/4 *	146/3	108	229	57	16	35	0.62	35	0.62	
138.014	146/3	145/48	76	229	68	16	35	0.51	35	0.51	
138.015	145/48	151/36	168	229	46	16	35	0.77	35	0.77	
138.016	151/36	151/37	114	305	118	16	35	0.30	35	0.30	
138.017	151/37	151/514	98	305	128	16	46	0.36	55	0.43	
138.018	151/514	151/29	200	305	89	16	68	0.76	93	1.04	s
138.019	151/29	151/30	71	305	150	17	73	0.49	103	0.68	
131.009	151/30	151/32	28	675	1921	26	167	0.09	265	0.14	
						Overflow Hydrograph from the following length stored					
131.010	151/32 *	151/35	5	229	269	26	34	0.13	34	0.13	

FIG. 5   ESSENTIAL FEATURES OF RESULTS

that drawing. Clearly each pipe connects two manholes and this is the reason for double records on the manholes file. The sort key is Manhole Number within Drawing Number, thereby achieving an ordered arrangement of the records. Consideration of the TRRL numbering system shows that, at any manhole:

10.1.1.   The main line will have a lower integer part than any branch lines.

10.1.2.   The downstream section of the main line has a higher decimal part than the upstream.

10.2.   The sort key can be enhanced by these facts to place each pipe within a manhole into a logical sequence. This sorted file is then printed with a simple program so that the engineer looking at a particular manhole can turn straight to the appropriate page by reference to drawing number and manhole number. He would there find an entry giving details of all pipes connected to that manhole, together with cross-references to the appropriate flow calculation for each pipe.

10.3.   The Appendix attached to this paper comprises the instructions issued with the sewer records file and gives examples of its use.

11.   Actual Flow Calculations

11.1.   The examples quoted in this paper are based on a large analysis undertaken for the Borough of Weymouth & Portland.   Since the main calculations have been completed, a need has arisen to consider actual flows from the outfall in a typical year. This is part of an investigation we are carrying out into the possibility of predicting rates of flow into a large pumping station by the use of information received from a series of automatic rainfall recorders. Past recordings of rainfall are available, but unfortunately they do not consist of, or even approximate to, a series of British Standard Storms! The solution was relatively simple - Q_N is given by

$$Q_N = \sum_{t=o}^{NT} A_t \cdot r_t \quad \ldots\ldots\ldots \text{(iv)}$$

Clearly the TRRL method provides values of Q_N derived from a particular set of r_t. Items such as storm overflows would affect the result and a conventional area/time diagram would not contain such effects, and in any case it was not available. Consideration of equation (iv) shows that it has a matrix form:

$$Q = R. A \dots\dots (v)$$

where R is a lower triangular matrix with r_1 in all
leading diagonal positions, r_3 in the next diagonal, etc.,
since the hydrograph is in the form of average flows for
2 minute intervals.

11.2.   Solution of this system appeared simple, but in
practice proved not so as:

11.2.1.   R is very poorly conditioned.

11.2.2.   Q values are approximate, having suffered many
interpolations.

11.2.3.   Beyond a certain time A's are zero. This time
is not, in fact, the time of concentration as this is
based on full bore velocity. Q's are based on actual
part-full velocities.

11.3.   A solution was obtained using the Householder QU
factorisation and solving as an overconstrained system
with A 0. The time of concentration obtained by this
method was about 140 minutes as opposed to 121 minutes
based on the full bore velocity. The 1 and 5 year storms
gave very similar values of A.

12.   Summary

12.1.   The development work for handling input and output
data described in this paper has been used in practice
and examples given are extracted from the work carried out
for the Borough of Weymouth & Portland.   We hope to
make available a Datapoint mini computer during the
Conference in order to demonstrate the method of input data
handling.

12.2.   The method of dealing with sewers in a surcharged
condition has not, at the time of writing, been used in
practice but it is anticipated that when the paper is
presented results will be available from the use of the
surcharge routine.

12.3.   The authors wish to thank the Partners of L.G.
Mouchel & Partners for permission to present this paper.
We should like to express our thanks also to Mr. J.R.
Kemble, B.Sc.(Eng), M.I.C.E., F.I.Mun.E., F.Inst.H.E.,
Borough Engineer, Weymouth & Portland Borough Council, for
permission to reproduce documents arising from the
Weymouth study.

APPENDIX

Use of Manhole Schedule

Example 1   What information is available relating to the manhole outside 47 Old Castle Road?

The manhole is shown as No. 37 on square number 151. Therefore it has the number 151/37 (see Fig. 6). Reference to the manhole schedules gives the following information for manhole 151/37 (see Fig. 7)

Upstream Connection

Invert level in 151/37	8.580 m. AOD
Upstream length number	138.016
Pipe diameter	305 mm.
Upstream manhole	151/36
Invert level	9.455 m. AOD
Area number	131

Downstream Connection

Invert level in 151/37	8.580 m. AOD
Downstream length number	138.017
Pipe diameter	305 mm.
Downstream manhole	151/514
Invert level	7.350 AOD
Area number	131

Use of Flow and Capacity Calculations

Example 2   What are the conditions in the combined sewer at the bottom of Old Castle Road?

The sewer in question runs downstream from manhole 151/37 (see Fig. 6). From example 1 it has the length number 138.017 and is in drainage area number 131. Reference to the calculations gives the following information for length 138.017 (see Fig. 5).

Manhole at top of pipe	151/37
Manhole at bottom of pipe	151/514
Pipe gradient	1 in 98
Pipe diameter	305 mm.
Full bore capacity	128 l.p.s.
Peak foul flow 3 DWF	16 l.p.s.
Peak flow in 1 year storm (including peak foul flow)	46 l.p.s.
Ratio to capacity of pipe	0.36, = 36%
Peak flow in 5 year storm (including peak foul flow)	55 l.p.s.
Ratio to capacity of pipe	0.43, = 43%
Surcharge	None

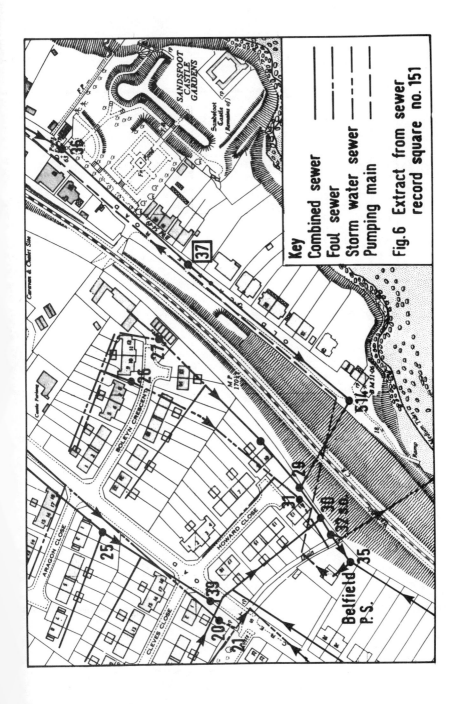

Fig.6 Extract from sewer record square no. 151

Key
Combined sewer
Foul sewer
Storm water sewer
Pumping main

Manhole Number	UPSTREAM CONNECTIONS						DOWNSTREAM CONNECTIONS					
	Invert Level	Length Number	Pipe Size	Connected Manhole	Invert Level	Area Number	Invert Level	Length Number	Pipe Size	Connected Manhole	Invert Level	Area Number
151/1	7.520	122.004	152	151/2	10.080	101	7.520	122.005	152	151/481	4.570	101
151/2	10.080	122.003	152	151/8	14.795	101	10.080	122.004	152	151/1	7.520	101
151/3	9.520	815.000	152	150/12	12.500	801	9.520	815.001	152	151/5	8.475	801
151/5	8.475	815.001	152	151/3	9.520	801	8.475	815.002	229	150/10	6.145	801
151/8	14.795	122.002	152	151/9	16.550	101	14.795	122.003	152	151/2	10.080	101
151/9	16.550	122.001	152	150/9	17.850	101	16.550	122.002	152	151/8	14.795	101
151/10	2.495	801.022	915	150/33	2.905	801	2.495	801.023	610	151/103	1.920	801
	2.785	816.003	305	154/320	4.490	801						
151/17	13.565	136.001	152	145/7	20.590	131	13.565	136.002	152	151/18	10.030	131
151/18	10.030	131.006	229	151/22 *	11.500	131	10.030	131.007	229	151/39	8.500	131
	10.030	136.002	152	151/17	13.565	131						
151/22 *	11.500	131.005	229	144/562	12.365	131	11.500	131.006	229	151/18	10.030	131
	11.500	134.000	152	151/23	13.145	131						
	11.500	135.000	152	145/1	16.150	131						
151/23							13.145	134.000	152	151/22	11.500	131
151/25	16.295	137.000	152	145/9	24.145	131	16.295	137.001	152	151/39	8.500	131
151/29	7.075	138.018	305	151/514	7.350	131	7.075	138.019	305	151/30	6.795	131
151/30	6.795	131.008	305	151/39	8.500	131	6.795	131.009	675	151/32	6.360	131
	6.795	138.019	305	151/29	7.075	131						
151/32 *	6.360	131.009	675	151/30	6.795	131	6.360	131.010	229	151/35	2.360	131
151/33							18.690	146.000	229	145/47	17.355	131
151/35	2.360	101.035	381	151/479	3.140	101	2.360	131.011	381	151/999	1.500	101
	2.360	131.010	229	151/32 *	6.360	131						
151/36	9.455	138.015	229	145/48	10.080	131	9.455	138.016	305	151/37	8.580	131
151/37	8.580	138.016	305	151/36	9.455	131	8.580	138.017	305	151/514	7.350	131
151/39	5.130	123.003	152	150/27	5.475	101	5.130	123.004	152	151/481	4.570	101
	8.500	131.007	229	151/18	10.030	131	8.500	131.008	305	151/30	6.795	131
	8.500	137.001	152	151/25	16.295	131						
151/103	1.920	801.023	610	151/10	2.495	801						

FIG. 7   MANHOLE SCHEDULE

Checking Available Capacity of a Sewer

Example 3   What possibility is there of increasing the
                load on the existing sewer in Old Castle Road?

The manhole numbers are found on the record drawings (see
Fig. 6). The sewer runs from manhole 151/37 to Belfield
pumping station. The length numbers and Area numbers are
found from the manhole schedule (see Fig. 7). The data is
given in the Flow and Capacity Calculations (see Fig. 5)

| From Record Drawing | From Manhole Schedule | | From Flow & Capacity Calculations |
Manhole No.	Downstream Length No.	Area No.	Downstream Manhole No.	
151/37	138.017	131	151/514	43% of Capacity on 5 year storm.
151/514	138.018	131	151/29	4% Surcharge on 5 year storm.
151/29	138.019	131	151/30	68% of capacity on 5 year storm.
151/30	131.009 additional branch enters number 131.008	131	151/32	14% of capacity on 5 year storm.
151/32	131.010	131	151/35	Overflow manhole 3 DWF = 26 lps overflows at 34 lps.
151/35	131.011	131	151/999	151/999 is pumping station sump.

Length 138.018 is at present overloaded on a five year
storm. This should be investigated further before any
additional loads are imposed on the sewer. The overflow
at manhole 151/32 operates at a very low level of dilution,
34 l.p.s., which is much less than 6 x DWF or 'Formula A'.
This also should be investigated. Finally, the capacity of
the pumping station should be examined to see if it is
able to handle any additional flows.

A METHOD OF ASSESSMENT OF PIPED DRAINAGE SYSTEMS TAKING ACCOUNT OF SURCHARGE AND OVERGROUND FLOODING

J.L. Thompson and A.R.R. Lupton

Chief Engineers, Binnie & Partners, London

SYNOPSIS

One of the major drawbacks in the design and analysis of drainage systems is the lack of a convenient method for taking account of surcharge and overground flooding as it occurs without modifying the carrying capacity of the pipes involved. This paper describes a method that has been developed to cover these situations as well as the normal free surface flow conditions. The method has been written into a program suitable for use in a digital computer and in a form easily understood by those with a basic knowledge of the design of drainage systems. The program simulates the hydraulic conditions which exist in a system of sewers as a result of a rainstorm or any other type of inflow. It predicts those conditions of particular interest throughout the period under consideration viz. the flow in the pipes, the water levels in manholes, the degree and extent of surcharge and flooding measured as the volume of spill-out through manhole covers and the time when peak hydraulic conditions would occur. In addition, the program will produce outfall hydrographs from any pipe or all pipes or parts of a system and it will cater for overflows from one part of a system to another part, from one system to another or by complete removal from a system either over or underground. Although the program is arranged to accept inflow hydrographs selected by the user a modified hydrograph taking account of the overland routing of rainfall is suggested. The program was developed, initially, for the analysis of a stormwater drainage system but its subsequent use has shown that it may be adopted for any gravity drainage system and may be used in an equally powerful manner as a design aid.

INTRODUCTION

The method was developed during a study which included the assessment of the capacity of the surface water drainage

ystem of the London Borough of Harrow.   The assessment was
eing carried out in order to pinpoint existing and potential
rouble spots resulting from existing and future developments
nd so permit an orderly programme of sewer augmentation and
emedial works to be arranged.   A companion paper (6.4) to be
resented to the Conference sets out further the need for the
ssessment and the benefits that have been realised to date
y this comprehensive approach.

ssentially, the area served by the Borough's surface water
rainage system covers roughly 51 km^2 and contains about 500km
f pipes which drain into five main open watercourses radiating
utward from Harrow Weald in the north of the Borough (Fig. 1).
lthough extensive, about 5000 lengths of piped and culverted
ections, it was soon apparent that the drainage system was
enerally undersized giving rise to much surcharge and over-
round flooding that could not readily be computed by the
echniques available.

new method for carrying out the complex calculations to
eal with these problems was therefore required but before
mbarking on such a task it was clearly necessary to set out
he desirable objectives.   These were formulated as follows :

i) The method of computation shall take account of surcharge
nd overground flooding as it occurs without modifying the
arrying capacity of the pipes involved,

ii) the method shall be capable of being written into a
rogram suitable for use in a digital computer,

iii) the method, so far as is practicable, shall be universal
n application,

iv) the method shall be in a form suitable for use in a
esign office by engineers or others with a basic knowledge
f the design of drainage systems,

v) the program output shall facilitate the rigorous data
hecking that is required with any computer program and shall
raw attention to areas where confirmation or correction is
esirable,

vi) it would be helpful if the results produced could be
eadily compared where applicable with other accepted methods,
n the case of Harrow, the TRRL method was chosen.

lthough the objectives are obvious, their re-statement was
ssential from time to time during the development period to
nsure that the resulting computer program, which was neces-
arily complex, could be easily understood and readily used
y those with little or no computer knowledge.

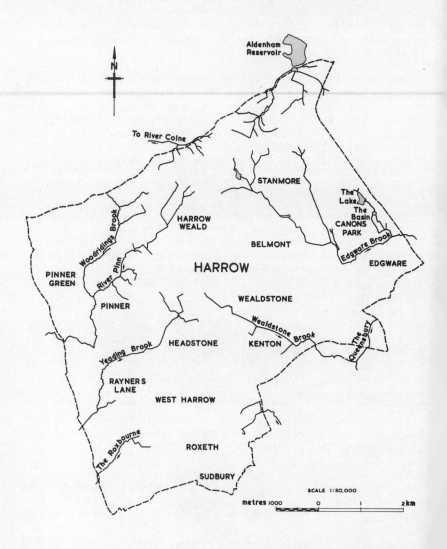

DRAINAGE PLAN

FIGURE 1

In view of the many variables and unknowns associated with catchment modelling, it was decided at an early stage that account of the overland routing of flows to the drainage system would be made by manual adjustment of the rainfall hydrograph.

The computerised part of the new method, code named SPARROW, deals solely with the hydraulic aspects of the drainage system from the time that flows enter the system until the time that they are discharged.

The data required for computation purposes fall into two distinct parts: that which describes the physical and hydraulic characteristics of each section of pipe and its catchment area and the hydrological data such as rainfall, runoff and impermeability.

PIPE AND CATCHMENT DATA

The detail and form of the pipe and manhole data are laid down by the requirements of the program and include cover and invert levels of the sewer system, pipe diameters, pipe lengths between manholes, pipe type and conditions for estimating roughness, catchment areas for each pipe section, estimates of dry weather flow and plan area of manholes, including allowance for gullies and house connections. When this information has been collected the system is separated into convenient lengths by nodes which must be at manholes but not necessarily at each manhole. The unit of calculation in the computer program is a pipe length which consists of a pipe and its upstream node, terminating just upstream of the downstream node. Each pipe length must be numbered, preferably by the decimal classification recommended by the Road Research Laboratory (1963)[1].

It is recommended that the large amount of manhole and pipe data which generally arises is handled by a framework based on natural drainage catchments. Within the framework it is found helpful if principal catchments, groups of sewer systems and, where appropriate, individual sewer systems are distinguished by a numbering or lettering system.

Data Preparation
The lengths of pipes which are coded (i.e. the length between nodes) should be kept reasonably consistent and nodes should be placed at changes in diameter and marked changes in gradient. Furthermore, the time of flow through a coded length at full flow velocity should be roughly equal to, or slightly larger than, the time interval between successive calculations in the computer program (usually 1 minute in urban drainage systems). Within an urban system the consistent coded length is generally found to be about 150 metres but in those cases where the time of flow through a pipe length is markedly different

from the time interval between successive calculations in the computer program the coded length may be allowed to vary from 15 metres to 250 metres.

Pipe roughness should be determined for each pipe length but examination of the sewer system may indicate that a single value for a whole system may suffice. In the London Borough of Harrow, the condition of the system was similar to that found in slimed foul sewers. For those conditions Ackers (1969)[2] recommended a roughness coefficient of 1.5mm and this value was adopted throughout the study. Where it is necessary to simulate ditches or other means of conveying the surface water, roughness coefficients of up to 40mm or more may be required.

The invert level of the manhole at the upstream and downstream nodes and the manhole cover level at the upstream node, complete the pipe data sufficiently to run the computer program. Where small differences exist in the incoming and outgoing invert levels in a manhole, the average invert level may be adopted. Special provisions are made for those locations where a drop or a step-up occurs.

Although the manhole area and the number of manholes in a pipe length may vary, it has been found that the manhole areas in surcharged or spilling conditions appear to have only minor local effects on the results.

The impermeability factor represents that proportion of the catchment area which drains to the sewers. Within urban areas it has been found sufficiently accurate to assume that all paved and roofed areas drain to the sewers and that non-paved or undeveloped areas make no contribution. Exceptions to this rule apply in less dense residential areas.

The choice of impermeability factor and the extent of the area contributing to a pipe length are probably the most subjective part of any assessment but a reasonable degree of consistency and accuracy results if determination of the catchment area for each pipe length follows the method recommended by the Road Research Laboratory[1] whereby property fence lines are taken as the catchment boundary, the boundaries plotted on large scale maps and the resulting area measured. Also, by typifying different densities of development within an area a range of characteristic impermeability factors is produced.

Non-Standard Data

The straightforward method of data preparation set out above is adequate for simulation of most sewer systems. However, there are instances where facilities are required to simulate non-standard features which cannot be dealt with satisfactori in the manner so described. Non-standard features which are allowed for are given below.

(a) <u>Short pipes</u> A short pipe is defined as the length of pipe between manholes where the time of flow when the pipe is running full is significantly less than the time interval (usually 1 minute) between successive calculations in the computer program. The inclusion of a short pipe results in mathematical instability in the computer program. For this reason, short pipe lengths (generally less than 15 metres) are not coded but are incorporated as a pipe of equivalent capacity in an adjacent pipe length. For completeness the short pipe data is included in the computer print-out. Short pipes are generally found at road junctions where confined conditions near other services may have required a sudden change in pipe gradient or direction for a short distance; or where additional sewers for a new road or housing development are linked to an existing sewer system.

(b) <u>Diverging pipes</u> There are two different situations which arise as a result of diverging pipes :

(i) a sewer bifurcates and continues as two separate systems,

(ii) a sewer bifurcates and the two limbs come together again as a single pipe further down the system.

When a sewer bifurcates and continues as two separate systems one of the pipes is selected as the main branch and continues as part of the sewer system under consideration. The other branch is treated as an overflow and dealt with separately; the volume of water passing down the overflow branch being determined by the overflow (water transfer) facility described below. When a sewer bifurcates and the two limbs or branches come together again as a single pipe further down the system, a pipe of equivalent carrying capacity is entered to simulate the twin pipe section.

(c) <u>Non-circular pipes</u> Non-circular pipes are dealt with by substituting circular pipes of equivalent capacity and roughness.

(d) <u>Overflows</u> When a sewer system is overloaded, relief is provided by

(i) water spilling out of a manhole and subsequently returning to the system through the same manhole,

(ii) water spilling out of a manhole and travelling overland to another entry point in the same sewer system or to another sewer system and

(iii) water travelling through an underground connection

(pipe) to another sewer system or to a stream.

Case (i) is dealt with separately. Cases (ii) and (iii) are treated as overflows and are dealt with by the water transfer facility.

Under the water transfer facility, the overflow is treated as a separate system arranged to operate when the water level in the loaded system reaches a predetermined level. The position of the outfall of the overflow system (or the position of the return manhole to a sewer system) is specified together with the transit time of flow. To ensure a satisfactory driving head in an overflow system the simulation requires that the overflow level is set artificially high. The choice of over-flow level is therefore subjective but in practice it has been found to be largely dependent on the relative carrying capacities of the main and overflow pipes. When the over-flow pipe is short, water may be transferred from one manhole to another by using a fixed time delay only.

(e) <u>Ditches</u> Consideration of ditch flow is necessary where ditches form

(i) the heads of pipe systems or

(ii) part of the sewer system.

When a ditch forms the head of a sewer system it has to be ascertained whether it carries a flow at all times and/or whether it is likely to affect the time of entry to the sewer system. Where it is found that there is always a flow this is estimated during a dry period and entered in the calculations as a dry weather flow into the sewer system. In those instances where the flow regime in a ditch is likely to modify the time of entry into the sewer system the appro-priate extra time is entered in the calculations.

In those areas where a ditch forms part of a sewer system, usually interlinking one part of the system with another, this may be accounted for by treating the ditch as a pipe of a diameter and roughness which would have similar flow chara-cteristics. To allow for water which can back-up in a ditch before flooding occurs, the node downstream of the simulated ditch is allocated an increased manhole area.

HYDROLOGICAL FACTORS

As explained in the introduction the computer program now proposed for the assessment of drainage systems starts with sewer inflow represented in the form of a modified rainfall hydrograph. Conversion of rainfall to sewer inflow is carrie out mainly outside the computer by manual methods. The

hydrological data or information required to convert the rainfall patterns of the area under consideration into a sewer
inflow which makes a realistic allowance for flow routing and
other conditions on the catchment is carried out by the procedures outlined below.

Storm severity

Storm severity is defined by the frequency or return period
with which a storm of a given rainfall depth and duration is
likely to occur. As three rainfall events are usually
sufficient to analyse and test the response of a surface
water drainage system one of the main concerns is, therefore,
the choice of severity of the rainfall events that should be
adopted. When a sewer system is full, additional rain produced by a storm of increasing severity by-passes the system,
either overland or by other means, from the time that the
system is full. Thus, the additional rain is of no assistance
in an assessment. The criterion for the choice of severity
of the rainfall events is therefore the maximum flow capacity
of the individual sewer system. Accordingly, one of the selected storms must be sufficiently severe to show the maximum
capacity of the drainage systems and the other two storms
less severe. From a questionnaire carried out by Colyer[3] in
1975 it would be reasonable to assume that in the U.K., the
1 year, 2 year and 5 year return period storms will be sufficient to enable comprehensive conclusions to be drawn.
Elsewhere, guidance should be obtained from a study of local
practice and events. Additional information arising from
the survey by Colyer indicated that flood events of 25 to 30
year return period are often associated with storms having
return periods as short as 5 years, but only when the catchment is fully saturated.

Storm rainfall

In the U.K., storm rainfalls will generally be estimated from
Bilham rainfall tables or the modified version of the tables
published in CP 2005 (1968)[4] or the methods outlined in
Vol. II (Meteorological Studies) of the NERC Flood Studies
Report[5] which is the most recent authoritative work on this
subject. The methods outlined in the Flood Studies Report
take account of regional variations in storm rainfall and
permit statistical estimates of storm severity to be made.
Comparison can be made with the most severe rainfall and
storm patterns measured at rainfall stations within the area
under consideration. The Meteorological Office is currently
engaged on a project to microfilm and digitise the autographic
records within the U.K. and this will assist in refinement
of chosen rainfall patterns for future work, in this country.
Elsewhere, similar but reliable information should be adopted.
In ascertaining storm rainfall due regard should be paid to
those areas large enough for there to be a significant point-
to-area reduction in rainfall intensity.

For the chosen return periods and adopted storm durations, total storm rainfalls are estimated and translated into appropriate storm profiles.

Storm duration

The choice of storm duration is governed by estimates of the time of concentration in the sewer networks. Calculations by the methods now proposed have shown that, for storm durations greater than twice the time of concentration, there is little variation in peak flow rates as storm duration increases. Storm durations not exceeding twice the time of concentration are therefore generally suitable for assessment purposes.

Storm profile

Storm profile is closely linked with storm duration. Storm profiles should be prepared which incorporate rainfall intensities during the chosen storm durations that are likely to produce the highest flow rates in the sewer system for the chosen return periods. These should be related to the most intense storm types expected in the area under consideration. As an example, Keers et al (1975)[5] stated that typically in the south-east of England 40% of all summer and autumn rainfall, which results in the highest flow rates in the sewer systems, is produced by convective storms with the associated thundery rain. For these reasons, a summer storm profile for those areas is required. Maximum rainfall intensities during such storms vary according to the profile which is likely to be exceeded by 50% of the storms and events have shown that there is no significant increase in peak sewer flows by using higher percentiles.

Design storms

After the design storms have been prepared for the chosen return period in the form of hyetographs, these are then modified so that they can suitably represent inflow to the sewers. These modifications take account of surface retention storage (which allows for the accumulation of storage as rainfall intensity increases and its subsequent release) by transferring rainfall from the rising limb of the hyetograph and adding it to the latter part of the storm. At this stage the hyetograph models the modified rainfall as entered in the computer program. The final modification of the hyetograph necessary to convert rainfall into sewer inflow is carried out within the computer program by making allowance for surface depression storage (part of the overall areal loss factor) and the time/area build up of run-off to the sewers. These modifications are illustrated in Figs. 2 and 3.

It is sometimes necessary to take account of storm movement over a catchment. Provision has therefore been made in the program for the simulation of a moving storm by translation of the hyetograph along the time axis.

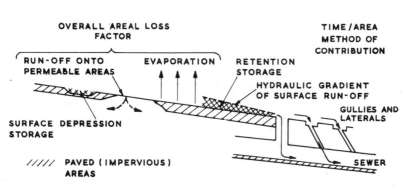

MODIFICATION OF THE STORM HYETOGRAPH

FIGURE 2

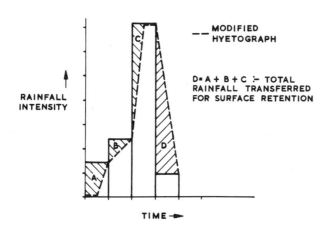

—— MODIFIED HYETOGRAPH

D = A + B + C := TOTAL RAINFALL TRANSFERRED FOR SURFACE RETENTION

MODIFIED HYETOGRAPH

FIGURE 3

Loss factor
This factor, usually referred to as the "overall areal loss factor", is incorporated in the computer program and defines that portion of the run-off from the impervious or contributing part of a catchment which reaches the sewer. The remainder which does not reach the sewers is lost to surface storage, soaking into paved areas and road surfaces, run-off from paved on to pervious areas and a small amount of evaporation.

In the study for the London Borough of Harrow, the value for this factor was derived from the Roxbourne catchment for notable flood events, using continuous water level records on the Roxbourne converted to flows by a stage-discharge relationship based on spot gaugings, and autographic rainfall records at two points within the catchment area. The results showed that, when there had been significant antecedent rainfall, the run-off, considered as coming wholly from the impervious area, reached 66% to 79% of rainfall. Studies carried out by the Road Research Laboratory[7] on a similar catchment in neighbouring Oxhey gave a run-off factor of 77% for 17 summer storms over the period 1953-59. On the basis of these studies, an overall run-off coefficient of 75% was assumed for each of the three design storms (1 year, 2 year and 5 year return period) considered in the analysis. For more severe storms and other catchment configurations the factor may need to be increased or decreased as conditions dictate.

Dry weather flow
Where it is known that a steady or dry weather flow exists it is recorded as a constant inflow against the pipe where it enters the sewer system.

THE COMPUTER PROGRAM

The analysis of the behaviour of a complex drainage system during storm events involves a large amount of data and repetitive calculations, and is eminently suitable for solution by digital computer. The calculation falls naturally into several activities which, though they occur simultaneously, can be conveniently considered separately as follows.

Rainfall and overland flow
Following adjustment of the rainfall profile as described above, a modified rainstorm is presented as a series of rainfall rates in discrete equal time intervals. This time interval must be short enough to give realistic modelling.

The modified rainstorm has been modelled to take account of the effects of terrain, soil moisture deficiency, etc. as previously described by adjustment to the rainstorm profiles;

profile shape being adjusted to allow for temporary retention
storage and factors representing permeability and depression
storage applied to the ordinates of each profile in the program.

To determine the variation of inflow rate to the sewer system,
a time/area diagram is employed. In the time area build up
of run-off to the sewers the velocity of flow over the catchment towards the point of entry into a pipe length is assumed
to be equal to the full-bore velocity in the pipe. The latter
assumption is reasonable where the catchment areas run roughly
along the pipe lengths, as in most urban street drainage,
because most of the water enters a pipe through gullies distributed along its length; furthermore, the mean propagation
velocity of kinematic waves along a pipe as it gradually fills,
Is similar to the full-bore velocity. The time of travel to
the sewer may be considered in two parts: one is concerned
with flow along a catchment parallel to the pipes; the other
with lateral flow across the catchment towards the sewer.
The first of these times is calculated from the mean kinematic
wave velocity and the catchment length; the second, which is
usually of smaller importance, cannot be calculated by the
program on the information available and is read in with the
catchment data for each pipe length.

The main objections to the simple time/area approach have
been overcome by the use of modified rainfall profiles incorporating the hydrological factors which take account of conditions on the catchment.

Pipe routing
There are three flow conditions under which water is routed
through sewer pipes; "free surface", "surcharged" or "partially surcharged". All three conditions may occur simultaneously in different pipes of a system. The concept of a
delay between the time at which water enters a pipe from the
preceding pipe and the time at which its influence reaches
the pipe outlet to be passed down the system has been incorporated in the program. The delay applies only to water
which is already in the pipe system and is flowing freely,
i.e. with no surcharge, towards the outfall. Provided
each pipe is treated as downhill from its associated catchment, no problems arise over this transit delay, although
the size of the delay may need adjustment.

When surcharging occurs, the response at the lower end of a
pipe length is almost immediate and the delay time is correspondingly reduced to zero. The effects of fluctuation of
inflow on the free surface at intermediate gullies are usually
damped-out within the calculation time interval, accordingly,
no adjustment of the time delay (zero) is required on this
account.

(a) <u>Free surface flow conditions</u>   For free surface flow conditions, a balance is struck for each pipe during each calculation time interval, between the average outflow rate, change
in storage and the inflow from its catchment and feeder pipes.
Since both the storage and free surface flow rate are functions of the water depth, a single expression can incorporate
them both leading to a simple linearised solution. Friction
losses are calculated from the Colebrook-White equation.

For the hydraulic calculations, it is necessary to adjust
values of hydraulic pipe diameter to avoid the instability
resulting from the equal hydraulic radii of circular pipes
running full and half full. This is done by adjusting the
relationship between flow and hydraulic pipe diameter as shown
on Fig. 4. Thus, the hydraulic diameter for maximum flow is
taken as 1.22 true diameters, but the velocity calculation is
based on the full bore flow rate. Fig. 4 also illustrates
that change of roughness has little effect on the shape of the
flow calculation curve.

In addition to the head loss in a pipe caused by surface
roughness, a further loss occurs at a manhole when the water
level rises above the pipe centreline due to the different
shape of the manhole benching (waterway) and the pipe bore.
This is represented by a factor, which rises steadily from
zero at half depth to a suitable number of velocity heads
chosen by the user, when the pipe runs full, and has the
effect of increasing the total upstream heads.

(b) <u>Surcharge conditions</u>   When the water level reaches the
soffit of a pipe full bore flow is initiated, and the storage
in manholes provides the only volumetric freedom in the system. The same basic flow balance still occurs; flow through
the pipes is calculated from the full bore flow, but no
allowance is made for the change in Reynolds number. The
program accepts a sudden reduction in delay time as conditions
change from free surface to surcharged but the sudden increase
in delay time on reversion from surcharged to free surface
conditions results in mathematical instability which would
appear as a temporary reversal of the sequence of outflow
rates from a pipe. The increase from zero delay is therefore
phased; the length of the transition period being at the choic
of the user.

When a pipe surcharges at its outlet end, because of conditior
downstream, it will become only partially surcharged. Under
this situation it is assumed that the upper part of the pipe
runs normally and that the lower part runs surcharged: the
position of the discontinuity between the two flow phases
allows a volumetric balance to be struck. It will be apparent, however, that the storage effectively available to the
downstream (fully surcharged) pipe is substantially greater

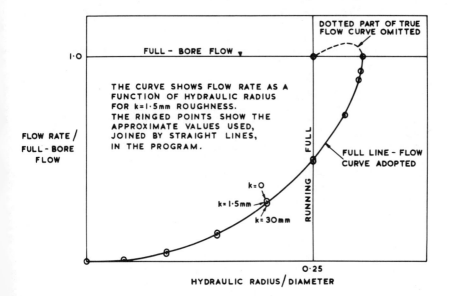

FLOW RATE /
FULL - BORE
FLOW

THE CURVE SHOWS FLOW RATE AS A
FUNCTION OF HYDRAULIC RADIUS
FOR k=1·5mm ROUGHNESS.
THE RINGED POINTS SHOW THE
APPROXIMATE VALUES USED,
JOINED BY STRAIGHT LINES,
IN THE PROGRAM.

FULL - BORE FLOW

1·0

DOTTED PART OF TRUE
FLOW CURVE OMITTED

FULL LINE - FLOW
CURVE ADOPTED

RUNNING   FULL

k=0
k=1·5mm
k=30mm

0·25

HYDRAULIC RADIUS / DIAMETER

ILLUSTRATION OF APPROXIMATION
USED IN FLOW CALCULATION

FIGURE   4

than its upstream manhole since water will simultaneously
back-up in the partially surcharged pipe above it. The horizontal free surface area in the upstream pipe is therefore
temporarily attributed to the downstream pipe. To alleviate
the additional mathematical problems which would arise if
storage were suddenly abolished or made available, an exponential smoothing factor is provided to effect the change
gradually.

(c) <u>Information transfer</u> Information on the flow regime is
transmitted from pipe to pipe in an upstream direction. For
the purposes of calculation a pipe length consists of a pipe
and its upstream node. At each round of calculations, the
condition in the downstream node (i.e. the upstream node of
the adjacent pipe downhill) is assumed to remain unchanged
at the value determined one time interval previously. However,
since this leads to rather violent changes in flow and to arith
metic instability, the downstream level is assumed to be
intermediate between the values calculated at the beginning
and the end of the previous time step. At the choice of the
user, the intermediate value may be weighted toward the older
or the newer level by use of the factor provided. Generally,
an equal weighting has been found satisfactory as it reflects
the average water level during each calculation interval.

Surface flooding
When the manhole water level reaches the ground surface,
flooding occurs. Such floods are treated as local, with insignificant further changes in water level until the whole
flood volume has returned to the manhole from which it came.
In some locations it may be that water spilled onto the surface will flow overland to another entry point. Provision is
made to simulate this by the water transfer facility.

Rectangular conduits
A rectangular conduit, like any other non-circular pipe,
can be represented by a circular pipe of equivalent carrying
capacity and storage. However, a special situation arises
in the case of a rectangular conduit because its carrying
capacity is suddenly reduced when the flowing water comes
into contact with the upper surface (conduit full or surcharged). Provision is therefore made in the program to deal with
the different flow conditions of rectangular conduits running
full and up to nearly full.

COMPUTER RESULTS

The computer calculates conditions of water levels and flows
throughout the sewer system, at the end of each time interval of the computer simulation. The number of time intervals is made sufficient for peak water levels and flows to
be reached in each pipe length, and to give a measure of the

apacity of the whole sewer system considered. Wherever
ncapacity of a part of the sewer system occurs, free-surface
low conditions change, first to partially surcharged conditions
nd then, when the water level in the upstream manhole begins
o rise, to fully surcharged flow conditions. If the manhole
epth from cover to the soffit of the pipe does not permit a
arge enough driving head to push the water through the pipe,
ater spills out of the manhole, and flooding occurs.

he predictions indicate that the capacity of a sewer pipe is
overned not only by its unrestricted full bore flow, but also
y changes in the hydraulic gradient due to rising water levels
n upstream and downstream manholes above the soffits of the
ipes. A frequency analysis of a sample of seventy seven
29mm pipes in the Harrow study showed that for the 50% which
ere predicted to surcharge or flood, the average ratio of
eak flow to unrestricted full bore flow was about 0.75 and
.00 during the 1 year and 5 year storms respectively. The
esults also indicated that the ratio could not be taken as
  useful guide as to where surcharging or flooding is likely
o occur.

low hydrographs can be produced for any pipe length in a
ewer analysis by the SPARROW program. Their main use is
n the determination of the magnitude and pattern of flow in
ny part of a system. At the outfall pipes, for which it is
ecommended that hydrographs should always be obtained, the
nformation provides evidence of the performance of the sewer
ystem as a whole and input data for the study of open water-
ourses.

ime of occurrence of maximum volume of spillage from a man-
ole, or the time to maximum water level in a manhole which
nly surcharges, is denoted as "peak time" in the computer
esults. Wherever "peak times" are found to be prolonged
eyond the typical range, severe constriction of the sewer
ystem may be suspected. Increase in "peak time", together
ith comparison of potential run-off volume, total outflow
olume and volume of water stored on the ground or in the
ewer system may also indicate that increased computer run
ime is necessary.

o verify the results of the computer program it is necessary
o compare them with all available information on the inci-
ence of flooding in the area under consideration and on the
torms which caused them.

n making a study of the records it is necessary to differen-
iate between complaints about the surface water and foul
ewer systems and to put more weight on recent flooding re-
ords as subsequent improvements to sewer systems often make
arlier flooding complaints irrelevant.

Whenever significant spill volumes are predicted, site visits are recommended to determine the consequences of the flooding which is expected. Site visits combined with local knowledge and experience should be used to improve the input data for future analyses. By this means, variations in local imper- meability due to changes in contributing areas, and the fate of spillage from manholes may be properly accounted for.

The quantity of information derived from the computer analysis was such that it called for a comprehensive though simple method of summarising the results.

CONCLUSION

The possible improvements to an existing sewer system are numerous. The benefits of this method are that the various solutions may be tested and properly appraised before any plans for construction work are formulated.

ACKNOWLEDGEMENTS

The Authors gratefully acknowledge the assistance of their colleagues and the encouragement of the Partners of Messrs. Binnie & Partners in the preparation of this Paper.

REFERENCES

Road Research Laboratory (RRL). "A guide for Engineers to the Design of Storm Sewer Systems. Road Note 35". Depart- ment of Scientific & Industrial Research, 1963.

Ackers, P. "Charts for the Hydraulic Design of Channels and Pipes". Hydraulics Research Paper No. 2. Hydraulics Research Station, HMSO London 3rd Ed. 1969.

Colyer, P.J. "Research in storm drainage design". Institute of Public Health, May 1975.

British Standards Institution (B.S.I.). Code of Practice CP 2005 "Sewerage" 1968.

Natural Environment Research Council. Flood Studies Report: N.E.R.C. London 1975.

Keers, J.F. and Rodda, R.C. "The variability of precipita- tion and evaporation in Engineering Hydrology Today". Proceedings of ICE/HID and IOH Conference. February 1975. Institution of Civil Engineers, London 1975.

Watkins, L.H. "The design of Urban Sewer Systems". Road Technical Paper No. 55. Dept. of Scientific and Industrial Research, London, HMSO 1962.

THE DESIGN OF SURFACE WATER SEWER SYSTEMS IN THE TROPICS

L H Watkins, Head of Environment Division, TRRL

D Fiddes, Environment Division, TRRL.

INTRODUCTION

1. The TRRL hydrograph method is a surface water sewer design method widely used in the United Kingdom, the country for which it was developed. It is simple and economic to apply requiring relatively few input data. For this reason attempts have been made to apply it either in its original form or with various modifications in other countries. In this paper a simple modification is described which has been developed at the Transport and Road Research Laboratory, UK, to enable the method to be applied in East Africa where, unlike the United Kingdom, run-off from unpaved areas can make a significant contribution to peak flows. This simple modification should make the method generally applicable in the tropics when suitable data are available. The method still requires few and simple input data, which is an immense advantage in regions where methods requiring far more information are impracticable. The application of the method is therefore examined to arrive at recommended values for the necessary parameters in parts of the tropics for which experimental verification is not yet available.

THE TRRL HYDROGRAPH METHOD AS APPLIED IN THE UK

2. The principle of the TRRL hydrograph method has been fully described elsewhere (1,2). The rainfall input to the computer program by means of which the calculations are carried out can be either a recorded storm or a theoretical rainfall profile. The surface hydrograph is calculated using the area/time diagram for the area, and surface effects are allowed for by adding the time of entry to the time of concentration. This latter approximation has sometimes been criticised by hydrologists who argue that it can introduce significant errors into the calculation. This is however not the case, as in the United Kingdom's summer storm conditions

the significant run-off in a normal sewered area is produced
by only the paved surfaces, the time of flow across which is
measured in minutes. It would be valid criticism in conditions
where runoff from unpaved areas became significant, and this
is discussed later where the modifications of the method for
tropical conditions is described.

At this point in the calculations the method gives the
hydrograph at the point under consideration that would exist
if the sewer system had negligible volume. However, sewer
systems do have considerable volumes and the hydrograph is
then modified significantly as a result of the flow through
storage. The next stage therefore is to route the hydrograph
through this storage, the maximum storage being the volume
in the sewers occupied by water at the peak rate of run-off.

In the early research it was shown that a satisfactory
relationship between storage and rate of flow could be
obtained for some sewer systems by assuming that at any given
instance the ratio of the depth of the water to the maximum
possible depth is the same for all the pipes in the system,
ie the proportional depth of water for the whole system is
constant at any instant. Implicit in this assumption are the
following two conditions which define the "simple system".

a) The system is designed with a reasonable degree of taper.

b) All the pipes in the system are geometrically similar,
ie all corresponding dimensions are in the same ratio to each
other.

If either or both of these assumptions is not valid in the
particular case then the assumption of uniform proportional
depth breaks down. However, as virtually all new pipe
systems satisfy these requirements the method is reasonably
applied in its original form to their design.

As the method became accepted and widely used the
restrictions imposed by these assumptions became increasingly
inconvenient. Pressure increased for the use of the method
for the redesign of the existing sewer system, which
frequently did not accord with these assumptions. The routine
technique was therefore re-examined, and it was found that a
more general solution was possible if the assumption of
constant proportional depth was maintained over only the
single pipe. The method revised in this way has now been
incorporated into the program available from major computer
bureaux.

ADAPTION OF METHOD FOR EAST AFRICA

3.   In order to carry out the research required in East
Africa to apply the hydrograph method, five catchments were
instrumented, three in Nairobi and two in Kampala. Full
details of the catchments were given in the Proceedings of the
Institution of Civil Engineers in 1976 (3). The Nairobi
catchments were selected to give a range of soil permeability
and were all relatively flat. It is easier to find steep
catchments in Kampala, and one catchment there was therefore
selected for the study of the effect of catchment slope. The
other was a small catchment over which the drainage pattern
could be very closely defined.

The three small catchment areas in Nairobi were selected
to represent the types of development and soil types which
predominate in the city. The first - Bernhard Estate - is a
low density housing estate on red lateritic soil drained by
open stormwater drains, the second is a high density
industrial development on a black grumosolic clay, draining
into closed storm-water sewers, and the third - Ofafa Estate -
is a high density council housing estate on a stony grey
grumosolic clay overlying 'murram' (ironstone laterite)
drained by open stormwater drains, leading to a box culvert
outfall.

The first of the Kampala catchments is a low density housing
estate in Wampewo Avenue and the second a high density
housing estate in Kira Road both being on similar soil and
slope. The slopes of these catchments are much greater than
those of the Nairobi catchments.

Casella natural siphon daily rainfall recorders were readily
available and were installed on each of the catchments
except Wampewo Avenue, which was covered by three high speed
autographic recorders that were part of the East African
rainfall project, Kampala network. The air purge method was
used for flow measurement in all the catchments. This
allowed the recorders themselves to be placed some distance
from the control structure, only the dip tube needing to be
installed in the measuring section. The method was found to
work well in practice. An absence of moving parts in contact
with the water eliminated troubles due to float jamming,
corrosion and silting. The only moving parts were enclosed
in waterproof cases which were kept dry with silica gel even
under the most adverse conditions.

Two modifications are required in the use of the TRRL hydrograph method program under tropical rainfall conditions.
They are:-

a) a method of modelling the delay and attenuation of the rainfall profile by its storage on the ground surface

b) a means of predicting the proportion of rain falling on the unpaved areas that flows into the drainage system.

For a) a linear reservoir sub-model was used;   b) was incorporated using a contributory area factor.

In the TRRL hydrograph method the run-off from paved areas is taken to be the rainfall profile routed into the sewer inlets for each of the contributing paved areas. When the flow from unpaved areas is taken into account the inflow needs modification to allow for infiltration and for any ponding effects. Storage on the unpaved areas can be likened to storage within a reservoir. If outflow from such a reservoir is proportional to the average depth over the area this is termed a linear reservoir.

Assuming that the unpaved areas act as linear reservoirs the ordinates of the unpaved surface flow hydrograph for any reach can be calculated using the equation

$$q_n = \frac{C_a A}{K} h_n \qquad (1)$$

The continuity equation for storage can be written as

$$AR_n C_a - \frac{(q_n + q_{n-1})}{2} \Delta t = AC_a(h_n - h_{n-1}) \qquad (2)$$

where R_n is the nth rainfall ordinate and t is the time increment between ordinates.

The two constants C_a and K can be derived from catchment records where these are available. The process of calculating the hydrograph for unpaved areas is then followed using the TRRL hydrograph method with two changes: the rainfall profile is replaced by the q_n profile and the paved area/time diagram is replaced by the unpaved area/time diagram. The run-off from the whole catchment is the sum of the run-off from the paved and unpaved areas routed simultaneously through the sewer system.

Optimum values of C_a and K were derived for each catchment by a comparison of recorded and calculated hydrographs. These are shown in Table 1.

TABLE 1. Optimum values of Ca and K

Catchment	Ca	K(hours)
Bernhard Estate	0.4	1.15
Industrial Area	0.7	4.00
Ofafa Jericho	0.7	0.65
Wampewo Avenue	0.1	0.60
Kira Road	0	– *

*At Kira Road there was no run-off from unpaved areas and the results from this catchment cannot therefore be used for the purpose of this paper.

SENSITIVITY AND RANGE OF VALUES FOR VARIABLES

Storm profile

4. A storm profile is required as input to the program. Four typical profiles for point rainfall are shown in Table 2, one for a temperate latitude and three from the tropics. In the United Kingdom storm profiles are readily available either in published form or for any individual location by application to the Meteorological Office. In the tropics it is generally necessary to derive the profiles from the few data available. The profiles for Nairobi, Kampala and Seawell Airport were derived by fitting the intensity duration equation:

$$i = \frac{a}{(b + t)} n \qquad (3)$$

where i is intensity (mm/hour)
t is duration (hours)
and a, b and n are constants.

With three constants in the intensity duration equation data for at least three durations are required for fitting. The three most readily available are totals for 24 hours, 1 hour and estimates of instantaneous intensities. The four stations shown in the table have been selected to demonstrate how these intensities can vary in different locations. For example Nairobi and Kampala have very similar daily totals but intensities at Kampala for durations comparable to typical times of concentration of urban sewer systems are nearly twice as high as at Nairobi. Seawell Airport has a larger daily total and comparable short period intensities to Kampala but at typical times of concentration the intensity at Seawell Airport is very little higher than Nairobi. The London profile shows that high short period intensities do occur in temperate latitudes but the intensity falls off much more rapidly than in the tropics.

TABLE 2. 2 year point rainfall profiles

Time from storm peak (mins)	London England (mm/hr)	Nairobi Kenya (mm/hr)	Kampala Uganda (mm/hr)	Seawell Airport Barbados (mm/hr)
$\frac{1}{2}$	96.0	94.4	186.2	175.0
$2\frac{1}{2}$	42.0	70.1	130.6	91.8
5	26.8	52.5	91.6	56.4
10	12.7	33.6	51.8	38.4
15	8.5	23.6	33.6	24.0
30	4.4	12.3	13.1	12.6
45	2.5	7.7	6.3	9.6
60	1.5	5.5	4.2	7.8
1 hour total (mm)	16.1	32.5	50.2	36.4
Daily total (mm)	34	70	70	101

For all but the smallest areas the point rainfall profile must be adjusted to take into account the reduced intensities appropriate when rainfall is averaged over the whole catchment. Because of the difficulty in measurement, data to derive area reduction factors are even less plentiful than those required for point rainfall profiles. A major study has been completed in the United Kingdom as part of the ICE/NERC flood studies programme (4). In the flood studies report the area reduction factors are shown graphically, but these have been tabulated by Keers (5). For areas up to 30 Km^2 (those usually considered for urban sewer systems), the tabulated values can be expressed by the equation

$$ARF = 1-0.038T^{-0.38}A^{0.37} \tag{4}$$

the comparable equations for East Africa developed by TRRL are

$$\text{Nairobi ARF} = 1-0.02T^{-\frac{1}{3}}A^{\frac{1}{2}} \tag{5}$$

$$\text{Kampala and Dares Salaam ARF} = 1-0.04T^{-\frac{1}{3}}A^{\frac{1}{2}} \tag{6}$$

The E African equations assume that storms last for up to 8 hours and a value of T = 8 is substituted for all longer durations(for example for daily values).

A similar equation is available for rainfall in West Africa.[7]

$$ARF = 1- (9 \log r - 0.042 P + 152) \log S.10^{-3} \tag{7}$$

where r = recurrence interval (yrs)
      P = mean annual rainfall (mm)
      S = catchment area (Km^2)

This equation applies to the total storm and should therefore be comparable to the 8 hour equation for E Africa. An examination of the equation will show that results are relatively insensitive to recurrence interval as found in the UK and E Africa.

Typical area reduction factors are shown in Table 4 from which it will be seen that although few data are available they are consistent and the selection of an inappropriate area reduction equation should not introduce significant errors in the flood estimation.

It is often unavoidable to prepare a design storm profile using data from a point distant to the catchment being studied. To examine the sensitivity of catchment response to extreme examples of this the program was run for each catchment with each of the 4 profiles shown in Table 2. The resulting peak flows are shown in Table 3 (note that the two larger storm profiles caused surcharge in the industrial area). Obviously the volume of the storm profile will affect the peak but it should also be noted that the shape of the assumed profile can have a large effect on the resulting peak. For example an error of approximately 50% in estimate of peak flow would occur if the Kampala storm shape were applied in a district where the flatter Nairobi profile was required even though the storm total depths are the same.

Comparing Tables 2 and 3 it can be seen that the 1 hour total rainfall is a better guide to peak flow than either the storm total or maximum instantaneous intensity.

The proportion of the peak run-off attributable to unpaved run-off was found to be virtually independent of the storm profile used and was approximately 10 per cent, 30 per cent and 80 per cent for the Industrial Area, Wampewo Avenue and Ofafa Jericho, the three catchments with paved area components. For comparison the unpaved areas comprised 70 per cent, 93 per cent and 94 per cent of the total areas respectively.

Contributing areas
5. The paved area of a catchment may be estimated from a catchment survey for an existing catchment, or from plans of the various development areas for a proposed scheme. It is important to remember that all paved areas do not contribute directly to the sewer system. Whereas it is common practice in the UK for all roads and roofs and most other paved areas to be connected directly to gullies, elsewhere it is quite normal, except in highly developed city centre areas, to allow roofs, footpaths and some road areas to discharge on to

grassed areas thereby reducing the peak flow to be designed for, both because of the reduced volume of run-off and by the delay in flowing over the rough unpaved surfaces. Highway engineers frequently debate the wisdom of providing sufficient gullies to divert the flow from tropical storms into the sewer system rapidly, arguing that fewer gullies could act as a constriction to to the flow, ponding it on the road surface, and hence reducing the size of sewer required. The argument that this would result in traffic delays is countered by the observation that during tropical storms of comparable size to the sewer design storm, traffic is effectively stopped by the lack of visibility due to the intensity of rainfall. If such a practice became common it would significantly affect sewer design techniques. Engineers would also have to ensure that the smaller, less numerous road gullies were well maintained and kept free from obstructions. The TRRL East African catchments may be considered as typical of current practice in the developing countries of the tropics in that the drainage systems were all designed to remove water as quickly as possible whilst the areas of paved surface ranged from 16 to 37 per cent of the total areas, and the percentages of this paved area directly connected to the sewer systems varied from 0-90 per cent, averaging 50 per cent.

In the application of the method, the unpaved area is taken to be the difference between the total area and the contributing paved area. Locally derived data for estimating contributing area coefficients are likely to be sparse or non-existent. Fortunately data from rural catchments, on similar soils, may be used suitably increased to allow for the generally more efficient drainage system in urban areas and large amounts of paved or bare areas considered as unpaved The contributing area coefficient for the East African catchments varied from 0-0.75 which was 1.5-2.2 times the equivalent factor for similar rural catchments depending on thepercentage of non-contributing paved areas and bare beaten-down soil.

Considerable care should be taken in assessing the contributing area coefficient, particularly on the more permeable soils where the possibility of error is much greater. It should also be borne in mind when using published tables of run-off coefficients that these tend to over-estimate the response of catchments generally, particularly so under small recurrence interval storms typically used for urban sewer designs and on the more permeable soils.

To examine the sensitivity of catchment response to the assumed contributing area coefficient the program was run for each catchment with the Nairobi storm profile (except for the industrial area where the London profile had to be used to

avoid surcharging) and three values of C_A (the measured value
and the measured value ± 50 per cent). The resulting peak
flows are shown in Table 5. Where there is no paved area
contribution, as at Bernhard Estate, the peak flow is directly
proportional to the assumed value of C_A. On the other
catchments the effect of an error in estimate of C_A is less,
depending on the percentage of contributing paved area and
lag time assumed and the initial value for C_A. (Compare the
response of the industrial area to the Ofafa Jericho catchment)

Lag times
6. This is the catchment parameter which can be estimated
with the least precision. A considerable amount of research
has already been undertaken on times of overland flow for flat
uniform small experimental plots, the work of Hicks([8]) and
Izzard ([9]) being the two most often quoted examples.

The Hicks equation for well-cared-for lawn is

$$T_C = \frac{9.34L^{0.298}}{B^{0.785}S^{0.302}} \qquad (8)$$

where T_C = time of concentration (minutes)
L = length of overland flow (feet)
B = supply rate of rainfall excess (inches/hour)
S = surface slope (%)

It will be seen that the time of concentration is
proportional to $L^{0.3} \times S^{-0.3}$. The length of overland flow
was not measured on the East African catchments but should
be directly proportional to the contributing area coefficient.
Lag time and time of concentration should also be related.
Tests were therefore run to determine how much of the
variation in K could be explained by the grouping:-

$$\left(\frac{CA}{S}\right)^{0.3} \qquad (9)$$

The equation resulting from the line fitting exercise was:

$$K = 42.5 \left(\frac{CA}{S}\right)^{0.3} - 24.1 \quad \ldots\ldots(r = 0.58) \qquad (10)$$

Thus the equation only explains 34 per cent of the variation,
but this is not too surprising when the different surfaces
of the catchments and the high sensitivity to surface type of
the coefficient (9.34 for clipped grass in the Hicks equation)
are considered. The line should pass through the origin and
allowing for the effect of surface type a generalised
equation may be assumed of the form

$$K = C \left(\frac{CA}{S}\right)^{0.3} \qquad (11)$$

where C is a coefficient whose value depends on the surface type.

The calculated values for C for the four catchments are Bernhard 30, Industrial Area 60, Ofafa Jericho 10 and Wampewo Avenue 30. These are in line with the expectation from the Hicks equation where the coefficient increases with surface roughness. Thus this equation may be generally applied with values of C of 10, 30 and 60 for high density housing, low density housing and industrial areas respectively.

The data from which the recommended values were derived are few, so where local information is available to estimate K values (by a comparison of simulated and recorded hydrographs) the locally derived values should be used. As before the effect of errors in estimation of this variable were tested by running the program for each catchment with a range of lag times. The resulting peak flows are shown in Table 6. It will be seen that on flat, rough surfaces, where large K values are appropriate, the predicted flood peaks are relatively insensitive to quite large errors in estimate of K. For example an increase in K from 3 hours to 4 hours on the Nairobi industrial area catchment reduced the predicted peak flow by only approximately 3 per cent. For smoother, steeper catchments the peak flow becomes much more sensitive to errors in estimate of K. Measurements of flow and rainfall on selected local existing catchments at an early stage in the use of the method to confirm the appropriateness of the values recommended above is therefore advised.

CONCLUSIONS

7.   A simple modification to the TRRL Hydrograph method of storm water sewer design has been developed for use in East Africa which can be applied elsewhere in the Tropics. Errors in estimates of the required sewer sizes will result if inappropriate storm profiles are used, or inappropriate values of the two parameters defining the unpaved area response (contributing area coefficient and lag time). However, with careful use of the methods described above designs within normal engineering accuracy are possible even in areas where little or no flow gaugings on urban sewer systems are available.

TABLE 3.   Effect of storm profile on peak flow (ft^3/sec)

Storm Profile	Bernhard	Industrial Area	Wampewo Ave	Ofafa Jericho
London	15.83	65.49	16.08	55.83
Nairobi	31.65	123.69	28.96	107.46
Kampala	46.92	– *	44.13	161.86
Seawell Airport	36.17	– *	34.13	124.04

* System surcharged

TABLE 4.   Typical area reduction factors

		Area (mk^2)			
		1	5	10	30
UK	24 hour	0.99	0.98	0.97	0.96
	1 hour	0.96	0.93	0.91	0.86
	5 min	0.90	0.82	0.76	0.65
Nairobi	8 hour +	0.99	0.98	0.97	0.95
	1 hour	0.98	0.96	0.94	0.89
	5 min	0.95	0.90	0.86	0.75
Kampala	8 hour +	0.98	0.96	0.94	0.89
	1 hour	0.96	0.91	0.87	0.78
	5 min	0.91	0.80	0.71	0.50
W Africa (2 yr 1000 mm)	8 hour +	1.00	0.92	0.89	0.83

TABLE 5.   Effect of contributing area coefficient on peak flow (ft3 sec)

Adjustment to C_A	Bernhard	Industrial Area	Wampwo Ave	Ofafa Jericho
– 50%	15.82	62.85	24.48	68.47
0	31.65	65.49	28.96	107.46
+ 50%	47.49	68.14	33.68	147.31

TABLE 6. Effect of lag time on peak flow (ft^3/sec)

Lag time (h)	Bernhard	Industrial Area	Wampewo Ave	Ofafa Jericho
0.6	48.79	91.82	28.96	112.43
1.15	31.65	78.51	25.46	76.01
3	15.38	67.40	22.38	43.34
4	12.19	65.49	21.83	37.79

ACKNOWLEDGEMENTS

8.   The work described in this Paper formed part of the programme of research of the Transport and Road Research Laboratory and the Paper is published by permission of the Director.

REFERENCES

1.   WATKINS, L.H. (1962). The design of urban sewer systems HMSO London DSIR Road Research TP55.

2.   Transport and Road Research Laboratory (1976). A guide for engineers to the design of sewer systems. HMSO DOE Road Note RN35 2nd Ed.

3.   WATKINS, L.H. (1976). The TRRL Hydrograph method of urban sewer design adapted for tropical conditions. Proc Inst Civ Engrs Part 2 61 Syst pp 539-566.

4.   National Environment Research Council (1975). Flood Studies Report: Volume 11: Meteorological Studies. NERC.

5.   KEERS J.F. (1977). Rainfall criteria for urban drainage design. Met Mag Vol 106 No 1257, pp 117-126.

6.   FIDDES, D. (1975). Area reduction factors for East Africa. Proc Flood Hydrology Symposium. Nairobi. TRRL SR 259.

7.   BRUNET-MORET, Y. (1975). Rainfall mapping and design storm forecasts in West Africa. Proc Flood Hydrology Symposium, Nairobi. TRRL SR 259.

8.   HICKS, W.I. (1944). A method of computing urban runoff.
     Trans ASCE Vol 109, pp 1217-1253.

9.   IZZARD C.F. (1946). Hydraulics of runoff from developed
     surfaces. Proc 26th Annual Meeting US Highway Research
     Board. Vol 26, pp 129-146.

MODIFICATIONS TO THE STORM WATER MANAGEMENT MODEL AND
APPLICATION TO NATURAL DRAINAGE SYSTEMS.

E. V. Diniz

Sr. Staff Engineer
Espey, Huston & Associates, Inc.

SCOPE OF STUDY

Increased runoff rates and increased pollutant loads are two of
the major effects of urbanization on the hydrologic regime of a
previously undeveloped watershed. The increase in impermeable
areas due to urbanization results in high velocity surface
flows which tend to increase the potential for capture of pollutants by the stormwater and reduce natural infiltration
processes.

The planned new community of the Woodlands near Houston, Texas,
is designed to minimize the detrimental effects of urbanization
upon runoff characteristics of Panther Branch Watershed in
which it is located. The U. S. Environmental Protection Agency
Stormwater Management Model (SWMM) was selected to model stormwater runoff characteristics at the Woodlands; but, due to the
innovative drainage methods implemented in this community,
several extensive changes had to be performed to the SWMM. The
necessary changes include modified computations of infiltration
volume and pollutographs and three new subroutines to develop
normalized area discharge curves for natural channel sections,
to model baseflow conditions, and to model runoff from porous
pavements. This paper discusses the changes that were performed to the SWMM program code, the new subroutines that were
developed, and the concurrent modeling effort in the Woodlands.
An urban Houston watershed, Hunting Bayou, was also modeled
because its drainage characteristics are similar to those of
the Woodlands.

Description of Study Areas
The Woodlands is a planned urban community being developed in
Montgomery County approximately 45 kilometers (28 miles) north
of Houston, Texas, in a heavily forested tract. The soils at
the Woodlands are highly leached, acid in reaction, sandy to
loamy in texture, and low in organic content.

256

Panther Branch, an intermittent tributary to Spring Creek, is
the major drainage channel as shown in Fig. 1. The drainage
channels tributary to Panther Branch and Spring Creek are generally broad and shallow swales having very mild slopes. Consequently, the dense vegetation, sandy soils, and mild slopes
result in high retention and infiltration losses from rainfall
on the watershed.

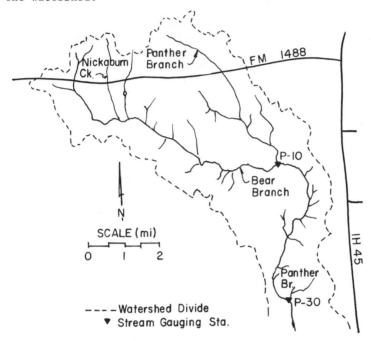

FIG. I   PANTHER BRANCH   WATERSHED

The existing drainage channels in the Woodlands are utilized to
the fullest extent possible and any new channels are constructed as wide, shallow swales lined with native vegetation to
emulate the existing channels. Storm sewers and drains are
used in high density and activity areas to conduct the excess
runoff to the nearest drainage channel with sufficient capacity
to safely carry the design flows. To minimize increases in
runoff volumes and peak flows, retention ponds are utilized
whenever practical. The net effect of this "natural" drainage
system is an increase in infiltration and storage capacity in
the channels, thereby reducing the effects of urbanization upon
the runoff regime.

There are two stream stage recorders located on Panther Branch:
Panther Branch near Conroe (P-10) and Panther Branch near
Spring (P-30) as shown in Fig. 1. Station P-10, located below
the confluence of Panther and Bear Branches, measures runoff
from 65.0 square kilometers (25.1 sq miles) of undeveloped

forest land. Station P-30 has a drainage area of 87.5 square
kilometers (33.8 sq miles) with the developing areas of the
Woodlands (Phase 1) immediately upstream.

The Hunting Bayou study area, located northeast of downtown
Houston, is within the metropolitan confines of that city. As
seen in Fig. 2, there are two gauging stations: Hunting Bayou
at Cavalcade St. (H-10) and Hunting Bayou at Falls St. (H-20).
The drainage areas of stations H-10 and H-20 are 2.7 square
kilometers (1.03 sq miles) and 8.0 square kilometers (3.08 sq
miles), respectively. Land use is primarily residential with
some commercial and industrial areas. There are very few storm
sewers and the major portion of the drainage system is made up
of grass lined swales comparable to those of the Woodlands.

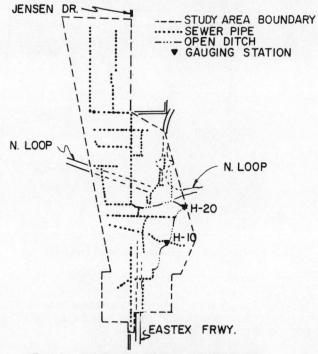

FIG. 2 HUNTING BAYOU WATERSHED

During storm events, stream flow quality sampling was conducted
in conjunction with flow gauging. The samples were analyzed
for a large number of parameters including suspended solids,
COD, nitrates, phosphates, and Kjeldahl nitrogen. Reconstitu-
tion of the observed hydrographs and pollutographs to calibrate
the SWMM were attempted in the modeling effort described in
this paper.

MODIFICATIONS TO THE SWMM

The SWMM was originally developed to model the hydrologic effects of older urban areas where an artificial drainage system was imposed upon, and in most cases entirely replaced the original drainage system. In the application of SWMM to the Woodlands, several defficiencies in the model were uncovered. The resulting modifications are discussed in the following sections.

Modification to Infiltration Volume Computation
Infiltration rates were originally computed in the RUNOFF BLOCK of the SWMM by means of Horton's equation defined as follows:

$$f_c = (f_i - f_o)e^{-kt*} + f_o$$

where f_c is infiltration rate at time t*, inches per hour,
       f_i is initial infiltration rate, inches per hour,
       f_o is final infiltration rate, inches per hour,
       k is the decay coefficient, per hour,
       t* is time from start of rainfall to the midpoint of the
          time interval Δt, or t* is equal to t + 0.5 Δt, hours.

The RUNOFF BLOCK of the SWMM was structured such that Horton's time dependent infiltration rate decay equation would become operative from the start of modeling time. Consequently, if the time of start of rainfall did not coincide with the start of modeling time, the infiltration rate would have decayed to a lower rate by the time rainfall had begun. This may be one reason why early investigators determined that the starting infiltration rate was not a significantly sensitive parameter (Metcalf and Eddy, 1971). The second problem with the computation of infiltration volume resulted from the input of two or more high intensity rainfall events separated by time periods of zero or low intensity rainfall that was not capable of satisfying the infiltration rate. The infiltration rate would decay without regard to the availability of rainfall for infiltration. Modeling runoff under these conditions was difficult and consequently the infiltration computation method in the SWMM was modified as follows.

The new computation scheme uses an integral form of Horton's equation and a time parameter to monitor the progress of infiltration only. The integral of Horton's equation is

$$M = f_o t + \frac{(f_i - f_o)}{k} (1 - e^{-kt})$$

where M is the accumulated infiltration volume at the end of time t, inches. The other variables are as defined previously.

During each time interval (Δt), the volume of water capable of infiltrating ($M_{t + \Delta t} - M_t$) is calculated and compared to the total volume of water available for infiltration determined as

$$D_t = S_t + R_t \Delta t$$

where D_t is water volume after rainfall during the time interval Δt, inches,
   S_t is water volume remaining from the previous Δt, inches,
   R_t is intensity of rainfall during Δt, inches per hour.

When the available volume is greater than the infiltration volume, the excess is calculated as the volume of water available
for runoff. The results are comparable to those previously
computed by the SWMM.

If the infiltration volume is greater than the available volume,
the time increment, $\Delta t^* < \Delta t$, is computed such that the infiltration volume is equal to the available volume.

$$M_{t + \Delta t^*} - M_t = D_i$$

where $M_{t + \Delta t^*}$ is volume of infiltration at time ($t + \Delta t^*$),
      inches,
   M_t is volume of infiltration at time t, inches,
   D_i is volume of water available for infiltration, inches,
and no runoff is generated for that Δt.

The infiltration rate at time ($t + \Delta t^*$) then becomes the starting infiltration rate for the next computational time interval
beginning at time ($t + \Delta t$). Therefore, the elapsed time for
infiltration rate decay by Horton's equation will not necessar
ily coincide with the elapsed runoff computation time.

Subroutine NATSEC
In the TRANSPORT BLOCK of the SWMM, normalized area discharge
curves are required for flow routing. Thirteen uniform channe
shapes (circular, trapezoidal, etc.) have their respective
curves preprogrammed to Block Data, but the curves for natural
sections have to be independently computed and input to the
model. Because of the large volume of manual work required i
preparing these curves for a "natural" drainage system, Subroutine NATSEC was written and incorporated into the SWMM.
This subroutine generates normalized area discharge curves fo
irregularly shaped cross sections and for cross sections with
varying values of Manning's roughness coefficient 'n'. The
cross section is input to the subroutine by means of a two-
dimensional linear coordinate system. Three Manning's 'n' va
ues, one for each overbank and one for the channel, may also
used. Depth increments for equal increments of area are calculated by an iterative process. When the depth of flow is
below bank elevations, a single application of Manning's

equation is sufficient. If the channel capacity is exceeded, the flows in each overbank as well as flows in the channel are computed by independent applications of Manning's equation to each flow area. The total discharge is equated to the sum of the individual discharges.

The output from Subroutine NATSEC is a tabular version of the normalized area discharge curves for natural channels and is comparable to the other area discharge curves in Block Data of the SWMM.

Subroutine BASFLO

The SWMM computational scheme considers all infiltration volume as permanently removed from the runoff volume. The volume of rainfall that soaks into vegetation debris and surface soils and which drains out at a delayed rate is not accounted for becasue in most urban areas this interflow volume is negligible. But, again in the "natural" drainage system of the Woodlands, interflow does become a significant factor. In vegetated areas, the volume of infiltration as computed in the SWMM includes evapotranspiration losses and losses to groundwater. Interception losses may be accounted for in either the infiltration or surface depression storage parameters.

That portion of the hydrograph beyond the point of inflection (where dq/dt approaches infinity) is generally considered as depletion of runoff volume stored in the drainage system of the watershed. As for most depletions in nature, the rate of depletion approximates an exponential decay. It is often referred to as baseflow recessions (Onstad and Jamieson, 1968) of the form:

$$Q_{t + \Delta t} = Q_t e^{-k\Delta t}$$

where $Q_{t + \Delta t}$ is flow at end of time interval Δt, cfs,
Q_t is flow at start of time interval Δt, cfs,
k is the recession coefficient.

The recession coefficients and their associated flow ranges are user supplied to Subroutine BASFLO. One theory of varying recession coefficients for a single hydrograph is the concept of drainage of different storage units of the hydrologic system (Holtan, and Lopez, 1973). In the Woodlands, both stations P-10 and P-30 exhibit two recession ranges. The recession coefficients (determined from observed hydrographs) are plotted against the flow at start of recession and the corresponding regression equations are derived. The coefficient of the regression equation are input to Subroutine BASFLO. All flow rates beyond the point of inflection are determined by consecutive applications of the recession coefficient regression equations and the baseflow recession equations.

Subroutine BASFLO also provides for inclusion of the groundwater component of runoff. Groundwater flow rate may be input
as a constant, linearly varying, or logarithmic function. All
computed groundwater flow rates are added to the runoff hydrograph with respect to time resulting in a corresponding upward
shift in the runoff hydrograph. The baseflow rates are substituted into the runoff hydrograph prior to addition of groundwater flow rates. The specific water quality loading rates are
applied to the new flow rates and the corresponding pollutographs are computed.

Subroutine PORPAV
The Woodlands Development Corp has envisioned an extensive use
of porous pavements in the place of conventional impermeable
basins. Subroutine PORPAV was developed to model the effects
of porous pavements on runoff volume and peak flows becasue the
SWMM did not have this capability.

The modeling scheme consists of delineating the porous pavement
and the subgrade as two hydraulically connected control volumes
for which the inflow and outflow conditions are established by
the equation for continuity or conservation of mass.

$$\frac{ds}{dt} = I - O$$

where ds/dt is the change in storage during the time interval,
    dt,
    I is the time average inflow,
    O is the time average outflow.

Inflow to the porous pavement area is determined as the sum of
direct rainfall onto the pavement and the overland flow hydrograph is computed by Izzard's method (1946). The outflow is
the sum of vertical seepage losses, horizontal seepage losses,
surface runoff when the porous pavement storage capacity is ex
ceeded, and evaporation losses. Vertical seepage losses are
computed by the variable head permeability equation. The modi
fied Darcy equation is used to model the horizontal seepage
losses (Diniz, 1976) and Manning's equation is used to establish the surface runoff rates. The instantaneous evaporatior
loss rate is computed from a time-lagged sine curve approximation of diurnal evaporation loss rates (Claborn, 1971).

Unfortunately, no comprehensive runoff data on existing porou‹
pavements are available. Therefore, only limited testing of
Subroutine PORPAV could be conducted, and this subroutine has
not been sufficiently verified.

STORMWATER QUALITY MODELING

A thorough analysis of the available water quality data from
the Woodlands and Hunting Bayou was conducted in an attempt t

define a methodology to predict runoff quality, specifically suspended solids, nitrates, phosphates, Kjeldahl nitrogen, and COD. The present version of the SWMM considers these as percentages of the dust and dirt volume. Recognizing that in the Woodlands the dust and dirt generation rates are not typical of older urban areas, new relationships between quantity and quality of flow were sought. Significant correlation was determined between the logarithmic transforms of cumulative volume of pollutant and cumulative volume of flow. An example of these correlations for COD is shown in Fig. 3. Also, it was determined that total pollutant loading in units of pounds per acre is a function of total inches of runoff as shown in Fig. 4, also for COD. The average slopes of the correlation line tend to increase or move towards the vertical with increasing urbanization, indicating an increase of pollutant loading for the same volume of runoff. Therefore the relative magnitude of urbanization effects may be determined by the increase in slope.

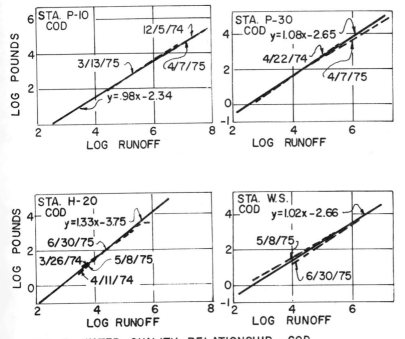

FIG. 3   WATER QUALITY RELATIONSHIP - COD

The pollutant loading relationship may be used to determine total pollutant mass and the cumulative pollutant mass relationship may provide a flow dependent mass transport rate. A combination of these two functions can be used to develop a pollutograph. This methodology to determine quality of runoff is totally unrelated to the dust and dirt accumulation

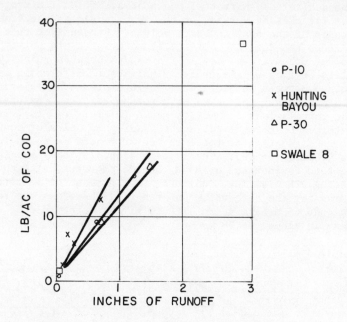

FIG.4 TOTAL LOADINGS OF COD AS A FUNCTION OF
RUNOFF.

approach as used in the SWMM. Consequently, the quality of
runoff computations in the SWMM would have to be completely re-
written to incorporate the new methodology. This was beyond
the scope of the current project, and therefore, a simpler
modification involving the input of user supplied pollutant
loading rates was developed. This approach was utilized durin,
the modeling phase of the study reported herein.

The original SWMM predicted the concentrations of suspended
solids, TOD, total coliform, COD, settleable solids, nitrites,
phosphates, and grease transported by storm runoff. The basic
theory used to predict these constituents assumes that the a-
mount of pollutant washed off at any time interval is a function of an availability factor and is proportional to the
amount remaining on the ground. This results in a first order
differential equation which integrates to the following form:

$$P_o - P = A_o P_o (1-e^{-kt})$$

where P_o is the initial amount of pollutant per unit area,
   P is the remaining amount of pollutant per unit area at
      time t,
   k is the decay rate, and
   A_o is an availability factor defined as a percentage of
      the pollutant P_o that is available for capture by
      storm runoff.

A detailed discussion of the overall procedure for predicting water quality constituents in the SWMM is provided in the program documentation (Metcalf and Eddy, et al. 1971).

The value of 4.6 for the runoff exponent (kt) implies identical rainfall intensities and washoff rates for all storms that are modeled. This is not always true, especially in Texas, where rainfall intensities have a very wide range. Ficticious curb miles, based on average feet of curb per acre of drainage area, were utilized because of the lack of curbs throughout the Woodlands. Also it was desirable to have more than five land uses to adequately describe a watershed. Consequently, the new water quality model was developed to ameliorate these problems.

Pollutant buildup is not considered in the new water quality model. Input data on dry days, street cleaning frequency, land use, or curb miles, is also no longer required and the effects of these parameters upon the pollutant loading rate have to be externally determined by the user. Also the transposition of data to different geographical areas becomes a user option. The pollutant loading factors in lbs/acre are input for each land use specified for all pollutants being modeled. Also the pollutant removal factor or decay exponent, kt, is an input parameter. This allows for model flexibility in the case where pollutants wash off a subcatchment at different rates for the same rainfall or runoff conditions.

The loading rates determined from analyses such as depicted in Fig. 4 are only approximate because of minor variability in the data caused by factors not considered in these analyses, for example, soil characteristics, flora and fauna, etc. The approximate loading rates determined from the figures can be verified for specific application by comparison of the modified SWMM output to observed data; this procedure was followed in all applications of the new water quality modeling version conducted during this study.

Calibration of the new water quality model is a relatively simple task. Modeling results based on initial loading rates and removal coefficient estimates are used to refine subsequent loading rates and removal coefficients, in a trial and error procedure, until the observed pollutograph is reproduced. In applying the model it was determined that pollutograph peak and loading rate were directly proportional while pollutograph shape and removal coefficient were inversely proportional. The pollutograph peak was raised in conjunction with an increase in the removal coefficient but a general relationship to quantify the rise could not be determined. It must be emphasized that the loading rate and removal coefficient thus derived are valid only for the storm used for calibration.

Application of these results to other storms is possible only
if prevailing antecedent conditions and rainfall runoff intensities are similar for both times and the study areas are
identical or at least homogeneous.

In analyzing the water quality modeling results, it became
evident that although pollutograph peaks could be reproduced
appropriately, the computed and actual mass transport graphs
did not correspond. It was determined that this condition resulted from slight variations in the runoff quantity modeled
results. For example, a slight difference between observed
and computed runoff rates concurrent to a high rate of mass
transport results in a large increase or decrease in pollutant
concentration. Consequently, pollutographs are highly dependent on the accuracy of the hydrograph model output. To improve the modeling of total mass transport of a pollutograph
the modified version was used to model pollutant mass flow
rates. This new approach identifies the need for the user to
be completely aware of the modeling objectives. If pollutant
mass flow rate or total pollutant loading are desired, water
quality modeling is essentially independent of water quality
modeling; but if pollutant concentrations are desired, then
both quality and quantity modeling results determine the
accuracy of the pollutant concentrations. Detailed analyses
of three storms modeled during this study are included in the
following section.

RESULTS OF MODELING WITH THE MODIFIED SWMM

The 87.5 square kilometer (33.8 sq mile) of drainage area upstream of Station P-30 was divided into 57 subcatchments with
an average size of 154 hectares (380 acres). The Hunting
Bayou watershed upstream of Staton H-20 was divided into 24
subcatchments with an average size of 37 hectares (91 acres).
Physical parameters were determined from topographic maps obtained from the U. S. Geological Survey and the Woodlands
Development Corp. Certain parameters such as infiltration
rates, width of subcatchment, and retention depths cannot be
directly determined for natural watersheds. Therefore, calibrated parameters were used. Initial infiltration rates,
which are a function of antecedent soil moisture, were found
to vary considerably in range from 0.51 to 4.1 cm/hr (0.2 to
2.0 in/hr) on the Panther Branch watershed and from 0.76 to
6.35 cm/hr (0.3 to 2.5 in/hr) on the Hunting Bayou watershed
But, the final infiltration rate in both areas was found to
be 0.25 cm/hr (0.1 in/hr) and the decay rate was 0.00115 per
second for Panther Branch watershed and 0.0005 per second fo
Hunting Bayou watershed. Width of subcatchment was first
estimated using the method described in the SWMM users manua
These values had to be reduced by approximately 40 percent
because overland flow would not occur as sheet flow over the
entire subcatchment.

Water quantity modeling results are presented in Tables 1 and 2 which list the observed and modeled peak flows and volumes at Stations H-10, H-20, P-10 and P-30. Loading rates and removal coefficients for the two study areas for three land use conditions and four selected pollutants are shown in Table 3. Water quality modeling results for the storm of 5-08-75 on Hunting Bayou and 12-05-74 on Panther Branch are presented in Table 4. Sample results are also graphically depicted in Figures 5 and 6.

Table 1
HYDROGRAPH MODELING RESULTS, HUNTING BAYOU WATERSHED

Date of Storm	Location	Total Runoff $(ft^3 x10^6)$ Observed	Computed	Peak Flow Rate (cfs) Observed	Computed
9/8/68	H-10	$1.43x10^6$	$1.85x10^6$	121	160
	H-20	$4.48x10^6$	$4.69x10^6$	325	355
9/17/68	H-10	$2.82x10^6$	$2.24x10^6$	144	155
	H-20	$8.37x10^6$	$6.02x10^6$	333	365
11/9/70	H-10	$1.5x10^6$	$0.97x10^6$	85	125
	H-20	$3.5x10^6$	$2.65x10^6$	161	220
3/26/74	H-20	$1.46x10^6$	$1.93x10^6$	40	38
5/8/75	H-20	$1.38x10^6$	$1.32x10^6$	73	80

Table 2
HYDROGRAPH MODELING RESULTS, PANTHER BRANCH WATERSHED

Date of Storm	Location	Total Runoff $(ft^3 x10^6)$ Observed	Computed	Peak Flow Rate (cfs) Observed	Computed
10/28/74	P-10	24.40	29.03	342	360
	P-30	39.34	36.16	376	410
11/10/74	P-10	64.48	53.44	979	600
	P-30	72.87	73.61	897	705
11/24/74	P-10	52.24	57.72	680	645
	P-30	73.70	78.97	774	735
12/05/74	P-10	36.06	32.66	273	315
	P-30	45.52	48.55	329	370
12/10/74	P-10	44.42	33.61	464	380
	P-30	51.73	43.02	517	425

The computed flow peaks and volumes agree well with the observed flows; the average absolute error in the volume of runoff was 15 percent and 26 percent of the observed value for Panther Branch and Hunting Bayou, respectively, while the average error in peak flow prediction was 20 percent of observed.

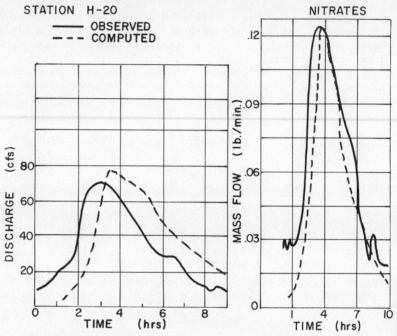

FIG. 5   MODELING RESULTS, STATION H-20, STORM OF
5/08/75

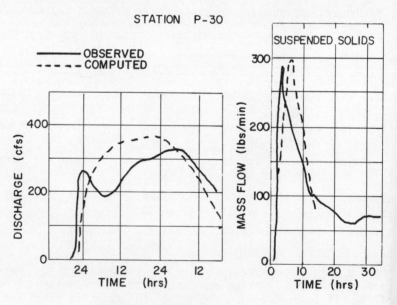

FIG. 6. MODELING RESULTS, STATIONS P-10 AND P-30,
STORM OF 12/05/74.

Table 3
POLLUTANT LOADING RATES

Hunting Bayou Watershed

Suspended Solids	4.00	5.000	0.500	35.0
Total COD	7.800	5.100	5.100	30.0
Nitrates	0.007	0.003	0.003	20.0
Total Phosphorous	0.020	0.004	0.003	23.0

Panther Branch Waterhsed

Suspended Solids	35.0	45.0	2.2	7.0
COD	14.0	7.0	7.5	3.5
Nitrates	0.090	0.015	0.004	5.3
Phosphates	0.022	0.004	0.009	10.0

Note: (1b/ac) x (5.45) = (kg/ha)
Land uses are defined as follows:
Residential - areas with single and multi-family homes
Construction - areas with construction activity
Undeveloped - open undistrubed land

Table 4
WATER QUALITY MODELING RESULTS

	Observed			Computed		
	Peak Conc. mg/L	Peak Mass 1b/min.	Total lbs. x10^2	Peak Conc. mg/L	Peak Mass 1b/min.	Total lbs. x10^2

Station H-20, Storm of 5-08-75

Suspended Solids	430.	114.	162.	1456.	109.	231.
COD	630.	166.	140.	2478.	166.	374.
Nitrates	.66	.124	.40	1.55	.125	.31
Phosphates	1.68	.34	.84	5.01	.304	.75

Station P-10, Storm of 12-05-74

Suspended Solids	130.	27.		75.	27.	
COD	60.	66.		113.	66.	
Nitrates	.092	.045		.105	.043	
Phosphates	.20	.19		.433	.115	

Station P-30, Storm of 12-05-74

Suspended Solids	670.	290.	296.	814.	296.	205.
COD	63.	73.	196.	169.	71.	149.
Nitrates	.276	.13	.08	1.67	.38	.24
Phosphates	.53	.18	.33	.723	.114	.24

Note: (1b/min) x (7.56 = (g/sec)
(1bs) x (2.2) = (kg)

The new water quality modeling scheme previously described was tested on Swale 8 which is a tributary to Panther Branch and the most affected developmental area within the Woodlands. The total Swale 8 drainage area of 194.0 hectares (479.3 acres) was divided into ten subcatchments ranging from 9.3 hectares (23 acres) to 26.71 hectares (66 acres) as shown in Fig. 7. Seventeen drainage system elements were used to model the entire area. Of these, two elemets were storage units, Lakes A and B, (construction of Lake C had not begun). All six channels were trapezoidal in shape as a result of channel enlargement. Land uses for the most extreme subcatchments were classified as undeveloped whereas the last three downstream subcatchments were designated as multifamily residential and commercial.

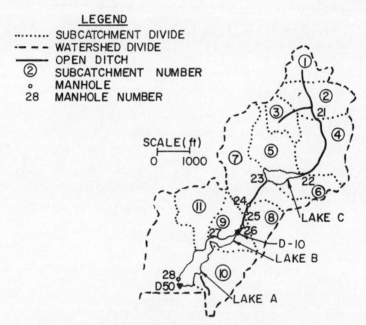

LEGEND

········· SUBCATCHMENT DIVIDE
— — — WATERSHED DIVIDE
———— OPEN DITCH
② SUBCATCHMENT NUMBER
o MANHOLE
28 MANHOLE NUMBER

SCALE (ft)
0   1000

FIG. 7   SUBCATCHMENTS AND DRAINAGE NETWORK – SWALE 8 WATERSHED

The lack of elevation-area-capacity data for the lakes, the ur recorded pumpage of well water into the lakes, and the complex outflow structure for Lake A resulted in extremely difficult modeling conditions at Station D-50 and the modeling results were unsuccessful. Consequently, all further modeling was con ducted only on that drainage area of Swale 8 upstream from Lake B (Station D-10). The pollutant loading rate curve previously discussed was used to derive each of the desired load ing rates listed in Table 5. The storm of 4-08-75 was used t verify the selected loading rates. Observed and computed

Table 5
POLLUTANT LOADING RATES FOR SWALE 8 WATERSHED

Pollutant	Undeveloped	Residential	Construction	Removal
	Loading Rates (lb/ac)			Coefficient
Suspended Solids	31.	101.	123.	4.6
COD	3.3	6.2	3.1	3.0
Nitrates	.009	.016	.006	5.0
Phosphates	.0082	.020	.0037	10.0

Note: (lb/ac) x (5.45) = (kg/ha)
Land uses are defined as follows:
Residential - areas with single and multifamily homes
Construction - areas with construction activity
Undeveloped - open undistrubed land

hydrographs and suspended solids mass transport rates for the
storm of 4-08-75 are compared in Fig. 8. Because the total
hydrograph is composed of two hydrographs resulting from two
distinct periods of rainfall, only the first period of rainfall, runoff, and water quality was modeled.

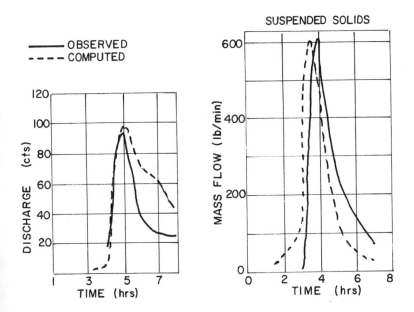

FIG. 8   MODELING RESULTS, STATION D-10, STORM
OF 4/08/75

In comparison to the loading rates determined for Station P-30,
it is evident that the loading rate estimates for developed
areas in Swale 8 would be too low indicating the extreme effects

of lake and golf course construction, as well as channel improvement. These activities were concentrated in the Swale 8
watershed and were diluted in the Panther Branch watershed.
Also, the areas already developed have not been stabilized.
When rainfall intensities are sufficiently high, even the
freshly sodded areas will erode severely; consequently, the
loading rates that were developed for the construction areas
in Swale 8 were much closer than expected to rates for developed areas. The water quality modeling results at Station
D-10 for the storm of 4-08-75 are presented in Table 6. As
can be seen, the peak mass transport rates are adequately
reproduced but the total volumes of mass transport are quite
divergent.

Table 6
WATER QUALITY MODELING RESULTS FOR STATION D-10
Storm of 4-08-75

	Peak Conc. mg/l	Peak Mass lb/min.	Total lbs.	Peak Conc. mg/l	Peak Mass lb/min.	Total lbs.
S. Solids	2152.	608.	1891.	6348.	609.	65290.
COD	87.	31.5	248.	217.	31.3	3539.
Nitrates	2.105	.113	.73	.93	.108	9.
Phosphates	.359	.130	.21	2.32	.129	9.

Note: (lb/min) x (7.56) = (g/sec)
       (lbs) x (2.2) = (kg)

The SWMM was also run for three phased development scenarios
for Swale 8. The storm of 4-08-75 was used to provide a basis
for comparison between future and existing conditions. As
anticipated, the modeling of subcatchment areas 3, 4, and 5
as construction areas, changes the pollutant loads considerabl·
the changes range from an increase of 70 percent for suspended
solids to a decrease of 8 percent for nitrates. After the
construction phase of the development has been completed, the
peak mass flow and total pollutant loads do not decrease as
might be expected, because additional developed areas would
contribute pollutants. These dramatic environmental effects o
construction activities are illustrated in Table 7.

CONCLUSION

The SWMM has undergone extensive evaluation and modification.
It has proved to be applicable in most cases; the only limita·
tions being areas with a transient land use and other areas
where extremely high suspended solids concentrations are gene
rated. Water quality modeling results are much more dependab
and observed events can be adequately simulated. Each of the

Table 7
MODELING RESULTS FOR FUTURE DEVELOPMENT
UPSTREAM OF STATION D-10

	Peak Conc. mg/l	Peak Mass lb/min	Total Pounds
Existing Conditions			
Suspended Solids	6348.	609.	65290.
COD	217.	31.	3539.
Nitrates	.93	.108	9.
Phosphates	2.32	.129	9.
Construction Conditions			
Suspended Solids	5713.	1080.	97163.
COD	231.	35.	4232.
Nitrates	.98	.099	11.
Phosphates	2.34	.133	12.
Developed Conditions			
Suspended Solids	5706.	1289.	95415.
COD	231.	68.	5862.
Nitrates	.98	.214	16.
Phosphates	2.34	.289	21.

storms used to test the new SWMM version were selected to
present a range of flows, water quality, and land use data;
thus the model was tested over a range of different conditions.
The new version of the SWMM can be applied universally but
the modeling results are highly dependent upon the availability
of local data.

The modifications and additions to the SWMM which are discussed in this paper indicate that the modeling of stormwater
quality for the SWMM has been considerably improved. The new
infiltration and baseflow models allow a closer parallel to
the observed hydrograph.

The modeling of smaller subcatchment areas with more definitive
hydrologic regimes will provide a method of evaluation of the
capabilities of the new subroutines and computational methods
in the SWMM. This type of data are presently being accumulated in the Woodlands. The present size and structure of the
SWMM limits the application of the new water quality modeling
methodology described in this paper. Therefore, it is expected that any significant improvement in water quality modeling
by the SWMM will necessitate a complete revision of the

present methodology.

In summary, the SWMM has been a valuable tool in determining the stormwater runoff characteristics of the Woodlands. The quantity of flow has been predicted satisfactorily and the quality of flow is also satisfactory if the local data are available for model calibration. The modeling of quality of flow from areas in a transitional phase of land use is very complex and further detailed data and study are necessary.

ACKNOWLEDGMENTS

This study was supported by Grant Number 802433 from the Storm and Combined Sewer Section, EPA. Partial funds for data collection were provided by the Woodlands Development Corporation.

REFERENCES

Claborn, B. C. and W. L. Moore (1970). Numerical Simulation of Watershed Hydrology, Report No. 54, Center for Research in Water Resources, Austin, Texas

Diniz, E. V. (1976). Quantify the Effects of Porous Pavements on Urban Runoff, Proceedings National Symposium on Urban Hydrology, Hydraulics and Sediment Control, p. 63-70, Lexingto Kentucky.

Izzard, C. F. (1946). Hydraulics of Runoff from Developed Surfaces, Proceedings Highway Research Board, Vol. 26, p. 129-150.

Holtan, H. N. and N. C. Lopez, (1973). USDAHL-73 Revised Mode of Watershed Hydrology, USDA Agricultural Research Service, Plant Physiology Institute Report No. 1.

Metcalf and Eddy, Inc., (1975). University of Florida, and Wa Resources Engineers, Inc. Storm Water Management Model - User's Manual, Version III, U. S. Environmental Protection Agency, Report No. EPA-670/2-75-017, Washington, D. C.

Onstad, C. A. and D. G. Jamieson, (1968). Subsurface Flow Regimes of a Hydrologic Watershed Model, Proceedings Second Seepage Symposium, ARS 41-147, Phoenix, Arizona.

RESULTS OF INVESTIGATION FOR URBAN HYDROLOGY IN HUNGARY

I.Wisnovszky

National Water Authority,Hungary

Urban experimental catchment in Hungary

The accelerating rate of urbanisation has prompted
the establishment of an urban experimental catchment
in 1970. An area of 74.4 hectares magnitude in the
town of Miskolc, Hungary, was selected for this purpose, where the storm runoff is collected by a network of storm sewers. Within the catchment the modern housing development with tall apartment houses,
the district center, the home-and-garden area and
the agricultural meadow area are clearly distinguishable.

From the steeply sloping meadow part the runoff is collected by an open intercepting canal and
conveyed to the storm sewer system with circular
cross-sections ranging from 0.3 to 1.3 m diameter.
A TDG-200 type telemetric rain gage is located close
to the geometric center of the area,whence the rainfall depth in millimetres is transmitted to a 16-
-channel point recorder.

Float controlled, potentiometer type level
gages mounted in four shafts of the storm sewer system transmit the electric signals through cables to
the same point recorder. The gaging system is controlled automatically to start operating whenever
the rainfall depth exceeds 1mm. The system operates
for a preset period, but is started again by a subsequent rain.

The printing and chart forewarding mechanism
is also controlled automatically. Point data are recorded at 12 sec intervals of time, permitting even
extreme variations in rainfall and water levels to
be registered.

The data for the experimental catchment and the four subcatchments have been compiled in Table 1. The storm sewer network and the gaging system are shown in Fig.1.

Table 1. Principal data of the Miskolc experimental urban catchment

	Runoff gages			
	1	2	3	4
Sub-catchment (ha)	26.75	22.66	11.85	13.53
Cover (%)				
Impervious	44.2	20.7	32.5	2.8
Grass,shrubs	53.1	75.1	58.1	97.2
Gravel road	2.7	2.5	0.9	–
Paved road	–	1.7	–	–
Barren	–	–	8.5	–
Total catchment area (ha)	74.68	48.03	25.38	13.53
Cover (%)				
Impervious	27.6	18.5	16.6	2.8
Grass,shrubs	68.6	77.2	79.1	97.2
Gravel road	1.9	1.4	0.4	–
Paved road	0.6	0.8	–	–
Barren	1.3	2.1	3.9	–
Sewer diameter (cm)	130	110	100	60
Sewer slope (0/00)	9.25	10.67	7.86	63.3

The runoff coefficient as a random variable

Completely synchronized rainfall and runoff time series have been recorded regularly at the gaging stations 1, 2 and 3. At the fourth gaging site flow rates sufficiently large to be measured were conveyed in the steep sewer at very violent rainfalls only.

Owing to the sensitivity of the telemetric system several records were discontinuous. Most records have been collected for the subcatchment No.3 having an area of 25.4 hectares. Processing these records on a computer permitted the runoff coefficient to be studied as a random variable. In the course thereof the mean intensity i_a of storm rains and the corresponding peak runoff discharges q_p have

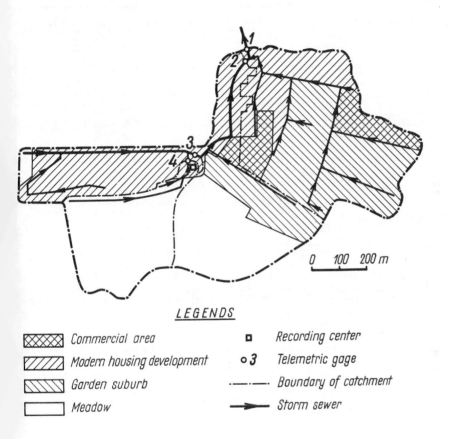

LEGENDS

⊠ Commercial area		▫ Recording center	
⟋⟋ Modern housing development		○3 Telemetric gage	
⟍⟍ Garden suburb		─·─·─ Boundary of catchment	
▭ Meadow		➤ Storm sewer	

Fig. 1.   Urban experimental catchment in Hungary

been determined for the sub-catchment pertaining to
the gages site No.3. The runoff coefficient and the
mean rainfall intensity relationship have been derived therefrom by regression analysis (Fig.2). Assuming runoff to have the same probability of occurrence as the rainfall triggering it, the runoff coefficient as a random variable has been related to
the probability of the mean rainfall intensity in
the next step. Regression analysis has resulted in
the expression

$$C = 0.117 + 0.0686\ i_a + 0.0344\ i_a^2 \tag{1}$$

for the runoff coefficient vs. mean intensity of the
triggering rainfall. Here C (dimensionless) is the
runoff coefficient and i_a (mm/min) the mean intensity of the triggering rainfall.

The correlation coefficient is 0.556. The low
value of the latter is explained by the fact that
the mean intensity of the actual rainfall can be
interpreted in different ways, depending on the duration of the rainfall regarded as the triggering
one.

The relationship shown is considered valid in
the range $0.2 < i_a < 2.0$ mm/min at the experimental
site,but the trend thereof seems to be general. With
accumulation times increasing in length the trend
will be evidently less pronounced than in the relatively small subcatchment of the experimental area.

In Hungary, just as in several other countries, the rational approach has been specified in
standards for the hydraulic dimensioning of storm
sewers. Storms recurring at 1, 2 or 4 year interval
on the average are specified as the basis of computation. However, according to the present investigations the method recommended for estimating the runoff coefficient is suited to the computation of the
runoff coefficient pertaining to the rainfall occurring on the average once every 20 years. This implies further that in the rational approach a rainfall of some probability is transformed into runof
by using a runoff coefficient of a different proba
bility. This is evidently illogical.

The unit hydrograph as a random variable

The rainfall-runoff records from sub-catchment No.
of the urban experimental area in Hungary have be
processed successively to determine the unit hydro
graphs resulting from effective rainfalls of 1 mm/m
intensity. These hydrographs showed marked differ
ences which could be characterized by the peak di

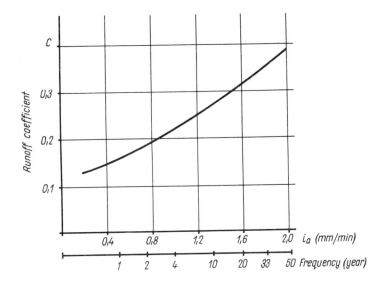

Fig. 2.   Runoff coefficient as random variable.

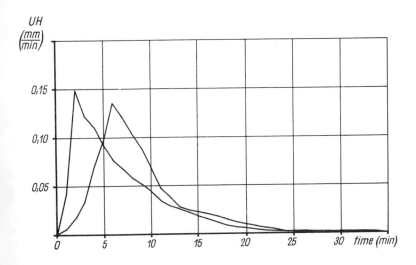

Fig. 3.   Unit hydrographs in the section No. 3.

charge of the hydrograph and the location of the mass center on the time axis, viz. the attenuation. Two of the unit hydrographs pertaining to $i_a = 1.75$ and 0.75 mm/min intensities are shown in Fig.3.

The number of records available is insufficient to arrive at general conclusions therefrom, but is suited to demonstrate the substantial differences in the unit hydrographs due to the mean intensities of different triggering rainfalls in a catchment of 25.4 hectares area and t = 20 minutes accumulation time. The differences are of interest, since according to Fig.3 the unit hydrograph is also a random variable. Attention is called further to the problem of assuming linearity in small catchments.

Model rainfalls in Hungary
Model rainfalls of constant intensity are specified in Hungary for storm sewer dimensioning. The intensity of the model rain is given as

$$i_a = a\ t^{-m} \tag{2}$$

where i_a is in mm/hour units, t (10 min) is the time of accumulation, while a and m are constants, whose values are given in Table 2.

Table 2. Constants for computing the intensity of model rainfalls

Frequency (years)	a (mm/hour)	m
1	47.9	0.69
2	73.1	0.71
4	97.2	0.72

The time series of model rainfalls of variabl intensity has also been determined by the author fo Hungary. The peak intensity thereof was shown to oc cur over 35 per cent of the period examined (1). Thu referring to the period before the onset of the pea intensity as antecedent rain duration and denoting it by t_e, whilst the duration examined is t, the tin ratio of the antecedent rain duration is

$$r = \frac{t_e}{t} = 0.35 \tag{$}$$

Conclusions

Processing the data of storm rainfall and runoff observations in Hungary has demonstrated both the runoff coefficient and the unit hydrograph to be random variables in a sub-catchment having a time of accumulation of t = 20 min. Quantitative and qualitative characteristics have been derived for the variations of the runoff coefficient and the unit hydrograph, respectively, in Hungary.

The $i_a(t)$ relationship of model rainfalls of variable intensity has been explored for Hungary.The results resemble those obtained by Papadakis and Preul (2).

References

1. Wisnovszky,I.: Rainfalls of variable intensity in Hungary (in Hungarian). Hidrológiai Közlöny, No.11, 1977

2. Papadakis,C.N.,Preul,H.C.: Development of design storm hyetographs for Concinnati, Ohio. Water Resources Bulletin, AWRA, April, 1973

HYDRAULIC INSTABILITIES OF STORM SEWER FLOWS

Ben Chie Yen

Professor of Civil Engineering
University of Illinois, Urbana, Illinois   61801   USA

SYNOPSIS

Storm sewer flows are inevitably time-varying (unsteady).
The flow depth may range from dry bed to pressurized pipe
flow. Many stability problems have been observed in routing
storm runoff in sewers. They can be classified into hydraulic instabilities and numerical instabilities. The latter
depend heavily on the routing models (flow equations) and the
numerical schemes used. They are not discussed here. Conversely, the former are the natural stability problems that
are a part of the flow characteristics. The five types of
hydraulic instabilities that have been observed in storm sewer
flows are (1) a near dry-bed flow instability which is a
Weber number flow phenomenon; (2) instability at the transition between subcritical flow and supercritical flow; (3) roll
wave instability which is a Froude number phenomenon occurring
when Froude number approaches or exceeds 2; (4) instability
at the transition between full conduit flow and free surface
flow; and (5) surge instability for full conduit flow. None
of these hydraulic instabilities has been investigated in
details for unsteady flows and they are not very well understood by many engineers. It is not intended to offer
theoretical solutions to these problems in this paper.
Instead, the present understanding of these instabilities are
reviewed in hope of stimulation of interests and possible
further research.

INTRODUCTION

From the hydromechanics viewpoint, flow in storm sewer networks is one of the most complicated hydraulic problems. The
flow is inevitably unsteady, usually nonuniform and turbulent
sometimes with a free surface and sometimes full-pipe
pressurized flow. The sewer pipes and junctions of a network
are hydraulically mutually affected. In the past, steady

uniform flow equations together with a number of other
assumptions were used to describe approximately the flow in
sewers. Stability problems that actually occur in sewer flows
are not reflected in these simplified models.

Recently, because of the increasing expenditure involved in
and pollution control associated with urban storm drainage, a
number of sophisticated mathematical models have been proposed to simulate the storm runoff in sewer systems (e.g., see
Chow and Yen, 1976; Colyer and Pethick, 1976; Yen, 1977).
These models represent more faithfully the physical phenomena
of the flow than the steady uniform flow approximations. But
they are also more difficult to solve. One of the difficulties is the stability problems involved in using these models.
The stability problems can be classified into hydraulic
instabilities and numerical instabilities. The latter are
relatively better known and widely but inadequately studied.
They depend heavily on the flow routing models (mathematical
equations) and the numerical schemes used. For example, the
dry-bed condition of zero depth and velocity imposes a
computational singularity in the finite difference schemes.
Also, by using finite differences to represent the partial
differential equations of the flow, it is well known that
explicit schemes are inherently numerically unstable whereas
certain implicit schemes and method of characteristics are
usually superior.

Conversely, the hydraulic instabilities are a part of the
flow characteristics. Five types of hydraulic instabilities
have been observed in storm sewer flows. They are

(1) A near dry-bed flow instability which is dominated
    by the surface tension effect.
(2) The instability that occurs at the transition
    between subcritical and supercritical flows.
(3) Water surface roll wave instability which is
    dominated by gravity effect and usually associated
    with steep sewer slopes.
(4) The instability at the transition between open
    channel flow and full conduit flow of the sewer.
(5) The surge instability of pressurized pipe flow
    due to interaction between the full sewer
    conduit and connecting manholes or junctions.

None of these hydraulic instabilities has been investigated
in details for unsteady flows, and perhaps except the second
type instability they are not very well understood by most
engineers. Although they may be important only under special
circumstances of sewer flows, they have been experienced by
the author during his research on sewer hydraulics.

In this paper only the hydraulic instabilities are considered.
Zovne (1970) suggested that hydraulically unstable supercritical flows are also numerically unstable. However,
numerical instabilities are not discussed because of the
broadness of the subject and the active on-going researches
that are being conducted by many investigators. No attempt
is made here to offer any theoretical solution to the hydraulic stability problems. Instead, the major purposes of the
paper are to call attention to express the existence of these
stability problems, to review present understanding of the
subject, and hopefully to stimulate interests and future
researches.

RUNOFF IN STORM SEWER NETWORK

Suppose runoff due to a heavy rainstorm of long duration
starts in a circular sewer. Initially the depth is small and
the flow may be laminar or turbulent, depending on the
Reynolds number of the flow. Since initially the velocity is
small, the flow is subcritical, unsteady, nonuniform, open-
channel flow. If the flow changes slowly with time steady
flow backwater curves may be adopted to approximate the free
surface profile. For higher discharges, the flow becomes turbulent if it was originally laminar, and a transition from
subcritical to supercritical flow may occur if the sewer slope
is steep. Conversely, if the flow is supercritical, hydraulic
jump may occur. Qualitatively, sewer flow is similar to culvert flow described by Bodhaine (1968) except that the unsteady effect is more important for sewers. A useful experience for anyone who attempts to model sewer flow is to
construct a small laboratory pipe-manhole unit and observe the
unsteady flow behavior.

Mathematically the open channel phase of the sewer flow can be
described by a pair of partial differential equations of
hyperbolic type (Yen, 1973, 1975)

$$\frac{\partial A}{\partial t} + \frac{\partial Q}{\partial x} = 0 \tag{1}$$

$$\frac{1}{gA} \frac{\partial Q}{\partial t} + \frac{1}{gA} \frac{\partial}{\partial x} \left(\frac{\beta}{A} Q^2\right) + \frac{\partial}{\partial x} (Kh \cos\theta) + (K-K')h \cos\theta \frac{1}{A} \frac{\partial A}{\partial x}$$

$$= S_o - S_f + \frac{1}{\gamma A} \frac{\partial T}{\partial x} \tag{2}$$

The first is a continuity equation whereas the second is a
momentum equation. In these two equations x is the direction
along the sewer invert, t is time, Q is the discharge, A is
the flow cross sectional area perpendicular to x, h is the
flow depth measured normal to x, θ is the angle between the
sewer bottom and a horizontal plane (Fig. 1), $S_o = \sin\theta$ is

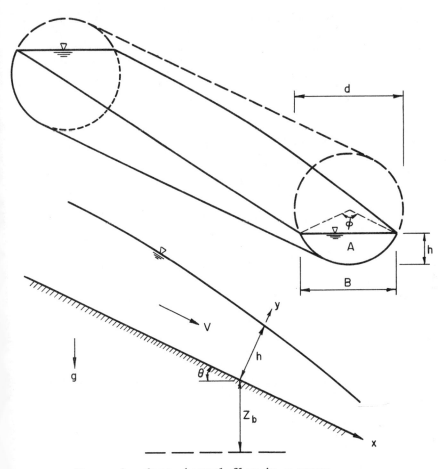

Figure 1. Open-channel flow in a sewer

the sewer slope, S_f is the friction slope (not energy slope),
γ is the specific weight of the fluid, g is gravitational
acceleration, β is a momentum flux correction factor, K and K'
are nonhydrostatic pressure distribution correction factors,
and T represents the force due to internal stresses acting
normally on A. In Equations 1 and 2 the liquid in the sewer
is assumed to be incompressible and homogeneous. For nonhomogeneous fluids the corresponding equations are considerably more complicated and they can be found in Yen (1973,1975).

For the pressurized conduit flow of the sewer, the corresponding continuity and momentum equations can be written as

$$Q = AV \qquad (3)$$

$$\frac{1}{g}\frac{\partial V}{\partial t} + \frac{\partial}{\partial x}\left(\frac{\beta V^2}{g} + \frac{P_a}{\gamma}\right) = -S_f + \frac{1}{\gamma A}\frac{\partial T}{\partial x} \qquad (4)$$

in which V is the cross sectional average   velocity and P_a/γ is the cross sectional average piezometric pressure head with respect to a horizontal reference datum. Equations 3 and 4 can be obtained from Equations 1 and 2 by noting that for a sewer pipe of constant cross section, $\partial A/\partial t = 0$, $\partial A/\partial x = 0$, and $P_a/\gamma = Kh \cos\theta + Z_b$ where $S_o = -\partial Z_b/\partial x$ (Fig. 1).

Theoretically, Equations 1, 2, 3, and 4 together with the appropriate continuity and dynamic equations for the junctions are adequate to describe the sewer flow and sufficient for the investigation of the flow stability problems, provided the coefficients and friciton slope are properly represented. Unfortunately the complete form of the open-channel momentum equation, Equation 2, was not derived until recently and no attempt has been made to investigate the possibility and characteristics of its solutions. Instead, currently most researchers are still investigating the solutions of the Saint-Venant equations which are

$$\frac{\partial h}{\partial t} + D \frac{\partial V}{\partial x} + V \frac{\partial h}{\partial x} = 0 \tag{5}$$

$$\frac{1}{g} \frac{\partial V}{\partial t} + \frac{V}{g} \frac{\partial V}{\partial x} + \frac{\partial}{\partial h} (h \cos\theta) = S_o - S_f \tag{6}$$

in which D is the hydraulic depth equal to A divided by the water surface width. The Saint-Venant equations are approximations of Equations 1 and 2 with the assumption that the flow is gradually varied so that the pressure distribution is hydrostatic leading to $K=K'=1$, and that the gradient of the force due to internal stresses, $\partial T/\partial x$, is relatively negligible. Consequently, their usefulness in hydraulic stability analysis is rather limited.

DRY BED INSTABILITY

This hydraulic instability occurs when the channel bed is nearly dry and the surface tension effect becomes important. It may occur at the beginning or the end of the runoff. Consider, for simplicity, a clean smooth flat plate with a gentle slope. When a small amount of water is spread on it, the water will not form a thin uniform film and flow. Instead, isolated small pools of water will form due to surface tension between water and the plate surface, holding the water from flowing down the slope. As water is supplied continuously onto the plate, the size of the pool increases, relatively the gravity force increases while the surface tension decreases. Soon the pools join each other, a thin film flow starts, the surface tension effect diminishes and the pools disappear.

Equations 1 and 2 are unsuitable for the analysis of this instability because the assumptions involved in deriving

these equations are violated. Particularly the surface
tension is not included. The thin-film flow instability is
important to many chemical and mechanical engineering problems. Hence research work in this topic has been done
mostly by investigators in these two areas. For instance,
one may refer to Javdani and Goren (1972) for film flow
instabilities. Obviously, the Weber number is the most
important parameter whereas the Reynolds and Froude numbers
of the flow also have their roles. The mechanism of formation of the pools is similar to that of breaking up of a
liquid jet into drops. Discussion of the effect of surface
tension on film flow stability can also be found in Yih (1965,
pp. 180-194). However, no results exist that can be adopted
directly to the present problem of sewer flow instabilities.
Fortunately, this instability bears little practical importance for sewer flow partly because it occurs only at the
initial and end of runoff when the discharge is insignificant,
and partly because of the rough sewer pipe surface after a
period of service making the surface tension effect relatively
small and negligible.

SUPERCRITICAL-SUBCRITICAL FLOW TRANSITION

The instability that occurs during the transition from supercritical flow to subcritical flow is perhaps, among the five
types of instabilities discussed in this paper, the most
familiar one to hydraulic engineers. It is commonly known as
a hydraulic jump. Conversely, the transition from subcritical
to supercritical flow is usually called hydraulic drop, and
for unsteady flow it is also called a surge or bore. One can
refer to standard open-channel flow reference books (e.g.,
Chow, 1959; Henderson, 1966) to find the characteristics of
steady flow hydraulic jumps and surges that move at a constant celerity. Most of the information is for flow over wide
channels and not for circular channels.

The subcritical-supercritical transition front of a sewer
flow is not stationary. It may move from upstream toward
downstream or vice versa, depending on the slope of the sewer,
the discharge, and the conditions at the upstream and downstream junctions. If it is moving at a constant speed it can
be analyzed by superimposing a constant velocity equal but
opposed to the celerity of the transition front. However,
usually the celerity varies with both time and space. No
theoretical solution has been obtained for such a transition
flow with a front of variable speed. Rigorously speaking the
Saint-Venant equations are not applicable to this type of
flow because the pressure distribution in the transition region is not hydrostatic. However, existing numerical results
seem to indicate that this region is limited to a small part
in the immediate neighborhood of the transition. Previous
studies on aerodynamic shocks and wave shoaling on beaches may

lend useful information for future analyses. Moreover, for
hydraulic jumps there is a possibility that the sequent subcritical depth is greater than the pipe diamater.

ROLL WAVE INSTABILITY

For fast flowing supercritical open-channel flow, the liquid
near the free surface is moving considerably faster than that
near the bed, generating a free surface instability known as
roll waves in hydraulic engineering and slug flow in chemical
and mechanical engineering. A sketch of a roll wave train is
shown in Fig. 2. Roll waves are caused by the relatively
strong gravity force pulling the liquid to flow rapidly down
slope and by the bed resistance exerted on the viscous liquid.
Therefore, they depend on the Froude number and resistance coefficient or Reynolds number of the flow.

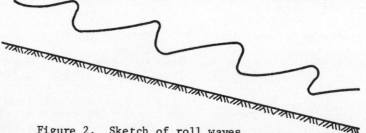

Figure 2. Sketch of roll waves

For a uni-direction flow down a wide plane, roll waves start
to occur at Froude number approximately equal to 1.6. For
circular, rectangular, and trapezoidal channels this stability
critical value is higher because of the sidewall effect.
Mayer (1961) and Brock (1969) described the generation and
propagation of different types of roll waves. Attempts to
establish the stability critical Froude number was reported as
early as half a centrry ago (Jeffreys, 1925). Based on
momentum consideration Iwasa (1954) proposed a stability
criterion for a constant discharge (quasi-steady) flow in a
smooth rectangular channel. Koloseus and Davidian (1966)
evaluated this criterion experimentally. Brock (1970) proposed to use the Saint-Venant equations together with a momentum consideration at the roll wave front to solve supercritical flow with periodic quasi-steady roll waves.

Brock's study demonstrates indirectly that the Saint-Venant
equations by themself are inadequate to represent the high
speed open-channel flow with roll waves. Zovne (1970, p. 54)
also showed that these equations are inapplicable when roll
waves occur. The coefficients β, K, and K' in Equation 2

become highly variable with both space and time, the $\partial T/\partial x$ term is no longer negligible, and the friction slope S_f is altered considerably and unknown at present. For sewer flow the problem may be further complicated by the possibility that the height of roll waves exceeds the size of the pipe, inducing a full-pipe flow.

Roll wave problems are also of great interest to chemical and mechanical engineers. In fact, studies on roll waves in circular pipes have been reported by them but not by civil engineers. They refer to roll-wave flow in pipes as slug flow. Most of these studies consider only laminar flow by solving the Navier-Stokes equations. The gravity effect is expressed in terms of the channel angle θ (Fig. 1) instead of Froude number. Representative literature include those by Dukler (1972), Javdani and Goren (1972), Yih (1965) and Lin (1974).

TRANSITION INSTABILITY BETWEEN OPEN CHANNEL FLOW AND PRESSURIZED CONDUIT FLOW

Sewers may carry pressurized conduit flow although usually they are designed on the basis of free surface flow (gravity flow). Pressurized full pipe flow occurs when the sewer pipe is under design, when the flood exceeds that of the design return period, or when pumping occurs. The major factors that affect this transition instability are the pipe discharge characteristics and the (lack of) air entrainment. Secondary factors include the pipe entrance or exit conditions, hydraulic jump, and the surface waves.

For illustration, consider the simple case of a circular pipe connecting two manholes. For steady uniform flow the discharge-depth and piezometric pressure gradient relationship is shown schematically in Fig. 3. Note that in the open-channel flow phase, the maximum discharge does not occur at depth h equal to the pipe diameter d. Instead, it is approximately at h = 0.94 d, depending slightly on the Reynolds number of the flow. For unsteady nonuniform flow the discharge-depth curve is modified but qualitative unchanged; i.e., maximum discharge of open channel flow occurs at h < d.

Assume that initially the upstream invert of the pipe is not submerged and the depth of the open-channel flow in the pipe is, say, 0.8 d. As the inflow rate into the upstream manhole increases continuously, the open-channel flow rate in the pipe also increases until the maximum discharge, Q_m, is reached with the upstream invert of the pipe unsubmerged. However, a small disturbance or slight increase in discharge will be reflected by a sudden increase in pipe flow depth not only filling up the pipe but also to establish the piezometric pressure gradient in accordance with the discharge.

Correspondingly, there is a rapid change in depth and amount
of water stored in the manholes. The disturbance can be surface waves, flow fluctuation, turbulence, or change of air
pressure in the pipe, and is usually random in nature. As
indicated by the curve HGFE in Fig. 3, for a discharge within
the range between Q_f and Q_m, the discharge-depth relationship
is nonunique, the possiblity of hydraulic instability exists.

Another case of instability is due to inadequate aeration in
the sewer. Consider the simple case of a constant discharge,
Q_e, into the upstream manhole which is equal to the constant
discharge taken out from the downstream manhole. Assume
that at a given instant the discharge into the pipe is
also Q_e corresponding to the water depths in the manholes and
the pipe water surface profile a shown in Fig. 4. Since the
sewer is not ventilated and its downstream part is sealed by
the hydraulic jump, entrainment of the trapped air into the
flowing water creates a low pressure air pocket (a situation
similarly to under ventilated weir or sluice gate). Subsequently, the discharge into the sewer increases while the
depth in the upstream manhole drops. One possibility is that
the higher discharge ($>Q_e$) pushs the hydraulic jump outside
the pipe, allowing air to enter, resulting in atmospheric
pressure for the air in the pipe. Consequently the discharge
drops ($<Q_e$) and the hydraulic jump occurs in the pipe again.
Another possiblity is that the depth in the downstream manhole rises because of the additional inflow, while the air
pocket eventually disappears, resulting in smaller head
difference between the manholes, and hence reduction in pipe
discharge ($<Q_e$). The discharge reduction reverses the change
of the depths in the manholes. If the downstream manhole
depth is reduced below the pipe invert and the upstream manhole depth is raised sufficiently, open-channel flow shown as
profile b in Fig. 4 will occur. If the flow air entrainment
is more than air supply due to restricted downstream opening,
hydraulic jump will reoccur in the pipe and the cycle will
repeat. Actural sewer flow is more complicated because of
flow unsteadiness. No theoretical work has been done on
either of the two types of instabilities just discussed.

SURGE IN FULL CONDUIT FLOW

The surges in full conduit flow of sewers are pressure waves
similar to waterhammer in pipe networks. The surges may be
due to meeting of flood waves from different sewer branches
at the junction, due to sudden surcharge of manholes or pipes,
or due to any other abrupt change of the flow. It has been
observed in many locations that surges of water spilled out
from manholes onto ground surface. The theory to analyze the
surge instability has been developed. It needs only to be
refined and applied to sewer networks. It should be reminded
that since pressure is transmitted immediately the surges

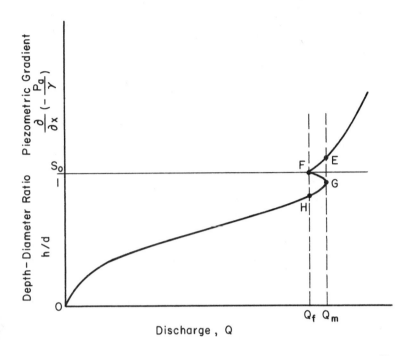

Figure 3. Discharge-depth and discharge-piezometric gradient
relationship for steady flow in circular pipe

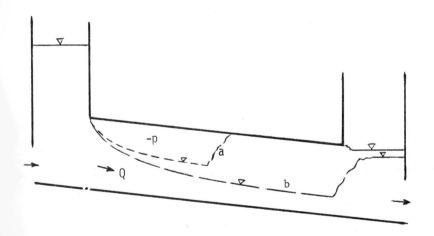

Figure 4. Sewer flow air entrainment instability

in the sewers and manholes are mutually related and there is the possibility of resonance.

ACKNOWLEDGEMENT

This paper is partially supported by U.S. Department of the Interior, Office of Water Research and Technology research project A-086-Ill, Agreement No. 14-31-0001-7029, as authorized under the Water Resources Research Act of 1964, P.L. 88-379, as amended.

REFERENCES

Bodhaine, G.L. (1968) Measurement of Peak Discharge at Culvert by Indirect Methods. Techniques of Water Resources Investigations U. S. Geological Survey, Book 3, Chapter A3.

Brock, R.R. (1969) Development of Roll-Wave Trains in Open-Channels. J. Hyd. Div. ASCE, 95, HY4:1401-1425.

Brock, R. R. (1970) Periodic Permanent Roll Waves. J. Hyd. Div. ASCE, 96, HY12:2565-2580.

Chow, V.T. (1959) Open-Channel Hydraulics. McGraw-Hill Book Co. New York.

Chow, V.T. and B.C. Yen (1976) Urban Stormwater Runoff-Determination of Volumes and Flowrates. Environmental Protection Technology Series EPA 600/2-76-116, U.S. EPA.

Colyer, P.J. and R.W. Pethick (1976) Storm Drainage Design Methods: A Literature Review. Report No. INT 154, Hydraulics Research Station, Wallingford, England.

Dukler, A.E. (1972) Characterization, Effects and Modeling of the Wavy Gas-Liquid Interface. Progress in Heat and Mass Transfer, 6, 207-234.

Goren, S.L. (1962) The Instability of an Annular Thread of Fluid. J. Fluid Mech, 12, 2:309-319.

Henderson, F.M. (1966) Open Channel Flow. The Macmillan Co. New York.

Iwasa, Y. (1954) The Criterion for Instability of Steady Uniform Flows in Open Channels. Memoirs of the faculty of Engineering, Kyoto Univ., 16, 6:264-275.

Javdani, K. and S.L. Goren (1972) Finite-Amplitude Wavy Flow of Thin Films. Progress in Heat and Mass Transfer, 6, 253-262.

Koloseus, H.J. and J. Davidian (1966) Free-Surface Instability Correlations. U.S. Geological Survey Water-Supply Paper 1959-C.

Lin, S.P. (1974) Finite Amplitude Side-Band Stability of a Viscous Film. J. Fluid Mech., 63, 3:417-429.

Mayer, P.G. (1961) Roll Waves and Slug Flows in Inclined Open Channels. Trans. ASCE, 126,Part I, 505-564.

Yen, B.C. (1973) Open-Channel Flow Equations Revisited.   J. Eng. Mech. Div. ASCE, 99,EM5:979-1009.

Yen, B.C. (1975) Further Study on Open-Channel Flow Equations. Sonderforschungsbereich 80 Report SFB80/T/49, Univ. Karlsruhe, Germany.

Yen, B.C. (1977) Hydraulics of Storm Sewer Design. Mini-Course Lecture Note, Proc. Internat. Symp. Urban Hydrology, Hydraulics and Sediment Control, Lexington, Ky. USA.

Yih, S.C. (1965) Dynamics of Nonhomogeneous Fluids. The Macmillan Co., New York.

Zovne, J.J. (1970) The Numerical Solution of Transient Super-critical Flow by the Method of Characteristic with a Technique for Simulating Bore Propagation. Ph.D. Thesis, Civil Eng., Georgia Inst. Tech., Atlanta, Ga.

COMPARATIVE STUDY OF CHARACTERISTIC METHODS FOR FREE SURFACE FLOW COMPUTATIONS

By K.Sivaloganathan

Senior Lecturer in Civil Engineering, Lanchester Polytechnic

INTRODUCTION

This paper deals with the scope and relative accuracies of the characteristic grid method, the rectangular grid characteristic method and a method developed by the writer, called the characteristic method of specified distances, for flow routing in channels of circular section. Past papers on flow routing by characteristic methods have used either the characteristic grid method (Amein 1966, Fletcher and Hamilton 1967, Liggett and Woolhiser 1967, Liggett 1968, Wylie 1970) or the rectangular grid characteristic method (Baltzer and Lai 1968, Harris 1970, Pinkayan 1972, Wylie 1970) and in general dealt with problems such as river flows or overland flows while relatively few have dealt with flows in channels of circular section (Harris 1970, Pinkayan 1972). In the above papers, in the case of river flows conditions have been subcritical throughout the domain of solution and in the case of free outfalls, conditions at the downstream boundary have been assumed to be critical throughout the flow duration. When the downstream boundary is a free outfall, conditions there could in certain cases change from critical to supercritical and vice-versa during the unsteady flow duration. The latter type of boundary condition is of particular interest in relation to flow routing in storm sewers and drains and merits its consideration. While comparison of numerical solutions with physically observed values is necessary to ascertain the validity of a mathematical model, the accuracy of the solutions must be sufficiently ensured before the comparison is made if a valid judgement is to be arrived at. Since the Saint Venant Equations do not in general have analytical solutions, the accuracy of a numerical solution of them cannot be precisely determined. However, useful insight can be gained into relative accuracies of the different methods by comparison of the computed depths at all grid points and by comparison of the extents to which quantity balance is satisfied by them. This has been done for a number of flow routing problems in channels of circular section and it has been possible to draw some general conclusions.

SAINT VENANT EQUATIONS

The Saint Venant Equations exist in various forms. Their derivations and basic assumptions made are available in the literature (Henderson 1966, Stoke 1957). The general form of the Saint Venant Equations may be stated as follows (Liggett 1968):-

$$A \frac{\partial v}{\partial x} + vB \frac{\partial y}{\partial x} + B \frac{\partial y}{\partial t} + v \left(\frac{\partial A}{\partial x} \right)_y = q \tag{1}$$

$$\frac{\partial y}{\partial x} + \frac{v}{g} \frac{\partial v}{\partial x} + \frac{1}{g} \frac{\partial v}{\partial t} = S_o - S - \frac{qv}{gA} \tag{2}$$

in which at a horizontal distance x from the origin at time t y = depth of water, A = area of the water section, B = width of the channel at the water surface, v = the velocity of the water, q = lateral inflow/unit length of channel, S = friction slope, $(\partial A/\partial x)_y$ = rate of variation of A with respect to x when y is held constant, S_o = bed slope at a distance x from the origin, and g = the acceleration due to gravity.

Characteristic Form

Equations 1 and 2 may be expressed in characteristic form as follows (Liggett 1968, Strelkoff 1970):-

$$\frac{1}{c} \frac{dy}{dt} + \frac{1}{g} \frac{dv}{dt} = S_o - S - \frac{q}{gA} (v - c) - \frac{vc}{gA} \left(\frac{\partial A}{\partial x} \right)_y \tag{3}$$

$$\frac{dx}{dt} = v + c \tag{4}$$

$$\frac{-1}{c} \frac{dy}{dt} + \frac{1}{g} \frac{dv}{dt} = S_o - S - \frac{q}{gA} (v + c) + \frac{vc}{gA} \left(\frac{\partial A}{\partial x} \right)_y \tag{5}$$

$$\frac{dx}{dt} = v - c \tag{6}$$

where $c = \sqrt{gA/B}$. The compatibility equation 3 holds along the +ive or forward characteristics given by equation 4 and the compatibility equation 5 holds along the -ive or backward characteristics given by equation 6. The right hand sides of equations 3 and 5 are hereafter referred to as C_f and C_b respectively for convenience.

INITIAL AND BOUNDARY CONDITIONS

It is assumed that there is steady flow in the channel initially. At the upstream boundary, in general the conditions may be subcritical, critical or supercritical. When conditions are subcritical or critical the upstream boundary condition may be given as a discharge hydrograph or a stage hydrograph. When conditions are supercritical both hydrographs are necessary. At the downstream boundary, in general the conditions may be subcritical, critical or supercritical. When conditions are subcritical or critical, a stage discharge relationship exists at the downstream boundary and has to be used.

SCOPE OF METHODS

Figure 1 shows the general pattern of characteristics for a case of flow routing in which conditions at the downstream boundary are always either subcritical or critical. Figure 2 shows the pattern of characteristics for a case in which conditions at the downstream boundary are to start with critical but turn supercritical subsequently. In the latter case, it may be observed that characteristic grid points start to fall beyond the downstream boundary, and computations by the characteristic grid method cannot proceed unless some channel properties are assumed beyond the downstream boundary too. It may also be observed that the number of grid points along forward characteristics originating from the upstream boundary decreases with successive forward characteristics, thus introducing a further difficulty. The characteristic grid method is thus unsuitable when the latter type of downstream boundary condition is involved. The rectangular grid characteristic method as normally formulated (Baltzer and Lai 1968, Streeter and Wylie 1967) is also unable to deal with the latter type of downstream boundary condition. However, the use of an algorithm introduced in this paper enables this method to deal satifactorily with the said boundary condition. The characteristic method of specified distances is able to deal with all types of downstream boundary conditions as will shown in subsequent sections.

FINITE DIFFERENCE SCHEMES

Reference is made to figures 1,3 and 5. LP is a +ive characteristic and RP a -ive characteristic intersecting at P. If the values of y and v at L and R are known, their values at P may be obtained by expressing equations 3 to 6 in finite differences. The following equations result from this exercise:-

$$\frac{(y_P - y_L)}{c_f} + \frac{(v_P - v_L)}{g} = G_f(t_P - t_L) \qquad (7)$$

$$x_P - x_L = (v_f + c_f)(t_P - t_L) \qquad (8)$$

$$\frac{-(y_P - y_R)}{c_b} + \frac{(v_P - v_R)}{g} = G_b(t_P - t_R) \qquad (9)$$

$$x_P - x_R = (v_b - c_b)(t_P - t_R) \qquad (10)$$

If c_f, v_f, G_f are evaluated at L and c_b, v_b, G_b are evaluated at R the resulting set of equations gives the explicit formulation These equations are linear and easily solved. If c_f, v_f, G_f are taken as evaluated using average values $(x_P+x_L)/2, (t_P+t_L)/2,$ $(y_P+y_L)/2$, $(v_P+v_L)/2$ and c_b, v_b, G_b are taken as evaluated using the corresponding average values between P and R, the resulti

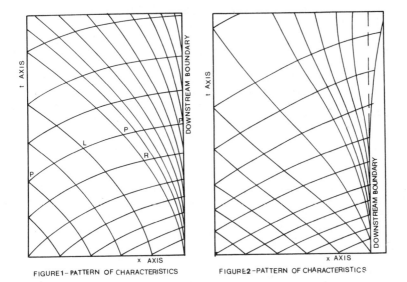

FIGURE 1 – PATTERN OF CHARACTERISTICS

FIGURE 2 – PATTERN OF CHARACTERISTICS

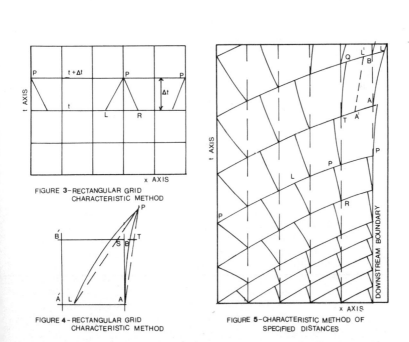

FIGURE 3 – RECTANGULAR GRID
CHARACTERISTIC METHOD

FIGURE 4 – RECTANGULAR GRID
CHARACTERISTIC METHOD

FIGURE 5 – CHARACTERISTIC METHOD OF
SPECIFIED DISTANCES

set of equations give the implicit formulation. These equations are nonlinear in x_p, t_p, y_p, v_p and an iteration procedure is necessary to solve them. The Newton Raphson iteration method can be used. At the upstream boundary, equations similar to equations 9 and 10 and another equation obtained by using the upstream boundary condition are solved for t_p, y_p, v_p. At the downstream boundary, if conditions are subcritical or critical always, equations similar to equations 7 & 8 and the equation obtained from the downstream boundary condition are solved for t_p, y_p, v_p.

CHARACTERISTIC METHODS

The characteristic grid method and the rectangular grid characteristic method, essentially use the finite difference schemes dealt with in the earlier section. These methods, with one boundary condition provided at each end, are well explained in the papers quoted in the introduction. Due to lack of space details of these two methods are excluded from this paper except for the algorithm that enables the rectangular grid characteristic method to deal with changes of the downstream boundary condition from critical to supercritical and vice-versa.

Characteristic Grid Method   This method yields values of y & v on an irregular network of grid points. A procedure for interpolation is necessary to obtain them on a rectangular grid.

Rectangular Grid Characteristic Method   This method yields values of y & v on a system of rectangular grid (figure 3), the time increments for which are obtained at each stage of calculation such that the Courant condition is satisfied at all points on the old time level. The values of y & v on a predetermined rectangular grid are easily obtained. If the downstream boundary conditions change from critical to supercritical and vice-versa during the flow duration, the algorithm that follows is to be used. Reference figure 4. A'A is the old time level and B'B is the new time level. The point L is the intersection of a straight line through B with slope that of the +ive characteristic at A, and the line A'A. x_L is easily calculated. y_L & v_L are calculated by interpolation from values at the grid points on the old time level. The curved line LP represents the actual +ive characteristic through L and the curved line AP represents the actual -ive characteristic through A. The values of x_p, t_p, y_p, v_p are now determined using the implicit formulation assuming the channel properties to be the same at P as at the downstream boundary. The points S & T are the intersections of the straight lines LP & AP with the new time level. The values of x_S, y_S, v_S are determined by linear interpolation from the values at L and P. The values of x_T, y_T, v_T are determined similarly from the values at A & P. The values of y_B, v_B at B are then determined by linear interpolation from the values at S & T.

Characteristic Method of Specified Distances   Figure 5 shows the scheme of characteristics used in this method. The calculations are conveniently performed along successive forward characteristics originating from the x-axis and the t-axis, using the difference schemes detailed in the earlier section. At each stage of proceeding from one forward characteristic to the next , interpolations are done to obtain the values of t,y,v at the specified distances and -ive characteristics are drawn from the latter points to intersect the next forward characteristic. The method yields depth and velocity hydrographs at predetermined distances from which values of y & v on a predetermined rectangular grid may be obtained by interpolation in time. Either the explicit or the implicit formulations may be used. The explicit formulation is recommended with the exception that when the downstream boundary is a free outfall, the last but one point be calculated using the implicit formulation, at each stage of the calculation. In the latter case if the downstream boundary conditions change from critical to supercritical and vice-versa the following procedure is to be adopted. The values of x,t,y,v at L are first calculated using the implicit formulation. A' is marked such that $x_A - x_{A'} = x_L - x_B$. Values of y,v,t at A' are obtained by interpolation from the values at T & A. A'L' is the -ive characteristic through A'. Now using the values of x,t,y,v at A' & Q, their values at L' are obtained. If $x_{L'} > x_B$, the procedure is repeated with the values at L' in place of those at L until $x_{L'} < x_B$. The values of t,y,v at B are now obtained by extrapolation from the values at Q & L'. If the distance $x_L - x_B$ is small, the values at B may be obtained directly by interpolation from values at Q & L.

ACCURACY OF METHODS AND COMPUTATION TIMES

The abbreviations that follow are used in the rest of the paper for convenience. The characteristic grid method is referred to by CGI when implicit formulations are used and by CGE when explicit formulations are used. The rectangular grid characteristic method using explicit formulations, Langrangian interpolation and the new algorithm whenever required at the downstream boundary, is referred to by RCE. Lagrangian interpolation is used, since it yields more accurate results than linear interpolation (Jolly & Yevdjevich 1974). The characteristic method of specified distances using explicit formulations is referred to by NCE. When the downstream boundary is a free outfall, methods CGE & NCE are to use the implicit formulation for calculation at the last but one point at each stage of calculation. A number attached to any of the above abbreviations is to be taken to denote the number of distance steps used in the calculation. For example, CGE40 means the characteristic grid method using explicit formulations and 40 distance steps.

Method of Assessment Values of depth have been obtained on a predetermined rectangular grid, by all the methods and with different numbers of distance steps and compared. In addition, quantity balance has been continuously checked throughout the computations. For this purpose, a parameter called the 'percentage quantity deficit' is defined as $100(W_{in} - W_{out} + V_o - V_t)/W_{in}$ in which W_{in} is the inflow volume and W_{out} the outflow volume up to time t, V_o is the volume of water in the channel initially and V_t the volume of water in the channel at time t.

Standard Method In problems in which the downstream boundary conditions are always critical, the standard for comparison has been taken as method CGI with the smallest practicable sizes of distance steps. Of the methods studied, CGI with small distance steps gives the smallest values for 'percentage quantity deficits'. Further, when the depth corresponding to any point in the x-t plane is plotted against the size of the distance steps used, the gradient of the curve so obtained is observed to be negligible corresponding to small distance steps, showing that the depths obtained for small distance steps are very close to the limits. For the same reasons, in the case of problems in which the downstream boundary conditions change from critical to supercritical and vice-versa, method RCE with the smallest practicable sizes of distance steps has been used as the standard.

Accuracy The percentage by which the depths in a hydrograph exceed the corresponding depths in the hydrograph by the standard method have been determined throughout the computations. Mean values, standard deviations, maximum and minimum values, of the 'percentage quantity deficits' and of the percentage excesses of depths at a number of sections have been computed and studied, and the following conclusions seem appropriate. When the downstream boundary conditions are always critical and the distance steps are small, all the methods yield solutions that are in close agreement. However, when the distance steps are large, method NCE gives the most accurate solution with the other methods following in the order RCE, CGI, CGE. If the downstream boundary conditions change from critical to supercritical and vice-versa, methods RCE and NCE yield solutions that are always in close agreement.

Computation Time In any of the methods, the time taken is appreciably longer if implicit formulations are used than if explicit formulations are used, and there seems to be no corresponding worthwhile improvement in accuracy, except in the case of methods CGI and CGE with large distance steps. In general therefore, explicit formulations are to be preferred. Comparison of the times taken by the explicit formulations of the different methods reveals that RCE takes appreciably longer than CGE or NCE. As an illustration of this, in example 1 in the next section, the times taken by CGI40, CGE40, RCE40, NCE40 are approximately in the ratio 2.37:1:2.34:1.5 .

APPLICATIONS

Two examples of flow-routing are dealt with in this section.

the first example, conditions at the downstream boundary are c-
ritical throughout the flow duration. In the second example, c-
onditions at the downstream boundary change from critical to s-
upercritical and vice-versa. In the first example, depth hydrographs obtained by CGI10,CGI5,CGE40,CGE10,CGE5,RCE40,RCE10,RCE5
,NCE40,NCE10,NCE5 are compared with depth hydrographs obtained
by CGI40. In the second example, depth hydrographs obtained by
RCE10,RCE5,NCE40,NCE10,NCE5 are compared with depth hydrographs
obtained by RCE40. The computations have been done in double
precision on an IBM360/65 machine.

<u>Example 1</u> The channel in which the flow routing is done is of
circular section of radius 1 metre, sufficiently open at the t-
op to permit lateral inflow. Its bed slope is 0.0005 and Manning's constant 0.01. The length of the channel is 20 metres and
it discharges freely at its lower end. A trapezoidal hydrograph
is superimposed at the upstream end on a steady flow of 0.8 cubic metres/second to give :-

$Q(0,t)=0.8$ $0 \leqslant t < 5$; $Q(0,t)=0.8+0.04t$ $5 \leqslant t < 30$
$Q(0,t)=1.8$ $30 \leqslant t < 55$; $Q(0,t)=1.8-0.04(t-55)$ $55 \leqslant t < 80$
$Q(0,t)=0.8$ $80 \leqslant t$

In addition, the following lateral inflow hydrograph is also s-
uperimposed over the whole length of the channel :-

$q=0$ $0 \leqslant t < 5$; $q=0.0024t$ $5 \leqslant t < 30$
$q=0.06$ $30 \leqslant t < 32$; $q=0.06-0.0024t$ $32 \leqslant t < 57$
$q=0$ $57 \leqslant t$

The computations have been carried out up to t=145 seconds. Tables 1 & 2 give comparisons between depth hydrographs by the
different methods. Fig. 6 gives the depth hydrographs by CGI40.

<u>Example 2</u> The channel in which the flow routing is done is th-
e same as in example 1. A trapezoidal hydrograph is superimposed at the upstream end on a steady flow of 0.8 cubic metres/se-
cond to give :-

$Q(0,t)=0.8$ $0 \leqslant t < 5$; $Q(0,t)=0.8+0.048t$ $5 \leqslant t < 30$
$Q(0,t)=2.0$ $30 \leqslant t < 55$; $Q(0,t)=2.0-0.048(t-55)$ $55 \leqslant t < 80$
$Q(0,t)=0.8$ $80 \leqslant t$

The computations have been carried out up to t=145 seconds. Tables 3 & 4 give comparisons between hydrographs by the relevant
methods. Fig. 7 gives the depth hydrographs by RCE40. In this
example, conditions at the downstream boundary are supercritical from approximately t=11 to t=52 seconds.

CONCLUSIONS

The characteristic grid method is unsuitable for dealing with
problems in which the downstream boundary conditions change from critical to supercritical and vice-versa. From the point of
view of reasonable accuracy and computing time over a wide range of sizes of distance steps, the characteristic method of specified distances using explicit formulations seems the best of
the three methods. However, in situations where it is necessary

to use the minimum possible memory storage, the rectangular grid characteristic method as presented herein might be the most suitable.

APPENDIX - REFERENCES

Amein, M. (1966) Stream Flow Routing on Computer by Characteristics. Water Resources Research, Vol.2, No.1: 123-130.

Baltzer, R.A. and Chintu Lai. (July 1968) Computer Simulation of Unsteady Flows in Waterways. Journal of the Hydraulics Division of the ASCE, HY4: 1083-1117.

Fletcher, A.G. and Hamilton, W.S. (June 1967) Flood Routing in an Irregular Channel. Journal of the Engineering Mechanics Division of the ASCE, EM3: 45-62.

Harris, G.S. (June 1970) Real Time Routing of Flood Hydrographs in Storm Sewers. Journal of the Hydraulics Division of the ASCE, HY6: 1247-1260.

Henderson, F.M. (1966) Open Channel Flow. Macmillan Company, New York.

Jolly, J.P. and Yevdjevich, V. (July 1974) Simulation Accuracies of Gradually Varied Flow. Journal of the Hydraulics Division of the ASCE, HY7: 1011-1030.

Liggett, J.A. (Aug. 1968) Mathematical Flow Determination in Open Channels. Journal of the Engineering Mechanics Division of ASCE, EM4: 947-963.

Liggett, J.A. and Woolhiser, D.A. (April 1967) Difference Solutions of the Shallow Water Equations. Journal of the Engineering Mechanics Division of the ASCE, EM2: 39-71.

Pinkayan, S. (Jan. 1972) Routing Storm Water through a Drainage System. Journal of the Hydraulics Division of the ASCE, HY1: 123-135.

Stoker, J.J. (1957) Water Waves. Interscience Publishers Inc., New York.

Streeter, V.L. and Wylie, E.B. (1967) Hydraulic Transients, McGraw Hill Book Co., New York.

Strelkoff, T. (Jan. 1970) Numerical Solution of Saint Venant Equations. Journal of the Hydraulics Division of the ASCE, HY1 223-252.

Wylie, E.B. (Nov. 1970) Unsteady Free Surface Flow Computation, Journal of the Hydraulics Division of the ASCE, HY3:2241-2251

APPENDIX - NOTATIONS

A= area of the water section; B=width of channel at the water surface; $c = \sqrt{gA/B}$; q=lateral inflow/unit length of channel; Q=discharge; R=hydraulic mean depth; S=friction slope obtained from the Manning formula; S_o=bed slope; x=distance from the origin; y=depth of water; v=velocity; t=time; g=acceleration due to gravity; N=Manning's constant; the subscripts and b denote the forward and backward characteristics respectively; G_f and G_b denote the right hand sides of equations 3 a 5 respectively.

Table 1 : Example 1 :

	Percentage excess of depths over CGI40 depths				
Method	Mean Value	Standard Deviation	Maximum Value	Minimum Value	Distance (metres)
CGE40	0.2576	0.1918	0.7762	-0.1135	
RCE40	-0.2734	0.1750	0.1299	-0.9920	
NCE40	-0.2167	0.1383	0.0850	-0.8624	
CGI10	0.2437	0.6951	3.5679	-1.2248	
CGE10	1.2708	0.8944	3.1375	-0.9938	x=0
RCE10	-0.9677	0.6122	0.1688	-1.9194	
NCE10	-0.4821	0.2950	0.6040	-1.0006	
CGI5	0.8344	2.0860	6.7586	-2.7185	
CGE5	6.6057	5.7375	17.1405	-1.4803	
RCE5	-1.8489	1.1078	1.3163	-3.3601	
NCE5	-1.4877	0.8221	0.5657	-3.5194	
CGE40	0.3166	0.1870	0.8243	-0.0326	
RCE40	-0.3651	0.1905	0.0164	-0.9283	
NCE40	-0.3302	0.1991	0.0792	-0.9465	
CGI10	0.1842	0.6390	3.1656	-1.2416	
CGE10	1.2874	0.8240	3.0712	-0.5384	x=12
RCE10	-1.3253	0.6810	-0.0118	-2.3644	
NCE10	-0.9021	0.3742	0.3708	-1.6136	
CGI5	0.4420	1.8939	6.1237	-2.7801	
CGE5	6.5358	5.4532	15.9064	-2.1144	
RCE5	-2.5327	1.1459	-0.0716	-4.4339	
NCE5	-2.2652	0.9095	-0.0493	-4.1725	
CGE40	0.0774	0.1073	0.7612	-0.2906	
RCE40	-0.1538	0.1370	0.0729	-1.0469	
NCE40	-0.1455	0.0984	0.0274	-0.9242	
CGI10	0.1888	0.8926	4.9584	-1.2740	
CGE10	0.3487	0.5511	2.1012	-1.4098	x=20
RCE10	-0.5255	0.3325	-0.0132	-1.4791	
NCE10	-0.3556	0.1959	0.0702	-1.0557	
CGI5	0.6365	2.1994	8.5391	-2.4550	
CGE5	7.6232	8.6385	19.6128	-2.0383	
RCE5	-1.0124	0.6377	0.5069	-2.9521	
NCE5	-0.8322	0.5318	0.8120	-2.2066	

Table 2 : Example 1 :

Percentage quantity deficits

Method	Mean Value	Standard Deviation	Maximum Value	Minimum Value	Distance (metres)
CGI40	-0.0490	0.1336	0.8091	-0.0581	
CGE40	0.2342	0.1057	0.8502	-0.0175	
RCE40	-0.3133	0.1092	-0.0583	-0.4158	
NCE40	-0.2698	0.2107	0.7767	-0.4034	
CGI10	-0.0231	0.4226	1.8845	-0.7055	
CGE10	0.4889	0.4385	2.2784	-0.6192	
RCE10	-1.2014	0.4065	-0.1967	-1.5826	x=20
NCE10	-0.9103	0.3290	0.4060	-1.2651	
CGI5	-0.4573	1.0870	3.7854	-2.1375	
CGE5	1.6179	2.1399	4.1700	-1.8969	
RCE5	-2.3422	0.6102	-0.4585	-2.9338	
NCE5	-2.3706	0.7113	0.1415	-3.4244	

Table 3 : Example 2 :

Percentage excess of depths over RCE40 depths

Method	Mean Value	Standard Deviation	Maximum Value	Minimum Value	Distance (metres)
NCE40	-0.0543	0.1952	0.6518	-1.0109	
RCE10	-0.2166	0.2476	0.1238	-0.8191	
NCE10	-0.3323	0.3408	1.2523	-1.2259	
RCE5	-0.5246	0.6007	0.7511	-1.7793	x=0
NCE5	-0.9557	0.8878	0.9810	-3.1018	
NCE40	-0.0599	0.1897	0.6076	-0.9060	
RCE10	-0.3858	0.3158	0.0004	-1.0018	
NCE10	-0.5105	0.2937	0.2383	-1.1317	x=12
RCE5	-0.9402	0.6781	0.3222	-2.1422	
NCE5	-1.2899	0.7972	-0.0491	-2.8796	
NCE40	-0.0811	0.2354	0.9095	-1.0890	
RCE10	-0.1643	0.1117	0.1298	-0.4075	
NCE10	-0.2524	0.3528	1.8296	-1.1190	x=20
RCE5	-0.3506	0.2489	0.0734	-1.9322	
NCE5	-0.5009	0.4440	1.7732	-1.4105	

Table 4 : Example 2 :

Percentage quantity deficits

Method	Mean Value	Standard Deviation	Maximum Value	Minimum Value	Distance (metres)
RCE40	-0.0536	0.0176	-0.0236	-0.0706	
NCE40	-0.2301	0.1408	0.4327	-0.4641	
RCE10	-0.2047	0.0654	-0.0830	-0.2706	x=20
NCE10	-0.3950	0.0705	-0.0297	-0.5691	
RCE5	-0.4092	0.2109	-0.1067	-1.1034	
NCE5	-0.6771	0.1163	-0.3908	-0.8474	

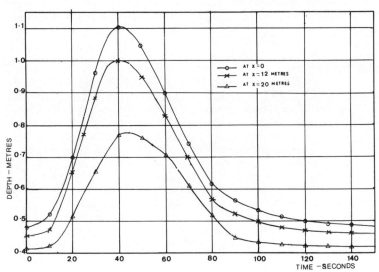

FIGURE 6 - DEPTH HYDROGRAPHS BY CGI40 - EXAMPLE 1

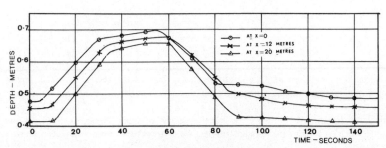

FIGURE 7 - DEPTH HYDROGRAPHS BY RCE40 - EXAMPLE 2

A SURCHARGING MODEL FOR STORM SEWER SYSTEMS

R. Bettess, Hydraulics Research Station, Wallingford
R. A. Pitfield, National Physical Laboratory, Teddington
R. K. Price, Hydraulics Research Station, Wallingford

SYNOPSIS

A simulation model is presented for flow in a storm sewer system. Unlike many previous models the extent of surcharging within the system is unrestricted. The equations of flow in part-full and surcharged pipes together with their appropriate approximations and solution techniques are discussed. Details of the program are described including the data handling and solution algorithms. Results are presented for an application of the model to a small catchment in Derby.

INTRODUCTION

Recently we have become aware of the need of design engineers for a method of simulating flow in storm sewer systems which is not restricted to part-full flow in the pipes but is capable of realistically modelling the system even when some of the pipes are surcharged. A significant proportion of the work of designers of storm sewer systems is involved with the improvement or extension of existing systems and it frequently happens that such systems are liable to surcharge. Hence, to be able to judge the performance of the system and the improvements necessary, the design engineer must have available a model capable of simulating surcharged flow. Also, in a period when more and more designs are having to be justified by a careful cost-benefit analysis the designer may have to consider the performance of his design for storms larger than the design storm. Using a model capable of calculating surcharged flow he may obtain estimates of the amount of surface flooding and associated damage resulting from storm with a larger return frequency than the design storm. We have, therefore, attempted to provide a model which, while remaining as simple as possible, is capable of realistically simulating surcharged flow throughout the system.

One may formulate models of storm sewer systems with varying degrees of complexity from the very simple, like the Rational method (Mulvaney, 1850; Lloyd-Davies, 1906) to the very sophisticated, like the SWM and ISS models (Metcalf and Eddy, 1971; Sevuk et al, 1973). Each model represents a balance between the quantity of data and computation required and the quantity and quality of the results provided. Hence, the choice of an appropriate model involves the consideration of the effort involved to and the necessary data and the nature of the corresponding results obtained. To be of practical use to designers, however, a model must be easy and cheap to use while providing sufficiently accurate and detailed results. We felt that this object has been approached by the TRRL method and our aim has been to provide a model of similar complexity which has comparable data requirements.

If the model was to simulate surcharged flow satisfactorily we considered that it must have certain essential features. We felt that it must be capable of handling any number and configuration of surcharged pipes within the system and, perhaps more importantly, that all the connected, surcharged pipes must be considered simultaneously. Thus the interaction of the flow in one pipe with the flow in other surcharged pipes can be fully accounted for. We believe that we have successfully incorporated these features into this model.

In the present model we have assumed that the pipe system has a tree structure, that is, there are no loops. Modelling of surcharged flow in pipes involves certain problems which are not present or may be ignored when considering only part-full flows, for example, surface flooding, energy losses at manholes and air entrainment in the pipes.

When a pipe becomes surcharged a head develops in one or both manholes associated with the pipe. As long as the water is contained within the manhole there is a straight-forward relationship between storage and depth of water. If water emerges from the manhole, however, the relationship is more problematical. To give a very crude indication of the effects and extent of surface flooding a large inverted cone was imagined super-imposed on every manhole, the vertex coinciding with the top of the manhole so that throughout the cross-sectional area of the manhole is a function of the depth of water. Thus one can calculate a depth of flooding once water has emerged from a manhole. This model assumes that no water is lost through surface flooding and that no water flows over the surface from one manhole to another.

Although some work has been done on energy losses at manholes (Ackers, 1959; Blaisdell and Manson, 1963; Taylor, 1944) the problem is a very complex one and much remains to be done before we can have a full understanding of the subject. The magnitude of the energy losses depends both on the flows involved and the geometry of the manhole. Laboratory experiments are currently being performed to obtain more information on this complicated problem. The full details of this investigation are not yet available but the interim results indicate that the energy loss for a surcharged, straight through, manhole is given by

$$0.2 \frac{v^2}{2g} \, ,$$

where v is the velocity in the downstream pipe. We appreciate that this can only be a provisional estimate for the energy loss and that it may need to be updated as more information becomes available.

Air may be entrained in surcharged pipes at junctions (Yevjevich, 1975). Sediment is frequently found in pipes and is transported by the flow (May, 1977). Both air entrainment and sediment may have the effect of reducing the discharge. We believe, however, that such effects are, in general, small and we have ignored both in the model.

The model of the flow in pipes may be considered in two parts, one part to solve the part-full flow and the other to solve the surcharged sub-systems. To enable the model to take rainfall as input a rainfall-runoff procedure has to be incorporated. We have used rainfall-runoff method developed at the Institute of Hydrology by Kidd (1976), and as applied by Price and Kidd (1978). The following sections describe the pipe flow model.

MATHEMATICAL DESCRIPTION

Equation for part-full flow

It has been shown that an equation of the form

$$\frac{\partial Q}{\partial t} + C\frac{\partial Q}{\partial x} + D\frac{\partial^2 Q}{\partial x^2} = 0 \qquad \qquad ....(1)$$

provides an approximation for slowly varying, unsteady flow in pipes (Bettess and Price, 1976), where x is the distance measured along the pipe, Q is the discharge and t is time.

Equations for surcharged flow

We must now consider the equations describing the flow in a surcharged pipe. Let the heads in the upstream and downstream manholes measured from the same datum be h and h* respectively, then the discharge through the pipe is given by

$$h - h^* = \frac{fL}{4gr}\frac{Q^2}{A^2} \quad , \qquad \qquad ....(2)$$

where r is the radius of the pipe, L the length, A the area and f is given in terms of the roughness coefficient k_s and the hydraulic radius R by the rough turbulent equation

$$\frac{1}{\sqrt{f}} = -2\log_{10}\left(\frac{k_s}{14.83R}\right) \quad ,$$

(Ackers, 1958). Equation (2) may be written as

$$h - h^* = \beta_i Q^2 \quad , \qquad \qquad ....(3)$$

where

$$\beta_i = \frac{fL}{4grA^2} \quad . \qquad \qquad ....(4)$$

Note that an asterisk denotes the value of a variable at the downstream end of the pipe the variable without an asterisk denotes the value at the upstream end. At each manhole we have a continuity equation of the form

$$\alpha_i \frac{dh}{dt} = \sum_{inflow} Q - Q_{outflow} \quad , \qquad \qquad ....(5)$$

where α_i is the cross-sectional area of the manhole.

From Equation (2) the discharge is a function of the square root of the head loss across the pipe, that is,

$$Q = (h - h^*)^{\frac{1}{2}} A\left(\frac{4gr}{fL}\right)^{\frac{1}{2}} \quad . \qquad \qquad ....(6)$$

Therefore, substituting for the discharges in Equation (5) we obtain, for a collection of pipes, a system of ordinary differential equations involving only the heads at each manhole. This system may be solved and then the corresponding discharges calculated using Equation (6).

A simpler system of equations for surcharged flow results if the storage effects of manholes are ignored. If the changes in discharge are on a time scale larger than the response time of the system then effectively the system progresses from one steady state to another. Thus the term $A\frac{dh}{dt}$ in Equation (2) can be assumed to be small and the equation becomes

$$\sum_{\text{inflow}} Q = Q_{\text{outflow}} \quad\quad\quad\quad\quad\quad ....(7)$$

The system of equations may then be solved more easily. Such an approximation, as previously stated, ignores the storage of water in the manholes and hence does not simulate the attenuation of the discharge caused by this effect.

Allowing parts of the system to be surcharged while the flow in the remaining pipes is part-full poses other problems. There is an essential difference in the mathematical equations which describe part-full and surcharged flow in pipes. The equations describing part-full flow are hyperbolic in character. These equations involve real characteristics along which information about the flow is propagated. Equation (1) has one set of forward characteristics and so, for this approximation, any changes in flow can only propagate down the pipe, that is, there can be no upstream influence or backwater effects. Information about changes in flow travels down the pipe at a finite velocity dependent upon the slope of the characteristic. Hence, one is able to solve the equations for part-full flow by going down the system pipe-by-pipe since the flow in any pipe depends only on the flow in the pipes upstream and not at all on the flow in the downstream pipes. In contrast the equations for surcharged flow, as described above, form a system of ordinary differential equations and are more akin to an elliptic system. Any change in one part of a surcharged system is propagated instantaneously throughout the system. Thus a surcharged group of pipes cannot be solved pipe-by-pipe but must be solved simultaneously.

If part of a pipe system is surcharged we must solve one set of equations for those parts of the system where the flow is part-full and another set of entirely different equations where the pipes are surcharged. The extent of the surcharged sub-systems continually changes in time as the discharges vary and is dependent upon the solution obtained. The problem thus has certain similarities with free-surface problems where different equations are applicable on either side of the boundary.

Transition between part-full and surcharged flow
We assume that initially the flow in all the pipes is part-full and the heads at all the manholes are zero. Discharges are then calculated throughout the system at successive time steps. At every time step the upstream discharge of each pipe is inspected and if it exceeds the pipe-full discharge that pipe is assumed to be surcharged. When a system of surcharged pipes has been isolated, the appropriate equations for surcharged flow are solved for the system.

When one pipe surcharges the effect of setting a depth in the upstream manhole may cause the next pipe upstream to surcharge. Hence, on traversing down a branch, when the first surcharged pipe is found the pipes immediately upstream are inspected. If

$$Q^2 \frac{fL}{4\pi^2 gr^5} - h^* > 0 \quad\quad\quad\quad\quad ....(8)$$

for the upstream pipe that pipe is assumed to be surcharged and the head upstream is taken to be

$$h = \frac{Q^2 fL}{4\pi^2 gr^5} - h^* \qquad . \qquad \qquad ....(9)$$

Thus the effect of a blockage or restriction in the system can gradually propagate up the system.

When the head at the upstream end of a surcharged pipe becomes less than or equal to zero, the flow in the pipe is assumed to be part-full.

NUMERICAL APPROXIMATIONS

Solution of equation for part-full flow

It was decided to use the Muskingum-Cunge finite difference scheme for calculating part-full flow. This is a scheme first developed by the U.S. Army Corps of Engineers (U.S. Army Corps of Engineers, 1940) and McCarthy (McCarthy, 1938) for flood studies in the Muskingum River Basin and incorporates improvements suggested by Cunge (Cunge, 1969) In a study at the Hydraulics Research Station this scheme was compared with a number of other finite-difference schemes for calculating flow in pipes and it provided the most acceptable combination of speed and accuracy (Bettess and Price, 1976).

The Muskingum-Cunge method uses a finite difference scheme of the form

$$\frac{a(Q_{j+1}^{n+1} - Q_{j+1}^n) + (1-a)(Q_j^{n+1} - Q_j^n)}{\Delta t} + \frac{C}{\Delta x}\{b(Q_{j+1}^{n+1} - Q_j^{n+1}) + (1-b)(Q_{j+1}^n - Q_j^n)\} = 0, ...(10)$$

where Q_j^n denotes $Q(j\Delta x, n\Delta t)$, Δt and Δx being the time and space steps respectively and C is the wave speed. This may be regarded as a finite difference approximation to Equation (1) if the parameters a and b are chosen appropriately (Cunge, 1969; Bettess and Price, 1976).

Solution of equations for a surcharged sub-system

We shall now extend Equations (2) and (5) to a set of ordinary differential equations describing the flow for a surcharged sub-system of m pipes; that is, a connected system of pipes which are all surcharged and for which the pipes immediately upstream and downstream of the sub-system are only part-full.

First we consider the storm sewer system as a tree structure of n nodes, where each node is defined as a manhole and the pipe immediately downstream of that manhole. We now insist that the nodes are given in end-order, (Knuth, 1972 and Fig 1), and define the vectors $\mathbf{Q}$, $\mathbf{Q^*}$, $\mathbf{h}$ and $\mathbf{h^*}$ to contain the elements Q_i, Q_i^*, h_i and h_i^*, as used in Equations (2) and (5), respectively. Using this same ordering, we denote by $\mathbf{P}$ the vector whose elements are the discharges from the area contributing directly to each manhole.

Next we define the system structure matrix, E_{nn}, by:—

$$E_{ij} = \begin{cases} 1 & \text{if node j is immediately upstream of node i} \\ 0 & \text{otherwise} \end{cases} \qquad ....(1$$

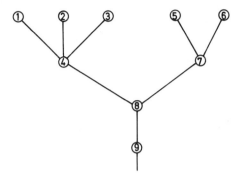

An example of a storm sewer system of 9 nodes
with the nodes numbered in end-order

FIG 1

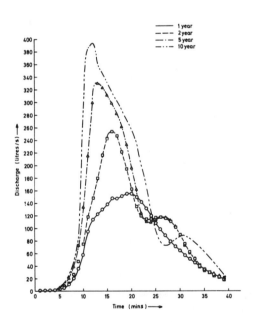

Discharge hydrographs at outfall
for 1,2,5 & 10 year return periods

FIG 2

Thus the discharge into manhole i is given by:—

$$P_i + \sum_{j=1}^{n} E_{ij}Q_j^*. \qquad ....(12)$$

The elements of a surcharged sub-system of m nodes are described by the matrix M_{mn} where

$$M_{ij} = \begin{cases} 1 \text{ if pipe j is the ith surcharged pipe when the sub-system is traversed in end} \\ \quad \text{order.} \\ 0 \text{ otherwise.} \end{cases} \qquad ....(13)$$

Using this notation we may rewrite Equation (5) for a sub-system of surcharged nodes as

$$M\alpha\frac{dh}{dt} = MP + MEQ^* - MQ \quad , \qquad ....(14)$$

where α is the diagonal matrix whose diagonal elements are α_i. The first term on the right-hand side represents the input from each manhole's own contributing area, the second term is the discharge into each manhole from the upstream pipes and the final term is the discharge from each manhole into the downstream pipe. The second term contains discharges from both surcharged and part-full pipes and may be written as

$$MEQ^* = ME(I_n - M^TM)Q^* + MEM^TMQ \quad , \qquad ....(15)$$

where the first term is the discharge from the part-full pipes and the second term is the discharge from the surcharged pipes; note that $Q_i = Q_i^*$ if pipe i is surcharged.

Finally, if we use Equation (6) to eliminate Q, and define $y = Mh$ and $y^* = Mh^*$, the set of ordinary differential equation (14) becomes

$$\frac{dy}{dt} = \alpha^{-1} [MP + ME(I_n - M^TM)Q^* - (I_m - MEM^T)z] \quad , \qquad ....(16)$$

where z is the vector whose elements are given by

$$z_i = [(y_i - y_i^*)/\beta_i]^{1/2}. \qquad ....(17)$$

Equation (16) is a set of first-order ordinary differential equations whose solution at time t is known and for which we require the solution at time $t + \Delta t$. Since the range of integration is small we have used the Runge-Kutta-Merson algorithm (Fox, 1962) to calculate the solution at $t + \Delta t$, because, as it is a single step algorithm, it does not require any special starting procedure. The computation speed is improved by making the transformation $t = u^2$.

The steady state solution of Equation (16) is

$$(I_m - MEM^T)z = MP + ME(I_n - M^TM)Q^* . \qquad ....(18$$

This solution is frequently attained before time $t + \Delta t$ and since it requires very little computation we compare the current solution after each step of the Runge-Kutta-Merso algorithm with the steady state solution to determine whether we need proceed with the integration.

Lastly we note that the discharges Q_i for the pipes in the surcharged sub-system are given in the vector z of Equation (16).

PROGRAM DATA STRUCTURE

In order to perform the numerical computation we require the following fixed data for each node:—

(i)   the three Muskingum-Conge coefficients;
(ii)   the coefficients α_i and β_i of the surcharged equations;
(iii)   the invert level of each manhole relative to a fixed base height;
(iv)   the pipe-full discharge; and
(v)   the surface area contributing directly to the pipe.

Using this data the program computes the following for each node:—

(vi)   the discharges at the upstream and downstream ends of the pipe; and
(vii)   the height of the water surface level in each manhole.

These data and results are stored in nine one-dimensional arrays arranged so that the elements are in end-order.

The program requires a one-dimensional work space and an integer array of pointers giving the address of the parent of each node, that is, the node immediately downstream. We also use a dummy node at the end of the system which represents a manhole that is never allowed to surcharge.

The program uses a stack, (Knuth, 1972), to store temporarily the addresses of nodes in a surcharged sub-system. The bottom element of the stack is the fixed constant $n + 1$, where n is the number of nodes in the storm sewer system.

METHOD OF COMPUTATION

At the beginning of every time step, the discharge into each manhole from its own contributing area is evaluated and stored in the work array. This array is later used to store the total discharge into each manhole. We then process the data for each node sequentially thus traversing the storm sewer system in end-order.

The node is first tested, using the inequality (8), to see if it is affected by surcharging in the node immediately downstream, whose address is given in the integer array of pointers. If it is affected we use Equation (9) to compute the new water level in the manhole. By performing this operation at the start of each time step we avoid having to traverse the storm sewer system in an upstream direction.

The program then determines whether the current node is surcharged; that is, either the discharge through its pipe exceeds the pipe-full discharge, or there is a head of water in the upstream manhole. If it is not surcharged we use the Muskingum-Cunge method to evaluate the discharge at the downstream end of the pipe. This discharge is then added to the element in the work array for the downstream node to give the discharge into the next downstream pipe. The discharge at the upstream end of the pipe is set to the total discharge into the upstream manhole. This value is available in the work array.

If the node is surcharged we put its address on the top of the stack and inspect the downstream node to test whether it is also surcharged. If it is we proceed to the next node in the list.

If we have just put the address of a surcharged node on the stack and the next downstream pipe is not surcharged, the addresses of an entire surcharged sub-system are available at the top of the stack. We can discover the number of nodes in this sub-system by inspecting the pointer of each node in the stack, starting at the top of the stack. If the sub-system contains m nodes the (m + 1)th node in the stack will be the first whose pointer is greater than the address of the top node in the stack.

Before describing the next two operations we consider Equation (16) in more detail. The vector $MP + ME(I_n - M^TM)Q^*$ is the discharge into the sub-system from all the pipes immediately upstream of the sub-system, and from the runoff into each manhole from its own contributing area. The values of these elements are given in the work array. The elements of the matrix $I_m - MEM^T$ are given by:—

$$(I_m - MEM^T)_{ij} = \begin{cases} 1 \text{ if } i = j \\ -1 \text{ if } y_i \equiv y_j^* \text{ as defined by Equation (17)} \\ 0 \text{ otherwise.} \end{cases} \qquad ....(19)$$

If $y_j^* \equiv y_i$ we know that $j < i$, and hence the matrix is unit lower triangular.

The steady state solution, given by Equation (18), is computed by performing a forward substitution to determine z and then using Equation (17) to evaluate y, and hence h. When executing the forward substitution we traverse the sub-tree, corresponding to the surcharged sub-system, in end-order. This is done by starting at the mth node of the stack and proceeding to the top node of the top node of the stack. To evaluate y from Equation (17) we start at the first element in the stack and proceed to the mth element.

Finally, the Runge-Kutta-Merson algorithm for solving the set of ordinary differential equations requires repeated evaluations of the right hand side of Equation (16) for given vectors y. These are evaluated by traversing the tree once again in end-order. We first start with the vector $F = MP + ME(I_n - M^TM)Q^*$, then calculate each element of z, adding or subtracting these elements from the elements of F as required.

RESULTS

The model was run on the St Mark's Road catchment, Derby. This is a small, urban catchment of about 10.4 ha with an average slope of 1/200. The pipe network consists of 81 pipes, varying in diameter from 152 mm to 533 mm. The pipe slopes are, in general, small and the surcharging of pipes is comparatively frequent. Since the system is flat, surcharging in one pipe may influence flow in many other pipes and the amount of the system that is surcharged may be extensive.

The rainfall inputs used were 15 minute, 50 percentile storm profiles obtained from the Meteorological Office. Storms of return periods 1, 2, 5 and 10 years were studied employing a time step in the model of 1 minute. The corresponding peak flows calculated in the pipes need not, however, have the same return periods. The relationship between the return periods of flow and of the input variables (rainfall depth, duration, profile and catchment wetness) is the subject of current research; see Price and Kidd, 1978. Figure 2 shows the discharge hydrographs at the outfall. It will be noticed that the time to peak decreases as the return period increases. For storms with larger return periods the rapid rise of the hydrograph poses certain problems. There is a sharp increase in the number of surcharged pipes, the number increasing from 0 to 42 in 5 minutes for the 10 year storm. In such cases it may be better to use a shorter time step though this would naturally involve more calculations. The problem does not occur for longer storms where the system does not change so rapidly. The bump on the recession of three of the hydrographs at a discharge of about 100 litres/s may be due to technical problems associated with the cessation of surcharging in the system.

This is only an interim report and we are actively engaged on improving this and other models. We envisage the model forming part of a comprehensive design-simulation package. After a system has been designed using methods such as those proposed by Price and Kidd (1978) or Price (1978), the present model would be used to indicate the frequency and extent of surface flooding. On the basis of such results and associated construction and damage costs the designer would have the option to redesign the system for alternative return periods. This would give the planning authority a more rational basis for the selection of an appropriate design.

ACKNOWLEDGEMENTS

The work described in this paper formed part of the research programme of the Hydraulics Research Station, and is published with the permission of the Director.

REFERENCES

Ackers, P. 1958 Resistance of fluids flowing in channels and pipes. Hydraulics Research Paper No 1, Hydraulics Research Station, Wallingford. HMSO.

Ackers, P. 1959 An investigation of head losses at sewer manholes. Civil Engineering and Public Works Review **54** pp 882-884 and pp 1033-1036.

Ackers, P. and Harrison, A. J. M. 1964 The attenuation of flood-waves in part-full pipes. Proc. I.C.E. **28** pp 361-381.

Bettess, R. and Price, R. K. 1976 Comparison of numerical methods for routing flow along a pipe. Hydraulics Research Station Report IT 162.

Blaisdell, F. W. and Manson, P. W. 1963 Loss of energy at sharp-edged pipe junctions in water conveyance systems. Agricultural Research Station, U.S. Dept. of Agriculture, Technical Bulletin No 1283.

Cunge, J. A. 1969 On a flood propagation computation method. J. Hyd. Res. 7 No 2, pp 205-230.

Fox, L. 1962 Numerical solution of ordinary and partial differential equations. Pergammon, New York.

Kidd, C. H. R. 1976 A non-linear urban runoff model. Institute of Hydrology Report No 31.

Knuth, D. E. 1972 The art of computer programming, Volume 1 Fundamental algorithms. Addison-Wesley.

Lloyd-Davies, D. E. 1906 The elimination of storm water from sewerage systems. Proc. I.C.E. **164** No 2 pp 41-67.

McCarthy, G. T. 1938 The unit hydrograph and flood routing. U.S. Engineer School, Fort Belvoir, Virginia, 1940. Presented as paper at Corps of Eng. Conference 1938.

May, R. W. P. 1977 The transport of sediment in sewers. WRC Conference on Opportunities for Innovation in Sewerage, Paper 17, Reading.

Metcalf and Eddy Inc et al 1971 Environmental Protection Agency storm water management model. Environmental Protection Agency, Washington. Report 11024.

Mulvaney, T. J. 1850 On the use of self-registering rain and flood gauges in making observations on the relation of rainfall and of flood discharges in a given catchment. Trans. I.C.E. Ireland **4** No 2, p 18.

Price, R. K. 1978 Design of storm sewers for minimum construction cost. International Conference on Urban Storm Drainage, Southampton.

Price, R. K. and Kidd, C. H. R. 1978 A design and simulation method for storm sewers. International Conference on Urban Storm Drainage, Southampton.

Sevuk, A. S. et al. 1973 Illinois storm sewer system simulation model — user's manual. University of Illinois Water Resources Centre, Research report 0073.

Taylor, E. H. 1944 Flow characteristics at rectangular open channel junctions. Trans. A.S.C.E., v, 109, pp 893-903.

U.S. Army Corps of Engineers 1940 Flood control. Eng. Construction Text X-156.

Yevjevich, V. 1975 Storm drain networks in Unsteady flow in channels. Volume II. Edited by K. Mahmood and V. Yevjevich, Water Resources Pub. 1.

CAN DETAILED HYDRAULIC MODELLING BE WORTHWHILE WHEN HYDROLOGIC
DATA IS INCOMPLETE ?

G. Chevereau,   SOGREAH, Grenoble, France
F. Holly,   SOGREAH, Grenoble, France
A. Preissmann,   SOGREAH, Grenoble, France

INTRODUCTION

In the course of developing calculation schemes for unsteady
free surface flow in rivers, SOGREAH anticipated from the
start the application of these methods to urban sanitary and
storm drain systems. In view of the physical difficulties
posed by conduits which can pass from free surface flow conditions into pressurised flow and back again, we planned on
taking this phenomenon into account (Cunge and Wegner, 1964).

For a decade or so the commercial development of drainage
system calculations using digital computers was considered unwarranted. This was especially true since, in contrast to most
of the river systems modelled up to that point, sewer systems
are in general looped (multiply connected), a feature which
requires rather delicate programming of the implicit scheme
used by SOGREAH.

It was only after observing that certain other organisations
had developed sewer system calculation programmes that we
decided to develop our own, a procedure known as CAREDAS,
which calculates unsteady free surface and pressure flow in
looped sewer system networks using the full de Saint Venant
flow equations.

The calculation algorithm, based on an implicit finite difference method (Cunge and Wegner, 1964) is without restriction
as to the number of nodes and direction of flow, and accepts
hydraulic features such as flooded and free-flowing weirs,
pumping stations, pressure-actuated flap gates, retention
basins, etc. The development effort has been focused on a
correct representation of flow in the conduits rather than
on the determination of the hydrographs entering the network
following rainfall on a runoff basin.

317

In the early applications of CAREDAS we became aware that :
. the authorities responsible for drainage systems do not in
  general know the discharges entering the system or the discharges in conduits within the system following a rainfall
  event;
. the determination of elementary runoff basin hydrographs in
  urban areas is of sufficient interest to have been the subject of numerous recent publications. For a given rainfall
  event, the application of the rainfall-runoff relationship
  proposed by different authors leads to quite different
  hydrographs (Brandstetter, 1975).

Thus we found ourselves in the situation of having a sophisticated and precise tool for the calculation of flow in drainage
network conduits, but lacking equally precise information on
the discharges entering the network.

In the first application of CAREDAS in France we adapted a
proprietary hydrologic procedure for the sizing of pipes, the
use of which was required by the client. This method, which
yields the maximum discharge leaving an elementary runoff
basin for a predetermined rainfall was applied to several rainfall events observed in the past. We were fortunate enough to
have had a means of approximate verification of the assumed
input hydrographs, in that during severe rainstorms drainage
system overflow zones were identified; it was a question of
seeing if the mathematical model was capable of approximately
reproducing the location and extent of these overflow zones.
It was apparent from the first trial calculation that the
rigorous application of the imposed hydrologic method yielded
input discharges which were too great, causing excessive overflow. By reducing the runoff coefficients, we were able to
obtain a quite satisfactory reproduction of the observed flooded zones. One may ask if it was not the hydraulic model, rather
than the hydrologic input, which should have been adjusted; but
the invert elevations, roughness coefficients, and hydraulic
characteristics of the network were known with sufficient
accuracy to preclude any significant adjustment of the hydraulic mathematical model.

We do not yet have a definitive procedure insofar as the rainfall-runoff transformation in urban watersheds is concerned.
We use procedures which are accepted or imposed by the client
and we try insofar as possible to reproduce observed events.

Thus the question to be answered is the following : given the
uncertainty in the urban rainfall-runoff transformation, can
it still be useful to make use of a sophisticated mathematical
model of unsteady flow in the sewage network? Before attacking
this question, it may be of some interest to examine the
influence of the various rainfall-runoff transformation parameters on the behavior of a drainage network.

BRIEF REVIEW OF THE RAINFALL - RUNOFF TRANSFORMATION

Elementary runoff basins (that is to say, basins which are
drained by a single inlet to the drainage network) are generally relatively small, the order of 1000 m2 to 10,000 m2. The
rainfall-runoff transformation can be described by, among other
methods, the unit hydrograph approach, as sketched below :

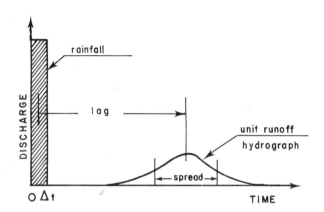

Figure 1. Rainfall-Runoff Transformation

The procedure consists in adopting a unit hydrograph which
results from a rainfall of short duration (one minute, for
example), and then calculating by convolution the hydrograph
which results from a given rainfall distribution. It is, then,
the characteristics of the unit hydrograph which determine
the rainfall-runoff transformation.

The most important characteristics of the unit hydrograph are :
. the runoff coefficient : this is the ratio of the volume of
  rainfall falling on the basin to the volume of runoff;
. the lag : this the time lapse between the centres of mass
  of the unit rainfall distribution and the outflow hydrograph;
. the spread : this is a measure of the "width" of the unit
  hydrograph (for example, twice the standard deviation).

It is evident that the runoff coefficient is of utmost importance. Its value depends on the nature of the terrain in the
basin, but also on the antecedent meteorological conditions.
Rainfall events which are liable to cause a drainage system
to function under pressurised conditions are generally of
high intensity but short duration, such that water retention

on supposedly impermeable surfaces (roof tiles, streets, etc.)
and in puddles can be significant. Thus it is necessary to
know if this retention capacity is in fact available, and this
depends on the preceding meteorological conditions.

On the other hand, it is not immediately evident that it is
necessary to precisely reproduce the lag and spread of the
unit hydrograph, at least for the dimensioning of large pipes
in a drainage system. In fact, the spread and lag of the hydrograph from an elementary basin are the order of a few
minutes, such that their effect on the hydrograph resulting
from a 30-minute rain, if perceptible, is not of great importance.

In the study of large drainage systems one does not try to
model all the pipes and elementary drainage basin inflows
One models the larger pipes, grouping the elementary runoff
basins into so-called computational sub-basins whose runoff
is introduced at one point on a pipe which is represented in
the mathematical model. In these computational basins, in
contrast to elementary basins, the runoff occurs simultaneously as overland flow and flow in smaller pipes which are not
represented in the mathematical model.

Due to the fact that the lag and the spread depend on the
distance of each elementary runoff basin from the sub-basin
outflow point, the rainfall-runoff transformation for the
sub-basin is characterized by a spread and lag which are greater than those for an elementary basin. The runoff coefficient,
on the other hand, is the same for the two cases. For a sub-
basin which is composed of thirty elementary basins or so,
one can expect that the spread and lag of the output unit
hydrograph will be increased by only a few minutes compared
to an elementary basin. Therefore in order that the overall
drainage system calculation results be relatively independent
of errors in the sub-basin unit hydrograph spread and lag, it
is necessary that :
. the spread error be small compared to the rainfall duration,
. the lag error be small compared to the time of propagation
  of discharge in the drainage network model.

We have performed two example calculations using CAREDAS to
illustrate some of the above points. Figure 2a shows the network adopted; a drainage surface of 300 hectares is divided
into 48 computational sub-basins, pairs of which feed surface
runoff into collector pipes at 24 points as shown. The collectors A-01 to JUNC, B-01 to JUNC, and JUNC to WEIR have an invert slope of 0.1 %, with diameters of 2.0 and 2.5 meters as
shown. Storm waters leave the system by passing over a free-
flowing rectangular weir with a four meter crest at elevation
10.5 meters.

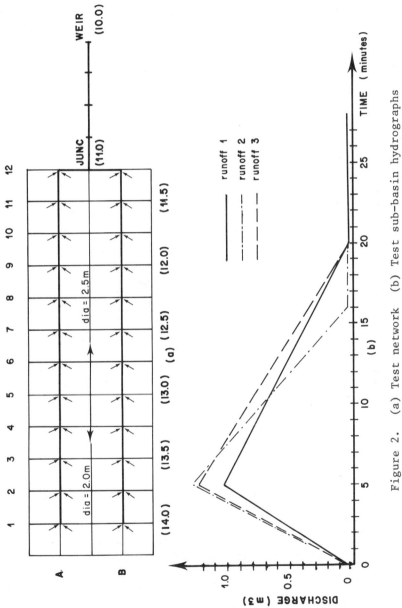

Figure 2.   (a) Test network   (b) Test sub-basin hydrographs

For the demonstration calculations we made no attempt to apply a rainfall-runoff transformation to the runoff basin, choosing instead an assumed hydrograph entering the network at each of the 24 points. Figure 2b shows the simplified triangular hydrograph adopted for the base calculations, labeled "runoff 1".

The first comparative calculation shows the effect of a change in the shape of the individual hydrographs, preserving the same total inflow volume. The hydrograph labeled "runoff 2" on Fig. 2b is constructed from the base hydrograph by increasing the peak discharge 25 % and reducing the spread from 20 minutes to 16 minutes. The effect of this change is seen on Fig. 3, where the water level and discharge at the point JUNC are plotted for the base calculation (runoff 1) and the modified hydrograph (runoff 2). It is evident that neither the water level nor the discharge at the point JUNC are significantly changed by the change in shape of the elementary input hydrographs.

The second comparative calculation shows the effect of a change in the overall runoff coefficient. The hydrograph labeled "runoff 3" in Fig. 2b is the base hydrograph with the peak discharge increased by 20 %; thus the runoff volume, or equivalently the runoff coefficient, is increased by 20 %. The curves labeled "runoff 3" on Fig. 3 show that at the point JUNC the peak discharge is increased by nearly 20 % and the peak water level is increased about 20 cm. Thus from these calculations it is clear that the water level and discharge in the downstream portion of the drainage network are much more sensitive to the runoff coefficient than to the shape of the hydrograph.

STUDY OF EXISTING NETWORKS

If improvement of an existing network is contemplated, by replacing certain pipes with larger ones, linking different parts of the network, creating overflow basins, etc., it is usually either because the network has not performed adequately during certain rainfall events or because extensions of the network are needed to accomodate a growing urban population. In this situation the jurisdictional authority usually undertakes a programme of observation and data collection in the network, the goal of which is to provide a data base on the functioning of the system. These observations, which consist usually of measurements of water level, pressure and, less frequently, discharges, can be analysed in parallel with observed rainfall data for the same period. Based on these data, one can adjust, or "calibrate", a mathematical model of the network by carefully varying the runoff coefficients in computational sub-basins so as to reproduce the observed water levels, pressures, discharges, and overflow volumes.

To test the efficiency of a proposed network addition or modi-

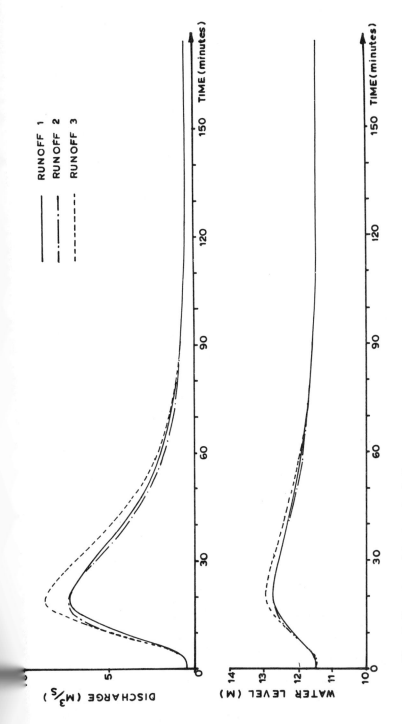

Fig.3 Results of test calculations

fication for a given rainfall event, one can compare the results
of mathematical model calculations with and without the improvement, using for each the same sub-basin input hydrographs.
The comparison is legitimate whether or not the sub-basin
discharges are precisely known. Differently stated, the comparison is made with respect to a rainfall event slightly different from the one observed or taken as a design standard.

One of the objections which can be made to the procedure just
described concerns the calibration of the mathematical model.
If in fact one has at his disposal only measurements of water
level and pressure and if discharge input points are infrequent, one can choose sub-basin input hydrographs so as to
reproduce approximately the observed levels and pressures.

But if discharge input points are spaced more closely than the
water level measuring points, the local discharges cannot be
estimated with a uniform-flow calculation, leaving considerable
uncertainty insofar as the input discharges are concerned. Thus
pipe and channel roughness coefficients, as well as discharge
coefficients for valves, weirs, etc., cannot easily be verified
when discharge measurements are sparse.

Therefore to be of real value, the procedure described above
demands at least a minimum of discharge observations in the
network. It is our judgment that, in any case, the network
authority should maintain a program of systematic network
observation and data collection so as to understand how the
system functions and to be able to identify its deficiencies.

A problem which arises in the study of proposed drainage network improvements is that of the uneven distribution of short-
term rainfall on runoff basins which have a surface area of
several tens of square kilometers. As the most dangerous
rainstorms for a drainage network are those of high intensity
and short duration, the intensity will not be the same over
the entire runoff basin. Therefore it would seem appropriate
to test the performance of network improvements for a range
of space-time distributions of rainfall. (Uniformly distributed rainfall is not necessarily the most dangerous situation). Given our poor understanding of the details of the
space-time distribution of rainfall intensity, it would seem
sufficient to assume several different distributions of
sub- basin input hydrographs. Under certain conditions one
observes that conduits linking large collectors can prevent
overflow damage when the rainfall is concentrated on a part
of the runoff basin, whereas the utility of these same condui
may not be apparent when the rainfall is uniformly distribute
We do not believe that, for this sort of investigation, a pre
cise hydrologic mathematical model is absolutely necessary.

The above remarks are especially applicable to the use of a

mathematical model of unsteady flow in a pipe network to investigate the performance of automatic valve or pump control systems. The opening or closing of valves is determined from data on the state of the drainage network at a given moment. Thus the calculation can reasonably be based simply on discharges entering the network from partial basins.

DIMENSIONING OF NEW DRAINAGE NETWORKS

In the dimensioning of new drainage networks, whether for new urban areas or for cities not previously having a subsurface drainage system, the rainfall statistics and the rainfall-runoff transformation must be known in some depth. But here again it is not so much the details of the rainfall-runoff transformation, as the runoff coefficient for rainfall events of duration less than one hour which must be known. It must not be forgotten that this coefficient depends on the soil moisture conditions at the beginning of the rainfall event in question, and that the rainfall may not be distributed uniformly over the entire runoff basin.

In the event that a network must be dimensioned by traditional methods (perhaps imposed by the responsible authorities), a detailed mathematical model of flow in the network can be useful as a check and/or verification of the dimensions chosen.

CONCLUSIONS

. Even if one does not have access to a precise hydrologic mathematical model, a mathematical model of unsteady flow in a pipe network can be useful in the determination of the relative efficiency of proposed improvements to existing networks, provided that a minimum of water level, pressure and discharge observations are available.

. Experience using mathematical models of flow in drainage networks has shown, at least for large networks, that the essential hydrologic parameter is the runoff coefficient.

. Data collection campaigns in existing networks should be systematically carried out so as to
    . better understand the hydraulic and hydrologic phenomena in play,
    . detect deficiencies,
    . determine overall runoff coefficients.
These observations should be for the most part confined to water levels and pressures in pipes, but it is essential, in order to avoid certain ambiguities, that pipe discharge measurements be taken in spite of the cost. Random discharge measurements are useful for the calculation of local pipe roughness coefficients, whereas systematic inflow/outflow measurements permit the overall water volume balance of a

network to be checked. The drainage network mathematical model is a useful tool for the planning and interpretation of such data collection programs.

. The inhomogeneous distribution of rainfall over urban watersheds during storms of less than one hour duration is a phenomenon of importance in the study of drainage systems. Unfortunately, one does not usually have at his disposal sufficient instantaneous rainfall records to properly take into account this uneven distribution.

. Multiply connected (looped) drainage networks can be more efficient than simply connected (branched) networks when rainfall distribution is non-uniform. Designers should not limit themselves to branched networks for lack of a calculation tool, since procedures such as CAREDAS permit the solution of the full de St Venant equations in looped networks.

REFERENCES

Brandstetter,A (1975) Assessment of Mathematical Models for Storm and Combined Sewer Management, Environmental Protection Agency (U.S.), Cincinnati, Ohio, 45268.

Cunge,J.A., and Wegner,N. (1964) Intégration Numérique des Equations d'Ecoulement de Barré de Saint-Venant par un Schéma Implicite de Différences Finies - Application au cas d'une Galerie Tantôt en Charge Tantôt en Surface Libre. La Houille Blanche, No. 1, pp. 33-39.

A DESIGN AND SIMULATION METHOD FOR STORM SEWERS

R K Price, Hydraulics Research Station, Wallingford, UK
C H R Kidd, Institute of Hydrology, Wallingford, UK

ABSTRACT

The paper describes a design and simulation method for storm sewers of similar complexity to the current TRRL method but different in its more realistic modelling of surface runoff and pipe flow routing. The surface runoff model is based on a non-linear reservoir model in which paved and roof areas are treated separately. Pipe flow routing is done using the Muskingum-Cunge method and a proper allowance is made for any surcharging of a pipe at its upstream manhole. The method is developed for observed storm events on catchments at Derby and Stevenage. Pipe diameters are designed as the smallest diameters to convey flows from a set of rainfall events with different durations for a prescribed return period without surcharging. An example of the redesign of an existing system and its construction cost is included.

INTRODUCTION

The significance of urban drainage in the economy of the UK has been highlighted by a recent report on the capital investment in sewerage; see DOE/NWC Standing Committee on Sewers and Water Mains report, 1977. The magnitude of this investment, reputedly of the order of £28,000 m, together with doubts based on experimental and theoretical evidence concerning the adequacy of existing design methods, has led to an increasing research effort into the development of more reliable and accurate methods for urban drainage design and flow simulation. In their review of storm drainage design methods, Colyer and Pethick (1976) conclude that a single design method suited to all situations does not exist. This is due to the very complex natural phenomena occurring in the runoff from an urban area, particularly in the overland flow phase. Consequently the number of possible modelling simplifications which are necessary in practice has led to a corresponding variety of methods, each with its own advantages and disadvantages. This fact, together with the demands of the practising engineer for design and flow simulation methods which meet varying requirements of accuracy, adequacy, convenience and economy of use, implies that there is a trend towards the adoption of an integrated package of several methods, each with its own particular advantages. The remainder of this paper is devoted to the description of one component of such a package.

The most widely used methods in the UK are the Rational or Lloyd-Davies method Lloyd-Davies, 1906) and the TRRL method (Watkins, 1962; Transport and Road

327

Research Laboratory 1976); see also Colyer and Pethick (1976). These methods are appreciated partly because of their simplicity compared with other more recent and complex methods, such as the Hydrograph Volume Method (Koniger and Klym, 1972), the Illinois Storm Sewer System Simulation Model (Sevuk et al 1973) and the Storm Water Management Model (Metcalf and Eddy Inc et al 1971), and also because the TRRL method, despite its inadequacies, produces results under simulation testing which compare favourably with the more complex methods. In fact Colyer (1977) observed that the TRRL method, together with several other methods, is able to simulate observed events with an accuracy which approaches the probable accuracy of the observations. However the TRRL method has been criticised on several theoretical grounds, in particular concerning the assumptions that (a) all reservoir-type storage in the catchment occurs in the pipe system; (b) the return period of the flow is the same as that for the rainfall event; (c) there is 100% runoff from the impermeable area; and (d) there is no contribution from the permeable area. In addition the method is unable to provide a realistic or accurate simulation of surcharged flow. It should be added that assumption (b) is made by all existing design methods and that assumptions (c) and (d) tend to compensate each other.

Any design and flow simulation package for urban drainage would benefit by including a method with the comparative simplicity of the TRRL method. Consequently the Hydraulics Research Station in co-operation with the Institute of Hydrology and the Meteorological Office is developing a new method as a counterpart to the TRRL method, and which meets some of the objections raised above. In particular the new method separates the above and below ground phases in the catchment runoff, and includes surface storage in the runoff from the individual contributing areas to each pipe and dynamic storage in the pipes.

A limited account is taken of surcharging by assuming that surcharging at the upstream manhole of a pipe does not affect pipes upstream. This assumption is necessary in a method which is to be used for both design and simulation and which has the simplicity of the TRRL method. An alternative method devoted solely to the simulation of flows in a sewer system is described by Bettess et al (1978).

The current version (1977) of the new method uses either a recorded rainfall hyetograph for simulation or rainfall events of prescribed return periods as recommended by the Meteorological Office (Folland, 1978) for the particular location. In the design case it assumed that the return period of the flow is the same as that of the rainfall. Research is currently underway to explore the relationship between the return period of the flow and the return period of the input variables (rainfall depth, duration, profile and catchment wetness). When this relationship has been established it will be incorporated in the method. When designing a system the method can use several events with different durations for a prescribed return period such that the system is designed for a uniform risk of failure throughout the catchment.

The modelling of the surface runoff and pipe routing phases is now described.

SURFACE RUNOFF

The surface runoff from the area contributing to a pipe depends on the spatial and temporal variation of rainfall and the irregular topography, surface roughness and varying permeability of the surface. Most contributing areas are extremely complex and an attempt to adopt a physical approach to the modelling of the processes imposes a se penalty both in terms of effort in data collection and computing cost. Therefore, surface runoff models as used by storm sewer design or simulation methods generally adopt a humped-parameter hydrological approach based on simplifications of the rainfall distribution and the nature of the surface. The particular model used in the present method is an adaptation of the non-linear model developed by Kidd (1976).

Each contributing area is divided into three components according to the surface types; paved, roof and permeable. A paved or roof area is divided into a number of equal sub-areas, each feeding a road gully or roof down-pipe respectively. Runoff from the permeable area is regarded as a uniformly distributed input to the paved sub-areas.

Three separate sub-models are used to calculate the runoff. A depression storage depth is uniformly distributed over a given surface type and rainfall has to fill this storage before runoff begins. A percentage runoff, r, is determined for the whole catchment either from data observed from a particular event to be simulated or from an equation derived at the Institute of Hydrology and based on a regression analysis of data from a number of UK catchments (Stoneham and Kidd, 1978; Kidd 1978). Thirdly, the discharge, Q (in l/hr), off an individual area, A (in m^2), is defined by the non-linear reservoir equations

$$\frac{dS}{dt} = i - q \qquad \qquad ....(1)$$

$$S = K q^n \qquad \text{and} \qquad ....(2)$$

$$Q = Axq \qquad \qquad ....(3)$$

Here S is the storage (excluding depression storage), i is the excess rainfall intensity in mm/hr, q is the discharge per unit area, K and n are the two parameters of the model for the appropriate surface type, and Ax is the actual area contributing to the flow. The factor x is determined for each surface type from the percentage runoff, r, according to the equation

$$(1 - \frac{d_P}{P})x_P + \frac{\overline{A}_R}{\overline{A}_P}(1 - \frac{d_R}{P})x_R = 100r(1 + \frac{\overline{A}_R}{\overline{A}_P}) \qquad ....(4)$$

where P is the total rainfall depth, $\overline{A}_P$ and $\overline{A}_R$ are the total paved and roof areas for the whole catchment, and d_P and d_R are depression storage depths for the paved and roof areas. It is assumed that

$$\frac{1 - x_R}{1 - x_P} = \nu \qquad \text{if} \quad \frac{x_P}{x_R} > \nu \qquad \qquad ....(5)$$

$$\text{otherwise} \quad \frac{x_P}{x_R} = \nu \qquad \qquad ....(6)$$

Here ν is a constant which is less than unity.

It remains to determine the model parameters. The application of the Manning equation to a flow situation where the width is large compared with the depth, such as encountered in overland or gutter flow, leads to a value for n in Equation (2) of 3/5. This value was adopted. The storage routing constant K varies with catchment characteristics and has units of mm (hr/mm)^n. For the work described here the analyses described by Kidd (1978) were not available and therefore K has been treated as a calibration parameter.

PIPE ROUTING

Flow in pipes is more amenable to a physically-based modelling approach than runoff from contributing areas because of the simpler geometry. However, complications can arise at pipe junctions and also where backwater effects are important or where surcharging occurs. If backwater effects are small, such as is generally the case for catchments with slope greater than 0.002, flow routing can be confined to discharge alone coupled with a measure of manhole water levels to determine surcharged flow. One of the simplest and better discharge routing models available is the Muskingum-Cunge method with fixed parameters; see Bettess and Price (1977). Unlike the kinematic wave method this method has the advantage that it includes an allowance for dynamic storage in the pipes. The marginal improvement in accuracy using the variable parameter Muskingum-Cunge method (Price and Mance 1978) is not justified because of larger computing costs.

The Muskingum-Cunge routing method is defined by the equations

$$\frac{dS}{dt} = Q_{up} - Q_{dn} \qquad \qquad ....(7)$$

$$S = \frac{L}{\omega} \left\{ \epsilon\, Q_{up} + (1-\epsilon)Q_{dn} \right\} \qquad \qquad ....(8)$$

$$\epsilon = \frac{1}{2} \left\{ 1 - \frac{\overline{Q}}{\overline{B}s L \overline{\omega}} \right\} \qquad \qquad ....(9)$$

where L and S are the length and gradient of the pipe respectively, Q_{up} and Q_{dn} are the upstream and downstream discharges, and $\overline{Q}$, $\overline{B}$ and $\overline{\omega}$ are mean values for the discharge, surface breadth and wave speed along the pipe. A value for $\overline{\omega}$ can be found from the equation

$$\overline{\omega} = \frac{1}{Q_{fb}} \int_{o}^{Q_{fb}} \frac{dQ}{dA}\, dQ = \frac{1}{Q_{fb}} \int_{o}^{d} \frac{1}{B} \left(\frac{dQ}{dy}\right)^{2} dy \qquad \qquad ....(1)$$

with Q defined by the normal depth relationship of the Colebrook-White equation:

$$Q = A(32gRs)^{\frac{1}{2}} \log_{10}\left(\frac{14.8R}{k_s}\right) \qquad \qquad ....(1)$$

Here g is the acceleration due to gravity, R is the hydraulic radius, d is the pipe diameter, k_s is the roughness height and Q_{fb} is the full-bore discharge. For typical values of k_s, $\overline{\omega} = dQ/dA$ for a water depth of 0.3 d. This value for $\overline{\omega}$ is used in Equation (7). However, $\overline{Q}$, $\overline{B}$ and $\overline{\omega}$ in Equation (8) are evaluated for a water depth 0.5d as these give an appropriate mean value for ϵ.

To avoid difficulties with the routing method when L is small and ϵ becomes negative the discharge hydrograph is routed along a virtual pipe of length $L_v = \Delta t / \bar{\omega}$ if $L_v > L$, and the instantaneous discharge is interpolated linearly to find the value at the downstream end of the original pipe.

A limited account is taken of surcharging at the upstream manhole of a pipe using the equation

$$h = \frac{Q^2 L}{2gA^2} \left\{ [16R\log_{10}{}^2(\frac{14.8R}{k_s})]^{-1} + \alpha \right\} - sL \qquad \qquad ....(12)$$

where h is water depth in the manhole above the soffit level of the outgoing pipe and $\alpha Q^2/2gA^2$ is the additional headloss in the manhole. α is a coefficient which can take a value between 0.2 and 1.5 depending on the geometry of the manhole and the alignment of incoming and outgoing pipes. By insisting that all runoff from the contributing area to the pipe is introduced at the upstream manhole, surcharging is only allowed to occur when the upstream discharge is greater than the pipe-full discharge.

METHOD STRUCTURE

The general format of the new method is similar to that of the TRRL method: the branch pipe nomenclature has been retained, and hydrographs are routed through each pipe in turn. By combining the hydrographs from the upstream pipes and contributing area the pipe diameter can be designed on the resulting peak discharge before the hydrograph is routed through the pipe. Facilities are included in the computer program to design each pipe for a set of design events of the same return period but of different durations. This ensures that a uniform design criterion is achieved throughout the catchment. A facility to evaluate the cost of the pipe system is also included.

DATA REQUIREMENTS

In its simplest form the method uses no more data for the pipes and contributing areas than is required by the TRRL method. Improved accuracy can be achieved by including more details of the contributing areas and the runoff parameters. Any cost evaluation requires additional data on pipe depths and surface and ground conditions; see Farrar and Colyer (1978).

COMPUTER PROGRAM

The present version of the computer program is written in standard FORTRAN and requires 14K words of store to simulate storms of up to 120 min duration. It takes 24 s of computing time on an ICL 1904S computer to design a pipe system of 87 pipes for two design storms of 15 and 30 min duration.

MODEL CALIBRATION

The model on which the design method is based was developed for limited sets of storm events on two experimental catchments: St Mark's Road, Derby, and Shephall in Stevenage. A more extensive simulation test of the model is reported by Pethick (1978).

The Derby system contains 87 pipes which collect runoff from a housing area of 10.4 ha. The catchment is relatively flat and surface runoff over the paved area is controlled primarily by the artificial camber. Rainfall is recorded at a gauge just to the west of the catchment and flow is gauged at three manholes by Arkon depth gauges which have been calibrated to 0.4 proportional depth by dilution gauging. Difficulties have been experienced in resolving the rainfall record to 1 min intervals and these have affected the results of the simulation reported below.

The Stevenage catchment is 143.6 ha and has an overall slope of about 0.011. Rainfall is recorded by two Dines autographic gauges at the top and bottom of the catchment. Flow is gauged using an Arkon depth recorder at the first manhole upstream of the outfall. Dilution gauging has confirmed the calibration of the flow up to 0.24 proportional depth. More recently a flume has been constructed at the outfall so that higher flows can be calibrated with greater confidence.

From the data for the Derby and Stevenage catchments two sets of 7 and 6 storm events respectively were selected for the calibration of the model. These events were selected as being typical of the range of recorded events on each catchment. The model was then run for each event with different values of the storage constant K. It was assumed that the K value for a standard roof area of 35 m^2 per down pipe was the same for both catchments and that a standard area per road gully of 300 m^2 could be used on the Stevenage catchment. The area per road gully on the Derby catchment was calculated separately for each contributing area, though the average area per road gully for the whole catchment is only marginally greater than the standard Stevenage value. k_s was defined as 0.0003 m for each pipe and the percentage runoff was calculated from the observed values of total rainfall and volume of flow so that volumes were forced.

Measures of goodness of fit between individual predicted and recorded events included the ratio, λ, of calculated to observed peak flow, Pearson product moment, integral square error and residual mass curve coefficient; see Pethick (1978). The overall selection of K values for the two catchments was made subjectively on the basis of the mean ($\bar{\lambda}$), standard (σ), and average ($\bar{\sigma}$) deviation of λ for each catchment.

Following a number of runs with the model K was related to A and x for each surface type by

$$K = K' \, {}'A^{0.3} \, x^{0.6} \qquad\qquad ....(1$$

It is emphasized that this relationship was obtained subjectively and will be superseded by the relationship derived by the Institute of Hydrology.

Tables 1 and 2 summarise the results obtained. The K' value for roofs was 0.07 and t paved area values for Derby and Stevenage were 0.085 and 0.12 respectively. The resu for Stevenage give $\bar{\lambda}$ = 0.98, σ = 0.13 and $\bar{\sigma}$ = 0.12 and the corresponding results for Derby are $\bar{\lambda}$ = 0.95, σ = 0.22 and $\bar{\sigma}$ = 0.18. There is more scatter in the Derby resul though in each of the Derby events there is doubt about the definition of rainfall for 1 min time step. The accuracy of the Derby gauges for peak flows is questionable.

APPLICATION TO DESIGN

The model described above was used to re-design the Derby system for existing pipe gradients. One year 50 percentile storms of 15 min and 30 min duration were obtain

...mated and recorded data for Stevenage catchment

Event number	Number of data points	Difference in time-to-peak (min)	Peak flow (l/s)		Volume of runoff (litres x 10^6)		λ	Pearson product moment	Integral square error	Residual mass curve coefficient
			obs	calc	obs	calc				
8	250	−1	257.48	232.14	0.817	0.808	0.902	0.953	0.025	0.792
9	281	17	370.51	337.38	1.764	1.752	0.911	0.921	0.026	0.700
10	163	0	158.72	125.60	0.715	0.685	0.791	0.900	0.076	0.724
14	91	3	185.23	197.26	0.361	0.348	1.065	0.885	0.167	0.794
16	200	−1	745.05	782.33	1.964	1.963	1.050	0.874	0.057	0.800
17	100	8	421.42	497.66	0.477	0.462	1.181	0.836	0.282	0.812

TABLE 2 Simulated and recorded data for Derby catchment

Event number	Gauge number*	Number of data points	Difference in time-to-peak (min)	Peak flow (litres/s) obs	Peak flow calc	Volume of runoff (litres x 10^6) obs	Volume of runoff calc	λ	Pearson product moment	Integral square error	Residual mass curve coefficient
2	1	241	1	66.08	74.83	0.099	0.108	1.132	0.989	0.035	0.765
2	3	218	-1	98.00	105.76	0.168	0.166	1.079	0.983	0.035	0.913
9	1	430	-4	87.78	128.33	0.304	0.351	1.462	0.948	0.047	0.922
9	3	408	-11	259.56	184.08	0.527	0.526	0.709	0.715	0.084	0.920
16	1	419	9	56.11	51.64	0.269	0.189	0.920	0.721	0.168	0.949
16	2	421	-7	77.00	59.97	0.299	0.233	0.779	0.732	0.154	0.964
21	2	302	-37	107.58	76.86	0.410	0.436	0.714	0.749	0.117	0.979
21	3	362	-37	137.35	95.26	0.550	0.546	0.694	0.772	0.092	0.985
27	1	481	6	62.43	58.52	0.123	0.086	0.937	0.774	0.422	0.991
27	3	241	4	77.98	77.77	0.125	0.110	0.997	0.781	0.414	0.995
29	2	101	3	80.67	94.54	0.102	0.098	1.172	0.783	0.515	0.998
32	2	340	-2	62.68	53.43	0.181	0.184	0.852	0.784	0.295	0.998

*... numbered in order down the catchment.

from the Meteorological Office for the catchment location and used in the design.
Table 3 summarises the designs of the main branch for two cases: the first allowing the
pipe diameters to decrease and the second with the non-decreasing diameter constraint.
Note that some of the pipes in the first case are designed on the second event.

TABLE 3

Pipe number	Original pipe diameter* (m)	Redesign 1 pipe diameter (m)	event number	Redesign 2 pipe diameter (m)	event number
1.0	0.229	0.075	1	0.075	1
1.1	0.229	0.150	1	0.150	1
1.2	0.229	0.150	1	0.150	1
1.3	0.229	0.225	1	0.225	1
1.4	0.229	0.300	1	0.300	1
1.5	0.229	0.300	1	0.300	1
1.6	0.305	0.375	1	0.375	1
1.7	0.305	0.450	1	0.450	1
1.8	0.305	0.375	1	0.450	1
1.9	0.305	0.375	1	0.450	1
1.10	0.305	0.375	1	0.450	1
1.11	0.305	0.450	1	0.450	1
1.12	0.381	0.450	2	0.450	1
1.13	0.381	0.375	1	0.450	1
1.14	0.381	0.525	1	0.525	1
1.15	0.381	0.450	2	0.525	1
1.16	0.457	0.525	2	0.525	1
1.17	0.457	0.525	1	0.525	1
1.18	0.457	0.375	1	0.525	1
1.19	0.457	0.450	2	0.525	1
1.20	0.457	0.375	1	0.525	1
1.21	0.457	0.375	1	0.525	1
1.22	0.457	0.450	1	0.525	1
1.23	0.457	0.450	1	0.525	1
1.24	0.533	0.450	1	0.525	1
1.25	0.533	0.450	2	0.525	1
1.26	0.533	0.525	2	0.529	1

Existing pipe diameters converted to metres from feet.

Using the costing subroutine the cost of the original Derby system (at January 1975 prices) is £70 426. The costs of the redesigned systems are £69 997 and £73 226 respectively. The larger cost for the second re-design is due primarily to the small gradien of pipe 1.14 which affects all downstream pipes in the main branch. It is problems like this which make pipe diameter—gradient optimisation for a minimum construction cost an attractive alternative; see Price (1978).

CONCLUSIONS

The design method based on the model described in this paper has been demonstrated as a potential design tool. Refinements are being made to the surface runoff model and the storage parameter K for paved areas is being related to catchment characteristics (Kidd, 1978). The method will undergo extensive trials in design offices during 1978. It is anticipated that the complete method should be available for general release by 1979.

ACKNOWLEDGEMENTS

The work described in this paper was carried out as part of research programmes into urban drainage at the Hydraulics Research Station and the Institute of Hydrology. The paper is published with the permission of the Director of Hydraulics Research and the Director of the Institute of Hydrology. Work at the Institute of Hydrology has been undertaken as part of a project commissioned by the Department of the Environment (Project No DGR/480/38).

REFERENCES

Bettess, R. and Price, R. K. (1976) Comparison of numerical methods for routing flow along a pipe. Hydraulics Research Station, Report IT 162.

Bettess, R., Pickfield, R. and Price, R. K. (1978) A surcharging model for storm sewer systems. International Conference on Urban Storm Drainage, Southampton.

Colyer, P. J. (June 1977) Performance of storm drainage simulation models. Proc. Instr Civ. Engrs. Part 2, **63**, pp 293-309.

Colyer, P. J. and Pethick, R. W. (March 1976) Storm drainage design methods — a literature review. Hydraulics Research Station, Report INT 154, p 85.

Cunge, J. A. (1969) On a flood propagation method. J. Hydr. Res., 7, no 2, pp 205-2

Department of the Environment/National Water Council (June 1977) Sewers and Wate Mains — A National Assessment. Standing Committee on Sewers and Water Mains, NWC, p 34.

Farrar, D. M. and Colyer, P. J. (1978) A procedure for calculating the cost of storm water sewer construction. International Conference on Urban Storm Drainage, Southampton.

Folland, C. K. Rainfall profiles recommended in Road Note 35 (Second edition). To published.

Kidd, C. H. R. (1976) A non-linear urban runoff model. Institute of Hydrology, Rej No 31.

Kidd, C. H. R. (1978) A calibrated model for the simulation of the inlet hydrograph for fully sewered catchments. International Conference on Urban Storm Drainage. Southampton.

Koniger, W. and Klym, H. (1972) Nichtlineare hydrologische modelle in der Stadtentwasserung Wasser/abwasser 113, pp 430-435.

Lloyd-Davies, D. E. (1906) The elimination of storm water from sewerage system. Proc ICE, Vol 164(2), pp 41-67.

Metcalf and Eddy Inc et al (1971) Environmental Protection Agency. Storm Water Management Model, Vols I, II, III and IV. Report Nos 11024 DOC07/71, 08/71, 09/71 and 10/71, Environmental Protection Agency, Washington.

Pethick, R. W. (1978) The testing of three storm sewer simulation methods. International Conference on Urban Storm Drainage, Southampton.

Price, R. K. (1978) Design of storm sewers for minimum construction cost. International Conference on Urban Storm Drainage, Southampton.

Price, R. K. and Mance, G. (1978) A suspended solids model for storm water runoff. International Conference on Urban Storm Drainage, Southampton.

Sevuck, A. S. et al (1973) Illinois Storm Sewer System Simulation model – User's manual. Univ. Illinois Water Resources Centre, Research Report No 0073.

Stoneham, S. M. and Kidd, C. H. R. (1978) The prediction of runoff volume from fully sewered urban catchments. Institute of Hydrology, Report No 41.

Transport and Road Research Laboratory (1976) Road Note 35. A guide for engineers to the design of storm sewer systems. HMSO, 2nd Edn., p 30.

Watkins, L. H. (1962) The design of urban sewer systems. HMSO, RRL Technical Paper 55.

COMPARATIVE INVESTIGATION OF COMPUTER METHODS

J. Keser

Institut für Wasserwirtschaft, Hydrologie und land-
wirtschaftlichen Wasserbau, Technische Universität
Hannover, West Germany

1. INTRODUCTION

More and more urban runoff calculations are put
through with the aid of digital computers. It must
be accepted that the calculations and results are
checked by the ordering institution or another
responsible authority.
There are less problems in testing the results of the
so-called rational methods. These methods and their
numerous modifications are all using the conception
of concentration time. The way of calculation is
straight forward and is accomplished without difficulties.
On the contrary the checking of so-called "urban
runoff simulation models" and the resulting output
is often particularly troublesome and difficult.
The aim of these methods is to reproduce the transformation of rainfall to runoff true to nature. For
that purpose surface runoff and sewer runoff are
mostly calculated by separate model sections. Each
section consists of physically founded formulations
The mathematical solutions are often quite non-
transparent. The checking of the results by hand is
seldomly possible. In addition the internal steps
of calculation for special problems are not always
obvious and transparent.
Nevertheless computations of existing sewer systems
which are overcharged and influenced by backwater
effects must be carried out for economical reorgani
zations. Consequently, physically founded computer
models must be used which take into consideration a
elements of influence to rainfall-runoff relationship. The problem itself often requires the application of sophisticated models which on the other
hand must be subsequently checked by the ordering

side.

2. OBJECTIVES

A joint working committee of the "Kuratorium für
Wasser und Kulturbauwesen" (KWK) has to work out a
manual of practise (guideline) for checking the
computations for urban sewer systems.
The KWK-Association is a central authority for
water and sewerage technology and law. Its purpose
is to solve common problems and to issue recommendations.
The guideline for checking urban runoff computations
should take into consideration the computer models
which are presently into operation.
To establish a basis for further work a comparative
investigation of computer methods was carried out.
The institute for water economics at the Technical
University of Hannover was appointed to conduct
these investigations.
The variety of the methods and their principal
distinctive marks were to be pointed out. The
causes of differences in the results should be investigated carefully. It was expected, that with
the aid of the results of this investigation it
would be possible to discover crucial points for
the guideline.
The following examinations were to be carried out
by the investigater:
- survey of distinctive features of computer methods
- comparison of results for two catchments
- comparison of measured data with results
- comparison of choosen parameters for fixed catchment characteristics.
Furthermore the following points were of interest:
Which methods will accept non-uniform precipitation
distributions as input?
Is it possible to calibrate the parameters of the
method for any corresponding measurement data of
precipitation and runoff?
Are the results improved if the computations were
put through with calibrated parameters?

3. CONTENT OF INVESTIGATION

The basis of the comparative investigation was the
runoff-computation results for two urban catchments. The catchments were of different size,
imperviousness and land slope.
The investigations consisted of two separate sections.

Section 1 - named relative comparison - corresponds
to the case of design computation. Input parameters
are hypothetical rainstorms with uniform intensity
and non-uniform intensity (hyetograph). For fixed
points of the sewer systems the results of the different methods were to compare with each other.
In section 2 - named absolute comparison - measured
rainstorms with varying intensities were given as
input functions. For these rainfall data corresponding measurement data of water level and runoff were
available at certain systempoints. Hence the
computed results could be compared with measurement
data.
Before computing section 2 measurement data for one
rainfall-runoff event were given, but only for
catchment No. I. By this the modelparameters could
be calibrated. It should subsequently be investigated if the computations with calibrated parameters
lead to better correspondence of calculated and
measured data for different rainstorms.
The following tables and figures 1 and 2 give some
detailed information.

Table 1: Characteristics of investigated catchments

Specification	Catchment No. I (see Fig. 1)	Catchment No.I (see Fig. 2)
drainage area in acres (hectares)	173 (7o)	91 (37)
imperviousness in %	51	39
average land slope in %	o,5	3,4

Table 2: Sewer system and input rainfalls

Specification	Catchment No. I	Catchment No.]
sewer system	combined	storm
investigated system point	side weir	fall structu:
rainfalls computed for relative compa- rison	rainfalls with uniform distri- bution , hyetograph	
rainfalls computed for absolute compa- rison	three different recorded hyeto graphs for each catchment	

FIG. 1.- DRAINAGE MAP ,   CATCHMENT I
COMBINED SEWER SYSTEM

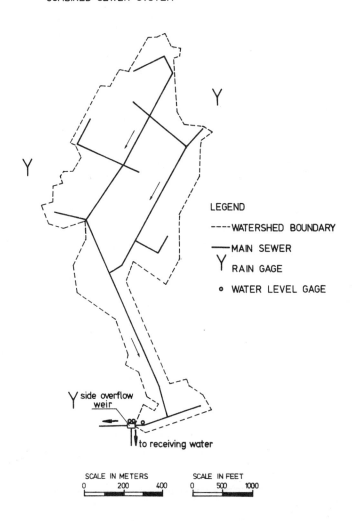

LEGEND

----WATERSHED BOUNDARY

——MAIN SEWER

Y RAIN GAGE

o WATER LEVEL GAGE

Y side overflow
  weir

to receiving water

SCALE IN METERS   SCALE IN FEET
0   200   400   0   500   1000

FIG. 2. - DRAINAGE MAP , CATCHMENT Ⅱ
STORM SEWER SYSTEM

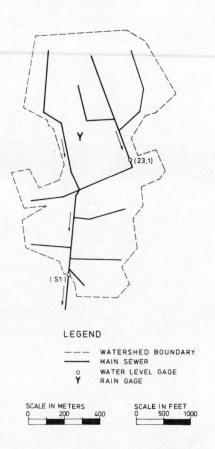

LEGEND

— — —	WATERSHED BOUNDARY
————	MAIN SEWER
○	WATER LEVEL GAGE
Y	RAIN GAGE

SCALE IN METERS
0   200   400

SCALE IN FEET
0   500   1000

4.   CLASSIFICATION OF PARTICIPATING METHODS

Ten engineering offices were invited to participate
in the investigation. All of them are using compute
programs for urban runoff calculations. With this
selection it was achieved that very different metho
could be investigated. The following methods took
part in the investigation: (the respective quantity
is given in parentheses)
- rational formula and modification (2)
- modified Unit-Hydrograph-Methods (2)
- methods solving the dynamic equations of flow (6)

Due to the differences in these methods a classification was proposed. By this methods were classified with regard to their application field.
Distinctions should be made between
- hydrologic methods with or without the possibility
  of calibrating the parameters with measurement
  data, and
- hydrodynamic methods with or without the possibility of calibration.
The term "hydrologic" implies that calculations
from rainfall to runoff are done by fixed formulas
or functions (rational method, unit-hydrograph).
The routing in sewers is done by algebraic functions
or simple routing-storage procedures.
Computations are characterized to be "hydrodynamic" if the routing procedure of sewer flow is
based on the dynamic equations. This denotation is
valid even when the terms of acceleration of the
momentum equation are neglected (momentum equation →
frictional equation). The possibility of calibration
implies that non-uniform rainfall distribution can
be accepted as input function. Otherwise model
parameters can be calibrated or tested insufficient-
ly.
All hydrodynamic methods consist of separated surface and sewer models.

5.   RESULTS

For given rainfalls maximum discharges and maximum water levels for the manholes in the main sewers
were to be calculated. For fixed system points hydrographs were to be computed. Furthermore maximum
velocities, total runoff volumes and - for catchment No. I - overflow volumes and hydrographs were
to be output.
From the immense number of results some essential
values are subsequently presented.
Results of the relative comparison:

   The peak discharges of hydrographs at the side-
   overflow weir of catchment No. I vary - for the
   uniform 3o-min. rainfall which is the critical
   rainfall duration for this catchment -
   from   51.2 cfs to 84.4 cfs (1.45 to 2.39 m^3/s)

   - see Fig. 3 -

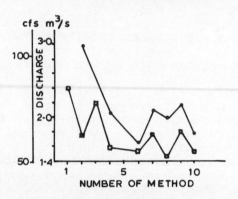

• = 60 min HYETOGRAPH
□ = 30 min UNIFORM RAINFALL DISTRIBUTION

Fig.3 Computed peak discharges
relative comparison, catchment I

and - for the 6o-min. hyetograph -

from   57.2 cfs to 1o4.9 cfs (1.62 to 2.97 m^3/s)

- see Fig. 4 -

Referring to the minimum values the deviations amount to 165 % and 183 % respectively. Deviations in total runoff volume amount to 143 % and 154 %. The calculated overflow volumes to the receiving water vary

- for the 3o-min. rainfall -

from   2163 m^3 to 3o34 m^3,   and

- for the 6o-min. hyetograph -

from   2492 m^3 to 3149 m^3.

The differences in the results are quite considerable for catchment No. I.
Nevertheless they were exceeded by the results for catchment No. II. At system point S1 peak discharges vary

- for the 5-min. rainfall -

from   28.6 cfs to 116.5 cfs (o.8 to 3.3 m^3/s)

- see Fig. 5 -

and - for the 6o-min. hyetograph -

from   52.3 cfs to 1o2.4 cfs (1.48 to 2.9 m^3)s).

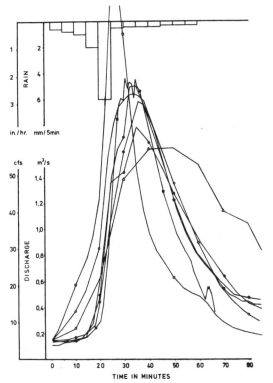

FIG. 4.- RAINFALL HYETOGRAPH , COMPUTED HYDROGRAPHS
RELATIVE COMPARISON , CATCHMENT I

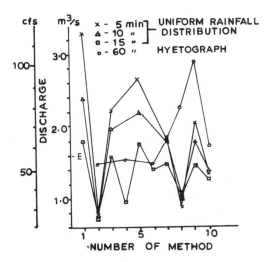

FIG.5 - COMPUTED PEAK DISCHARGES
RELATIVE COMPARISON, CATCHMENT II

Here the maximum deviations range from 4o7 % to
196 %. The differences in the results for system
point 23,1 are corresponding.
- see Fig. 6 -

FIG. 6.- RAINFALL HYETOGRAPH, COMPUTED HYDROGRAPHS
RELATIVE COMPARISON, CATCHMENT Ⅱ (23,1)

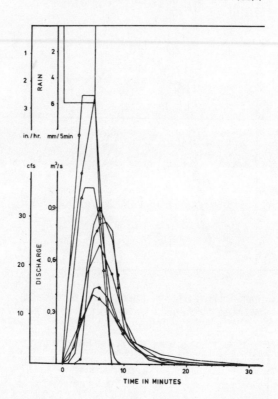

It should be noted that the maximum values for uniform rainfall distribution is always produced by
methods applying the rational formula.
Results of the absolute comparison:

As described before, recorded rainstorms with
varying intensities had to be used in this
section. Consequently only the simulation models could take part.
For catchment No. I it was possible to calibrate the model parameters with the aid of
one given rainfall-runoff event.

Fig. 7 shows the computed discharge hydrographs for
the calibration event compared with the recorded

hydrograph.

FIG.7. - CALIBRATION - RAINFALL HYETOGRAPH, COMPUTED
AND RECORDED HYDROGRAPHS, ABSOLUTE COMPARISON
CATCHMENT I

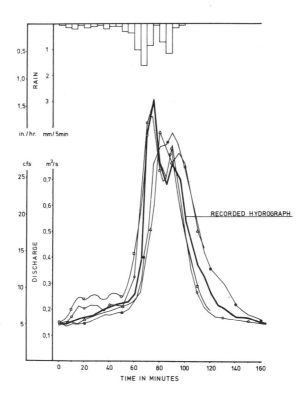

Fig. 8 gives the comparison of another recorded
hydrograph with computed discharge hydrographs for
calibrated model parameters.
In Fig. 9 the differences between computed and
recorded peak discharges for three rainfall events,
catchment No. I, are collectively presented. Due to
possible inaccuracies a tolerance range of ± 1o %
is marked for critical examination.

It is obvious that the runoff characteristics for
catchment No. I are well reproduced. Again it
should be emphasized that the model parameters were
previously calibrated.
For computing runoff characteristics for catchment
No. II it was not possible to calibrate the model
parameters. Corresponding to the results of the

FIG. 8. - RAINFALL HYETOGRAPH, COMPUTED AND RECORDED
HYDROGRAPHS, ABSOLUTE COMPARISON, CATCHMENT I

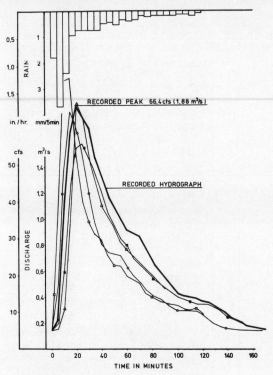

FIG. 9. - RECORDED AND COMPUTED PEAK DISCHARGES
ABSOLUTE COMPARSION, CATCHMENT I
-CALIBRATED PARAMETERS

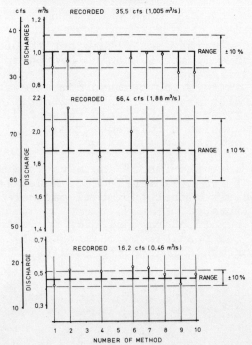

relative comparison, great differences in the results of the absolute comparison were computed.
In Fig. 1o computed and recorded discharge hydrographs of one event are plotted.

FIG 10. - RAINFALL HYETOGRAPH, COMPUTED AND RECORDED
HYDROGRAPHS, ABSOLUTE COMPARISON, CATCHMENT II

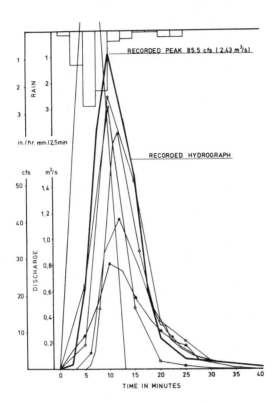

Fig. 11 shows the differences between computed and
recorded peak discharges of all three events.

FIG. 11. - RECORDED AND COMPUTED PEAK DISCHARGES
ABSOLUTE COMPARSION, CATCHMENT II

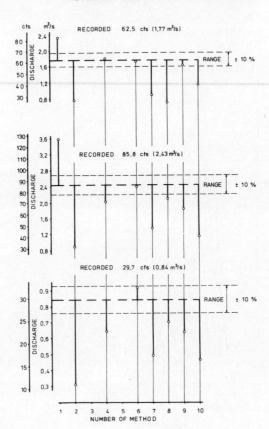

6. CONCLUSIONS

The aim of this investigation was to disclose differences in the computer methods for urban runoff
calculations. It was planned to achieve this by
comparing computed and recorded results for two
urban catchments.
The differences in computed runoff volumes however
point out the fact, that the deviation in peak
discharges depend not so much on the respective
model structure.
The reasons are rather caused by different surface-
loss assumptions for clearly described catchment
characteristics.

Consequently the problem of urban sewer computations lies not only in the modelling of the rainfall-run-off process but in the fixing of surface losses or specific rainfall-runoff ratios.
To underline this statement the computed discharge hydrographs were altered. The ordinates of the hydrographs were modified by the respective ratio of recorded to computed total runoff volume. In this way the volumes of the computed hydrographs correspond to the recorded volumes while the shape of the hydrographs remains unchaged (see Fig. 12).

FIG. 12. - ADAPTION OF RUNOFF VOLUMES (LOSSES) AND
MODIFICATION OF COMPUTED HYDROGRAPHS

COMPUTED RUNOFF VOLUME

—·—·— RECORDED RUNOFF VOLUME
DIFFERENCE VOLUME

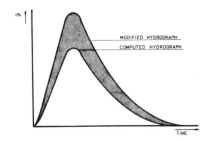

This adaption of runoff volumes demonstrated that computed peak discharges as well as hydrograph shapes correspond closely with recorded data. Of course this manipulation may be subject to criticism due to the different structures of the models.
The differences in fixing surface losses or runoff coefficients are demonstrated by the following example:
Supposing an urban catchment with an imperviousness coefficient of 5o % and flat land slope. For an initial rainfall rate of 3 mm/5 min the runoff coefficient is to compute. According to the statements of the engineering offices participating in the comparative investigation the effective rain-

fall varies from o,5 mm to 1,7 mm which is equivalent to runoff coefficients between o.17 and o.57.
It is quite remarkable that informations about the
quantities of surface losses are at present unsatisfactory. This requires the intensified collection of corresponding rainfall-runoff data in
existing sewer systems and appropriate research in
this field.
At present the accuracy of input parameters seem
to be of much greater importance than possible
further improvements of computer models.

URBAN RUNOFF AND DESIGN STORM MODELLING

Desbordes Michel

Laboratoire d'Hydrologie Mathématique. Université des Sciences
Montpellier - France.

INTRODUCTION

Rapid urbanization makes sewage problems more and more difficult. They cannot be accurately solved using classical ponc-
tual methods, deriving mostly from the so called rational formula. So, since 1972, the Laboratory of Mathematical Hydrology
(L.H.M.) of the Montpellier Sciences University, has been studying new design and management methods for complex sewerage
systems. The modelling is composed of a link of sub-models ;
each of them achieves a step of the transformation. Figure 1
shows that link of sub-models studied by the L.H.M. (Desbordes
M. and als 1976). Some of them are connected with hydraulic
civil engineering or economy domains. The L.H.M. has more particularly studied runoff and storm design models. The hydrological processes of the rainfall-runoff transformation over an
urban watershed, are complex. Many of the physical variables
of these processes are not easy to measure. So the L.H.M. has
used the methods of systems analysis as to build urban runoff
and design storm models. These models are described hereinaf-
ter.

URBAN RUNOFF MODELLING

The model gives a runoff hydrograph of a given storm at the
outlet of a small homogeneous catchment (10 to 100 hectares),
equipped with an elementary drainage system. The catchment
and its sewage network is taken as a transformation system in
which rainfall is the input and the discharge at the outlet,
the output.
The definition of the model was made by means of the conceptual approach of the system of the rainfall runoff transformation. That approach was prefered to that of the numerical identification of systems. That las one is probably more accurate,
but its interpretation in term of a general model is often
more difficult, as a consequence of its accuracy.

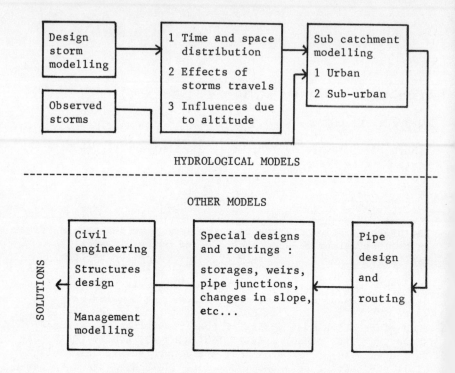

Fig. 1 - Urban runoff modelling : the L.H.M. system

Description of the model

The study of data observed on small experimental urban catchments (Normand,D. 1971) led to retain a model with a storage
effect. Such a model results from the association of two equations. The first one, the storage equation has the general
form :

$$S(t) = \sum_{j=0}^{j=m} A_j(i,Q,t) \frac{d^j Q}{dt^j} + \sum_{k=0}^{k=n} B_k(i,Q,t) \frac{d^k i}{dt^k} \qquad (1)$$

$S(t)$ is the storage at the time t in the system, $i(t)$ is the
net rainfall and $Q(t)$ the discharge. The second one is the
classical continuity equation :

$$d\,S(t)\,/\,d\,t = i(t) - Q(t) \qquad (2)$$

When j and k are greater than 2, it is difficult to find a
physical meaning to equation 1. The model will be as more accurate as it will have more numerous parameters ; but its fitting will need as more data and the extrapolation of the model

will be as more difficult. At last the model will be useful in the case of ungaged catchments (which is the general case in the design of urban sewage systems) if it has a minimum number of parameters. After some tests, the simplest model has been retained. Its storage equation is :

$$S(t) = K Q(t) \tag{3}$$

The K model parameter is the lag time between the mass centers of net rainfall and discharge. When K being a constant, the association of equations 2 and 3 leads to :

$$Q(t) = Q_0 \, e^{-(t - t_0)/K} + 1/K \int_{t_0}^{t} i(u) \, e^{-(t - u)/K} \, du + Qb \tag{4}$$

Qb is a steady state discharge, Q_0 the unsteady state discharge at the initial time t_0. The model pulse response or Instantaneous Unit Hydrograph is given by :

$$h(t) = 1/K \, e^{-t/K} \tag{5}$$

The numerical integration of equation 4, for a given input $i(t)$ may be done by means of recurrent use of equation 6, for each time interval dt, for $Q_0 = Qb = 0$:

$$Q(n\,dt) = e^{-dt/K} \, Q(n-1)\,dt + (1 - e^{-dt/K}) \, i(n\,dt) \tag{6}$$

K may be considered as a constant during a storm only for typically urban catchments (i.e. more than 20 % imperviousness). For semi-urban watersheds, K is a function of time ; the general solution becomes :

$$Q(t) \, e^{\int_{t_0}^{t} du/K(u)} = \int_{t_0}^{t} \frac{i(u)}{K(u)} \, e^{\int_{t_0}^{u} dv/K(v)} \, du + Q_0 \tag{7}$$

The numerical integration of equation 7 may be done more easely using an analytical approximation of K(t). One may also solve the differential equation of the process by means of a finite difference scheme.

Estimation of the K model parameter

From rainfall and runoff data of 21 french and american experimental urban catchments (Normand, D. 1971) (Sarma,P.S., Delleur J.W. 1970), the equation 8 has been established by means of multiple regression :

$$K = 5.1 \times A^{0.18} \times p^{-0.36} \times \left[1+(IMP/100)\right]^{-1.9} \times TE^{0.21} \times L^{0.15} \times$$

$$\times HP^{-0.07} \tag{8}$$

(multiple regression coefficient : 0.95).
The K parameter is in minutes, A is the catchment area in hectares, p its mean slope as a percentage, IMP is the imperviousness percentage, TE the duration in minutes of the heavy part
of rainfall, L is the catchment length in meters and HP the
rainfall depth in millimeters during TE. The variation range
for these parameters was :
$0.4 \text{ ha} \leqslant A \leqslant 5000 \text{ ha}$; $15 \% \leqslant IMP \leqslant 100 \%$; $110 \text{ m} \leqslant L \leqslant 17800 \text{ m}$;
$0.4 \% \leqslant p \leqslant 5 \%$; $5 \text{ mn} \leqslant TE \leqslant 180 \text{ mn}$; $5 \text{ mm} \leqslant HP \leqslant 240 \text{ mm}$.
Because K changes with the HP and TE storm values, the model
is pseudo-linear. The tests of the model for experimental watersheds (Desbordes, 1974) have shown quite good hydrograph
regeneration in almost all cases (less than 20 % error in
peaks discharge and 10 % in times to peak). For semi-urban
watersheds, equation 8 is also used. But IMP is replaced by
the instantaneous runoff coefficient. It may be obtained by
subtracting runoff losses from total runoff. These losses may
be found using for example an Horton's infiltration equation.

Study of the sensitivity of the model to K parameter
In equation 8, the K value is only statistical parameter which
has a sampling error. The best interest of this equation is
that it can be used in the case of an ungaged catchment, the
most usual case in sewer design projects. The study of the
sensitivity of the model to the K parameter can lead to answer
to the following question : with what precision the K parameter is known for a good definition of the hydrograph resulting
from a given storm ?

Data used in the study The rainfall used in the study are
known by their duration t, their mean intensity $\bar{i}$ on that duration, their maximum instantaneous intensity iM and the position tp of iM over t. Three simple geometric shapes have been
retained : rectangular, triangular and exponential. The $\bar{i}$ and
variables have been related by the so-called intensity-dura-
tion-frequancy curves of a type :

$$\bar{i} = a \, t^b \tag{9}$$
or
$$\bar{i} = a \, / \, (t + b) \tag{10}$$
in order to have physically probable rainfall volumes on a
given duration. The a and b coefficients corresponded to frer
decennial return periods.

The main hydrograph parameters were the peak discharge Qp the
time to peak Tp and the runoff volume HR. That last parameter
is not influenced in case of linear models, for the equations
4 and 7 are conservative.
As rainfall have geometric shapes, it is possible to know by
the mean of equation 4, the analytic form of each resulting
hydrograph and so to make easy the sensitivity study.

Estimation of the measurement and sampling errors   The measurement error is implicitly included in the sampling error of
equation 8. Using the plausible combinations over the variation range of the parameters, it was found that the 85 % confidence limit allowed resolution of K with a precision less than
40 %. It is hoped that with a greater sample of catchments,
this precision level would be quite a bit higher, but not less
than 20 %.

Sensitivity to the K parameter   This study was done with K
changing from 5 to 150 integration time units, for storms of
5 to 360 minutes. Incremental changes in K of $\pm$ 10 % to $\pm$ 40 %
have led to the following effects on resulting hydrographs :
- A 40 % error of K leads to changes in Qp less than 20 %. For
a given rainfall shape and a given K, the model gives a duration tM corresponding to the maximum peak runoff QpM. For the
critical pair (K, tM) which would be used in design, the maximum error is smaller.
- For a given t and K changes in peak runoff are not very influenced by the rainfall shape and the tp value. For example,
there are differences of less than 3 % between peak runoff
changes resulting form a fully advanced rainfall (tp = 0) and
those resulting from a fully delayed one (tp = t).
- A 40 % error of K leads to changes in Tp less than 15 %. For
the critical pair (K, tM) these changes are less than 7 %.
- The changes in Tp are more important for advanced rainfall
shapes which lead to smaller peaks runoff than delayed ones.

Utilization of the model
For gaged catchments, the K value may be optimized. In that
case the accuracy of the model may be very good in case of
small urban watersheds (Desbordes,M. 1974).
For ungaged catchments (storm sewer design) it has been seen
that peak runoff may be known with a precision less than 20 % ;
that is to say an error in pipe diameters less than 8 %. The
model is not very sensitive to the K value and has an adequate
precision level for the design of storm sewers projects, if
used with an adequate rainfall.

DESIGN STORM MODELLING

For design, the use of the model needs the knowledge of i(t).
As rainfall is essentially random, it is not very easy to define a design storm by only studying rainfall. As i(t) is trans-

formed by the model, one can ask for that question : for a
given K value, what are the influences of the rainfall parameters on the hydrograph parameters ? The study of these relative
influences may lead to retain only the most important parameters which may be used in a simple design storm definition.
Using the sensitivity study of the model to the rainfall parameters, the L.H.M. has answered to that question and built up
a new design storm pattern (Desbordes,M., Raous P. 1976).

Study of the sensitivity of the model to the rainfall parameters
The same rainfalls shapes were used as in the sensitivity study to the K parameter. In order to have probable storm, a random component was added to the geometrical shapes. The effects
of time discretization have also been studied.

Effects of storm duration   As it has been said, the model
shows that there exists a critical duration tM leading for a
given K and a given storm shape to the maximum peak discharge.
The model is not very sensitive to tM. Beyond t = 3 or 4 hours,
the changes in Qp or Tp may be neglected. The design storm
could have such a duration.

Effects of storm shape   Triangular and exponential shapes gives approximately the same model responses, even with differences in iM of more than 75 % and for the same tp values.
These effects on Tp may be neglected.

Effects of maximum intensity position   They are very important for Qp and Tp. For a given rainfall shape, the differences in Qp reach 20 % between a fully advanced and a fully
delayed storm. The tp rainfall parameter seems to be an important parameter of a design storm.

Effects of rainfall time discretization   It has no important
effect when the rainfall duration exceeds 6 to 8 time steps.

Effects of random component   The superimposition, on the geometrical storm shapes, of log-normally distributed variances
(as observed on real storms) with very important coefficients
of variation (reaching 300 % beyond 30 minutes from each side
of iM) have led to maximum changes in peak discharge of about
5 to 10 % for 75 % of the 600 simulations made. Greater changes, about 25 %, were observed for very short duration rainfalls, as the tp values were greatly modified by the importan
coefficients of variation.

Design storm definition   As it can be seen above, a design
storm would be accurately defined, for urban hydrology, by
means of three principal random variables :
- The mean maximum intensity $\overline{I}M$ (tM) during the intense rainfall period, tM, critical for a given small urban catchment,

and whose duration varies from 15 to 60 minutes in most cases
(definition of peak runoff).
- The mean maximum intensity $\overline{iM}$ (4 h - tM) during the complement to 4 hours of the tM period (hydrograph volume).
- The time position tp of the intense rainfall period over the
4 hours storm (time to peak to peak runoff).
- A double-triangular storm pattern as shown by the studies of
the effects of rainfall shapes and of a random component added
to these shapes. The design storm pattern is given in figure 2.

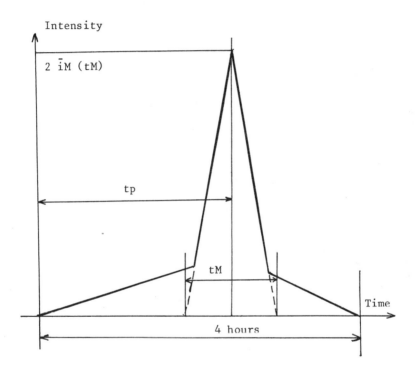

Figure 2 - Design storm pattern

Study of a long period rainfall data

The study of the statistical relationships between three random variables, has been made using the observed rainfall data
from 1921 to 1971, at the Montpellier - Bel Air Station
Desbordes,M., Masson, J.M., Ruiz, L. 1975). The storms were
sorted by means of intensity levels over different time dura-
ion, in order to have an homogeneous sample of some interest
or urban hydrology. About 250 storms were so retained. It has
een observed that the three variables were statistically in-
ependent variables. The rainfall depthes HM (tM) and

HM (4 h - tM) are exponentially or log-normally (3 parameters) distributed. The tp variable is uniformly distributed. As an example for tM = 30 minutes :

$$HM (30') = e^{0.45 u + 2.68} + 6.9 \qquad (11)$$

$$HM (4 h - 30') = e^{0.46 u} + 2.96 \qquad (12)$$

with HM(tM) = $\overline{I}M$ (tM) x tM . In equations 11 and 12, u is a standardized normal variable and the HM values are in millimeters. Moreover it was found that the storm number over a given period was controled by a Poisson's law. For the storms sample above, it has been observed that the annual storm number, N, agrees very well with :

$$Prob \left[N = n \right] = \frac{4.73^n}{n !} e^{- 4.73} \qquad (13)$$

Design storm modelling
It may be done by many ways. One can first make a general simu lation over a given period, using independent random samplings in the statistical relationships (water management problems, economical calculations, definition of the optimal frequency of storm overflow incidents,...). In that case the simulated rainfalls may be transformed by hydrological models and one can study the statistical distribution of special parameters of the resulting hydrographs in special points. This will be done independently of the statistical distributions of the rainfalls parameters.
On an another hand, in first approximation, it is also possible to give to the design storm and to its consequences, the return period T of the $\overline{I}M(tM)$ variable (peak runoff) or of the $\overline{I}M$ (4 h - tM) variable (volume runoff). The analysis of a long period of rainfall data may give the approximative return period T' of the associated $\overline{I}M$(4 h (tM) or $\overline{I}M$ variables. That last method may be used in storm sewer design with a sufficie precision.

CONCLUSIONS AND FUTURE RESEARCH

The generalized modelling of hydrological urban phenomena is very interesting way to find new calculating and management methods for complex sewerage systems. However, many other pr blems must be solved. They are for example : a better knowledge of spatial distributions of rainfall, the influences d to altitude variations, the importance of runoff or economic data associated to storm overflow incidents,...

These problems cannot be solved without an important experim al help. The L.H.M., the Ministry of Equipment and the Mini try of Interior, have started during 1976 an important measu

rement program for 11 new experimental urban catchments
(Desbordes,M., Normand, D. 1976).

Desbordes, M. (1974). Study of methodes used in storm sewers
design. Ing. Dr. Th. - University of Sciences. Montpellier.

Desbordes, M., Masson, J.M., Ruiz, L. (1975). A new definition
of a design storm. Ministry of Equipment. Paris. Report L.H.M.
14/75. 77p.

Desbordes, M., Normand, D. (1976). Urban hydrological modelling
and catchment research in France. T.M. n° IHP 8. A.S.C.E.
New-York. 57 p.

Desbordes, M., Raous, P. (1976). An example of the interest in
sensitivities studies for hydrological models. La Houille
Blanche. Grenoble. n° 1. 37-43.

Desbordes, M. and als. (1976). Study of a computer program for
sewage networks. Final report. Ministry of Equipment. Paris.
Report L.H.M. 21/76. 33p.

Normand, D. (1971). A general study of urban runoff. Ministry
of Equipment. Paris. Report SOGREAH R 10737.

Sarma, P.S., Delleur, J.W., Rao (1970). A program in urban
hydrology. Part II. Technical report n° 9. Purdue University.
Indiana.

THE DEVELOPMENT OF STORM DRAINAGE MODELLING IN CANADA

Alan R. Perks

The Proctor & Redfern Group

INTRODUCTION

The design and analysis of urban drainage systems has recently undergone a significant reorientation in Canada, as it has in many other highly urbanized countries. This has been brought about partly because of the increasing complexity and scope of urban water problems, as well as the growing trend towards systems analysis and watershed planning in all surface water administration.

In past, there has been a reliance upon rather simplistic, empirical means for hydrologic analysis in urban areas, as exemplified by the Rational Formula. Such techniques have be applied to the planning of small watershed drainage systems with little or no consideration given to the underlying hydro logic assumptions nor to the true nature of the urban hydrolc cycle. In many cases the result of this restricted drainage outlook can be measured in inadequate and inefficient drainac systems, disruption of natural water balance and supplies, costly downstream flood-control works, and pollution of numer lakes and streams in urban areas.

The administration of surface waters in Canada has historica reflected the strong political regionalization in the countr A variety of water administration agencies, with varying responsibilities, has evolved. At present, the basic sharin of these responsibilities is outlined in the following table

TABLE 1 - WATER RESPONSIBILITIES

ederal	Provincial	Shared
avigation	Water Supply	Agricultural
isheries	Hydroelectric Power	Interprovincial
		Projects
oundary Waters	Irrigation	
ederal Properties	Drainage	
	Reclamation	
	Pollution	
	Licensing	
	Development	
	Regulation	(after ref.[4])

he responsibility for urban drainage, although provincial, has
n fact been delegated to municipal governments over the years.
n the face of increasing concern about drainage problems
quantity and quality), the practical responsibility was there-
ore with the level of government least able to finance and
onduct the necessary research, and develop guidelines and
riteria applicable on a watershed basis.

he passage of the Canada Water Act (1970) signified renewed
fforts to integrate water-related planning and administration
etween the various levels of government. Under its initiative,
mproved approaches encompassing watershed-wide concerns have
volved, and are now finding their way into urban drainage
ractice.

tormwater Management represents the focus of this new planning-
riented approach on specifically urban water problems.

TORMWATER MANAGEMENT

tormwater Management may be defined as the *planning, analysis,*
nd *control* of stormwater runoff to achieve specified objec-
ives. Many of the principles and procedures encompassed by
he term are not new and have been applied for many years in
arger, rural watersheds. They are now being considered and
mplemented in smaller, urbanized watersheds because they offer
 logical approach to solving complex water problems associated
ith these areas.

omprehensive *planning* is essential to co-ordinate stormwater
acilities with transportation, land use, and other master plans.
ithout this co-ordination costly remedial works are sometimes
ecessary, and opportunities for economies of scale are lost.
lanning is the framework within which all efforts are directed
owards achieving stormwater management objectives.

Analytical methods are changing to keep pace with greater a
more detailed information needs. New hydrograph techniques
and hydrologic simulation models have been developed especi
for highly impervious areas. These are quickly displacing
traditional methods, such as the Rational formula, which ha
been over-used for many years. A hierarchy of methods has
evolved which may be selectively applied to problems of dif
ing complexity and scale.

The entire process of stormwater *control* is undergoing a si
ficant redirection. Conventional philosophy sought the rap
removal of surface water from each individual site. Althou
this ensured maximum local convenience and access, the cumu
tive effect on the watershed was increased flooding and erc
and massive downstream control and conveyance works. New
emphasis has now been placed on controlling stormwater runc
in upstream areas by storing and detaining runoff near its
source. Solutions incorporating this idea can have benefic
effects both at the individual sites and in downstream area
where reduced facilities would be required. Each problem i
different, however, and no single panacea can be relied upc
to the exclusion of all others.

URBAN DRAINAGE PROBLEMS AND NEEDS

The demand for better analytical methods was created by awa
ness of the growing magnitude of urban drainage problems ar
the inability of traditional methods to fully deal with the
At the root of most drainage problems are basic changes to
hydrologic cycle brought about by urbanization. Balanced h
logic relationships which have developed over centuries may
quickly and significantly disrupted as an area undergoes de
ment. Removal of vegetation, increased imperviousness, anc
efficient conveyance channels all serve to increase the amc
of surface runoff and increase flooding conditions at downs
locations. The causes and effects of urban water problems
summarized in the following table.

TABLE 2 - STORMWATER PROBLEMS

Causes	Effects
Removal of vegetation	Increased runoff volume and erosion.
More imperviousness	Increased runoff volume and rate, flooding, structural damage, pollutant discharge. Less groundwater and baseflow reduced natural storage.

TABLE 2 - STORMWATER PROBLEMS (cont'd)

Causes	Effects
More efficient channels	Higher velocities, increased erosion. Decreased concentration times.
Grading and filling	Increased runoff volume. Reduced natural storage, ground-water, and baseflow.
Construction activities	Increased erosion, sediment and turbidity in streams.

Traditionally, Canadian engineers have relied upon relatively simple empirical methods for designing drainage systems. The Rational Formula, primarily because of its ease of application, is still the most popular runoff estimating method. It is reasonably well suited to providing conservative peak design flows for drainage areas where the objective is to remove the water as rapidly as possible. For larger areas, however, and in situations where storage and channel routing are of concern this method has several inadequacies. These include the facts that only peak flows are produced, rainfall patterns are ignored, runoff factors must be selected subjectively, concentration times are subject to uncertainty, and the theory is truly applicable only for very small areas. Many engineers were quick to realize that this method would not be adequate to deal with very large drainage systems, combined sewer flooding, complicated systems with inter-connections, and renewed interest in storage and detention of stormwater.

The same need was recognized by the Governments of Canada and Ontario, who were increasingly concerned with the contributions of urban runoff and combined sewer overflows to the pollution of the Great Lakes. A joint Urban Drainage Subcommittee[1] was formed to investigate this problem and develop a program of investigation with 3 basic goals:

1. To define the problem
2. To develop solution capability
3. To develop a strategy for implementing solutions.

Figure 1 shows the key program activities and some of the projects undertaken in each. The program has achieved considerable success over the last 4 years, and is presently completing its final tasks. A number of these projects were undertaken by consulting engineering firms.

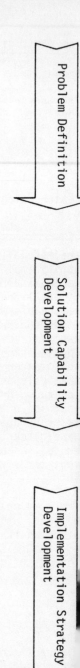

Problem Definition

Typical Projects:

● Monitoring

● Great Lakes
Loads evaluation

● Stormwater
Bacteriology

Solution Capability
Development

Typical Projects:

● Treatment studies

● Urban drainage
modelling

● Modelling demonstration
studies

Implementation Strategy
Development

Typical Projects:

● Domestic and foreign
practices review

● Seminars

● Manual of Urban
Drainage Practice

● Policies and
guidelines

FIGURE 1 - URBAN DRAINAGE PROGRAM

THE CANADIAN STORMWATER MANAGEMENT MODEL

A key study in the Urban Drainage Subcommittee's program was
the development of a Canadian Stormwater Management Model[5].
The approach taken was the modification of an existing runoff
model, the U.S. Environmental Protection Agency STORMWATER
MANAGEMENT MODEL, rather than the development of an entirely
new model which might be more costly and not offer any significant advantages over existing ones.

A previous study[2] had been carried out which consisted of a
review of Canadian design practice and an assessment of available runoff models. This study found that, although the
Rational Formula was almost universally applied in Canada,
there was widespread interest in Stormwater Management concepts
and models. Three models were selected for detailed comparison
and evaluation; the U.S. Environmental Protection Agency Stormwater Management Model (SWMM); the University of Cincinatti
Urban Drainage Model (UCUR); and the British Road Research
Laboratory Model (RRL). The study concluded that runoff models,
in general, were sufficiently well developed for practical
application. It also concluded that the Stormwater Management
Model (SWMM) gave the best overall performance, and had further
advantages of the most sophisticated Transport model, greater
flexibility, a water quality modelling capability, and was
continuously being updated through research and development.

Based upon the results of this previous study, and considering
differences in the Canadian climate, engineering practices, and
costs, a second study was commissioned with the following
objectives:

1. To review and modify the U.S. EPA SWMM for Canadian
    conditions.

2. To select and modify a submodel for the simulation
    of snowmelt runoff quantity and quality.

3. To develop a submodel capable of extracting model
    input data from existing data banks.

4. To select and modify a high-speed model for simulation
    of the frequency of storm discharges and overflows.

The Urban Drainage Subcommittee contracted the firm of Proctor
& Redfern Limited to conduct the study, which was completed in
December 1975. The major components of the study are described
in the following sections. A brief description of the SWMM is
provided to introduce the study.

Brief Description of the Basic SWMM

The STORMWATER MANAGEMENT MODEL is a detailed, single event
model which simulates the quantity and quality of urban storm
runoff in a drainage network, resulting from a real or design
storm. The SWMM accepts as input a detailed description of th
urban drainage system broken down into its component subcatchment areas and drainage conduits, as well as a rainfall intensity distribution, or hyetograph. The model then performs an
accounting of the movement of the resulting stormwater through
the drainage system, including overland flow routing, hydrauli
routing of subcatchment hydrographs through the sewer system,
flow and quality routing through any storage or treatment
facilities, and simulation of the receiving water body effects
The model is structured in 4 main computational blocks, each
representing a major phase of the drainage cycle.

The RUNOFF Block computes an overland flow hydrograph for eac
specified subcatchment. Each subcatchment is characterized b
its size, imperviousness, slope, infiltration potential,
depression storage, and several factors relating to the
accumulation of surface pollutants. For each time interval,
a sequential computation is made of rainfall, infiltration,
depression storage, net rainfall excess, and subcatchment
outflow rate according to Manning's equation and the continui
equation.

The TRANSPORT Block accepts as input a geometric description
the sewer system, and combines and routes the various subcat
ment hydrographs to the outlet using a modified kinematic wa
approach.

The STORAGE Block simulates the effect of any storage or tre
ment facility on the runoff hydrograph. Treatment processes
are simulated by user-selected strings of formulas represent
the desired processes.

The RECEIVING Block represents a river or lake system by a
figuration of channels and nodes. Water level and pollutan
concentration variations are solved using the equations of
continuity, motion, and conservation of mass.

Activities in the Canadian Study

Basic Data New urban runoff data was collected and reviewe
from 9 different Canadian sites. Three areas were selected
use in the study, ranging in size from 48 acres to 2,330 ac
None of the areas provided data of completely acceptable qu
but together were sufficient for use in various parts of th
study.

Debugging At the time this study was commenced, the basic
SWMM version obtainable from the U.S. Environmental Protec
Agency was still undergoing rapid modification and develop

The Canadian study team corrected numerous program inconsistencies, which were also incorporated into the U.S. version. Included in this category were modifications to the overland flow routines, gutter routing routines, pollutant washoff routines, and storage simulation routines. Considerable efforts were required in this area which were not fully anticipated at the start of the study.

Flow and Quality Simulations   Flow and quality simulations were performed using the new Canadian data. The flow results were very encouraging under a variety of different storms and watershed conditions. Sensitivity analysis confirmed most of the default values in SWMM RUNOFF and TRANSPORT Blocks. Under surcharged conditions, the SWMM results were found to be generally unsuitable for detailed design or analysis, as compared with simulations by the proprietary Dorsch Consult HVM and the Water Resources Engineers SWMM, both of which simulated surcharged conditions.

The quality simulation routines of SWMM were found to be less accurate than the flow routines. In most simulations only a general "order-of-magnitude" correlation was achieved with observations. Sensitivity analysis indicated that the detailed quality routing and decay routines were also not significant in terms of the computed results. Because of these factors, a simplified approach to quality simulation was suggested.

The simplified quality model was developed using the basic SWMM theory but eliminating the concepts of quality routing and decay, and sub-catchment discretization. This model was tested successfully using data from several catchments and also against the detailed SWMM results. This model was intended for use in cases where numerous quality simulations were required, or for use with any other hydrologic model. The model may be easily calibrated and applied to numerous areas and events at relatively little cost.

Snowmelt Simulation   One of the major modifications was the development of a routine for the simulation of snowmelt quantity and quality. This represented one of the first attempts to simulate snowmelt in urban areas, which may be the critical runoff event in some Canadian cities.

Anderson's snowmelt model was selected following an extensive search of available literature, and was incorporated into the SWMM as an optional feature. Additional input data required by the new routine were a description of the snowpack distribution and its physical parameters, hourly air temperature, and wind speed. Snowmelt quality was simulated by extending the list of SWMM quality constituents to include chlorides and lead. Only very preliminary verification of the snowmelt model was possible due to lack of suitable data.

Simplified (Planning) Simulation Use of the SWMM as a multi-event, or continuous simulation model was investigated by reducing the number of subcatchments and sewer elements include in the simulation. Little information was available regarding the effects of fewer subcatchments on simulation accuracy, and some conflicting results were reported in the literature. Simulation of several watersheds was carried out at varying levels of discretization, from 40 or more at one extreme to a single subcatchment at the other. Hydrologic parameters of the lumped subcatchments were defined as spatial averages.

It was found that by simulating fewer subcatchments, and con-sequently fewer pipes, a great deal of conduit storage was ignored which, through the routing process, created less attenuation of the runoff hydrograph. The procedure developed was to compensate for this neglected pipe storage with increas surface storage achieved by reducing the subcatchment "width" parameter of the simplified subcatchments.

Good results were achieved using both fewer subcatchments and larger time steps, which introduced the concept of using SWMM to simulate numerous events or for continuous simulation at reasonable cost. This latter use would require modification of SWMM, for moisture balance accounting between storm events which was not performed in this study.

Other SWMM Refinements A number of other modifications and refinements were made in this study. The Infiltration routin were reviewed and related to Canadian practice, the RECEIVINC Block was debugged, applied, and compared with other models, the TREATMENT Block was modified to reflect current Canadian treatment practices and costs, and the Storage routines were tested and debugged.

Continuous Simulation The available continuous simulation models were reviewed and compared. The U.S. Corps of Engine "STORM" model was selected and applied to Canadian watershed with very good results. STORM results for individual simula were compared with detailed SWMM results, and the idea of in facing the two models was developed, in which STORM could be used to identify critical events for subsequent SWMM simulat

Data Analysis Model The final study task was the developmer a data submodel to analyse meteorological data and provide ` to SWMM and STORM. This model was developed with capabilit for reading the Canadian Atmospheric Environment Service Climatological data bank, screening the precipitation and temperature data for errors, combining various records into single output file, summarizing storm events, and creating suitable model input data.

Distribution With the completion of this study, the Canadian
SWMM was made available through publication, and the staging of
several user seminars which were heavily attended.

OTHER DRAINAGE MODELS

In addition to SWMM and STORM, several other urban drainage
models are available and widely used in Canada. Two of these
constitute, with SWMM and STORM, what is regarded as a reasonable hierarchy of models for selective application to all types
of drainage problems.

The Illinois Urban Drainage Area Simulator (ILLUDAS)[6]
The Illinois State Water Survey Urban Drainage Area Simulator
is a single-event urban runoff model based upon the British
Road Research Laboratory Model. It differs from that model,
however, in that it computes runoff from pervious areas and
has the capability of sizing circular pipes or retention basins.
ILLUDAS does not model water quality.

Surface runoff hydrographs are computed for each specified
subcatchment, composed of grassed area runoff, directly connected impervious area runoff, and indirectly connected
impervious area runoff. These surface hydrographs are then
accumulated and routed downstream to the outlet. Storage may
be specified as a detention volume at any point, in which case
the decrease in peak flow will be reported, or alternatively a
limiting discharge rate may be specified, in which case the
required storage volume will be reported. ILLUDAS is relatively easy to use, has been well tested, and is well documented.

The U.S.D.A. Hydrologic Model (HYMO)[7]
HYMO is based upon a synthetic unit hydrograph procedure, and
was designed for use in flood control studies and general
watershed hydrology investigations. The model computes flood
hydrographs from specified subcatchment areas, and routes
these hydrographs downstream through channel and reservoir
systems. It was designed primarily for larger natural watersheds, but has more recently been applied in urban areas.

The net rainfall excess is determined by the U.S. Soil Conservation Service procedure. The basic unit hydrograph for each
subcatchment can be calibrated to measured hydrographs by
directly inputting the time to peak and a recession constant,
or these can be computed internally based upon regression
equations relating to watershed area, length, width, and slope.
Channel routing is by the Variable Travel Time Method, and
reservoir routing by the Storage-Indication Method. The model
does not consider water quality, although sediment yield can be
estimated.

HYMO is well documented and is relatively easy to use, although its applicability to small urban watercourses has not been established to date.

Use of Models

No single method for runoff analysis will satisfy the require-ments of all types of drainage studies. The best approach lie in selecting the model or technique whose capabilities best suit the requirements of the problem at hand. Figure 2 illus-trates how the models discussed in this paper form a hierarchy which may be used to solve various drainage problems.

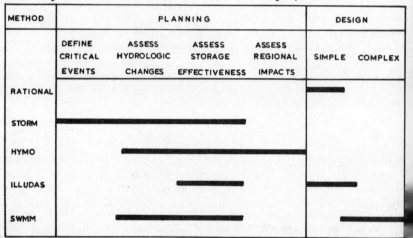

FIGURE 2 - USE OF DIFFERENT RUNOFF METHODS

MODELLING APPLICATIONS

A large number of modelling applications have been reported recently in Canada. Several applications carried out by the author's firm are briefly described in the following section These applications demonstrate the range of capabilities in the various models.

Hamilton - Combined Sewer Analysis   The SWMM was employed to analyse the behaviour of 2 combined sewer systems under desi storm conditions. Each area was modelled separately, includ most of the local sewers. Runoff data was collected with wh to calibrate the models. The problem of surcharging was ana ysed using the Extended SWMM TRANSPORT Block which simulates surcharge behaviour. The simulations revealed that replacem of the main trunk sewer would only partially relieve the flo ing problem, since the lateral sewers were also inadequate convey the storm runoff. An improved relief system was ana ysed utilizing a smaller, cut-off relief sewer running para to the trunk, and intercepting several of the laterals at t points where surcharging was first revealed.

The STORM model was used simultaneously to estimate the magnitude and frequency of combined sewer overflows to the receiving watercourse. No suitable long-term data was available with which to calibrate the model.

Hamilton - Combined Sewer Storage In another application, the STORM model was used to study the feasibility of storing combined sewer overflows from a large 3,000 acre combined catchment. A preliminary analysis of separation costs showed these to be far beyond available funds.

Flow records at the overflow chamber were used to calibrate STORM computations over one year. The computed flows were found to be within 6% of the measured total over this period. The model was then used to process 14 years of observed rainfall data and simulate the number and magnitude of overflows that would result from different sizes of storage. The study recommended a 20 MIG combined storage facility, and estimated that overflows would thereby be reduced by 90%.

Toronto International Airport A portion of the apron area was subject to frequent ponding occurrences, and a proposed extension to this apron threatened to create worsened conditions. The SWMM was employed to design a new trunk interceptor to serve the problem area plus a portion of a neighbouring drainage area. Simulations were carried out to determine levels of surcharging in the original area and in the new drainage system under severe rainstorms.

St. Catharines - Combined Sewer Analysis The City of St. Catharines is served by two major trunk systems which suffer basement flooding in times of wet weather. In a first stage of study, the SWMM was used to assess the trunk sewer capacity and performance under a variety of storms for which some flow records were available.

A second stage, currently underway, is analysing selected drainage areas where flooding problems exist.

Port Colborne - Storm Sewer Study The City of Port Colborne is situated on flat, low-lying land adjacent to Lake Erie. High lake levels combined with winds seriously reduce the capacity of existing storm sewers.

The SWMM was used to evaluate the capacity of the existing system, investigate the causes of street flooding, and design a new system to serve the entire city. Alternative drainage concepts utilizing existing systems and available storage areas were studied.

Burlington - Storm Sewer Study The SWMM is being used to
analyse the existing storm drainage system in the downtown
core area, made up of both pipes and open channels. This
analysis is part of an overall storm, sanitary, and water
system improvement study. For major storms when the surface
runoff predominates, both SWMM and the unit hydrograph model
HYMO are being utilized and results compared.

Mississauga - Assessment of Detention Measures Mississauga
is one of the fastest growing urban areas in Canada. Vast
tracts of land are being developed for residential uses, with
the resulting impacts of runoff rates.

The SWMM was used to demonstrate the effectiveness and benefi
of stormwater detention to a major developer. A particular
watershed, Loyalist Creek, was simulated under conventional
drainage designs, and also with disconnection of roof downspouts, roof storage in industrial areas, and detention
facilities at selected points. Comparative hydrographs were
produced illustrating the hydrologic benefits of these measur

Kitchener - Innovative Drainage Design Many Canadian municipalities have requested that stormwater management principles
be embodied in new development designs. In this study the
SWMM and ILLUDAS models were applied to study and compare
different drainage plans. The area in question is a small
watershed in which drainage problems were anticipated at dow
stream locations after development. The soil conditions wer
highly pervious, and it was desirable to promote infiltratio
as much as possible. Models were used to simulate configura
tions of rear-lot greenways leading to multiple stormwater
control areas. Land runoff and roof discharge are directed
to these rear-lot greenways where the stormwaters would be
slowly conducted to control areas. Only minimal roads sewer
would be installed to conduct runoff to the greenway. Under
major storms, greenways and selected roadways would function
as emergency flood channels.

Barrie - Stormwater Drainage Study This project concerned a
2,500 acre watershed, the upper portion of which was rural,
the lower portion highly urbanized. The lower channel was
severly constrained, and frequent flooding resulted.

The SWMM and HYMO were employed to identify the severity of
drainage inadequacy under existing and future development
conditions. Subsequently, the models were employed to asse
various stormwater management measures which might be appli
to make full use of downstream capacity but avoid any massi
re-channelization. Several alternative detention schemes w
evaluated.

Transport Canada - Guidelines for Airport Stormwater Management
Many municipalities and agencies who have not been involved
with modelling concepts are unclear as to the direction in which
urban drainage is evolving. This report was completed to serve
as a guideline for application of stormwater management at
airport properties. Large airports may significantly alter the
hydrology of small watersheds, and increasing problems are being
encountered with downstream property owners.

The report presented and discussed applicable stormwater management objectives, precipitation analysis, runoff analysis, storage,
and the conduct of different drainage studies. The SWMM, ILLUDAS,
STORM, and HYMO models were recommended for use.

Other Applications These are some of the modelling applications
undertaken by Proctor & Redfern in the last 2 years. Table 3
briefly summarizes other applications reported in the literature
[3]. The range and variety of applications indicates the wide
ranging interest in modelling techniques which has evolved in
Canada recently.

TABLE 3 - MODELLING APPLICATIONS

Location	Models Used	Objectives						
		Analyse Sewer Systems & Relief	Analyse Urban Drainage Systems	Assess Impacts of Development	Assess Storage & Detention Measures	Estimate Quality of Runoff & Overflows	Master Planning	Guidelines and Criteria
Halifax	ILLUDAS				X			
Ottawa	SWMM, STORM			X	X			
Midland	STORM			X		X		
Toronto International Airport	SWMM, STORM	X		X		X		
Port Credit	SWMM	X						
Toronto	SWMM, HYMO		X	X			X	
Hamilton	SWMM	X				X		
Ontario	STORM					X		
Winnipeg (A)	SWMM	X					X	
Winnipeg (B)	SWMM							X
Edmonton	SWMM	X						
Burlington (A)	ILLUDAS, HYMO		X	X	X		X	
Burlington (B)	SWMM, ILLUDAS, HYMO							X
Waterloo	HYMO		X		X		X	
Kitchener	SWMM, HYMO		X	X	X		X	

CONCLUSIONS

At present it is generally regarded that, for estimating runoff
quantity, sufficient reliable models exist for practical application in analysis and design. The emphasis is now upon applying the various models in appropriate cases.

A great deal of modelling activity is evident across the count'
from reported applications. There appears to be a general sat
isfaction with the performance of models among those responsib
for the administration of drainage systems, which is further
adding to the momentum.

A strong need has been perceived to establish guidelines and
criteria under which modelling applications may be carried out
At present only a few jurisdictions specifically mention model
in criteria manuals, which makes it difficult for those unfamiliar with the models to initiate new applications.

Provincial government policies are now being formulated which
encompass stormwater management. One of the difficulties lie:
in resolving administrative conflicts between urban drainage
and floodplain management, which must be considered simultaneously in watershed planning but responsibilities for which
remain separated in most areas of the country.

REFERENCES

[1] Environment Canada, "Report of the Urban Drainage Sub-committee Program", Environment Protection Service, Ottawa, 1976.

[2] James F. MacLaren Ltd., "Review of Canadian Design Practice and Comparison of Urban Runoff Models", Research Report 26, Canada-Ontario Agreement, Ontario Ministry of the Environment, Toronto, 1975.

[3] Marsalek, J., "Urban Hydrologic Modelling and Catchment Research in Canada", Environment Canada, Technical Bulletin No. 98, Inland Waters Directorate, Canada Centre for Inland Waters, Burlington, Ontario, 1976.

[4] Proctor & Redfern Limited, "Airport Stormwater Management", a report submitted to Transport Canada, Surface Structures Division, Ottawa, 1977.

[5] Proctor & Redfern Limited, and James F. MacLaren Ltd., "Storm Water Management Model Study", Research Report 47, Canada-Ontario Agreement, Ontario Ministry of the Environment, Toronto, 1976.

[6] Terstriep, M.L., and Stall, J.B., "The Illinois Urban Drainage Area Simulator", Illinois State Water Survey Bulletin 58, 1974.

[7] Williams, J.R., and Hann, R.W., "HYMO: Problem-Oriented Computer Language for Hydrologic Modelling", U.S. Department of Agriculture, Agricultural Research Service, 1973.

URBAN STORMWATER MANAGEMENT THROUGH LAND USE CONTROLS IN THE UNITED STATES

Robert L. Hiller[1], Eugene J. Riordan[2], Neil S. Grigg[3]

INTRODUCTION

Rapidly developing areas in the U.S. are experiencing significant alternations in watershed hydrologic response, both in
quantity and quality. The urbanization-induced reduction in
pervious land area and increase in stormwater removal efficiency (lined channels, concrete drains, etc.) are causing
measured increases in peak runoff rate and total runoff volume.
The quality of this "urban" stormwater runoff is significantly
lower than its "rural" counterpart due to loss of the watershed's natural filtration system and the accumulation of urban
residues (oil, gas, etc.) being deposited within the developing watershed.

Due to upstream growth patterns which are predominant in the
U.S., the impacts of these alternations are generally not expressed within the developing area itself but rather downstream in previously developed sections of town. This relationship between cause and impact creates a need for proper
drainage-related control and management of the developing area.
Without such control, the developer or builder "sees" no
need to internalize the drainage impact costs generated by his
development. The burden of these development-induced impacts
(expressed either as downstream damages due to increased flooding, or costs associated with upgrading existing facilities

1. Assistant Professor, Natural Resources Law, Dept. of Agricultural and Chemical Engineering, Colorado State University, Fort Collins, Colorado, USA.
2. Graduate Research Assistant, Dept. of Civil Engineering,
   Colorado State University, Fort Collins, Colorado, USA.
3. Director, Water Resources Research Institute of the University of North Carolina, Raleigh, North Carolina, USA.

to accommodate the altered stormwater flows) are then unfair)
borne by existing developments.

Unfortunately, need does not necessarily guarantee a solution
The requirement for drainage-related growth management must b
satisfied within the confines of constitutional law, the law
of surface water; and just as important, within the social,
economic, and political desires of the community. In additic
a drainage-related growth management program must address al]
phases of development, from raw undivided land to established
commercial development. Control of drainage with respect to
new developments is extremely important but represents only a
part of a successful management program since all phases of
development, each with specific drainage problems that need t
be corrected, will exist when a developed program is implemented.

This paper discusses drainage programs currently enforced in
the U.S. It begins with a brief summary of the law of surfac
water drainage, and the authority by which local governments
in the U.S. are enabled to establish drainage control program
Next, court cases pertaining specifically to drainage ordinar
ces or drainage-related cost recovery programs are examined
for the necessary legal criteria upon which these ordinances
must be based, and then some of the typical U.S. drainage ordinances are categorized and discussed. Finally, the paper
closes with a discussion of what appear to be "optimal" criteria for a drainage program and the legal interpretations
needed to guarantee the successful implementation of this
program.

LAW OF SURFACE DRAINAGE

There are three different rules that govern the drainage of
diffused surface waters in the U.S. They are the "common
enemy" rule, the "natural flow" rule and the "reasonable use"
rule. The "common enemy" or "common law" rule is a doctrine
that suggests that surface runoff is a common enemy and can
be dealt with as one wishes in order to protect one's propert
regardless of any injury so caused. This doctrine, which
apparently grew from a desire to promote land development in
the 19th Century, has been tempered over the years to include
a requirement of reasonableness in one's actions to protect
himself against this "common enemy." Wilful, wanton and
malicious conduct concerning the disposition of surface water
can now result in liability of the landowner under this
doctrine. (Beck, 1972)

The "natural flow" or "civil law" rule is based on the aqua
currit maxim -- "water runs as it is wont to run, and ought
to run." Following the laws of nature, this doctrine assumes
the existence of a natural eastment or servitude over lower

lands within a drainage basin. The owners of these servient lands cannot obstruct the natural flow of water to the detriment of the dominant (or higher) landowner. It should be noted that if strictly construed, this doctrine does not require a servient landowner to accommodate development-induced changes (in quantity or quality) in stormwater flows, only natural flows. Like the "common Law" rule, this doctrine has been modified over the years to include a reasonableness criterion. Servient landowners may now modify existing drainage patterns as long as the modifications are reasonable, and dominant landowners may increase runoff above natural flows as long as the increase is not ruled unreasonable.

The "reasonable use" doctrine is concerned with the rights of individual landowners, be they dominant or servient. This doctrine states that any landowner may make reasonable use of his land even though such use may alter surface water flow to the injury of others. Three questions that recognize the similarity and equality of the rights of each land owner are basic to an analysis by this doctrine (Beck, 1972):

1. Was there reasonable necessity for the actor to alter the drainage to make use of his land?,
2. Was the alteration done in a reasonable manner, with due care to prevent injury to another's land?, and
3. Does the utility of the actor's conduct reasonably outweigh the gravity of the harm to others?

It is apparent that the "reasonable use" doctrine is not significantly different from the modified interpretations of the doctrines of "common Law" and "civil law." The only suggested difference among all three is in "the practical question of prediction and proof." (Beck, 1972) That is, the basic philosophy behind the doctrines is identical (reasonable use of one's own property) but each places the burden of proving unreasonableness on different actors (either the servient or dominant landowner) resulting in different outcomes.

It is important to know which of these doctrines, or modifications thereof, applies when developing drainage programs that attempt to recover the costs of drainage facilities from the beneficiaries of those facilities. For example, if the prevailing doctrine was the "common enemy" rule, it would be difficult to assign benefits to a dominant landowner in an effort to recapture costs of drainage facilities constructed on lower lands. On the other hand, benefits could be more easily assigned if the "reasonable use" or "civil law" doctrine prevailed; the dominant landowner enjoys the benefit of being able to collect and discharge stormwater in excess of natural flows or in an unreasonable manner. This collection and discharge of increased stormwater flows to a servient landowner is precisely the hydrologic outcome of the upstream development patterns in the U.S.

AUTHORITY TO ESTABLISH DRAINAGE CONTROL PROGRAMS

All of the modern drainage-related cost recovery programs find their origin in the early storm drainage and reclamation districts. These districts, affirmed in the courts, were based on the "police power" authority and the "power to tax" authority of the legislature. The "police power" authority is a right vested with a sovereign to require owners of property to use their property only to the extent that such use does not preclude a neighbor's reasonable enjoyment in his land. This authority allows regulation, management, and control of private property to promote the general health, safety, and welfare of the community. The "power to tax" is a right similarly vested with a sovereign to make exactions from the general public to pay for services that are rendered in the interest of the general health, safety, and welfare of the community.

Currently enforced municipal drainage ordinances account for the societal and environmental needs and desires that have grown over time. It cannot be ignored that these ordinances regulate and tax private property far more than the early drainage districts, yet they still rely on the "police power" authority for affirmation. The expansion of the "police power" authority required to support these ordinances has been upheld and the U.S. Supreme Court in Euclid v. Ambler, the case establishing the general constitutionality of zoning, suggests the necessity of such expansion.

> Regulations, the wisdom, necessity, and validity of
> which, as applied to existing conditions, are so
> apparent that they are now uniformly sustained, a
> century ago, or even half a century ago, probably
> would have been rejected as arbitrary and oppressive.
> Such regulations are sustained, under the complex
> conditions of our day, for reasons analogous to
> those which justify traffic regulations, which,
> before the advent of automobiles and rapid transit
> street railways, would have been condemned as
> fatally arbitrary and unreasonable. And in this
> there is no inconsistency, for, while the meaning
> of constitutional guarantees never varies, the scope
> of their application must expand or contract to
> meet the new and different conditions which are
> constantly coming within the field of their operation.
> In a changing world it is impossible that it should
> be otherwise.[1]

---

[1] See footnotes at the end of this paper for citations to legal references.

DRAINAGE-RELATED CASE LAW

The previous section charted the authority for drainage facility recapture programs. Unfortunately, this authority does not specify the necessary requirements of these cost recovery programs and a great deal of litigation questioning the equitability of some of them has resulted. The specific questions which the courts have tried to answer are whether or not a particular cost recovery program insures that:

1. The calculated benefits from a drainage project exceed the calculated costs,
2. The benefits from a drainage project accrue to the landowners in the area being assessed,
3. The assessment costs have been properly and fairly apportioned among the landowners in the assessment area, and
4. The benefits from, and costs of the drainage project have been properly calculated.

The following examination of court cases dealing with these questions attempts to establish the equitability criteria upon which a drainage-related cost recovery program should be based to withstand judicial review.

Benefits greater than costs
A common element noted in the various legislation that authorize the establishment of improvement districts (including drainage districts) is the requirement that the costs of constructing the improvement shall not exceed assessed benefits. The courts have consistently upheld this requirement.[2] Unfortunately, the term "benefit" is generally not specifically defined within the authorizing legislation, and the courts have had to interpret the legislatively implied definition of "benefit." The resultant broad range of interpretations makes it difficult to determine project benefits and costs. A standard interpretation is needed to insure the proper and consistent identification of the special and general benefits that can be included in a computation of project benefits and the relative weights of each. Such a breakdown has recently been developed (Shoemaker, 1974) but has not enjoyed widespread usage.

Benefits accruing to assessed area
In addition to demanding that the benefits of a project are greater than its costs, the authorized agency (district, municipal government, etc.) must insure that the proposed drainage project especially benefit the area to which the cost assessments are made. A California court in City of Buena Park v. Boyar upheld the collection of a drainage fee which was to be used expressly for a drainage project that would benefit the development to which the fee was assessed.[3] In addressing an earlier California case (Kelber v. City of

Upland)[4] where the court held a similar drainage fee invalid, the Boyar court differentiated the two cases stating that, "...in Kelber v. City of Upland, the city (of Upland) could use the collected fees anywhere in the city..."[5] (emphasis added).

This requirement for "special" benefit has similarly been upheld in a number of different situations. In Duncan v. St. John Levee and Drainage Dist. for example, the court found that deliquent payments on bonds issued for a certain portion of a drainage district--with separate and distinct benefits accruing to it--cannot be paid back with money collected from bonds issued over the other portions of the district[6]. The court stated that this would amount to taxing property for benefits that didn't exist, and that "Any attempt by taxing authorities to impose a burden without a compensating advantage is power arbitrarily exerted, amounts to confiscation and violates the due process provisions of the 14th amendment"[7]. This citing of Constitutional guarantees is prevalent in improvement district cases and appears in a case dissolving a drainage district. The court, in Thibault v. McHaney, was determining the amount of authorized claims against the district and states, "...from Kirst v. Street Improvement District No. 120:[8] 'Special assessments for local improvements find their only justification in the peculiar and special benefits which such improvements bestow upon the particular property assessed. Any exaction in excess of the special benefits is, to the extent of such excess, taking of property without compensation.' "[9]

Proper apportionment of costs

The law regarding the equitability of a particular cost recovery program is clear; the courts have universally maintained that an assessment program or cost recovery program is a matter under legislative control and not normally subject to judicial overrule. In Luckehe v. Reclamation District No. 2054, the courts affirmed an assessment for cost recovery and maintained that "...the formation of a reclamation district is a legislative act carried out in the exercise of the police or taxing power of the state."[10] The importance of legislative authority in this area is nicely expressed in Funkhouser v. Randolph in which the court found that a law providing for the organization of the Little Wabash River Drainage District was void because it directed the county court to "decide legislative questions," namely, the extent of the district, who benefits, etc.[11] In still another case, Reclamation Board v. Chambers, a California court in determining the legality of the state appropriating money from the general fund for payment to a reclamation district explained:

The method of paying for the same (drainage works)
is solely a matter of legislative discretion. The
state, if it elects, may pay all the costs, or
place the same upon the lands specially benefited
by the work, or it may in its discretion divide the
burden between the landowners and the state in such
proportions as the state may deem equitable[12] (emphasis
added).

The courts, however, have recognized the problems with this
"blanket" legislative authority and have warned that the
judiciary can invalidate legislative actions in this area if
there is a clear indication of legislative impropriety. Unfortunately, the courts have given varied interpretations of
legislative impropriety. On Hurley v. Board of County Commissioners of the County of Douglas, the Kansas Supreme Court
maintained that a sewer assessment scheme (equal acreage
charge throughout service area) based on "equal benefits" is
not proper since all lands within the district are simply not
benefited equally.[13] The assessment was ruled "unjust, unreasonable, discriminatory, and grossly disproportionate to
the benefits received." In this case the action by the administrative body that should have been "conclusive on property
owners and courts," was nullified because it was not "fair,
just, and equitable."

In contrast to the Hurley case, the Kansas Supreme Court in
City of Wichita v. Robb upheld a state law which apparently
taxed landowners for costs associated with drainage works
within the Arkansas River basin who would not enjoy any direct
benefits.[14] The court ruled that "...the legislature may
exercise its discretion in fixing a taxing district for drainage or flood control projects, and its action in so doing is
not open to judicial inquiry unless it is wholly unwarranted
and a flagrant abuse and by its arbitrary character is a mere
confiscation of particular property."[15] The court relied
heavily on the "general" benefit principal stating that, "...
benefits to a taxpayer conferred under a drainage and flood
control project may be direct or tangible, or they may be indirect and intangible where they redound to the benefit of the
whole taxing district in which he is a taxpayer."[16]

A Florida case further illustrates the variety of interpretations of legislative impropriety. In Board of Supervisors of
South Florida Conservancy District v. Warren, Governor, the
Florida Supreme Court affirmed the assessment of benefits to
the plaintiff who claims that his lands were not benefited in
any way by the reclamation project.[17] The court acknowledged
that the plaintiff had constructed, on his own, certain on-
site structures for reclamation but that the benefits of the
assessment project in question go beyond simply direct benefits
to particular parcels. The dissenting opinion, however places

significantly more weight on the criterion of "special" benefits, and disagrees with the reasonableness of the assessment
to plaintiff's property. The opinion feels that the facts of
the case support a holding that the assessment was an abuse of
legislative discretion![18]

Proper computation of benefits and costs

The presumption of validity of legislative action has seemed
to limit the scope of court review of drainage programs to
questions of cost apportionment. There has been virtually no
action taken questioning the technique used to calculate the
hydrologic information necessary to compute project benefits
and costs. This failing is dosconcerting because all of the
drainage programs with cost recovery provisions in the U.S.
seem to be based on a Master Plan of Drainage--a plan that
requires high initial costs. The question facing young, rapidly urbanizing communities is whether or not they can collect
drainage fees based on a "simplified" drainage plan that does
not require severely high initial costs. Would the courts
rule this simplification a "gross abuse of legislative discretion?" We can guess from Kelber v. City of Upland[19] that
without any sort of drainage plan, the courts will have a
tendency to invalidate a fee collection program, but the question of reasonableness of a "simplified" plan has not been
addressed.

U. S. DRAINAGE ORDINANCES

The drainage ordinances in the U.S. have grown from "police
power" and "power to tax" authorities and from the court
rulings regarding other cost recovery programs. The previous
sections of this paper have illustrated how the "police power"
authority and the case rulings can, and will, change over time;
it should not be surprising that this variability has resulted
in a variety of approaches to drainage-related cost recovery.
At one end of the spectrum, there exist drainage ordinances
that deal only with new development and at the other end there
exist drainage ordinances that permit drainage control and
management over all phases of development (from raw undivided
land to existing populated areas). A review of all of these
ordinances suggests dividing a particular ordinance into two
portions for ease of discussion: one portion that deals ex-
clusively with requirements imposed upon developers and builders (New Development), and another that deals with requirement
imposed upon subdivided land (or home) owners (Existing Development). As implied above, not all drainage ordinances will
necessarily contain both portions.

New development

All of the existing ordinances reviewed set some requirement
for drainage within a proposed new development. Some of them
(Tampa, Florida) address the satisfactory drainage within the

new subdivision only, and make no mention of where the drainage waters collected within the development should be discharged. Other drainage ordinances are more specific requiring, in the case of Colorado Springs, Boulder, and Arvada, Colorado, the developer to insure that all his storm runoff waters and those draining onto his property are properly conveyed to a designated outfall--the costs of this conveyance to be borne by the developer.

In other instances, the ordinances confine themselves to on-site drainage, but attempt to insure against drastic alterations in hydrologic response due to development by requiring detention of stormwaters. This detainment of water is accomplished by either on-site ponds or regional ponds as determined by the local authorities. The Dekalb County, Georgia drainage ordinance is a good example of this approach, and the Metropolitan Sanitary District of Greater Chicago accomplishes similar objectives through a sewer permit issuance program that essentially mandates on-site detention of stormwater runoff.

In addition to on-site drainage facilities the more "advanced" drainage ordinances provide for off-site drainage improvement fee collection. These fees are generally referred to as Drainage Fees and are based on the rationale that upstream developers are impacting downstream drainage facilities (even with the installation of their on-site improvements). The fees are collected to either upgrade inadequate drainage facilities, or construct new downstream drainage facilities. This thinking is explicit in the Fairfax County, Virginia's zoning ordinance, Chapter 30 entitled, "Pro-Rata Share of Costs for Drainage Facilities," wherein they state:

> The purpose and intent of this section is to
> require a subdivider or developer of land to pay
> his pro-rata share of the cost of providing
> reasonable and necessary drainage facilities,
> located outside the property limits of the land
> owned or controlled by the subdivider or developer,
> but necessitated or required, at least in part,
> by the construction or improvement of his sub-
> division or development.

The apportionment of these drainage fees that "specially" benefit landowners within a particular basin have varied from ordinance to ordinance--the two most popular being the "Acreage Fee" and the "Land Use Fee." The "Acreage Fee" assumes equal benefit throughout the drainage basin and is calculated by dividing the total cost of the proposed basin improvement by the total land area within the drainage basin. The Arvada, and Colorado Springs, Colorado drainage ordinances use this 'Acreage Fee" apportionment method.

The "Land Use Fee" is computed in essentially the same way as
the "Acreage Fee" except that land use is considered. This
apportionment approach recognizes that single-family development does not alter the hydrologic response of a watershed to
the extent that a shopping center does and hence should not be
assessed the same acreage fee. Fairfax County, Virginia has
used this rationale to develop a system of graduated drainage
fees based on land use. Des Moines, Iowa, has gone further
than simply differentiating among land uses. They have developed a fairly complete set of variables (including area, runoff coefficient, distance to outlet, slope, etc.) that should
be used to graduate the fee schedule in a way that best reflects the hydrologic impact of a specific development.

Existing development
Fees exacted from existing developed areas have taken three
forms that are not necessarily a part of the drainage ordinance per se: Bond Issue, Assessment District, and Utility
Fees. The Bond Issue requires a referendum and has not been
very successful in recent years because of the unwillingness
on the part of voters to vote themselves a higher tax. One
of the last major bond issues was voter approved in 1964 and
generated funds to support a sizable flood control agency to
deal specifically with drainage and flood control in the
greater Los Angeles, California area.

The assessment district, too, is not extremely successful.
Part of this is due to the extra tax it imposes on landowners
and part is due to the tremendous support required to institute and administer the district. Another disadvantage of the
assessment district is that it tends to be "piecemeal" with
a number of small, non-cooperative districts that have no
authority or desire to address basin-wide drainage problems.
The apportionment approach of the existing districts is similar to the drainage fee assessed to new developments as described previously.

The "Utility Fee" is a relatively new approach in assessing
general off-site drainage costs and has been implemented in
Boulder, Colorado. This city created a "Storm Drainage and
Flood Control Utility," similar to a water utility, whose
task is to provide city-wide storm drainage services. The
utility collects a monthly fee from each landowner based on
the use of the land (here again, the hydrologic impacts of
different land uses are recognized). This technique affords
a comprehensive city-wide approach to storm drainage control
but relies heavily on the "general" rather than "special"
benefits created by the control facilities.

DISCUSSION

In order to develop an "optimal" drainage ordinance, it is appropriate to compile the best attributes of the reviewed ordinances into one program. However, the development of this composite ordinance should be guided by the following criteria; the final drainage program must be:

a) legal - The "police power" authority has expanded with the requirements of modern drainage management, and extensive control over private property in the interest of the general health, safety, and welfare of the community has repeatedly been upheld. A drainage program would have to be substantially confiscatory to be deemed unconstitutional with this expanded authority.

b) equitable - The review of the drainage-related court cases indicates that a drainage program with cost recovery provisions must insure that the benefits of a project exceed their costs, and that at least some of these benefits accrue to those being assessed. These "benefits" can range from direct, tangible, and computed in great detail (Des Moines, Iowa), to indirect, intangible, and spread over the entire municipal area (Boulder, Colorado) for the courts to honor the program. Assessments based purely on indirect and intangible "general" benefits will suffer considerably more litigation, as the justifiable proportion of "general" benefit to "special" benefit has not been conclusively specified.

c) simple - All of the drainage programs reviewed were based on a master drainage plan, yet young urbanizing communities with a small tax base and considerable "growing pains" cannot afford the expenditure of substantial funds for these kinds of studies. The high initial costs associated with these studies need to be cut in order to insure implementation of a proposed drainage program by a growing community. This cut can be realized if the courts rule that "simplified" drainage studies represent a reasonable basis for cost assessment.

d) easy to administer - The equitability of a detailed benefit analysis and cost assessment for a drainage program is drastically diminished if this level of detail costs the general taxpayer more in city services to administer the program. Ease of administration must be tempered with equitability; a proper balance between the two must be developed.

FOOTNOTES

1. Euclid v. Ambler Realty, 272 U.S. 365 (1926)
2. See, for example, Thibault et.al. v. McHaney et.al. 177 SW 877 (1915), and McMahon v. Lower Baraboo River Drainage District 200 NW 366 (1924)
3. City of Buena Park v. Boyar, 8 Cal Rptr. 674 (1960)
4. Kelber v. City of Upland, 318 P2d 561 (1957)
5. City of Buena Park v. Boyar, 8 Cal Rptr. 674 (1960)
6. Duncan et.al. v. St. John Levee and Drainage District et. al. 69 F2d 342 (1934)
7. Ibid.
8. Kirst v. Street Improvement District No. 120, 109 SW 526 (1908)
9. Thibault et.al. v. McHaney et.al., 177 SW 877 (1915)
10. Luckehe v. Reclamation District No. 2054, 238 P 760 (1925)
11. Funkhouser et.al. v. Randolph et.al., 122 NE 144 (1919)
12. Reclamation Board v. Chambers, 189 P 479 (1920)
13. Hurley v. Board of County Commissioners of the County of Douglas, 360 P2d 1110 (1961)
14. Board of County Commissioners of Sedgewick County v. Robb, and City of Wichita v. Robb, 199 P2d 530 (1948)
15. Ibid.
16. Ibid.
17. State ex.rel. Board of Supervisors of South Florida Conservancy District v. Warren, Governor, et.al. 57 S2d 337 (1951)
18. Ibid.
19. Kelber v. City of Upland, 318 P2d 561 (1957)

BIBLIOGRAPHY

1. Beck, R. E. ed. (1972) The Law of Drainage. Water and Water Rights, 5, pp. 475-648.

2. Brower, D. J., et.al. (1976) Urban Growth Management through Development Timing, Praeger Publishers, New York, NY.

3. Dague, R. R. (1970) Storm Sewer Assessments--The Des Moine Plan. Public Works, August, p. 62.

4. Debo, T. N. (1975) Survey and Analysis of Urban Drainage Ordinances and a Recommended Model Ordinance. Georgia Institute of Technology, National Technical Information Service Publication No. PB-240-817.

5. Freilich, R. H., and P. S. Levi. (1975) Model Subdivision Regulations, Text and Commentary, American Society of Planning Officials.

6. Hagman, D. G. (1971) Urban Planning and Land Development Control Law, West Publishing Company, Minnesota.

7. Shoemaker, W. J. (1974) What Constitutes "Benefits" for Urban Drainage Projects. Denver Law Journal, 51, pp. 551-565.

8. Subdivision and Drainage Ordinances for the Following Municipal Governments:
    a. Arvada, CO, USA, City of
    b. Boulder, CO, USA, City of
    c. California, USA, State of
    d. Chicago, IL, USA, Metropolitan Sanitary District of
    e. Colorado Springs, CO, USA, City of
    f. Dekalb, GA, USA, County of
    g. Fairfax, VA, USA, County of
    h. Ingham, MI, USA, County of
    i. Lakewood, CO, USA, City of
    j. Larimer, Colorado, USA, County of
    k. Los Angeles, CA, USA, City of
    l. Los Angeles, CA, USA, County of
    m. Pueblo, CO, USA, City of
    n. Tampa, FL, USA, City of

ACKNOWLEDGMENTS

Financial support for this paper has been provided by the US Department of Interior, Office of Water Research and Technology, through the Environmental Resources Center, Colorado State University.

MANAGEMENT OF URBAN DRAINAGE AND FLOOD CONTROL IN BRAZIL

By Silvio A. C. Wille[1] and Neil S. Grigg[2]

ABSTRACT

Although design and planning procedures are important, drainage and flood control management questions continue to be of concern, particularly in urbanizing areas. As part of a larger research program, case studies of several countries have been initiated. This paper describes the Brazilian case, considered to be somewhat characteristic of Latin America. Apparently, the federal government is most influential, with cooperative federal-local programs being the rule. Land use and drainage management practices at the municipal level appear to be widely varied and remain a question mark. Modern planning procedures are being implemented due to an increasing base of technology and engineering manpower.

INTRODUCTION AND OBJECTIVE

Urban drainage and flood control (UDFC) management programs are normally implemented at the level of local government. In most countries, we also find state and federal governments involved and, in some cases, regional management agencies. The principal causes of urban drainage and flood control problems are urbanization and industrialization. The United Nations is very interested in this problem as evidenced by the International Hydrological Program (IHP) and a recent UNESCO conference in Amsterdam on the subject.

This paper is from a research project which seeks to develop improved management methods for urban drainage and flood control with emphasis on economic and financial aspects. The work program includes case studies of several other countries including Britain, Australia, and Brazil at the present time. They serve to supplement a series of international studies coordinated through UNESCO as part of the IHP subproject on urban hydrology coordinated by Murray B. McPherson. To date, over ten studies have been completed as part of this program.

The objective of the present paper is to describe the Brazilian case, considered to be somewhat characteristic of Latin

---

[1] Graduate Student, Colorado State University, Ft. Collins Colorado.
[2] Director, Water Resources Research Institute of The University of North Carolina.

America, and to present information which will be useful for
drainage planners to compare the Brazilian case to those of
other countries.

PROBLEMS OF URBANIZATION IN BRAZIL

Brazil has high growth rates in most every category. Currently, the population growth rate is near three percent with urban population growing at approximately 5.15 percent (Costa,
1973). Economic growth has averaged 10.4 percent since 1970
(Figbe, 1976), and Brazil recently passed the 50 percent urbanization mark meaning that more than half the people live in
the cities (Costa, 1973). Migration from rural areas has been
the principal force causing urbanization. The eleven largest
cities of the country contained 33.7 percent of the total population by 1970 (Iorio Filho, 1973).

Brazil is seventh in world population having approximately 10
million in 1976 which represents half the population of South
America. The urban population is projected to be about 67 percent of total population by 1980 (Costa, 1973). Most of the
urban population is concentrated in the large cities located
on the coast, including the megalopolis area of Sao Paulo-Rio
de Janeiro. Several other cities exceed one million in population. Most Brazilian cities, except in the most depressed
areas, are undergoing rapid urbanization. Naturally, problems
of planning, finance, and growth control are very serious.
Although Brazil is developing rapidly, there is still a problem of income distribution and of financial capacity to develop infrastructure rapidly.

INSTITUTIONS FOR MANAGEMENT OF DRAINAGE AND FLOOD CONTROL

Several federal agencies have responsibilities in the area of
urban drainage and flood control. These include the Ministry
of the Interior which has a supervisory role. Included in its
other duties are regional development, control of floods and
droughts, and assistance to the population in the case of public disasters. Other important agencies include the National
Housing Bank (BNH) and the National Department for Works and
Sanitation (DNOS). It is normally difficult to split responsibilities for urban drainage and flood control, but DNOS is
primarily responsible for flood control and BNH for urban
drainage. This arises naturally due to the linkage between
urban drainage and development of housing. DNOS seems to cor-
espond closely to the U. S. Corps of Engineers and BNH to the
U. S. Department of Housing and Urban Development.

Recent policies developed in the country related to UDFC include the Second National Development Plan (NDP) and the Industrial Pollution and Preservation of the Environment Policy
(IPPEP) (Republic of Brazil, 1974). The NDP is very general

and concerned with the national structure of the urban system,
especially with infrastructure. The IPPEP calls for the development of basic environmental legislation for the institu-
tionalization of environmental issues and for the creation of
implementation mechanisms. These include standards and norms
for environmental protection. Storm water pollution loads are
apparently not yet widely recognized or considered to be significant in the federal policy documents.

DNOS works on implementation of national programs through the
lead of the Ministry of the Interior. They follow integrated
planning efforts, which include flood control, as part of the
national water management strategies. DNOS has a special program for flood control and rehabilitation of flood river valleys (FCRRV) which received 39 percent of funds available in
1976. It currently has fourteen projects underway, including
construction of dams, dikes, and other flood control works.
Most of the design and construction is done by private consultants and contractors. DNOS carries out management and
supervisory activities as well as management and operation.
Since 1975 DNOS has been sharing construction responsibilities
with municipalities where small projects are being built.
These are projects not exceeding about U. S. $200,000. Joint
DNOS-municipal projects are implemented through agreements between regional offices of DNOS and municipal mayors. State
and municipal cost sharing is common (DNOS, 1975).

Urban drainage and minor flood control activities are the only
water-related services provided by municipalities in Brazil.
This arose after the creation of statewide public companies
for water supply and sewage all across the country. Since
most drainage systems are built by private developers instead
of municipalities, we see that the relationships between the
municipal government and private developers are very important, the same as in the United States. Political arrangements
for urban drainage come through master plans, zoning codes,
and subdivision ordinances. Master plans and ordinances for
the cities of Manaus, Belo Horizonte, Santo Andre, and Sao
Bernardo do Campo were studied for urban drainage provisions
(Manaus, 1975a; Belo Horizonte, 1976; Santo Andre, 1976; S. B
Campo, 1972). In general, we found that urban drainage was
not addressed in the detail that it is in the United States.
Seven additional subdivision ordinances were studied to deter
mine the regulations concerning urban drainage and flood control (S. B. Campo, 1972; Nova Iguacu, 1976; Manaus, 1975b; B
Horizonte, 1935; Goiania, 1971; Foz de Iguacu; Ribeirao Pret
Most prohibited land development in the flood plains and in
swamps, but few controls were evident. All required that
final designs for storm water networks be presented for city
approval. Developers are required to build the system withi
two to three years and to pay for them completely. There ar
no national, regional, or locally accepted universal UDFC

design standards, although some municipalities have a few guidelines. Strips of land are sometimes reserved along streams and rivers to preserve drainage easements. These generally vary from about four meters to about 30-50 meters for streams and rivers. These are associated more with reserving open space and transportation arteries than due to flood plains.

There are serious problems between cities and developers. In 1973 BNH reported that: (1) growth often occurred without proper public supervision; (2) subdivision regulations were usually incomplete; (3) few cities had satisfactory levels of sanitation; and (4) land speculation is a widespread practice causing much idle urban land (BNH, 1973).

The financing of urban drainage and flood control facilities comes from several sources. Public projects are often financed through loans from BNH designated for improvement projects and sometimes from assessment programs. The improvement program through BNH has been most effective. It is carried out through a "financing of drainage" program (FIDREN). Up to 1976 FIDREN had financed nearly two hundred million dollars of drainage projects in twenty-nine cities. For the period of 1977-1979 the FIDREN budget is $185,000,000. No performance data is available on these projects (BNH, 1976 and 1977).

The basic land development procedure followed by Brazilian developers is as shown in Figure 1.

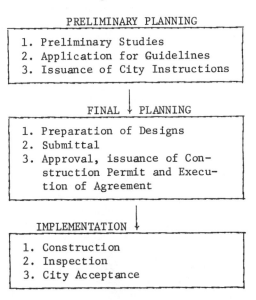

PRELIMINARY PLANNING

1. Preliminary Studies
2. Application for Guidelines
3. Issuance of City Instructions

FINAL PLANNING

1. Preparation of Designs
2. Submittal
3. Approval, issuance of Construction Permit and Execution of Agreement

IMPLEMENTATION

1. Construction
2. Inspection
3. City Acceptance

Figure 1. Land Development Permit Process in Brazil

LEVEL OF TECHNOLOGY

Apparently, the process of urbanization and rapid growth is inhibiting the construction of adequate urban drainage and flood control facilities in some cases. In other cases, particularly those where DNOS is active, adequate facilities to control major floods can be expected. The basic problem of controlling urban development in the face of land speculation and inadequate supervision remains serious.

The level of technology and availability of scientific and technical manpower in Brazil is increasing rapidly. Through international and national efforts, a number of Brazilian professionals have acquired the skills necessary to develop an adequate national program in urban drainage and flood control. The writers know of several instances where this is so. In the university system we find a number of scientists trained in the most modern aspects of urban hydrology, considering both quality and quantity problems. It only remains a matter of time until additional manpower and an adequate technological infrastructure is in place to overcome the apparent problems. Some of the Brazilian consulting firms have impressive qualifications as well. We anticipate that the urban drainage and flood control problems in Brazil are not insolvable but are symptomatic of the rapid process of urbanization. The development of adequate institutional and financial capabilities will eventually meet the needs.

CONCLUSIONS AND RECOMMENDATIONS

This paper represents a first attempt to deal with the perspective of UDFC in Brazil. It has covered a great deal of material and necessarily addresses problems at quite a general level. Some specific conclusions are: (1) there is no national policy on urban drainage and flood control in Brazil, but several agencies are aware of the problems and are working on them deliberately; (2) there is not a clear distribution of responsibilities for UDFC among the three levels of government; (3) nationally accepted design standards for flood control and urban drainage are needed; (4) as in the United States, the development of adequate urban drainage and flood control is inhibited by the lack of a national land use policy (5) little performance data on urban drainage and flood control is available for the proper evaluation of past programs; and (6) financial constraints inhibit the development of appropriate and adequate systems.

Some helpful additions to the current programs might be as follows: (1) a national information and data collection center for urban drainage and flood control in Brazil; (2) a national policy for UDFC including the distribution of responsibilities as well as objectives and design standards; and (3) adequate guidance from the appropriate levels of government on land use and financial aspects should be implemented.

REFERENCES

Belo Horizonte, City of 1935: Municipal Decree No. 54 of Nov. 4, 1935. (in process of being changed)

_____, 1976: Municipal Law No. 2662 of Nov. 29, 1976.

Banco Nacional de Habitacao (BNH), 1973: Resolucao do Conselho No. 7/73, Secretaria de Divulgacao do BNH, Rio de Janeiro.

_____, 1976: Linhas de Financiamento do BNH Consideracoes Gerais, Assessoria de Planejamento e Coordenacao, Secretaria de Divulgacao do BNH, Rio de Janeiro.

_____, 1977: Orcamento Plurianual, Secretaria de Divulgacao do BNH, Rio de Janeiro.

Costa, Rubens Vaz da, 1973: Urban Growth: The Foundation of Economic Development, BNH Information Office, Rio de Janeiro.

Departamento Nacional de Obras de Saneamento - DNOS, 1975: Resolucao No. 148/75, Presidencia do Conselho de Administracao Rio de Janeiro.

Foz do Iguacu, City of, Subdivision Ordinance.

Fundacao Instituto Brasileiro de Geografia e Estatistica - FIBGE, 1976: Anuario Estatistico do Brasil 1976, FIBGE, Rio de Janeiro.

Goiania, City of 1971; Municipal Law No. 4526 of Dec. 31, 1971.

Iorio Filho, Osvaldo, 1973: Crescimento Urbano Tema e Consideracoes, Secretaria de Divulgacao do BNH, Rio de Janeiro.

Manaus, City of 1975a: Municipal Law No. 1213 of May 2, 1975.

_____, 1975b: Municipal Law No. 1214 of May 2, 1975.

Ministry of Interior, 1977: DNOS - Desempenho em 1976 Programacao para 1977, 100 - CCMI, Brasilia.

Nova Iguacu, City of 1976: Municipal Decree No. 1478 of Jan 9, 1976.

Republic of Brazil 1967: Decree Law No. 195 of Feb. 24, 1967, Brasilia.

_____, 1974: II National Development Plan (1975-1979), Servico Grafico do IBGE, Rio de Janeiro.

Resende, Thierry C. and Sergio Bianconcini, Normas para Projetos de Drenegem Urbano, COPLASA, VI Congresso Brazileiro de Engenharia Sanitaria, Brazil.

Ribeirao Preto, City of _____: Project of law of Subdivision Ordinance. Santo Andre, City of 1976: Municipal Law No. 5042 of March 31, 1976.

Sao Bernardo do Campo, 1956: Municipal Law No. 533 of Dec. 5, 1956.

_____, 1972: Municipal Law No. 1980 of June 19, 1972.

Wilken, Paulo Sanipaio, 1974: Normas para Projetos de Sistemas Urbanos de Sancamento Basico - Drenagem Urbana (estudo preliminar), Escola Politecnica de Universidade de Sao Paulo, Sao Paulo.

ACKNOWLEDGEMENTS

Financial support for this paper has been provided by the US Department of Interior, Office of Water Research and Technology, through the Environmental Resources Center, Colorado State University.

STORM DRAINAGE IN INDIA

Avinash Chandra Chaturvedi

Secretary, Irrigation Commission, U.P. Lucknow, India

SYNOPSIS

Storm drainage has posed serious problems to Indian cities
where traffic has been held up from several hours to a couple
of days causing great dislocation. With the advent of Janta
(Peoples) government in India, the problem seeks an imminent
solution. The relevance of system concepts to urban, regional
and national planning with emphasis on planning in the context
of control systems theory is indicated. This has led to
evaluation of existing planning procedures and proposition of
new ones along with socio-economic modelling and forecasting.

The functioning of storm drainage combined with urban waste
water disposal needs rethinking. This has increased the load
on waste water treatment plants but reduced contamination from
waste material picked up by the storm run-off. Pumping plants
and gates to prevent back flows are recommended in areas where
storm run-off may coincide with high water levels in rivers
and estuaries. Hydrographs from small urban watersheds were
found to have very sharp peaks. Small amounts of storages
were found to substantially reduce flow peaks, storm sewer
cost (including required pumping cost) was substantially
reduced by detention storage in small sumps. The relationship
between sump capacity and storm sewer capacity for a given
storm frequency was approximated for an accumulated run-off
diagram. (Chaturvedi 1975)

Our newer cities need facilities to dispose of excess storm
water. Modelling is extensively used to fix the current size
and culvert capacity as the flow through the culvert during
the greatest head water which can occur without causing any
damage. We found that selecting culvert design capacity by
economic efficiency ignored the income redistribution effect
of the cost bearing. Damages from backwater flooding initially afflicted adjacent property owners. Where high

construction caused the backwater damage, it became incumbent
to compensate upstream property owners who were flooded.
Collection was made from downstream owners for the benefits
accorded to them from culvert dampened peaks. (Chaturvedi
1976)

INTRODUCTION

Ten percent of our land is affected by drainage and the
problem affects the entire urban areas of our country. The
worst sufferers are cities like Bombay, where millions of
people are stranded and work dislocated for periods on the
advent of monsoons. Structural measures for urban drainage
are basically the same as for flood control. We have employed
small open ditches or underground sewers to transport storm
water from low lying areas to a nearby major water course.
Storm drainage is combined with urban waste water disposal.
This has increased the load in waste water treatment plants
but reduced contamination from waste material picked up by
the storm run-off. The system provided for the collection and
disposal of storm water includes gutters, inlets, manholes and
outlet works. Pumping plants and gates are provided to prevent back flow in low areas where storm run-off may coincide
with high water levels in rivers and estuaries.

We found the hydrographs from small urban watersheds to have
very sharp peaks. Because small amounts of storage can
substantially reduce flow peaks, storm sewer cost (including
required pumping cost) was found to be substantially reduced
by detention storage in small sumps. The relationship between sump capacity and storm water capacity for a given storm
frequency was approximated from an accumulated run-off diagram.

OPTIMISATION TECHNIQUES

Graphic Approach
The economic evaluation of drainage projects may combine the
marginal benefit curves to obtain an aggregate drainage marginal benefit curve by the summation process. Likewise, the
drainage marginal benefit curves were combined to obtain the
system drainage benefit curve. The marginal cost curve for a
single drainage project was determined from the relationship
between drainage cost and drainage size. The system drainage
cost curves were determined by combining drainage marginal
cost curves to maximise drainage for a given cost. The optimum cost of the optimum system was determined from the point
of intersection of corresponding marginal benefit and cost
curves by marginal analysis.

The marginal analysis was found to work best for individual

multiple drainage projects or for simple drainage systems.
Simplifying assumptions were required to condense all information determining the merit of the drainage project into two
marginal curves. The more complex the prototype system is,
the more restrictive the assumptions become. Nevertheless
marginal analysis provides valuable insight into the economic
aspects of drainage development and in developing a starting
point for simulation iteration. (Chaturvedi 1977)

Data for sample study. The drainage project for Dhampur was
under study. The economic data required to optimise the
drainage were collected and summarised in curves under the
following procedure:

(1) By designing drainage works of a number of different sizes
for the site and estimating the total annual cost amortised
over a 50 years' life, a curve of annual drainage cost as a
function of active drainage was developed.

(2) Drainage operation studies were completed to determine the
annual yield developed by a range of conservation storm drainages.

(3) Prospective benefits and other pertinent factors were
analysed to develop a drainage demand curve.

(4) Drainage damage as a function of drainage flow was determined for the downstream drainage area. Based on a selected
amount of drainage, storms covering a range of frequencies
were routed through the area to develop a curve of residual
damage against frequency. The reduction in area under the
drainage frequency curve provided the annual drainage benefit
from selected drainage systems. The frequency was developed
for a number of drainages to develop a curve of annual
benefits against drainage.

(5) Damage concentration, flow rate concentration and flow
duration curves were developed from a damage survey and gauge
records. The three curves were combined to estimate drainage
damages without the project. By modifying the flow duration
curve by low flow augmentation expressed in units of maximum
flow allowed, residual damages were estimated as a function of
minimum flow. Subtracting residual from initial damages produced a curve relating benefits to minimum flow.

(6) The flow duration curve was analysed to determine the low
low augmentation. The flow gave augmentation required.

Analytical Optimisation techniques

The analytical approach. The many complex systems introduced by modern technology provide a major challenge in planning for optimum system design and operation. The disciplines of statistics, decision theory and operational research were developed with increasingly sophisticated methodology to meet this need, and the numerical work was carried out by the digital computer. The decision problem was converted to a standard mathematical model for the analytical approach. The value gained depended on the close approximation modelling of prototype by a solvable algorithm. (Chaturvedi 1977)

Conceptual models. The application of analogue computers to storm drainage was effected by computer simulation of conceptual models of drainage basins. The components of various models were represented by different analogue circuits. The operational amplifier in the circuit was a voltage transforming device which provided a means of interconnecting a series of drainages. This device served as a buffer and ensured positive direction of flow of current through series connections although it was not absolutely necessary. An indirect analogue circuit for the linear drainage had suitable computing elements connected to complete the circuit. By connecting the analogue circuits for various combinations of linear or non-linear drainages the conceptual models for drainage basins were simulated electrically. Improved models have been developed for practical applications. The model consisted of a linear channel of translation coefficient C and two linear reservoirs of different storage coefficients K1 and K2 in series. The linear channel was used to produce a time area diagram for the whole drainage basin with variable area distribution of the instantaneous effective rainfall. In applying this model to actual data, K1 was assumed to have a constant average value of about 0.25 and the time area diagram to be one of basic geometric forms such as a rectangle, triangle, trapezoid or sine curve instead of the actual diagram. We considered run-off and rainfall relationship by system analysis and a general equation of storage for a non-linear reservoir was proposed. By continuing this equation with the continuity equation, the resulting general differential equation was that of many mechanical and electrical dynamic systems.

Our storm drainages were represented by block diagrams. Each block or so-called operational block, represented the mathematical model of a reservoir. All blocks were connected in various ways in series and/or in parallel form, to represent different cases. Each block did not exactly simulate a physical linear reservoir, because the input and output to an operational block was made unequal. Three functions of equal magnitude were input to three blocks in parallel. The sum of these outputs from the blocks were automatically adjusted to be equal to the triangle input instead of three times input and this sum equalled the outflow. The circle or so-called junction in the block diagram represented the addition of the block output to produce the outflow and the positive sign or negative sign indicated that the inputs to the junction must be multiplied by the appropriate sign before being added. (Chaturvedi 1977)

For comparative purposes, flow duration curves were plotted in terms of ratios to the average discharge rather than the actual discharge units. The curves were also constructed on a lognormal probability paper with the flow on the logarithmic scale and percent of time on normal probability scale. The curve so plotted tended to appear as a straight line, particularly in the middle portion. We made such lognormal probability plottings of the flow records of a large number of storm drainages in the urban areas of Eastern U.P. From a flow duration curve thus plotted, the values of drainage at ten percent intervals from 5 to 95 percent of the time were read off. The logarithms of these discharges were then found and their standard deviation was computed. The index of the flow was worked which is essentially the standard deviation of the flow statistics. The correction due to various storm characteristics was found approximately. The approximate shape of a duration curve was attained by drawing a straight line on logarithmic probability with a slope such that the ratio of the discharge exceeded 15.87 percent of the time to the discharge exceeded 50 percent of the time is equal to the antilogarithm of the variability index. (Chaturvedi 1977)

BENEFITS IN URBAN DRAINAGE

Benefits from storm drainage measures result from the reduction in damage caused by ponded storm water. Increased protection was found to place a high nuisance value on ponded storm water while the expenses borne by individuals to drain storm water from private land indicated a nuisance damage of significant magnitude. Difficulty in direct quantification has caused alternate approaches to be used for selecting the optimum level of protection. This includes:

(1) Installation of storm drains with additional facilities provided when public pressure develops.

(2) Nuisance damage may be regarded as intangible. The planner may select design frequency with economic analysis being used only to determine the least cost design for accomplishing the objective.

(3) To provide economic justification of the measure a function is postulated to relate nuisance damage to the depth and duration of ponding and combine benefits from reduced damage

nuisance with flood damage.

(4) Storm drainage practices are surveyed to derive an implicit nuisance damage function for use in studying new measures.
The parameters of the derived nuisance damage function may be
related to the local land use.

The first two methods have been widely practised by us but
experience has shown them to result in a wide variation of
drainage designs among areas with identical drainage problems.
Such design diversity is not an economically efficient resource allocation. While the third method relies on an arbitrarily selected nuisance damage function it ensures constant evaluation of equivalent situations. Resources may be efficient-
ly allocated among drainage projects, but   a demonstration
of economic feasibility is provided.
The fourth method offers promising research possibilities as
private investment in drainage may offer the key to evaluating
nuisance damage. Once a method for evaluating nuisance damage
is selected, the demand curve is the customary plot of marginal benefits against storm frequency.

In the process of economic analysis benefits and costs were
identified and expressed in monetary terms as far as possible.
The total benefit equalled the measure of economic improvement
received with development. The total benefits and costs were
converted to an annual basis. All benefits were discounted
to the time when the project started to operate. The intangible benefits related to prevention of loss of life, personal
injury and sickness and maintenance of public morale.

Cost benefit
Sump storage was made expensive by the high cost of right of
way in our urban areas, but the portion of cost allocated to
drainage was substantially reduced by also using the area for
recreation and aesthetic pruposes.

Overall system cost was found to be largely governed by popul
ation density and design standards. The supply curve for
storm drainage was developed. The non-structural measures
used for flood control were unlikely to be economical where
the storm water rose very quickly, was shallow, and seldom
entered buildings. The least cost combination of sump storag
and sewer capacity for each of a series of selected design
frequencies defined a total cost curve. The marginal cost of
the supply curve was the slope of the total cost curve.

The direct benefits related to prevention or reduction of
direct physical damage to properties and land. The indirect
benefits related to reduction or prevention of indirect dama
ges such as loss of income, wages, goodwill or interruption

of services utilised and transportation, disruption of markets
temporary rental unusual expenditure. (Chaturvedi 1977)

Project composition

By applying the benefit cost analysis to each separate purpose
or service, it was possible at least to some degree to ascertain which of the purposes was to be given paramount consideration in formulation of the plan for development of Moradabad
drainage. The problem of choice of the purpose and of its
degree of service to be rendered was always present. Our
problem was to determine in a practical way how to combine and
compare the benefits and the cost. A practical cut-off line
was determined stochastically both for benefits and for costs,
and this cut-off line was an important one which gave equal
emphasis to benefits and costs. A small chart was used to
illustrate the use of benefit cost analysis in the determination of the size and characteristics in our projects and the
degree of service that any one function of the project might
best be planned to render. The benefit cost ratio was computed separately for each particular service to be rendered.
This was done for a number of projects of different sizes.
The points were then plotted for each service and a curve was
drawn through them. These individual service curves were used
to ascertain the best degree of development for each particular
service in relation to other possible project services. Having
determined the relative degree of services to be provided a
composite curve was drawn for sizes of completely formulated
drainage projects. (Chaturvedi 1977)

Research programme

As part of our research programme, seeking an improved methodology for the design of storm water drainage systems and
ascertaining of benefits, two linear programming models were
developed. The multistructural model sought optimum values
for the decision variables of active and dead storm run-off
from a hypothetical system of two storms and four drainage
systems. The model had three versions according to whether
the optimum solution was based on the storm during a typical
year in isolation on the inflow during the entire critical
period, or on a set of inflows drawn at random from an appropriate probability distribution. The first version was found
to be simple but the least accurate. The second version gave
better results because it made provision for the storm. The
third version recognised that the operator never knew future
storm flow in advance. All the three versions were limited
by the constricting approximation required to use the linear
rogramming algorithm.

n the second stochastic sequential model we sought to opti-
ise the storm drainage outflows and operating procedure
imultaneously using storms drawn at random from probability

distribution based on historical flow records. The amount of
storm drainage was determined as a function of the fraction of
the time it was available. The model optimised the target
outputs and operating procedure for a given reservoir size by
selecting the yield probability distribution with the highest
expected value. However, the process had to be repeated for
a series of storm drainages in order to optimise the variable.
(Chaturvedi 1977)

Value judgement
Engineering decisions on storm drainage of Lucknow area were
based on economic value and other values in economic terms.
When future decisions concerning the operation of storm drainage systems were made it became possible to adopt the value
system implied in past operations of the system to derive
optimal decisions. Social decisions concerning the management
of storm drainage systems made it possible to define and redefine the value judgements and to develop a value system for
future decisions concerning multiple purpose allocation of
drainage.

Water is an important and sometimes critical input for economic development and growth. Storm drainage systems are often
large and complex. They require large scale capital investment with a long gestation period. They are affected by the
three market imperfections - externalities, public goods and
merit wants. Hence regulations and management of storm drainage systems have often been in the public government sector
with a hierarchy of institutions and a number of decision
makers. Management of storm drainage hence requires the
specification of stated goals and priorities in terms of the
values associated with the level of achievement of each goal.
The importance of using welfare economics and in particular,
multiobjective (or multiple goals) evaluation is well known
in the state of U.P.

Multiobjective analysis
Multiobjective analysis comprises two steps: first the maximum levels of attainment along the several objectives were
determined. These defined the production possibility frontier or transformation curve which related the available
resources to possible alternatives such that objective combinations above the frontier were not feasible, and those below
were inferior, in that at least one of the objectives was
attained at a smaller level. Part of the surface defined the
para to frontier or non inferior set. The second step consisted in determining the optimal decision based on a suitab
value system.

Generally the functional relationships among the system para
meters and the social goals were not available. The social
goals were achieved through multiple purpose development, th

is through irrigation, flood control, recreation etc. From
techno-economic data and relationships the technological
relationships among system parameters and the level of attainment of each purpose were determined. Further planning in the
State of Uttar Pradesh is based on sectors like farm irrigation and tourism rather than explicit socio-economic objectives.
It must be noted that risk aversion in storm drainage is
provided. Thus for example, objectives for operation of storm
drainage system at the end of monsoon may be farm irrigation
and carry over storage. Using available technology it is
possible to derive the non-inferior set for the three objectives.

Value System

The preference information or the value system for different
combinations of the levels of attainment of the several goals
may be provided by the social welfare function in terms of in-
different curves. In its absence the preference curves of the
decision makers, which may be biased, may be used. The point
of tangency between the non-inferior set and the preference
curve identifies the optimal compromises, or satisfactory
solution. The slope of the tangent line through the point of
tangency was found to be the inverse of the ratio of the
weights of the objectives. These weights defined as the optimal weights indicated the marginal importance of the society
and the decision makers for the objective, and this naturally
depended on the degree of fulfillment of the objective. Thus
for optimal solution , it was sufficient when the decision
maker gave his relative preference for points on the non-
inferior set. As the trade-offs depended on the levels of
attainment of the objectives, it was possible to have interaction with the decision maker to facilitate redefinition of
relative values and an iterative convergence to the optimal
solution.

Planning Process

In reality, the problem of social decision making encompassed
the technology, economics and the political process, and was
represented by the mutual bargaining model. Paretian analysis
was used to illustrate such considerations in storm drainage
systems. In the above approaches, the value systems were
defined by the decision makers and this led to optimal decisions and they constituted top down planning. Generally the
value systems and trade-offs were not explicitly available.
Yet decisions concerning the operations of storm drainage were
made regularly. It was possible to study the past decisions,
infer and impute the trade-offs implicit to the past decisions
 evaluate the consistency of the implied value system and
make the trade-offs explicit to the decision maker. This was
the 'bottom up' planning procedure where past decisions were
used to infer the value system. This led to a better under
standing of the value system and eventually to 'top down'

planning when future decisions concerning the operation of
the same or similar systems were made,it became possible to
adopt the value system implied in the past operations to
derive optimal decisions.

Hydrologic and economic uncertainty are inherent in any storm
drainage system. Farm and industry competed with each other
for the limited storm drainage. Social decisions concerning
the management of storm drainage systems implied value judgements by the decision makers. It seemed necessary and possible to refine and redefine the value judgements and develop
a value system for future decisions concerning multiple purpose allocation of drainage in the face of hydrologic uncertainty.

Kanpur drainage

Industrialisation in the developing area of Kanpur inevitably
led to the expansion of the existing urban area and the
growth of townships in areas that were once sparsely populated
and in a state close to nature. Development of such industrial towns in India have altered the landscape and changed many
hydrologic parameters such as infiltration rate, percentage
of impervious area, channel roughness etc. Such changes were
found to lead to two main consequences (i) increase in the
volume of storm flow and peak discharge and reduction in the
time lag and (ii) increase in the pollution load in the storm
run-off. It was found that unless proper measures were incorporated in developing an appropriate storm water collection
system, there was considerable dislocation of the industrial
activities and consequent economic and other losses. Some of
the methods available to use are minimising the disruption due
to storm over-flow in an already existing drainage system and
provision of new systems. Such measures include separately
or in combination realignment of drainage networks, proper
placing of impervious areas, use of porous pavements etc.

System elements

The system elements suggested were classified under three
categories, measures to modify overland flow within channel
measures and off-channel measures. However the drainage systems often designed for conditions existing prior to the rapi
growth of the area were found to be inadequate after the area
developed. Due to this there was frequent flooding of the
urban area (recently for 10 days with storm density of 700 mm
in 36 hours) with considerable inconvenience to the public
because of flooded business areas. Further the ground water
supply to the city was affected by about 10% due to the rapid
urbanisation since the recharge to the aquifer was lessened
by the impervious cover that resulted from urbanisation. A
rational approach provided for methods which would reduce th

run-off and bring it as close to the pre-urban or pre-expansion
flows as possible in addition to the conventional method of
providing drainage capacity. Reduction of run-off was found
to be an attractive alternative to capacity expansion, at
least for the interim, since expanding the capacity of the
drainage systems in urban areas was found to be capital intensive, took considerable time and caused a lot of inconvenience.
Further steps taken to reduce run-off often resulted in increased infiltration and hence 10% more discharge to ground
water with consequent increase in ground water supplies. Another aspect of rapid urbanisation was the pollution potential
of storm run-off.

The aim was to increase the travelling time of water by increasing overland flow length, by decreasing the slope of overland
flow surface, increasing the roughness of overland flow surface and replacing conventional pavements with porous pavements. The within-channel measures included increasing the
capacity of the existing storm sewer systems, adding new sewer
lines in the existing system to handle the excess flow and
exploitation of underutilised sewers in the existing systems.
The off-channel measures provided included disposition of
impervious areas, discharge of roof top on to the pervious
areas including storage on areas such as playgrounds, parking
lots and roof top, retention basins and storing run-off in
underground tunnels and vaults. The magnitude of flooding was
found to vary with the type of locality flooded whether residential or industrial. Providing retention basins is a way of
reinstating the storage lost by urbanisation. The run-off from
the roofs was discharged in areas where soil conditions were
dry enough to absorb significant amounts of water. This measure was found most effective in areas which were predominent-
ly residential.

CONCLUSION

The simplification required to model a complex drainage storm
system in a form amenable to optimisation by the standardised
analytical techniques call for further research. We eventually used simulation to complete the optimisation. We have to
undertake many levels of sub-optimisation to precede optimisation of the overall system. Linear programming was used to
optimise drainage system for the storm water, dynamic programming to optimise a stage construction sequence and queuing
theory to select the optimum number of drains. The standard
analytical algorithms offer great promise as viable tools for
the overall optimisation of complex systems and are finding
increased application in the optimisation of system components.
The formation of the techniques of computing down urban storms
involved not only routing the upstream flow but simultaneous
expected contribution of the storm rainfall volumes from the

small intermediate subbasins. This study helped in the direction of establishing typical time and a real distribution of storm which could be realised in individual pockets that join the main run-off. We found that if the position and direction of movements of monsoon depression was received well in advance, the preliminary assessment of temporal distribution of the associated rainfall with respect to the smaller sub-areas could be made for predicting the storm drainage and designing economic systems with the help of system analysis and offer better design and management.

REFERENCE

Chaturvedi A.C. (1975) Flood Problems of U.P. Journal of South African Society of Civil Engineers.Vol. 30.
Chaturvedi A.C. (1976) National Reconstruction and the Engineer. Journal of Association of Engineers (I) Vol.51, No. 1-20.
Chaturvedi A.C. (1977) Designs for Low Cost Ground Structures. Proceeding, National Seminar of Indian Association of Geohydrologists, 24.
Chaturvedi A.C. (1977) Water Quality and Availability in India. Proceedings of Third International Hydrology Symposium. Fort Collins, Colorado, 24.
Chaturvedi A.C. (1977) Irrigation Modelling in India. Proceedings of International Conference System Science IV, Wroclaw Tech. Univ., Poland, 20.
Chaturvedi A.C. (1977) Water Resources Modelling in India. Proceedings of IFIP Working Conference on Modelling and Simulation of Land, Air and Water Resources System, Ghent, Belgium, 120.
Chaturvedi A.C. (1977) Water Resources Information. Proceedings of International Seminar, Agr. Res. Centre, State Univ. Colorado, Fort Collins.
Chaturvedi A.C. (1977) Optimisation in Data Processing. Proceedings, Fourth International Seminar, Sofia Verne, Bulgaria, 108.

URBAN STORM WATER MANAGEMENT IN THE U.S.

C.T. Haan, B.J. Barfield and T.Y. Kao

University of Kentucky, Lexington, Kentucky

ABSTRACT

For the past four years we have held National Symposiums on the topics of Urban Hydrology, Hydraulics and Sediment Control. The Symposiums have been attended by over 750 engineers from throughout the U.S. and more than 20 other countries. At these Symposiums a great deal of the current research and practice in the area of urban hydrology and urban storm drainage design in the U.S. has been presented. In this paper we will summarize these presentations and attempt to relate the current methodology that is being used as well as the current technology that is available and could be used in storm drainage design as it relates to the U.S.

This paper will review the more than 120 papers and mini-courses that have been presented at our Symposiums and relate our feelings as to the best ways for transferring new information to design engineers so that design practice can better keep pace with available technology.

INTRODUCTION

In the 1960's the citizens of the U.S.A. and many other countries began to become aware on a massive scale of the dangers of the then current rate of environmental degradation. As a result of this awakening new regulations and new federal and state agencies were created to "protect the environment". People began to realize that urban development could be done in a manner that did not create additional flooding and sedimentation problems. Many communities began instituting development regulations with respect to the allowable change in stormwater runoff rates brought about by the development process. More recently there has been a substantial movement toward ordinances aimed at controlling erosion and sedimentation from developing areas.

As a result of this surging interest in environmental quality and new ordinances regulating stormwater and sediment originating on newly developing land, many engineers found themselves faced with design decisions with which they were totally unfamiliar. No longer was the Rational Equation the sum total of the knowledge required to determine the hydrologic inputs into a storm sewer system. Now storages had to be incorporated into the design and this meant that knowledge of runoff volumes was required. The allowable release rate from these storages was specified by ordinances. The engineers now had to route stormwater runoff hydrographs through the storages to estimate the hydrograph of runoff leaving the development site.

Estimates had to be made of the volume of sediment that might be produced during the development phase and methods of preventing this sediment from leaving the site had to be devised. Again the engineers were generally not familiar with techniques for accomplishing this.

KENTUCKY SYMPOSIUMS

In recognition of the need for engineers to have additional training in the area of urban stormwater management, a series of short courses were initiated in 1971 at the University of Kentucky to provide some of the needed training. At first these short courses were local and regional in geographic extent. Because of the National interest that was apparent in the topics of short courses, in 1974 the first National Symposium on Urban Rainfall and Runoff and Sediment Control was held at the University of Kentucky. Since that time the Symposium has been an annual event generally held in the latter part of July on the campus of the University of Kentucky in Lexington.

Since 1974 the Symposium has continued to grow and attract a number of speakers and participants from outside the U.S. In 1977 the Symposium was called an International Symposium on Urban Hydrology, Hydraulics and Sediment Control. The 1978 Symposium to be held the last week of July is titled International Symposium on Urban Stormwater Management.

In the four-year period 1974-1977, 125 papers on the topic of stormwater management have been presented. Sixteen of these papers were authored by engineers from 10 countries other than the U.S. In addition to these 125 technical papers, 9 mini-courses have been presented. The mini-courses consist of 4 hours of lecture and problem solving with each being presented by one or two individuals. The mini-courses are intended to cover a limited aspect of urban stormwater in considerable detail and in a fashion that provides the practicing engineer with material he can put to use in the everyday performance of his job.

In the past four years, 795 engineers from 18 countries have attended these Symposiums. Of these engineers, 43 were from outside the U.S. The attendees have been fairly evenly divided between researchers, governmental workers and private consultants. The Symposiums have been aimed at updating the abilities of practicing engineers as opposed to a forum for the exchange of the latest developments in the research aspects of urban stormwater management.

The papers presented at the Symposiums have ranged from general presentations on procedures to follow in organizing an urban storm drainage project to theoretical treatments of overland flow. The papers have, to a great extent, covered the entire range of methodologies currently being employed by engineers, especially in the U.S., in their efforts to design stormwater management systems. It is the attempt of this paper to summarize some of the methodologies that have been discussed. Thus this paper is not a general review of urban stormwater management literature but is largely limited to papers presented at the past 4 University of Kentucky Symposiums. The Symposium papers have, however, been representative of a broad spectrum of current practice.

CURRENT HYDROLOGIC PRACTICE

The hydrologic procedures employed by many engineers are selected on the basis of several factors. Some of these factors are:

1. Training of the engineer.
2. Scope of the project.
3. Time available for analysis.
4. Availability of computing equipment.
5. Money.
6. Availability of data.
7. Regulations and/or organization of regulating bodies.

These factors work interactively in their influence on the selection of hydrologic methodologies. For instance if an engineer is extensively trained in computers but his firm has no access to computers or the project is a very small one, the use of any procedures that require a great deal of repetitive calculations would be precluded.

Until very recently most engineering curriculums in the U.S. did not include any formal training in hydrology. Even today hydrology is generally an elective subject. Most engineers working in the field of stormwater management specialized in highway engineering or sanitary engineering. In these areas their exposure to hydrologic techniques was largely limited to the Rational Method for computing peak rates of stormwater runoff. Indeed it has only been in the past few years where this single relationship has not been sufficient to handle a majority of the hydrologic problems encountered by design engineers. Even today when surveys are conducted, it is found that the Rational Equation is the single most widely used method of hydrologic analysis for urban areas.

This can be contrasted to a growing number of engineering firms that rely on the latest developments in hydrology for structuring models for simulating the hydrology of urban basins.

Since 1965 a large percentage of the engineering firms have moved from sole reliance on the Rational Equation to the adoption of procedures for developing runoff hydrographs for single storms. It is in the development of runoff hydrographs that the influence of local regulating agencies is most strongly felt. In many instances these agencies recognized that many engineers did not have the capabilities of developing runoff hydrographs. The agencies then would produce a set of guidelines specifying the type of analysis that should be done. This approach not only limits the engineers creativity but is a deterrent to the engineer in terms of upgrading his hydrologic knowledge.

Reasons that local regulating agencies tend to adopt a single, uniform methodology include their desire to have relatively uniform flow estimates regardless of the engineer doing the analysis, the lack of hydrologic training possessed by the practicing engineers, the lack of hydrologic training possessed by the staff of the regulating agencies, the ease of checking storm drainage plans if they are all based on the same methodology and the large number of storm drainage plans that must be processed by a small number of engineers on the regulating agencies staff.

Hydrology is a very inexact science at best. Different methods of analysis can produce widely varying flow estimates. The only way to know which method is producing the best flow estimates is to have some actual flow data to use as a basis of comparison. Urban governments in the U.S. have been especially lax in collecting long-term flow data on which to base their hydrologic procedures. In the absence of this data, the tendency has been to adopt a set hydrologic procedure. Calculations based on this procedure are taken as a basis on which to judge other procedures. In other words the adopted procedure is treated as though it were truth and any other procedure judged on the basis of its compliance with this "truth". This type of thinking is of course hydrologic nonsense but extremely convenient administratively. What is badly needed are actual hydrologic records.

HYDROGRAPH PROCEDURES

The most common methods of developing runoff hydrographs are based on variations of unit hydrograph procedures. The engineer first selects an appropriate time interval to work with based on the size of the basin and the time it is estimated for flow to be contributing from the entire basin. A rainstorm temporal pattern is developed in blocks of uniform intensity with a time base equal to the selected time interval.

Rainstorm Temporal Pattern

Several methods for developing the rainstorm temporal pattern are employed. One method employs rainfall depth-duration-frequency curves to estimate the intensity of rainfall for a given frequency and for various durations. Based on the intensity and duration, the depths of rainfall are computed. From the accumulated depths for various durations, blocks of incremental rainfall depth and intensities are determined. Finally, the intensity blocks are arranged to form a rainfall intensity temporal pattern. The U.S. Soil Conservation Service has developed some dimensionless curves that simplify and standardize this procedure. A major problem that is inherent in the method is that even though for any duration, the rainfall depth approximately corresponds to the chosen return period of the storm, the return period of the entire storm is not known. Since rainfall depth-duration-frequency curves are based on the greatest rainfall depths for various durations from a number of storms, the probability of receiving a single rainstorm with an intensity pattern as developed from the above procedure is much lower than would be deduced from the return period used in developing the storm.

A second common procedure for arriving at a rainstorm temporal pattern is to use a severe historical storm that has been experienced in the near geographic area. This procedure has the advantage of using a storm that engineers and the general public can relate to but again has the disadvantage of an unknown return period. As a matter of fact the return period of the storm will be different for different watersheds since the critical storm duration for different watersheds will vary. This means that the appropriate rainfall duration may vary from watershed to watershed. It is unlikely that the return period of the chosen storm will be the same for different durations within the storm. Thus drainage facilities on all watersheds will not be designed on the basis of a common frequency.

Rainfall Abstractions
Having estimated the rainstorm pattern, the next step is to estimate the losses from this rainfall to produce the effective rainfall hyetograph. Methods of loss calculation are widely varied. Some assume a constant loss rate depending on the type of surface, some use an empirical infiltration equation, and others employ the runoff curves developed by the U.S. Soil Conservation Service.

In the U.S., storm rainfall intensities generally exceed the infiltration rates of the pervious areas during part of the storm. This means it is not possible to simply assume no runoff from pervious areas. It is also common in residential areas for roof drains to empty onto lawns and other pervious areas. Thus the assumption of 100 percent runoff from impervious areas may be in error as well. The estimation of infiltration losses is a truly complex problem that currently has no satisfactory and practical solution.

Unit Hydrographs
Unit hydrographs themselves are developed in a wide variety of ways. Some methods estimate a peak flow rate and a time to peak and then "sketch in" the actual shape of the unit hydrograph. Some methods employ a mathematical function to define the shape and some simply use a triangular shape. It appears that if the time increment that is used in computing the runoff hydrograph (duration of the unit hydrograph) is relatively short in comparison to the total time involved in the effective rainstorm, the actual shape of the unit hydrograph will have minor influence on the runoff hydrograph as long as the same peak discharge and time to peak is used on the unit hydrograph. Of course, if only 2 or 3 unit hydrographs (multiplied by the appropriate runoff

volume factor) are summed, the shape of the resulting runoff hydrograph is strongly influenced by the shape of the unit hydrograph.

Another common procedure for developing unit hydrographs is through the use of dimensionless unit hydrographs. Unit hydrographs are determined by simply multiplying the ordinates of the dimensionless unit hydrographs by appropriate factors. Again methods for determining these factors are varied but generally are dependent on estimating the unit hydrograph peak flow rate and time to peak.

MODELING APPROACHES

Recent legislation in the U.S. concerning water quality and especially that legislation affecting water quality from non-point sources of potential pollution has greatly accelerated the use of hydrologic models especially in the area of regional planning. Many of the models in use have the capability of simulating not only the quantities of stormwater runoff, but its quality as well. Of course models of this type have rather demanding data input requirements. In applying these models there is no substitute for local data to use in parameter estimation and model validation. The models generally have default values for many of the parameters in the event that the user cannot provide estimates for them. There is accumulating evidence that the use of default values may lead to substantial errors.

In the design phase of stormwater management, the main tool continues to be empirical relationships, hydrograph procedures or simple hydrologic models. There is a growing recognition of the desirability of designing stormwater management systems on a regional basis. As this trend develops, more comprehensive hydrologic models will be used in the design phase. Currently the use of hydrologic simulation models for design is not as widespread as it is for planning.

The use of models for managing a stormwater control system once it is installed is in its infancy. These operational models will become more common as engineers become more comfortable with models and as equipment for the rapid accumulation and processing of real time data increases. Use of the current and past conditions of the hydrologic-hydraulic system to arrive at an optimum project management decision at the current time step can result in considerable improvement in the performance of the stormwater management system over what can be obtained by preset operating rules. Communities having large, integrated stormwater management systems will adopt models of this type in the future.

The hydrologic models currently in use range from relatively simple ones to the very complex. Some are proprietary and some non-proprietary. Because of the diversity of models, it is not possible to discuss or even list these models. Many of them are being discussed at this Conference. New models appear each year. It appears that a trend is currently developing toward fewer new models and more improvements in existing models. This trend is probably in the right direction but makes it difficult to keep track of the current status of a model as several organizations may be modifying a model at any one time without any apparent link among the modifications. Thus, we end up with many versions of the same model.

The application of simulation models involves many difficulties that often keep engineers from using them. Estimation of parameters for simulation models, so as to reasonably match expected or observed results, presents major problems. The estimation of rainfall abstractions such as infiltration and surface storage to produce an effective rainfall hyetograph in the face of various soils, covers and land uses scattered throughout a watershed is a major hurdle for consultants.

Lumb and James at the 1975 Kentucky Symposium presented a unique approach to overcoming this problem. For a given municipality they produce a runoff file or what is equivalent an effective rainfall file based on historical rainfall and four soil permeability classes ranging from very rapidly permeable to impermeable. They further produced a routing model UROS4 which routes rainfall excess from subareas through channels and storage segments.

The consultant takes the four runoff files and by properly weighing them, produces a single runoff file for his basin. He then uses UROS4 to produce runoff estimates for his study area. This procedure relieves the consultant of tasks he is not likely to be familiar with (estimation of rainfall excess) and provides him with a modeling approach that requires as data input only physically measurable features such as slopes, roughness, etc., that the consultant can readily determine.

EFFECTIVENESS OF KENTUCKY SYMPOSIUMS

As we have conducted our Symposiums we have experimented with various methods of helping the consulting engineer aquire new and useable information. We have used standard 20-minute papers with discussions, reporters with author replies, keynote or longer state-of-the-art papers, panel discussions and mini-courses. We have also tried to evaluate all of these procedures through questionnaires and formal and informal interviews of the conference attendees.

We have found that the least effective procedure is the longer lectures or state-of-the-art papers and the most effective procedure has been the mini-course concept. A mini-course as we use them is 4 hours of instruction and application on a narrow topic with classes of about 40 engineers. The instructor prepares a complete set of notes for distribution and then lectures on the topic for ½ to 2 hours. Following a break, the participants work on class problems using the technique being taught. By keeping the classes small it is possible to get instructor-participant interaction and for the instructor to spend some time with each participant as needed during the problem solving portion of the course. Different instructors use variations of the above procedure in terms of mixing lecture with problems.

In general we feel, and our evaluations have confirmed this, that the mini-courses are very effective in advancing the knowledge of consultants in areas they are interested in. We generally have 3 simultaneous mini-courses on different topics so that the participants can select the ones that really are of interest to them.

Our experience with shorter (20 minute) papers has shown them to be a moderately effective means of information transfer. They achieve their maximum potential only if accompanied by discussion periods and especially by opportunities for individual discussion between the speakers and participants at coffee breaks and social functions.

By employing a variety of methods of presentations at the Symposiums, we are able to maintain the interest of the conference participants at a high level for the duration of the Symposiums. We have employed an expert in the evaluation of adult educational experiences to assist us in evaluating our conferences and to improve them. This evaluation process has indicated to us that the consulting engineer is significantly increasing his knowledge of stormwater management techniques through conference attendance. Other less formal forms of feedback indicate that much of the material being covered by the conferences is being applied by the participants in the performance of their normal work assignments.

Based on this assessment, we feel the Symposiums have been an effective educational tool for a great many engineers.

SUMMARY

Engineering consultants are anxious to increase their capabilities in the area of stormwater management. They are willing to travel great distances and spend considerable time attending conferences that they feel will help them in their work. They are especially appreciative of in-depth treatments of topics of interest to them.

In the U.S. there is a growing trend toward the adoption of hydrologic models for both planning and design of stormwater management systems. The use of the models in planning is well ahead of their use in design. Currently the Rational Method is still widely employed but is being rapidly replaced as more and more communities adopt regulations concerning the allowable changes that development may cause in stormwater runoff. Hydrograph procedures are gaining wide acceptance.

There is a growing awareness that stormwater management based on small sub-basins is not only expensive but may lead to hydrologic solutions that are suboptimal. Regional stormwater management is being recognized as more economical and more effective than the sub-basin approach. The regional approach requires the use of a hydrologic model and will only reach its full potential as the capabilities of planning and regulating agencies and consulting engineers in the use of models grows.

ACKNOWLEDGEMENTS

The work on which this report is based was supported in part by funds provided by the Office of Water Research and Technology, United States Department of Interior, as authorized under the Water Resources Act of 1964 and in part by the Kentucky Agricultural Experiment Station and is published with the approval of the Director of the Station.

REFERENCES

Barfield, B.J. (editor) (1976). Proceedings National Symposium on Urban Hydrology, Hydraulics, and Sediment Control. July 27-29, 1976, University of Kentucky, Lexington, Kentucky. Report UKY BU 111.

Haan, C.T. (editor) (1975). Proceedings National Symposium on Urban Hydrology and Sediment Control. July 28-31, 1975. University of Kentucky, Lexington, Kentucky. Report UKY BU 109.

Kao, D.T.Y. (editor) (1974). Proceedings National Symposium on Urban Rainfall and Runoff, and Sediment Control. July 29-31, 1974. University of Kentucky, Lexington, Kentucky. Report UKY BU 106.

Kao, D.T.Y. (editor) (1977). Proceedings International Symposium on Urban Hydrology, Hydraulics, and Sediment Control. July 18-21, 1977. University of Kentucky, Lexington, Kentucky. Report UKY BU 114.

SOME ASPECTS OF THE DESIGN OF STORMWATER BALANCING PONDS FOR
CATCHMENT AREAS SUBJECT TO URBANISATION

M.J. Hall

(Sir William Halcrow & Partners, London, UK)

T.M. Prus-Chacinski and K.J. Riddell

(C.H. Dobbie and Partners London, UK)

INTRODUCTION

In recent years, the loss of agricultural land in the United
Kingdom to urban development has occurred at a rate of almost
20000ha per annum (Economic Development Committee for Agriculture, 1977). This average turnover of farmland to urban use,
which is equivalent to losing half the area of the Isle of
Wight every year, is expected to continue well into the next
decade.

Since urbanisation represents one of the more dramatic
examples of Man's interference with the hydrological cycle,
such rates of change of land use have important consequences
for the design of drainage works both within and downstream of
urban areas. The changes in flow regime which occur as a
catchment area undergoes urbanisation have been described by
several authors (for example. Leopold, 1968; Hall, 1973).
Assuming that storm conditions remain the same, owing to the
increased proportion of impervious area, greater volumes of
runoff are discharged from an urbanising drainage basin than
from the same catchment in its rural state. Moreover, since
the natural channel system is supplemented or even replaced
completely by stormwater sewerage, velocities of flow from an
urbanising area are higher and so time characteristics of the
hydrograph, such as the lag time and the time base, are short-
ened. The discharge of greater volumes of runoff within
shorter time intervals inevitably increases peak rates of flow.
Furthermore, since soil moisture recharge decreases as the
volume of runoff increases, low flows between storm events are
generally reduced.

Owing to the dearth of hydrometric records spanning the period
of development of urbanising catchment areas, little

quantitative information has become available on the magnitude
of the changes in both the frequency distribution of floods
and the shape of the flood hydrograph that are to be expected
at different stages of urban growth. This lack of information
has compounded the problems of protecting the areas downstream
of an urbanising catchment from flooding. In recent years,
many developments have proceeded only on the understanding
that measures are taken to ensure that the flow regime of the
receiving river system at some downstream point does not
change significantly. In practice, one of the most frequently
applied methods of meeting such a condition has been the
provision of temporary storage facilities within the urbanising catchment itself.

Recognising the lack of guidance available to engineers concerned with the design of stormwater balancing ponds, the
Construction Industry Research and Information Association
(CIRIA) in 1976 set up a Research Project on the Design of
River Works to Cater for Runoff from Catchments with a Degree
of Urbanisation. The principal objective of this project has
been the production of a Guide on the Design of Stormwater
Balancing Ponds. This Guide will contain the details of
procedures for both flood estimation for catchment areas undergoing urban development and the sizing of flood storage ponds.
These techniques will be presented as part of a step-by-step
design procedure, the details of which are described in this
paper.

DESIGN OF FLOOD RETENTION PONDS

The suggested sequence of steps which should be followed in
designing a flood retention pond is summarised in Figure 1 in
the form of a flowchart. This flowchart outlines the various
options which are available at successive stages of the design
procedure. The extent to which its provisions apply to any
particular scheme depends largely upon the hydrological
characteristics of the area of interest. The greater the
changes in the flow regime of a catchment area affected by
urbanisation, the more sophisticated the level of analysis
required to achieve an economical design solution. The flowchart applies specifically to the design of flood retention
ponds and does not allow for the analysis of alternative
solutions, such as flood alleviation schemes. For convenience
the design procedure may be considered in four parts, the
details of which are discussed below.

Preliminary analysis

The first four steps in the design procedure (indicated by the
numerals in Figure 1) enable the designer to form a preliminary assessment of the hydrological significance of urbanisation. This portion of the flowchart does not require a large
amount of detailed survey information and can be carried

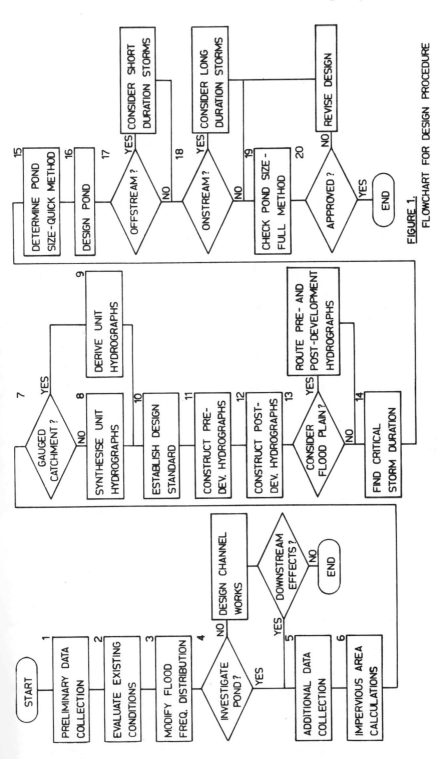

FIGURE 1.

FLOWCHART FOR DESIGN PROCEDURE

out relatively quickly without recourse to computer facilities.

Step 1 Initially, data are required on the topography, soils
and geology of the area of interest in order to construct a
frequency distribution of floods for the catchment in its
pre-urban state. The procedures outlined in Volume I of the
Flood Studies Report (Natural Environment Research Council,
1975) are recommended for this purpose. If the catchment is
gauged. the flood frequency distribution may be obtained from
either a partial duration series or an annual maximum series
of observed flood events depending upon the length of the
available record. However, where no records are available,
the ungauged catchment procedures from the Flood Studies
Report must be applied. which require a knowledge of catchment characteristics obtainable from either Ordnance Survey
maps to scales of 1:25000, 1:50000 and 1:250000 or maps
included in Volume V of the Report. Particular care is
necessary in determining values of the soil index from maps of
the winter rain acceptance potential for small catchment areas.

In order to assess both present and future flooding potential
downstream of the proposed urban development, representative
river cross sections are also required. If high flood stages
are to be considered, such cross sections should extend for
the width of the valley floor.

Step 2 The second step of the procedure involves the estimation of stage-discharge relationships for the chosen channel
cross sections. A site visit for the purposes of assessing
channel roughness and examining channel constrictions, such as
bridge openings, which may act as a control at high stages,
provides a useful complement to the survey information at this
stage of the investigation. An attempt should be made to
extend the stage-discharge curves above-bank-full conditions.
Information on the level and areal extent of inundation on
previous occasions. obtainable from local authorities, news-
paper reports and interviews with residents affected by the
flooding, is helpful in estimating the areas and velocities of
the valley flow. A site visit also provides a useful opportunity to assess the number and type of properties at risk and
the services likely to be affected at different flood stages.

Step 3 In Step 3 of the procedure, the flood frequency
distribution for pre-urban conditions obtained in Step 1 is
modified to reflect the effects of urbanisation. In general,
urban development may be expected to result in an increase in
the mean annual flood and a flattening of the growth curve,
since the influence of urbanisation is intuitively assumed to
diminish as the return period of the flood increases. The
magnitude of such changes may be predicted using an extension
to the procedure for ungauged rural catchments contained in
the Flood Studies Report suggested by the Institute of

Hydrology. In this extended method, which should be regarded
as an interim procedure pending the publication of further
work by the Institute under contract to the Department of the
Environment, the changes in the mean annual flood and the
growth curve are treated separately. The urban mean annual
flood is obtained from the rural mean annual flood using a
multiplier which is a function of catchment characteristics
including the proportion of the catchment area devoted to
urban land use. For return periods up to 50 years, the
appropriate growth factor (ratio of the urban T-year flood
to the urban mean annual flood) is read from the rural
regional growth curves using an "equivalent reduced variate",
which decreases from its rural value as the proportion of
urban area within the catchment increases. For return periods
greater than 50 years, the urban T-year flood is obtained by
multiplying the rural growth factor obtained from the
appropriate regional curve by the ratio of the urban growth
factor to the rural growth factor. This ratio is assumed to
decrease exponentially from a value of unity at a return
period of approximately seven years and approaches the
quotient of the rural mean annual flood and the urban mean
annual flood as the return period increases.

Step 4 Having completed Step 3 of the procedure, the pre-
development flood frequency distribution may now be compared
with that relating to post-development conditions. Given
the changes in flood magnitude relating to different return
periods, the stage-discharge relationships obtained in Step 2
may be used to evaluate the downstream effects of urbanisation
using indices such as the increased frequency of overbank
discharges. If the increase in flows downstream of the urban
development is tolerable or within the capacity of channel
improvement works that are required in any case, then the use
of flood retention ponds need be investigated no further.
Sufficient information is contained in the derived flood
frequency distributions for the design of works which do not
involve storage considerations. However, if channel improvements are insufficient to provide the necessary degree of
protection to downstream land users, then a flood retention
pond is worthy of further consideration. Since volumes of
inflow are as, if not more, important than peak flow rates
in determining the size of storage ponds, a more detailed
hydrological analysis in which the whole inflow hydrograph
is derived must now be applied.

Derivation of flood hydrographs

Step 5 The second part of the overall design procedure begins
with the collection of the additional topographical and
hydrological information upon which the more detailed investigation of a flood retention pond is based. Firstly, a more
detailed topographical survey of the development area is

required, which will both indicate the extent of the flood plain and assist in the choice of locations for storage ponds. For the larger developments, aerial survey provides perhaps the most economical and convenient method for obtaining contoured plans. Such a survey should also include flood-prone area downstream of the urban development. Even if aerial survey is used, additional channel cross sections will be required which must be obtained using conventional land survey techniques. In addition, the extent of both existing and projected future development should be assessed from local authority plans and records.

If the catchment of interest is gauged and the area of development is sufficiently large to merit such a detailed study, the flow record should be examined and the dates of significant flood events noted. An attempt should then be made to construct the associated rainfall hyetographs using both autographic charts and daily rainfall total from raingauges within the vicinity of the catchment. Even if these data are not used to derive unit hydrographs (see Step 9 below), the information on lag times and percentage runoff obtained from individual storm events may be used to improve estimates produced using the ungauged catchment procedures described in Step 8.

Finally, information on rainfall depth-duration-frequency relationships, storm profiles and areal reduction factors may also be compiled at this stage of the analysis. Such data may be produced for the catchment area of interest using the procedures outlined in Volume II of the Flood Studies Report. Alternatively, the same information is available on a repayment basis from the Meteorological Office, who have developed a suite of computer programs for the purpose (see Keers and Wescott, 1977).

Step 6 The proposed method of deriving flood hydrographs for urbanising catchment areas uses the percentage of impervious area as an index of urban development and not the proportion of the catchment devoted to urban land use employed in Step 3. Although the former is perhaps a more sensitive measure from the hydrological viewpoint, its computation can be time-consuming. In an attempt to reduce the expense of such analyses, Gluck and McCuen (1975) have suggested the use of prediction equations containing independent variables, such as population density and distance from the centre of the nearest business district, based upon representative catchment data. In addition, Jackson et al (1977) have reported favourably on the costs of estimating percentage impervious area from LANDSAT satellite imagery. For the purposes of the present design procedure, however, a sampling method has been found to be adequate. In this method, the impermeable areas and housing densities are measured on several small represent-

ative sections of the development using 1:2500 scale maps.
The date of construction of each section is also noted.
Using these data, a plot is prepared of percentage impermeable
area against housing density, with the points grouped according to the age of the property. Given such graphs, impermeable areas may be computed directly by counting houses,
given the approximate age of the development. The same
procedure is also applicable to new developments if the
proposed housing density is known.

Step 7 At this stage in the procedure, the unit hydrographs
which will later be used to compute flood hydrographs for
the chosen design storm must be either synthesised from
catchment characteristics (Step 8) or derived from observed
rainfall hyetographs and streamflow hydrographs (Step 9).

Step 8 Where no hydrometric data are available or the value
of the development does not justify the additional expense
of analysing the available records, a unit hydrograph must
be synthesised from a knowledge of the catchment geomorphology
and the extent of the urban area. In the present procedure,
a dimensionless one-hour unit hydrograph, derived from a
study of urbanising catchment areas in West Sussex and North
London (Hall, 1974, 1977), is employed. This hydrograph may
be made dimensional by multiplying the abscissae and dividing
the ordinates by a single scaling parameter, lag time, whose
value depends upon the length and slope of the main channel
and the percentage impervious area of the catchment.

The dimensionless one-hour unit hydrograph is available both
in the form of a table of ordinates for manual calculations
and as a polynomial function for computer usage.

Step 9 If the catchment area of interest is reliably gauged
and the additional expense is justified, unit hydrographs
may be derived from the available rainfall and runoff records.
In order to evaluate the effects of urbanisation, suitable
storm events are selected from throughout the period of record,
and the unit hydrographs obtained are grouped according to
their shape and date order. Smoothed hydrographs for each
grouping produced (say) by fitting a suitable function by
least squares, may then be taken to represent particular
states of development.

Step 10 The selection of design standards is a contentious
topic, with the majority of water engineers preferring to
treat each scheme on its own merits. As applied in the
present context, the design standard is the return period of
the storm up to which either no change in downstream flow
regime is discernible or peak flow rates do not exceed a
predetermined threshold of discharge.

Step 11 Design storm hydrographs for existing (pre-development) conditions may now be derived using an extended version
of the unit hydrograph plus design storm approach contained
in Volume I of the Flood Studies Report. The suggested
procedure differs from the latter in several important aspects.
Firstly, the dimensionless one-hour unit hydrograph described
in Step 8 is used in place of the triangular approximation
proposed in the Flood Studies Report. As noted above, this
dimensionless unit hydrograph is scaled by means of lag time,
and so the design storm duration also depends upon this
parameter and not time of rise as in the Flood Studies Report.
Secondly, a 50 per cent Summer rainfall profile is adopted in
place of the 75% Winter profile. Thirdly, the return period of
the total rainfall depth within the computed design storm
duration is assumed equal to that of the required peak flow
rate. Finally, rural percentage runoff is computed by means
of a regression equation based upon data from gauged catchments in the United Kingdom containing less than five per cent
of their area in urban land uses.

Application of this procedure for existing conditions serves
to define the limits which should not be exceeded when the
flow regime is altered by urban development. If an upper
limit on downstream discharges has already been established,
Step 11 can be omitted.

Step 12 The procedure described in Step 11 for the derivation
of storm hydrographs for existing (pre-development) conditions
may now be applied to synthesise a comparable set of hydrographs for the future (post-development) situation. These
calculations are based upon a revised unit hydrograph which
exhibits the shorter lag time characteristic of an urbanising
catchment area. In addition, the percentage runoff is
modified to reflect 80 per cent runoff from the proportion of
impervious area and the previously-computed rural percentage
runoff from the proportion of undeveloped area within the
catchment.

Post-development storm hydrographs should be derived for a
range of storm durations, since, when determining the size of
a flood retention pond, the critical event is that which
produces the maximum storage requirement and not necessarily
the largest peak flow rate. A range of durations of from two
to five times the lag time of the catchment is generally
sufficient for this purpose.

Step 13 Although Steps 12 and 13 are sufficient to evaluate
the changes in hydrograph shape that can be expected when the
catchment is urbanised, other aspects of the development which
may also influence the timing of flows should also be considered. For example, where the development encroaches on the
existing flood plain, the loss in over-bank storage may cause

an increase in peak flows which is comparable to, if not
greater than, that caused by the urbanisation. In these
circumstances, both sets of hydrographs must be routed downstream using an appropriate technique from Volume III of the
Flood Studies Report.

Step 14 The post-development hydrographs obtained from Step
12 (or Step 13) may now be used to estimate the duration of
storm which gives rise to the maximum storage requirement.
This critical duration will depend ultimately on the design of
the pond itself. If the proposed pond is onstream, i.e.
outlet-controlled, the required duration may be estimated by
drawing a straight line from the beginning of each hydrograph
to a point on the recession limb equal to the predetermined
peak outflow rate. The critical duration is that for which
the volume contained by the hydrograph above the straight line
is a maximum. For an offstream pond, which is usually inlet-
controlled, horizontal lines may be drawn across the hydrographs at the required outflow level. Again, the critical
duration is that associated with the largest volume under the
hydrograph above this line.

Design of retention pond

Step 15 Having obtained an estimate of the critical storm
duration from Step 14, preliminary estimates of the size of
pond or the configuration of the control structures, or both,
may now be obtained. For onstream ponds, several rapid
routing methods are available for this purpose, the majority
of which depend upon the use of geometrical figures such as
a trapezium (West, 1974) or methematical functions, such as
the Gamma distribution (Sarginson, 1973), to approximate the
shape of the inflow hydrograph. Using such methods, the
feasibility of several alternative schemes can be examined
both quickly and economically. In the case of offstream ponds,
which may involve up to three separate control structures,
each of which may have radically different characteristics,
schemes must be examined individually using finite-difference
routing methods.

Step 16 Sufficient information has now been obtained to undertake the detailed design of the pond. In particular, several
of the major assumptions employed in the preliminary analysis
of Step 15 must now be verified. For example, the pond may be
either onstream or offstream. With the latter, smaller
volumes of storage and therefore less land is required than
with an onstream pond. However, if an offstream pond is to
be evacuated by gravity, more fall is required between inlet
and outlet. If the difference in level is insufficient, the
depth of the pond may have to be reduced, and the saving in
land-take thereby lost. In addition, the pond may contain
water at all times of the year or drain completely between

major storm events. A 'wet' pond may be used as a local
amenity for boating and fishing, but the permanent storage
obviously cannot be used to reduce peak flow rates. A 'dry'
pond can also be managed as an amenity in the form of parkland
or playing fields, but some storage capacity may be lost if
the bed has to be graded for drainage purposes.

The preliminary estimates of storage capacity from Step 15
were also based upon general assumptions on the size and
configuration of control structures. More detailed
examination of the site may lead to some refinement at this
stage in the design procedure. Consideration should also be
given to the need for telemetry and the advantages in making
the most efficient use of the available storage that can be
afforded by automatic gate operation. Such refinements may
add significantly to the cost of small schemes, but may be
amply justified when several ponds are to be operated con-
junctively.

The detailed design of the pond should also include the provision of an emergency spillway. For ponds less than
25000 m^3 in capacity, such a spillway should be designed to
carry away the 150-year flood. Ponds of larger capacity fall
within the terms of reservoir safety legislation and are
therefore subject to statutory supervision and inspection.
In the latter case, a spillway of much higher capacity may
well be required.

For multi-purpose ponds, which serve both as an amenity and a
flood control device, the former usage may sometimes influence
the size of a pond more than the hydrology and hydraulics. For
example, the use of a 'dry' pond for football pitches may
dictate a minimum plan area of pond, which in turn may reduce
the depth of water to give the required storage capacity and
thereby alter the design of the control.

Check of pond design

The last four steps in the design procedure consist of a
series of checks on the scheme elaborated in Step 16 in which
selected storms are routed through the works as proposed.
The choice of appropriate storm durations is critical to the
success of each test.

Step 17 If an offstream pond has been selected, the proportion of the volume of the flood which is diverted into
storage depends upon the inlet control to the pond. In these
circumstances, short-duration storms will not create condition
that are critical to storage requirements, but they may yield
peak flow rates which overload the inlet, in which case the
scheme may not provide the necessary attenuation of the higher
rates of runoff. Onstream ponds will overdamp inflows from
short-duration storms and therefore do not require checking

for this condition.

Step 18   Although a flood retention pond may successfully
restrict the flows immediately downstream to predetermined
maxima, the proposed works may be only one element in a
larger scheme for which the situation at some point further
downstream may become critical with a longer duration storm.
If the pond is onstream, such a long-duration storm may
produce a condition in which inflow and outflow are comparable
in magnitude and the water level stabilises. The increase in
flow caused by urbanisation will therefore be passed downstream. Even though the increase will not be comparable in
size to that associated with the shorter duration design
storms, the effect may be undesirable. These longer duration
storms can be attenuated by increasing the storage capacity
and restricting the outlet control. In extreme circumstances,
the type of reservoir could be changed to offstream.

Step 19   A full level-pool routing technique may now be used
to check the performance of the works as designed.

Step 20   If the proposed flood retention pond does not meet
the design criteria completely, further adjustments to the
scheme must be made and Step 19 repeated. Satisfaction of
these criteria ends the procedure.

MULTIPLE BALANCING POND SYSTEMS

The procedure outlined above is intended to apply to the case
of a single flood storage pond, the purpose of which is to
mitigate the changes in flow regime that have occurred as a
result of urban development within its tributary catchment.
When the pond forms part of a larger scheme in which adjacent
catchment areas are also provided with some form of balancing
storage, the need arises to examine in addition the operation
of the whole system. Since a balancing pond both delays and
lowers the peak rate of runoff, the possibility that the
attenuated, post-development hydrograph will coincide with
that from a neighbouring catchment when the pre-development
hydrograph did not, cannot be ignored. Knowledge of such
behaviour is particularly important to those schemes in which
some form of remote control is employed to maximise the use of
the available storage volume.

The hypothetical example of a cascade of ponds, in which the
outflow from an upstream storage becomes the inflow to its
neighbour immediately downstream, has been examined by both
Wycoff and Singh (1976) and Mein and Woodhouse (1977). Their
results showed that the reduction in peak discharge at a
downstream point is greater for a single pond than for two
ponds each of half the storage capacity in series. Similarly,
two ponds are more effective than three ponds, each

providing one-third of the same storage capacity. In addition, one large pond often has greater potential for recreational use than two or more smaller ponds.

For the case of ponds in parallel, i.e. located on adjacent tributaries of the same river system, McCuen (1974) has drawn attention to the difficulties of phasing outflows so that increases in peak flow rates in the downstream reaches are avoided. This problem has also been investigated by Mein and Woodhouse (1977), who concluded that conditions under which the addition of a balancing pond had a detrimental effect on downstream peak flows were unusual. This apparent contradiction in opinion serves to emphasise the site-dependent nature of such effects. Additional case studies on multiple balancing pond systems would be particularly useful in clarifying this issue.

ACKNOWLEDGEMENT

The authors wish to thank the Construction Industry Research and Information Association for their permission to present this paper, and the members of the Steering Group for the Research Project referred to in the Introduction for their helpful comments during the drafting of the Guide on the Design of Stormwater Balancing Ponds.

REFERENCES

Economic Development Committee for Agriculture (1977) Agriculture into the 1980s. Land use. National Economic Development Office, London, 32pp.
Gluck, W.R. and McCuen, R.H.(1975) Estimating land use characteristics for hydrologic models. Wat.Resour.Res., $\underline{II}$, 177-179.
Hall, M.J.(1973) The hydrological consequences of urbanisation and introductory note. Construction Industry Research and Information Association, Research Colloquium on Rainfall, Runoff and Surface Water Drainage of Urban Catchments, Bristol, Proc.paper 10.
Hall, M.J.(1974) Synthetic unit hydrograph technique for the design of flood alleviation works in urban areas. Internat. Assoc.Scient.Hydrol. pubn. no. 108, 485-500.
Hall, M.J. (1977) The effect of urbanisation on storm runoff from two catchment areas in North London. Internat.Assoc. Scient.Hydrol. pubn. no. 123, 144-152.
Jackson, T.J., Regan,R.M. and Fitch, W.N. (1977) Test of LANDSAT - based urban hydrologic modelling. Proc.Am.Soc.Civ. Engrs., J.Wat.Resour.Planning and Management Div., $\underline{103}$, WR1, 141-158.
Keers, J.F. and Wescott, P.(1977) A computer-based model for design rainfall in the United Kingdom. Meteorological Office Scient. Pap.no. 36, H.M.S.O., London, 14pp.

Leopold, L.B.(1968) Hydrology for urban land planning - a guidebook on the hydrologic effects of urban land use. U.S. Geol.Survey Circular 554, 18pp.
McCuen, R.H.(1974) A regional approach to urban storm water detention. Geophys.Res. Letters, $\underline{1}$, 321-322.
Mein, R.G. and Woodhouse, M.P.(1977) Design of retarding basin systems. Instn.Engrs.Austr., Proc. Hydrol.Symp., Brisbane, 141-145.
Natural Environment Research Council(1975) Flood studies report, 5 vols., N.E.R.C., London.
Sarginson, E.J.(1973) Flood control in reservoirs and storage pounds. J.Hydrol., $\underline{19}$, 351-359.
West, M.J.H.(1974) Flood control in reservoirs and storage pounds - a discussion. J. Hydrol.,$\underline{23}$, 67-71.
Wycoff, R.L. and Singh, U.P.(1976) Preliminary design of small flood detention reservoirs. Wat.Resour.Bull., $\underline{12}$, 337-349.

HYDRAULIC ASPECTS OF BALANCING STORM WATER AT MILTON KEYNES

L.H. Davis, C.Eng., Area Engineer.
D.R. Woods, C.Eng., Principal Engineer.
Anglian Water Authority, Bedford Sewage Division.

INTRODUCTION

It is unusual to find an urban situation where the opportunity
has been taken to deal with additional storm water run-off
from new development by a comprehensive system of storm-water
balancing reservoirs incorporated into the surface water
drainage system.

This is what is being done at Milton Keynes New Town and,
whilst it is recognised that this solution is possible, mainly
due to the "greenfields" situation in which this New Town was
planned, it is nevertheless true that Water Authorities are
becoming increasingly aware that they can no longer permit
development, even on a much smaller scale, to take place
without at least some consideration being given to balancing
excess run-off, before discharge to the country's river
systems.

The permissions given for development of all types over many
years without this safeguard, have created increasing
problems with regard to the greater incidence of flooding of
agricultural land, as well as industrial and housing areas;
although the control of flooding in these developed areas may
have special problems of their own which could involve local
solutions of a different kind.

In this country, apart from inconvenience, flooding is mainly
an economic problem due to high densities of development.
Although it may not be possible in many cases to solve this
problem in such a comprehensive way as at Milton Keynes, this
does not change the principles involved with regard to the
need for storm water balancing, nor those associated with the
design of such facilities.

It is for the above reasons that the authors recommend that

more consideration should be given in urban situations to the use of some form of storm water balancing facility, both to protect development and agricultural land downstream and as a means of effecting, in some cases, considerable economies in the surface water sewerage system.

It is proposed in this paper, therefore, to set out the background requirements for storage of run-off from the development at Milton Keynes, to indicate some of the problems encountered in assessing run-off from partially urbanised catchments and to give a description in some detail of the solutions adopted and the hydraulic methods used for designing the various storage reservoirs already constructed in this rapidly growing New Town.

Topography & Geology
The New Town Designated Area is approximately 8900 ha with the River Great Ouse flowing from west to east on the northern boundary. The area is divided into two main catchments, the River Ouzel flowing south to north through the eastern half and the Loughton Brook flowing through the western half, both joining the River Ouse near the northern boundary.

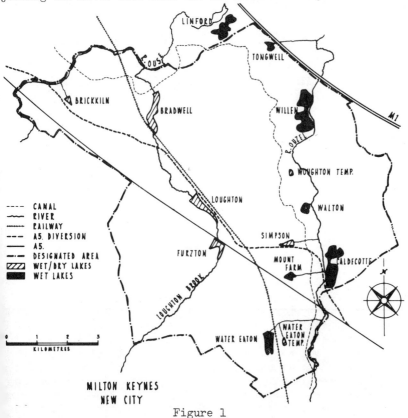

Figure 1

The western half forms nearly the total catchment of the
Loughton Brook, which is approximately 4800 ha, whilst the
eastern half, lying over the lower reaches of the River Ouzel,
is only about 14% of the total catchment of 28,000 ha draining
to this river. It was found necessary to provide different
solutions to the problems of flood control in these two areas.

The New Town area is low-lying with gently undulating
topography liberally provided with ditches and small watercourses flowing to the basic river system. Following early
feasibility studies to indicate the drainage system required,
it was decided to provide separate surface water sewerage
systems to drain development to the main watercourses, which
could be improved as necessary.

Due to the nature of the countryside and the subsoil geology,
rainfall tended to be collected and retained in large areas of
open pastureland. In the past this has no doubt helped
considerably to avoid excessive run-off rates in time of storm
and has relieved the river system but, nevertheless, widespread flooding was a frequent occurrence. It was recognised
that the construction of a New Town would considerably worsen
this situation.

The geology of the area was the subject of a special study by
the Institute of Geological Sciences (1970). The detailed
geology is of great importance related to the construction of
the dams and balancing reservoirs and a general knowledge,
especially of superficial deposits, is also of importance when
considering the run-off characteristics of undeveloped areas.

The geological formations are of the Jurassic and Cretaceous
series which are principally Oxford clays and Cornbrash.
These are partly overlaid by Pleistocene and recent deposits,
mostly Boulder clays and associated deposits, which cover some
two-thirds of the area.

Alluvium and first and second terrace gravels appear in the
river valleys and these of course are pervious formations, but
most of the area is covered by heavy clays which give rise to
a large percentage run-off during storm periods.

Planning
The River Ouse flood plain and both the Ouzel and Loughton
Brook valleys have flooded at regular intervals in past years
some of the flooding being very widespread. The conditions i.
1947 and 1968 are notable in this respect, when extensive
areas in the valleys were under water and existing areas of
development at that time, such as parts of Bletchley, Newport
Pagnell, Bradwell and two villages, Loughton and Simpson, wer
seriously affected. Both the residents and local farmers wer
very concerned that any new development, to say nothing of a

major New Town, would increase the maximum flood levels and also the frequency of smaller floods.

Following investigations, it was accepted that if the New Town were to be drained in the traditional way by constructing large surface water sewers discharging to the river system, future flooding would be considerably worse, and even if the new development was not affected, the problem would simply be transferred downstream.

Proposals were made therefore, to provide storage reservoirs within the development area to balance all additional storm water run-off brought about by the increased paved and roofed areas of the New Town and thus ensure that flooding would be no worse after completing the development than it had been in the past. This proposal is quite feasible, but it has posed the drainage engineers with a number of difficult and interesting design problems, which would not necessarily apply to developments of a lesser extent.

General design considerations

The worst recorded flood which has occurred in the area of the Ouse and Ouzel valleys was in 1947 and this was caused by a long period of rain associated with melting snow. This flood was specified by the Great Ouse River Authority, the previous Land Drainage Authority, as the design criterion to be adopted for any flood control works in the Ouzel catchment.

An original feasibility study of surface water drainage for Milton Keynes was carried out by Consulting Engineers, J.D. & D.M. Watson and submitted in 1969.

The report recommended that an allowance of 100% run-off from paved areas and 70% from unpaved areas, combined with a rainfall of 64 mm in 24 hours with half this quantity falling in the first 2 hours, be used for design purposes in the sub-catchments. It also indicated that conditions worse than these would be unlikely to be experienced more frequently than about once in 20 years.

The immediate reaction to this proposal was that 70% for unpaved areas was too high. However, it was shown that both Grendon Underwood, a nearby catchment being studied by the Institute of Hydrology, and the catchment at Harlow used by Watkins (1962) in his TRRL Hydrograph studies, had produced run-off from individual storms during the winter approaching and exceeding this figure. Grendon Underwood and Harlow are situated on heavy clay and, in this respect and others are similar to catchments in Milton Keynes.

Later studies by the design engineers showed that the suggested rainfall approximated to the Bilham formula (1936) for a return period of 10 years and it was decided to use this curve in design calculations for sub-catchments.

The very large catchment area of the River Ouzel compared with the much smaller Loughton Brook catchment meant that the control of flooding in these two areas of the New Town would involve very different solutions. There were also smaller catchments within the area such as Brickkiln, which drains directly to the River Ouse, and Simpson which lies upstream of an existing village and these required separate consideration.

These smaller catchments lie entirely within the New Town area and will be developed in the normal way. Due to this fact, the impermeability factors and times of concentration, within reasonable limits, were readily calculable and it was found quite satisfactory to assess the storage required by the 'Triangle' method proposed by Davis (1963) with peak discharge rates suited to the capacities of downstream watercourses.

In the Loughton Brook Catchment, also, the use of a 10 year Bilham storm was considered satisfactory providing, as with all balancing reservoirs, proper account was taken of the run-off from undeveloped areas within the development as well as the paved areas. Whilst the 'Triangle' method, which provides a very quick solution, could be of use here for checking, and particularly for indicating the likely duration of the worst storm, it was not considered to be a reliable method where large natural areas are associated with developed areas, mainly due to the fact that incorrect assessment of the time of concentration would seriously affect the results.

In the case of Loughton Brook where a series of on-stream reservoirs was proposed, these had to be considered as a whole and, apart from considerable areas of grass within the development, this catchment included some 890 ha outside the New Town which would stay as open countryside.

The available methods of assessing run-off from such areas involve the use of impermeability factors. At best these can give only very approximate results and at worst they are wildly inaccurate.

It was decided, therefore, that normal hydrographs would be calculated for the paved areas and a separate hydrograph constructed for all undeveloped areas based on the method proposed by Nash (1960) which was based on the characteristics of a large number of catchments. The two hydrographs were then combined to give the final inflow hydrograph to the balancing reservoirs.

Latterly of course the Natural Environment Research Council has produced the 'Flood Studies Report' (1975) which is a mor

factual and statistical approach to the assessment of floods
for a given return period and this is found to be very useful
in connection with the assessment of peak flows from
undeveloped catchments. Unfortunately, for two reasons, the
methods recommended are of little use in the solution of
storage problems because, firstly, it is not recommended for
use where more than about 20% of the catchment is urbanised,
and secondly, it is not possible to allow for changes in the
storm duration, and this is essential in the design of
balancing reservoirs where it is not peak flows that are
critical but volumes of water.

It should be emphasised also that the determination of the
worst storm in balancing reservoir design is directly related
to the allowable discharge rate from the reservoir. It is <u>NOT</u>
determined to any great degree by catchment characteristics
and is the reason why standard methods of sewer or open
channel design including those in the Flood Studies Report
should not be used.

The requirements of the major catchment in the Ouzel Valley
were found to be quite incompatible with the use of the above
design methods. As had been stated, the 1947 flood was caused
by the combination of a storm and melting snow. There were a
number of rainfall gauges in the area, privately maintained
for the meteorological office, which did give some background
information of total daily rainfall, but the only gauging weir
in the area measuring the flow in the River Ouzel at Willen
was not constructed until 1961 and this was overtopped in
storm periods. It did not, therefore, record peak rates of
flow even in the limited recording period available and, of
course, was of no help in establishing flows in 1947.

It was not possible, therefore, to relate rainfall to run-off
in this area during 1947 especially when associated with melting snow. Nor could the return period of such a flood be
assessed with any degree of accuracy although this is thought
to be in the order of 50 to 100 years.

There was, however, another method to achieve the criteria
laid down. If conditions were to be no worse than those
occurring in the past, then the required storage would be made
up of flood plain storage losses due to construction, sewage
effluent and the additional run-off due to development. The
latter is of particular interest and, briefly, was determined
by investigating two different approaches.

A) Consideration of the run-off hydrographs from each subcatchment in the New Town area before and after development.

B) Consideration of the increases in run-off coefficients of
the developed areas which would produce greater volumes of
run-off.

With the latter approach the coefficient for paved areas was
taken as 1.0 and those for natural areas were varied depending
on the storm return period used.

These two approaches, one of which is described in the later
sections, gave a good correlation and, because the second
method was much simpler, it was used to give the storage
volume required for additional run-off from the whole area.

Having determined the total storage volume required it was
necessary to establish a method of control which would provide
such storage.

The simple methods of control used in the sub-catchments,
usually a piped outlet giving a variable discharge rate depending on the head of water in the reservoir, was out of the
question in this main catchment where a much greater degree of
control was required.

It became clear that it would be necessary to provide flood
gates across the river which at suitable times could be partly
closed to divert flood water into off-stream reservoirs, thus
making maximum use of the storage to be provided.

In the event this method was adopted, but due to the
complications of multiple inflows from sub-catchments, the
need to gauge and record total rates of flow entering and
leaving the area and of linking up with flows in the River
Ouse it was decided that a telemetry and telecontrol system
linked to a master computer was essential to effect an economical storage system. This would not only automatically
control the storage of storm water in the most efficient way,
but also make possible the continuous updating of design parameters as the New Town develops and operating experience increases.

Finally a considerable amount of research work has been and is
quite rightly being directed towards the problems of rainfall
and run-off to effect economies in sewer design. However, if
the Authors' opinion is accepted that more regard should be
given to the balancing of storm flows for economic reasons,
then it is contended that more research time should be
specifically directed towards the very different requirements
of balancing reservoir design.

The foregoing gives a general picture of the balancing of
surface water run-off at Milton Keynes. There follows a more
detailed discussion of the design methods used for the
various types of balancing facilities being provided.

Sub-Catchments
For the smaller catchments, two types of reservoir have been

constructed, those built off-stream in flat areas where a
constant discharge control could be provided and those built
on-stream where an earth dam could be constructed across a
valley allowing for a variable discharge control. A cross
section through the former is shown in Figure 2.

DISCHARGE CHANNEL WITH
CONTROL DOWNSTREAM.

OVERFLOW WEIR

RETURN PIPES
WITH FLAPVALVES

Figure 2

The capacity of this type is the most straight-forward to calculate and a graphical method is shown in Figure 3. The total

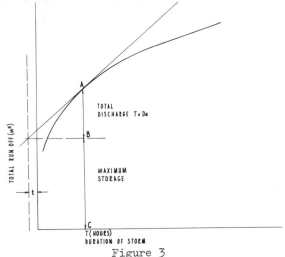

TOTAL RUN OFF (m³)

A

TOTAL
DISCHARGE T × De

B

MAXIMUM
STORAGE

t

C

T(HOURS)
DURATION OF STORM

Figure 3

rain in any time T is multiplied by the Impermeable area Ap
and a curve is plotted of total run-off against time. A
straight line representing the allowable constant rate of
discharge De is drawn from a line to the left of the origin
equal to the time of concentration (t) of the sewer system so
that it is tangential to the curve at A. The maximum storage
required is represented by BC.

Either Bilham rainfall tables or local rainfall predictions,
in this Country available from the Meteorological Office, may be
used in the calculation. If the unmodified Bilham formula is
used the required capacity may be determined by calculation
instead of graphically.

Tongwell Lake and two temporary reservoirs using this method
of design have been constructed.

Figure 4 shows a cross-section through an on-stream reservoir
where the discharge rate is variable, only reaching the maximum allowable rate De when the reservoir is full.

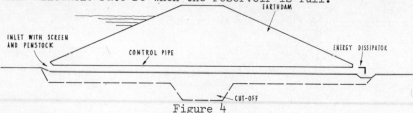

Figure 4

The problem presented here is solved simply by the use of the
Triangle method which, along with the previous method is
eminently suitable for comparatively small areas of Urban
development, providing there are no substantially undeveloped
portions. Referring to Figure 5, the main assumption is that
the area of triangle, height De and base (Tw + t) is equal to
the area DFH formed by the rising arm of the reservoir discharge
curve and the recession curve of the inflow hydrograph.

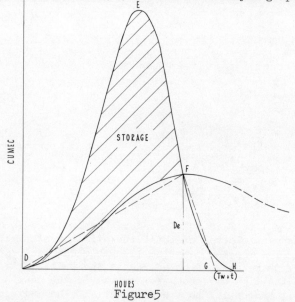

Figure 5

The errors which may occur from this assumption are small
compared with the variations resulting from differing assessments of the run-off factor which, to a large extent are at
the moment, bound to be subjective. In addition the unmodifie
Bilham formula is used, which does not concern itself with
intensities and is, therefore, much simpler in assessing
storage capacities than methods which include rainfall
intensity. It's use however is restricted to those parts of
this Country where 'Bilham' gives a reasonable approximation o
actual rainfall. The evidence shows this to be true in the
Milton Keynes Area.

The total intake to the reservoir is represented by the
hydrograph DEH which is equal to Ap x r for any storm length T.
The storage S is equal to the total intake less the area of
triangle DFG. By differentiating S with respect to T, the
worst storm length Tw for maximum storage Sm can be found.

By substituting Tw in the storage equation it can be shown that
for a 10 year storm $S_D = 56A_D^{1.3922} - 25A_D - 1800t$ where A_D is
the Area/Discharge ratio $\dfrac{Ap}{De}$

S_D is the Storage/Discharge ratio $\dfrac{Sm}{De}$

This equation may be tabulated or the most useful portion of it
can, for practical purposes, be plotted as a straight line on
log log paper for any values of S_D and A_D where t = 0 and this
is now shown in Figure 6 in metric units for a variety of
storm return periods. The correction for - 1800t may then be
applied and the resulting value for S_D multiplied by the
allowable discharge to give the volume of storage required.

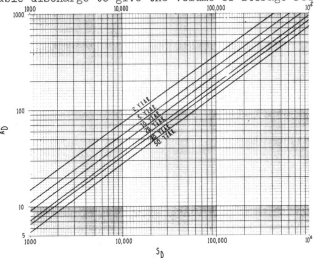

CORRECTION OF -1800t TO BE APPLIED WITH t IN HOURS
Figure 6

The following example for the Simpson Reservoir at Milton
Keynes shows the simplicity of the method:-

Gross area contributing   =   525 ha
apply areal reduction factor (Holland 1967) of 0.87   =   457 ha
34% of area paved @ 100%   =   155
66% of area unpaved @ 70% =   211
Impermeable Area Ap   =   366 ha

Maximum allowable discharge rate De   =   1.4 cumec
$A_D = \dfrac{366}{1.4} = 261$

From graph (Figure 6) for a 10 year storm S_D (t = 0) = 122,500
time of concentration t = 0.67 hours.

S_D (t = .67) = 122,500 - (1800 x .67) = 121,300
S^D = 121,300 x 1.4 = 170,000 m^3

Loughton Brook Catchment

Three on-stream wet/dry reservoirs were proposed along the
Loughton Brook, the first to be constructed being the Bradwell
Reservoir at the northern end near the confluence with the
River Ouse.

Due to staff shortages at that time this was designed by
consultants J.D. & D.M. Watson. Because no flow or rainfall
records were available, the design method used the assumed
similarity between the ungauged catchment and a nearby gauged
catchment for which a rainfall run-off correlation was already
established.

The hydrographs thus built up were routed through the reservoir
using a standard tabular method and the maximum capacity
required for the design storm determined.

Loughton Reservoir, the second to be constructed in the
Loughton Valley, immediately upstream of Loughton village, was
designed by the Authority.

The triangle method was not considered suitable for this much
larger catchment with an area of 2,380 ha, of which 890 ha is
outside the designated area of Milton Keynes. The method used
for Bradwell was not adopted since the procedure for obtaining
the hydrograph was based on comparison with only one other
catchment and, in addition, no distinction was made between the
hydrographs shapes from the paved and unpaved parts of the
catchment.

The need to provide for this omission led in 1972/73 to the
development of a method using concurrent hydrographs. The
essential principal of this method is that the hydrographs of
inflow from the unpaved and paved areas are found separately
and then the paved hydrograph superimposed on the unpaved to
give the full inflow hydrograph for the reservoir.

In determining the hydrograph from the unpaved area it was
assumed that the whole area was contributing and the vertical
ordinates were then reduced in proportion to the total area of
hard surfacing i.e. if the gross area of the catchment was
1000 ha and there were 280 ha of net paved surface then the
vertical ordinates were reduced by 28%. The rainfall was
assumed to have uniform intensity and the resulting hydrograph
for the paved area therefore assumed an inverted trapezoidal
shape with the rising arm reaching a constant discharge rate a

a time equal to the time of concentration for the sewer system.
The time of concentration was fairly easy to ascertain within
reasonable limits by using the rational method in a preliminary
design of the future surface water sewerage system.

The determination of the hydrograph shape for the unpaved area
was not so easy. Following discussion with the Institute of
Hydrology Nash's work on synthesising Unit Hydrographs for
ungauged catchments was adopted, being based on the results
from a large number of British Catchments. The catchment
characteristics in the form of area and overland slope were
used to determine the time to peak while the addition of stream
length and use of the incomplete gamma function provided the
shape of the S curve of the Instantaneous Unit Hydrograph. A
one hour Unit Hydrograph was obtained by subtracting the
vertical ordinates of the S curves set an hour apart. The
effective rainfall which, in this instance, as for the paved
area, was assumed to fall at a constant rate, was combined with
the one hour Unit Hydrograph to obtain the full hydrograph for
the selected storm length. The two hydrographs obtained were
summated as shown in Figure 7 and the resulting composite
hydrograph routed through the reservoir basin.

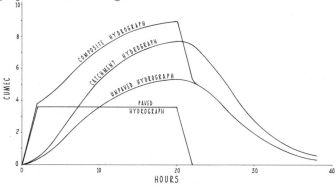

Figure 7

Various methods are available for flood routing, but unless a
computer is readily available or a considerable number of
exercises are to be done, the graphical method described by
Sorenson (1952), which is both reasonably accurate and quick
may be used.

The application of Nash's method to produce a Unit Hydrograph
is considerably longer and more complicated than that advocated
in the recently published Flood Studies Report. In addition
the assumption of uniform rainfall intensity for any given
quantity of rain was obviously wrong, although it was not known
if this would seriously affect the results.

A re-appraisal of the Concurrent Hydrograph method was therefore carried out using the 'triangular unit hydrograph' shape
proposed in the Flood Studies Report and the '75% winter'

profile for rainfall also proposed. In addition to the '75% winter' profile with the peak occurring during the middle of the storm, calculations of maximum reservoir capacity have been checked with the peak occurring at the beginning and end of the storm period and with a constant rate of rainfall.

In the sample catchment used, the '75% winter' profile set symetrically in time proved to require the greatest balancing capacity for any length of storm. The results are shown in Figure 8 and it will be noted that for the '75% winter' profile very little variation in capacity is required over a large number of storm lengths. This naturally simplifies the calculations since only a small number of trial storms at large time intervals need be worked through to determine the maximum capacity required.

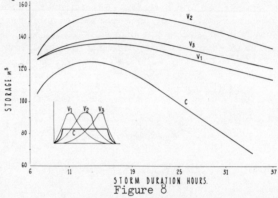

Figure 8

The same storm profile will be applied to the paved area for which a one hour Unit Hydrograph must also be constructed. Experience has shown that with long storms and a sewer system with a time of concentration of one hour or less (which will cover most cases) an isosceles triangle with a time to peak of one hour, as shown in Figure 9, is satisfactory.

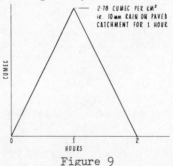

Figure 9

Loughton Reservoir was designed using the method in it's original form and its capacity has been checked with the revised Unit Hydrograph and rainfall profile. This resulted in a higher estimate of storage capacity. However, it was found in practice that due to the way the original method had

been applied, and to other factors, that sufficient storage
was available.

The design method has also been applied to Bradwell Reservoir
and resulted in a storage requirement lower than that provided.
On the basis that the method of assessing capacity should be
consistent throughout the Loughton Valley it was decided to
reduce the allowable discharge from this reservoir to make full
use of the storage available. As a result it will be possible
to decrease the scope and cost of improvement works downstream.

<u>The Ouzel Valley</u>
The methods of control in the sub-catchments and the Loughton
Brook were dictated by simplicity and consistency of operation
keeping maintenance requirements to a minimum. These reservoirs
have resulted in an improvement to the existing flooding situation, as well as alleviating future problems caused by
development.

In the Ouzel valley, with its large catchment, this principle
could not be applied due to the enormous capital cost involved.

The original assurance that "development would not make flood
conditions downstream of the new City worse than at present "
became the prime consideration.

The flood storage to be provided fell into two main categories:

1. Existing capacity lost in the flood plain due to the
construction of road embankments, the lakes themselves, and
the new Sewage Treatment Works.

2. The increase in run-off caused by impervious areas replacing green fields.

The assessment of (1) was fairly straight-forward and amounted
to 1,754,000 m^3 but (2) presented some problems. No information could be found on the <u>increase</u> in run-off from urban
areas.

On the premise that longer return period floods were caused
generally by a combination of lower moisture deficit (higher
run-off factor) and heavier rain, the following table of
factors was adopted for a 24 hour storm considered to be the
worst case.

Return Period in Years	Bilham Total Gross Rainfall r mm	Est. run-off Coef. for Unpaved Areas K
1	32	.15
2	40	.25
5	52	.50
10	64	.60
20	78	.65
30	88	.70

The method of deriving the volume of storage was as follows:-

Consider an unpaved area A (hatched) before development. A fraction of this area (y) is developed and has 100% run-off.

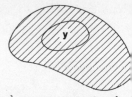

r   is the total rainfall
k   is the run-off coefficient for unpaved areas expressed as %

V_1 is the volume run-off before development
V_2 is the volume run-off after development
V_b is the volume run-off to be stored after development

$V_b = V_2 - V_1$
$V_1 = A.k.r.$
$V_2 = Ay.1.r. + (A - Ay) k.r.$
$\quad = Ayr + Akr - Aykr$
$\quad = V_1 + Ayr - Aykr$
$V_2 - V_1 = Ayr - Aykr$
Therefore $V_b = Ayr - Aykr$

Let f be the percentage increase in run-off factor due to development.

Then $V_b = A.r.f.$
Therefore $Arf = Ayr - Aykr$
Hence $f = y - yk$ which is of straight line form with f and k the variables.

when $k = 0$ $f = y$ (% of paved area) $V_b = A.r.y.$
and $k = 100\%$ $f = 0$ $V_b = 0$

By plotting f against k the additional percentage increase in run-off due to paving can be ascertained for any value of k. Figure 10 shows the percentage increase in run-off due to paving for that part of the Designated Area within the Ouzel catchment and it is interesting to note that the greatest difference is provided by the 2 year storm.

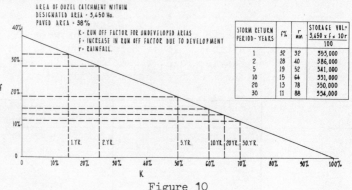

Figure 10

The figure of 386,000 m3 is less than 18% of the total storage
to be provided so that any rational variation in the assumptions made was not going to significantly alter the final
volume of storage required.

Acknowledgements:
The Authors wish to thank the Bedford Sewage Division of the
Anglian Water Authority for permission to present this paper
and to point out that any opinions expressed are their own and
not necessarily those of the Authority. They wish to acknowledge the work done by the staff at Milton Keynes over the
past few years with design and construction of the balancing
reservoirs and to express appreciation for the assistance
given in the preparation of the diagrams.

References:
Bilham E.G. (1936) Classification of Heavy Falls in Short
Periods. British Rainfall 1935, 75, 262-280.

Davis L.H. (1963) The Hydraulic Design of Balancing Tanks and
River Storage Pounds. Chartered Municipal Engineer 90, 1-7.

Holland D.J. (1967) The Cardington Rainfall Experiment Meteorological Magazine 96, 193-202.

Institute of Geological Sciences (1970) Regional Geotechnical
Investigation of the Milton Keynes Area, Preliminary Report

Nash J.E. (1960) A Unit Hydrograph Study with Particular Reference to British Catchments. Proceedings of the Institution
of Civil Engineers 17, 249-282.

Natural Environment Research Council (1975) Flood Studies
Report, London.

Sorensen K.E. (1952) Graphical Solution of Hydraulic Problems
Proceedings of the American Society of Civil Engineers 78,
Separate 116.

Watkins L.H. (1962) The Design of Urban Sewer Systems. Transport and Road Research Laboratory, Technical Paper No. 55.

DESIGNS AND ANALYSIS OF STORMWATER RETENTION BASINS USING QUALITY AND QUANTITY CRITERIA

Martin P. Wanielista, PhD, P.E. and
Yousef A. Yousef, PhD, P.E.

Professors of Engineering
Florida Technological University
Orlando, FL   32816

The goals of water pollution control in the United States were specified by the U.S. Congress (1). Interpretation of these goals have resulted in water pollution abatement programs which improve water quality of both the ground and surface waters. The sources of pollution have been classified into two general categories, point and nonpoint. Point sources are those discharges from controlled sewer systems, such as municipal and industrial water pollution control facilities. Nonpoint sources are those locations or land uses which discharge wastewater resulting from precipitation. Stormwaters are discharges from nonpoint sources. These stormwaters can contain pollutants equivalent to or greater than treated sewage from the same land area.

Treatment performance levels have been obtained for the more conventional processes, such as dissolved air flotation, high-rate filtration, microstrainers, and a newer treatment, swirl concentrators (2). Broom sweeping and sewer flushing performance efficiencies and cost also are available (3). The most commonly used treatment practice is sand filtration by existing soils. As areas become more impervious (streets, buildings, etc.) the existing soils are less available for recharge of groundwaters and purification. Direct surface water discharges are increased and surface waters receive greater quantities and concentrations of stormwater impurities. In this paper, results are presented from stormwater sampling before and after percolation/retention basins. These data on runoff hydrographs and loadographs (mass vs. time) are used as a basis for design criteria based on quality (performance criteria) and quantity of water diverted. Regulations (4,5) for control of stormwater results in designs based on a volume of stormwater runoff, but currently (1977) do not specify performance cri-

teria. This work will help establish performance criteria by
relating removal efficiencies to a first flush diversion volume.

PRECIPITATION

If all precipitation of a storm event resulted in runoff, this
could be the extreme case for design of a stormwater abatement
system. However, depression storage, infiltration, and evapotranspiration reduces runoff quantities. Thus the runoff/
precipitation relationship must be determined.

When using percolation for treatment, the antecedent rainfall
and watershed conditions are important. Storms of lesser
volume but of more frequent occurrence could require a larger percolation basin volume because the previous storm volume
has not had time to infiltrate or otherwise be treated. In
addition, the percolation volume removed is proportional to
the percolation area. Thus, shallow percolation basins will
drain faster than deep percolation basins.

The cumulative volume of rainfall produced for storms of less
than or equal to a stipulated rainfall event volume is shown
in Figure 1. If no treatment were specified, then the cumulative volume of runoff for a stated rainfall event volume
is as shown in Figure 1 (no diversion curve). However, if
a stated volume is treated no matter how large the storm, then
a greater volume is cumulated relative to the no diversion alternative. As an example, consider the stated event volume
equal to ½". This specifies that events up to ½" will be
treated and those events over ½" will have the first ½"
treated and the remaining amount will be directly discharged
without treatment. Illustrated on Figure 1 is the treatment
curve. If concentrations of pollutants were relatively constant
with time, then the mass of pollutants removed would be equal
to the fraction of yearly runoff treated. On very large watersheds, it was shown that mass was related to flow by a linear
relationship which implies that the random concentrations can
be represented by a constant volume (7). This is equivalent
to saying that on large watersheds, the first flush effect is
non-existent. Figure 1, therefore, can represent the average
yearly efficiency (% removal) of treatment given a stated
event level for treatment.

If the first flush events are present, then the average yearly
efficiency should increase because the concentrations or quantities of pollutants are greater during the early part of a
storm event. Thus, either the efficiencies expected from diversion systems will be accepted to be higher in smaller watersheds, or the stated volume of treatment can be reduced.

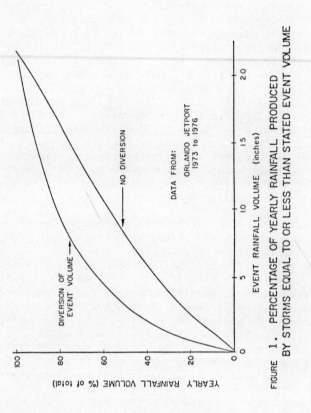

FIGURE 1. PERCENTAGE OF YEARLY RAINFALL PRODUCED
BY STORMS EQUAL TO OR LESS THAN STATED EVENT VOLUME

This analysis of precipitation assumes that the resulting run-off produces pollutant concentrations at a random rate which can be represented by an average concentration value. The efficiencies for the smaller watersheds using diversion with percolation have to be established. The minimum average yearly efficiency for larger basins has been established in Figure 1. It is diversion with treatment. However, the efficiencies must be determined as a function of watershed size, type of soil for percolation, travel time, land use, and volume of diversion.

FIELD INVESTIGATIONS

Fourteen field sites were established in the Orlando area. Four of these are considered to represent large watersheds; namely, a freshwater swamp, flatwoods, an improved pasture, and a mixed agricultural low density residential area. The remaining ten field sampling sites are characteristic of urban development on small watersheds, such as parking lots, low density residential, medium density residential, high density residential, motel parking lot, commercial/residential area, and apartment complexes.

The data from a 4.6 acre motel complex will be used to illustrate the first flush effects, calibration of a computer program model, and efficiencies of a diversion/percolation management practice.

Engineering drawings of the diversion box and the percolation pond are shown in Figure 2. Stormwater drains from the watershed (parking lot) and is diverted to a percolation pond. The soils in the vicinity of the pond are excellent for percolation. Once the first inch (2.54 cm) of runoff is diverted to the percolation pond, approximately 2½ feet of water are held in the pond, thus any other runoff water will flow over the diversion baffle and receive no treatment. Thus, there are no moving parts or mechanical failures.

Figure 3 illustrates a typical first flush with concentrations increasing during the early part of a storm event. Also shown are the typical loadographs(loads) which are a plot of the mass of pollutants with time.

Using watershed, pollutant removal, soils, and precipitation data, the diversion/percolation system was simulated over a 20 year period to determine average yearly efficiencies and range of efficiencies. These data were then compared to six storm events whose water quality and quantity were measured relative to the hydrograph. From these six events (about 8 quality determinations/event), the average yearly efficiency

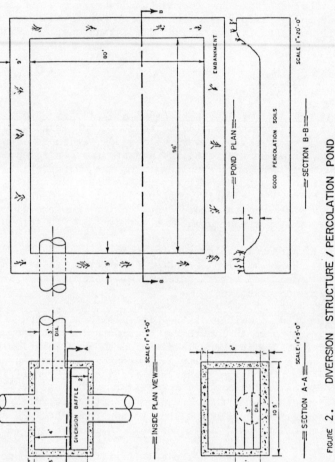

FIGURE 2.   DIVERSION STRUCTURE / PERCOLATION POND

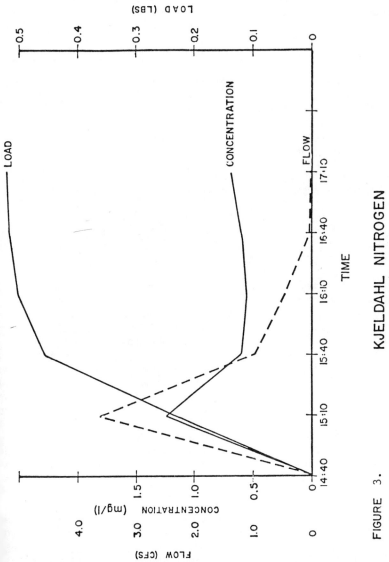

FIGURE 3.

KJELDAHL NITROGEN
4.6 Ac. MOTEL COMPLEX – 9/18/76 STORM

was calculated at about 99% for all water quality parameters.
This calculation was based on the average yearly volume of
rainfall.

GENERAL DESCRIPTION OF "BMP" COMPUTER PROGRAM

The Best Management Practices (BMP) computer program is used
to simulate runoff quantities and qualities with diversion
of a first flush of stormwater into a treatment facility. A
percolation pond is assumed but provisions are available for
a specified treatment rate. The program also is used for
simulation of runoff from parking areas and roof tops. Runoff from impervious and pervious areas are calculated for a
time interval. The time interval is chosen such that the
resultant hydrograph can be calibrated for a reasonable hydrograph shape. The time interval is usually less than 1/5 the
storm duration. Infiltration of pervious areas is calculated
using the Curve Number (CN) procedure. The CN is adjusted
for soil moisture conditions using a soil moisture budget over
time. Rainfall excess is considered as the excess from both
the pervious and impervious areas. Routing is performed by
means of the Muskingum method; therefore, a time of concentration for the watershed is required. Pollutant loadings at
the diversion point are calculated using average daily loading
on the watershed and a pollutant washoff function. If desired
site specific equations can be used and are available for the
Orlando area to calculate total loadings per storm event. The
watershed loadings are calculated for each storm event by considering the antecedent dry conditions and the quantity of pol
lutants removed during the previous storm event. The polluta
removal function assumes an exponential removal rate. One hu
dred sixteen storm events for a 100 acre watershed were simulated using a 10 minute time interval. This resulted in a
computer charge of less than $1.50 on the IBM 370 system. Th
program calculated the volume of treatment required for a spe
cified watershed. Average annual efficiencies are calculated
Efficiency is defined as the quantity of pollutants treated
relative to the total loadings expressed as a percentage.
The treated quantity of pollutants are output from the progra
In addition, treatment (percolation basin) sizes can be deter
mined using one storm event. Calibration was performed using
data from the Orlando (7) area and the verification process
used data from the Orlando and Winter Haven, FL, areas.

EVALUATION OF DESIGNS

Evaluation of the diversion/percolation system was conducted
using the computer program to simulate the rainfall/runoff
and percolation basin conditions (7).

Simulation was first done using 1975 rainfall data. Next, the
1960 rainfall data were used. The 1960 rainfall represented
the year of maximum record (68.74"). A volume of treatment
in units of inches were specified and twenty years of rainfall
data were simulated. The yearly efficiency curve did not significantly change (less than one percent) from one year to
the next. The water quality parameters used to calculate
the efficiencies were BOD_5, nitrogen, phosphorus, and suspended solids. The maximum variability per storm event occurred
with the 1960 rainfall data. The frequencies of the average
efficiencies are shown in Table 1. Average efficiency is defined as the average removal of total nitrogen, phosphorus,
BOD_5, and suspended solids. If the volume of storage is 0.5
inches, 92.9% of the time, the average efficiency for an event
will be 100%, 98.2% of the time, the average efficiency will
be greater than 88%, and at no time will an event efficiency
drop below 84%.

If the cost of the percolation basin is calculated for each
volume of storage and plotted against the resulting average
yearly efficiency, a cost-efficiency curve results. This
is shown in Figure 4. Essentially, one can estimate the cost
of treatment (construction) using Figure 4, if an efficiency
or volume of treatment is given. It should be noted that
these curves are relatively flat or do not increase rapidly
up to about 70% efficiency, then, for additional treatment,
the unit cost (marginal cost) increases rapidly. When treating one inch of runoff from 4.6 acres, the simulation reported
99% plus removal efficiencies. The unit cost data used to
construct this curve is considered low or related to rural
or large contract construction work. In addition, the diversion structure was not included in the cost calculation.
If small construction job unit costs for urban areas and the
cost of the diversion structure were calculated, the construction cost is estimated for a volume of one inch of runoff to
be $9,500.

The soil percolation rates at a larger watershed were considered excellent or generally over 10"/hour. Therefore,
the percolation pond almost drained before the next storm
event. However, a poor soil with a percolation rate of 0.2"/
hour was substituted to determine the increased cost of the
percolation basin. This comparison is shown in Figure 5. For
.5 inches of runoff, the basin size increased so significantly
that it became impracticable to build the basin because area
would not be available or the depth of the basin was below
the water table. It would almost appear likely that good
soils could be placed in the area to provide filtration of
the stormwaters at less of a cost than the differential shown
in Figure 5. From this comparison, it appears that the sequence of storm events is important for the sizing of the per-

TABLE 1

EFFICIENCIES FREQUENCY OF OCCURRENCE (%)
Area = 4.6 Ac, 85% Impervious, T_c = 20 min

Average Efficiency*	Volume of Storage (inches)			
	0.1	0.25	0.5	1.0
100	35.4	66.4	92.9	99.0
> 96	42.5	74.3	97.3	100.0
> 92	46.0	77.9	97.4	
> 88	47.8	81.4	98.2	
> 84	50.4	90.3	100.0	
> 80	56.6	92.9		
> 76	61.1	96.3		
> 72	66.4	97.3		
> 68	72.6	98.2		
> 64	82.3	100.0		
> 60	94.7			
> 56	100.0			

* Average Efficiency is the average removal of
BOD_5, SS, N, and P over a twenty year period.
Average rainfall events producing runoff per
year are 116.

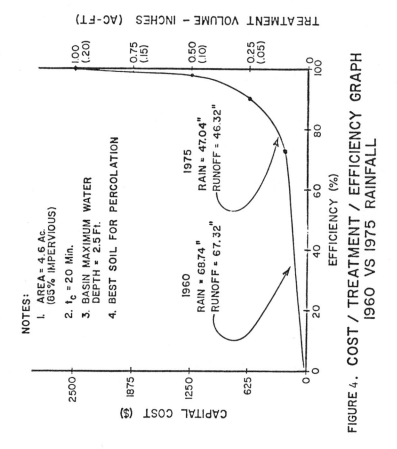

FIGURE 4. COST / TREATMENT / EFFICIENCY GRAPH
1960 VS 1975 RAINFALL

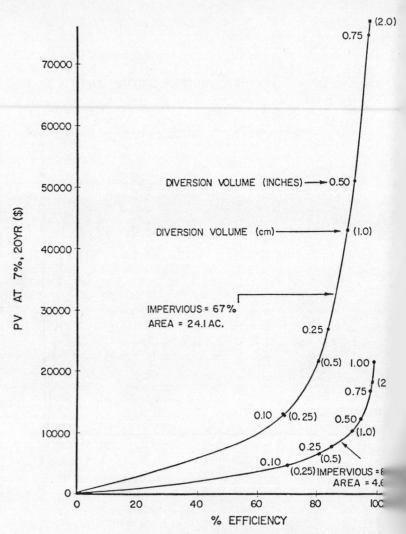

FIGURE 5. PRESENT WORTH OF PERCOLATION BASIN
FOR BEST PERCOLATION CONDITIONS

colation basins and soil percolation is an important variable for design.

The volume of treatment (percolation) increases as the size of watershed increases provided watershed and treatment basin soils remain the same. Basin size increases are reflected in the cost efficiency curves shown in Figure 6. These curves can vary depending on watershed runoff volumes. Procedures for sizing have been developed elsewhere (8).

GROUND WATER EFFECTS

If percolation basins are used for treatment after diversion, direct discharge of pollutants are minimized. However, storm-water pollutants may not be removed by soil filtration and exchange capacity. This can result in a deterioration of ground water or indirect discharge of pollutants into the surface waters by means of ground water transmission.

An underdrain system discharge and ground water in the vicinity of a percolation pond were measured to determine soil removal potential. Ground water samples collected at the bottom of a percolation pond and in the vicinity of the banks of the per-colation pond indicated higher concentrations of nitrogen and phosphorus in the water collected at the bottom of the pond relative to the water in the bank of the percolation pond. A comparison of selected water quality measures are shown in Table 2 (10). The upper Florida aquifer values were from a well within 6 miles of the percolation pond. Iron was one element that with high probability could violate drinking water standards.

PRACTICAL DESIGN CONSIDERATIONS

Percolation or infiltration in a holding basin produces an infiltration hydrograph (time-variable) which significantly drains the holding basin. From field tests (9) it was de-termined that the Horton equation form is applicable, or the volume of infiltration (inches) over a time period (t) is:

$$VDEP = BCI \ (\Delta t) + \frac{(BOI - BCI)}{BK} \ (1 - e^{-BK(\Delta t)})$$

where: VDEP = volume infiltrated between storms (inches)
       BCI = basin final infiltration rate (in/hr)
       Δt = time between storms (hours)
       BOI = basin initial infiltration rate (in/hr)
       BK = basin infiltration depletion rate (1 hr)

From the simulations using the "BMP" computer program, the maximum depth of basin based on the requirement for drainage

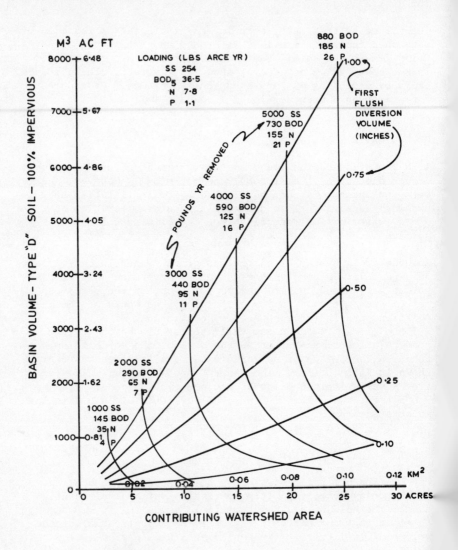

Fig. 6 Volume of retention / percolation as a function of
watershed size

within three days is 5 feet for Type A soils and 3 feet for
Type D soils. Type A soils are those classified by the U.S.
Soil Conservation Service (SCS) classification for soils with
percolation rates generally greater than 10 inches per hour,
and D is for soils with percolation rates generally around ¼
inch per hour. The limiting infiltration rates for both soils
was zero vertical percolation at saturation conditions; however,
side or horizontal percolation rates used were 1 inch/hour
for "A" soils, and ¼"/hour for "D" soils.

If the time variation of storm events and build-up of pollutants on the ground were not considered, a minimum basin size
could be calcualted using first flush volumes. This is parallel to an assumption that the basin will treat the stormwater
before the next runoff. Minimum volume is calculated as:

$$V_m = \frac{A \times DI}{12}$$

where: V_m = minimum basin volume (AC-FT)
 A = contributing watershed area (AC)
 DI = diversion volume (in)
 12 = conversion factor (in/ft)

For a watershed with 24 contributing acres, the minimum basin
size for one inch of diversion is 2 AC-FT. This volume is
the minimum volume and must be increased because of the time
variability of rainfall events, infiltration rates, and
depth of basin. Deep basins will take longer to drain relative to shallow basins. Thus deeper basins require larger
volumes relative to the shallow basins. Depth of water in a
basin is a varaible in design. To aid in the choice of
this variable, Figure 6 was constructed for basin depths of
5 feet. For less deep basins, the following equations were
developed for basin volumes.

Type A Soils

$$V_D = V_m + \frac{(V_5 - V_m)}{4} (D - 1), \quad 1 \le D \le 5$$

and

$$V_D = V_m, \quad D < 1$$

where: V_D = volume of basin at depth "D" (AC-FT)
 V_m = minimum basin volume (AC-FT)
 V_5 = basin volume at 5 foot depth from Figure
    6 (AC-FT)
 D = depth of basin water (feet)

Type D Soils

$$V_D = V_m + \frac{(V_5 - V_m)}{4.5} (D - 0.5), \quad 0.5 \le D \le 3$$

TABLE 2

GROUND WATER QUALITY UNDERLYING A PERCOLATION POND
COMPARED TO OTHER GROUND WATER DATA
(mg/l unless indicated otherwise)

Parameter	Percolation Pond		Drinking Water Std. Max. Contaminant Level	Upper Florida Aquifer[2]
	Average[1]	Maximum		
Chromium	0.005	0.006	0.05	-----
Copper	0.069	0.120	1.00	0.0025
Mercury	0.0005	0.0005	0.002	-----
Nitrate-N	0.14	0.48	10.0	0.25
Phosphorus-Ortho	0.102	0.204	-----	0.032[3]
Zinc	0.21	0.33	5.0	< 0.2
BOD5	6.5	20.0	-----	-----
Cadmium	< 0.003	< 0.004	0.01	< 0.03
Cyanide	< 0.004	< 0.004	-----	-----
Iron	0.7	0.8	0.3	0.048
Lindane (g/l)	< 0.01	< 0.01	4.0	-----
2-4-D	< 0.10	< 0.10	100.0	-----

[1] Three (3) sample average

[2] One (1) sample, 200 feet deep

[3] Deep well (1080 feet felow surface)

and

$$V_D = V_m, \; D < 0.5$$

For a given type of soil in the basin, Figure 6 is used to determine the volume at the 5 foot depth. Then adjustments to
volume are made for other than 5 foot depths. The choice of
depth and volume is primarily an economic and land availability
type decision once the above required volumes are calculated.
Shallow basins may be more aesthetic and serve as alternate
land uses during dry weather conditions. Some alternate land
uses are recreation facilities (sports or children's playgrounds) and green areas.

Knowing the contributing watershed area and the volume of diversion or yearly removal quantities, one can pick a size of
percolation basin.   Figure 6 is drawn to assume a curve number (runoff relation) close to 100 (completely impervious).
Adjustments for other curve numbers are made and presented
in other places (9). The choice of basin depth must be less
than the depth to the water table. The depth to water table
should be determined during the wet season.

AN EXAMPLE

Part A

An engineer wishes to size a diversion percolation system for
a 24.9 acre watershed (Figure 7). It's land use is classified
as commercial. There is no depression storage. He is designing for an area of type D soils and is working with a regulation specifying the removal of at least:

> 5000 lbs/yr of suspended solids,
> 760 lbs/yr of BOD5,
> 187 lbs/yr of nitrogen, and
>   26 lbs/yr of phosphorus

1. What volume percolation basin is required to meet regula-
    tions? The watershed is 88% impervious (assume a compo-
    site curve number of 97).

2. What area of percolation basin is required (in acres) if
    the pond will have 5.0 ft storage depth?

1. From Figure 6, at a contributing watershed of 24.9 acres
    and a diversion volume of 0.25 inches, the 5000 lbs/yr
    removal of SS and 760 lbs/yr of BOD5 removal are easily
    accomplished, but the N and P regulations require a larger
    basin size. With the limiting parameters of N and P, a

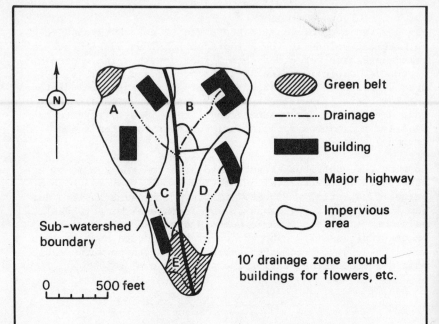

Green belt

Drainage

Building

Major highway

Impervious area

Sub-watershed boundary

0   500 feet

10' drainage zone around buildings for flowers, etc.

COMMERCIAL AREA EXAMPLE

SUB WATERSHED	TOTAL AREA (AC)	IMPERVIOUS AREA (AC)	PERVIOUS AREA (AC)	ROOF TOP AREA (AC)	PARKING & ROADWAY AREA (AC)
A	8.0	7.43	0.57	1.21	6.04 (parking) 0.18 (road)
B	6.0	5.35	0.65	0.83	4.20 (parking) 0.32 (road)
C	3.9	3.76	0.14	0.51	2.85 (parking) 0.40 (road)
D	5.0	4.86	0.14	0.52	4.22 (parking) 0.12 (road)
E	2.0	0.40	1.60	0	0.40 (road)
TOTALS	24.9	21.80	3.10	3.07	

IMPERVIOUS AND PERVIOUS AREAS
FOR EACH SUB-BASIN

Figure 7

basin size of 3.0 ac-ft if required. This corresponds to a diversion volume of 0.5 inches.

2. Knowing that a basin of 3.0 ac-ft will satisfy regulations, with a 5.0 ft storage depth this becomes 3.0 ac-ft/5.0 ft = 0.60 acre retention/percolation pond.

Part B

What size basin is required if a regulation specifies on inch of diversion treatment? What level of pollutants are removed? Express the answer in pounds and percentages.

If a stipulation of 1.00 inch of diversion treatment is included one must go to Figure 6 and recalculate. At 24.9 acres of watershed area and 1.00 inch of diversion treatment, the new volume of basin required is 6.0 ac-ft.

At this level of treatment the following pounds of pollutants are removed (obtained by linear extrapolation):

$$SS: 6100 \text{ lbs/yr}$$
$$BOD: 895 \text{ lbs/yr}$$
$$N: 188 \text{ lbs/yr}$$
$$P: 26.5 \text{ lbs/yr}$$

At a watershed size of 24.9 acres and one inch of diversion, the additional pounds removed per year are minimal with increasing treatment volume.

Compared to the basin volume of 3.0 ac-ft and approximately 0.5 inches of diversion treatment, this extra removal can be very costly. Costs were calculated assuming an interest rate of 6-3/8% for 20 years, excavation at $1.50/cu yd, cover crop at $0.12/sq ft, and a diversion structure cost proportional to flow rate (9).

Basin Size (ac-ft)	Capital Cost	PV	Pounds Removed/Year			
			SS	BOD_5	N	P
3.0	$31,000	$47,500	6050	890	187	26.0
6.0	$56,000	$90,000	6100	895	188	26.5

CONCLUSIONS

This work was limited to efficiencies calculated for the removals of impurities from direct discharge to surface waters.

Some ground water effects were noted. Watersheds which are not near impervious are treated elsewhere. The treatment of the first volume of runoff by diversion and percolation results in high removal efficiencies. The size of percolation basins can be determined from this work if a diversion volume and watershed size are known.

When designing a percolation pond, the percolation rate of the stormwaters is important because of the antecedent dry conditions between storms. If sufficient time is not available for percolation, a larger basin volume is necessary to treat the first flush volume from the next storm.

On large watersheds, the volume of runoff, and the relative lack of first flushes may cause excessive investments in a single diversion/percolation system. Possibly, a few systems may be more cost effective.

ACKNOWLEDGEMENTS

This work was funded in part by the 208 program in the Orlando, FL, USA, area through the East Central Florida Regional Plannin Council. The 208 program is administered through the Environmental Protection Agency.

REFERENCES

1. U.S. Senate and House of Representatives, Federal Water Pollution Control Act, Public Law 92-500, October 18, 1972.

2. Field, R., Tafuri, A., and Masters, H., Urban Runoff Pollution Control Technology Overview, USEPA, Publication #600/2-77-047, Cincinnati, OH, March 1977.

3. Lager, J. A. and Smith, W. G., Urban Stormwater Management and Technology, USEPA, Publication #670/2-74-040, Cincinna OH, December 1974.

4. Bateman, J. M., Orange County Florida Subdivision Regulation, Orange County, FL, p. 26-30, 1975.

5. Wanielista, M. P., Nonpoint Source Effects, Florida Technological University, Orlando, FL, January 1976.

6. U.S. Department of Commerce, National Oceanic and Atmospheric Administration, Local Climatological Data, Orlando, FL, by year 1957 through 1976.

7. Wanielista, M. P., Best Management Practices, Florida
    Technological University Report to the East Central
    Regional Planning Council, June 1977.

8. Wanielista, M. P., Surface Water Management, Textbook,
    under development.

9. Wanielista, M. P. and Shannon, E. E., Best Management
    Practices Evaluation, East Central Florida Regional
    Planning Council, 1977.

EVALUATION OF ON-SITE STORMWATER DETENTION METHODS IN
URBANIZED AREA

Akihiko Tsuchiya

Director of River Division
Public Works Research Institute
Ministry of Construction

1. INTRODUCTION

   Frequent disasters caused by flooding in small river
watersheds, particularly in urbanized areas are a remarkable
development in recent years. This can be explained by the
following facts.

(1) Stagnation of river improvement in urban areas. Urbanization close to a river course is quite usual in many cities,
and increases the difficulty of widening the river or the
construction of embankment and parapet walls on account of the
high expenditure involved in removing the inhabitants and
their strong resistance to such measures.
(2) Increase of outflow by the development of upper reach
watersheds. Urbanization of forests and farm lands causes an
increase in the impermeable area and the effective precipitation. Construction of gutter and drainage systems accelerates
the outflow and increases peak discharge, consequently promot-
ing inundation in urban areas down stream.
(3) Increase of damage potential along the river course.
Most lowlands along a river course are flood plains and have
the important role of retarding flooding. The rapid increase
of residential or industrial areas in such dangerous lowlands
is a recent trend. Most of the damage due to flooding is apt
to occur in these areas.
   An increase in flood damage along river courses is recognized in many cities and urbanized areas in Japan. It is
therefore a matter of urgency to suppress the increment in the
run-off due to the development of the watershed and to maintai
a sufficient inundation and infiltration capacity of lands in
the already developed areas.

2. RUN-OFF CONTROL BY DETENTION POND

2.1 Construction of detention ponds in Japan
    Construction of detention ponds is now obligated by the

local governments and authorities under the regulations or
administrative control when housing lots, factories and golf
courses are developed on a large scale. The scale of developed area under the application of regulation differs from 0.1
hectare to 20 hectares by different local governments, and
tentative or permanent use of detention ponds also depends on
the local regulations.

　　About 1900 detention ponds were constructed in Japan,
in which the ponds for permanent use (excluding the ponds for
golf courses) are counted as 235. Purposes and number of
ponds are shown in Table 1. The detention ponds, for the most
part, are for the development of housing lots. The relation
between area of detention ponds and development lands are
shown in Figure 1, in which the ratio of one to two percent
is dominant.

Table 1.　Purposes and number of detention ponds

Purpose	Number	%
Housing lots	144 (42)	62
Land readjustment	17 (6)	7
Trade and industry	32 (12)	14
Recreation	27 (9)	11
School	7 (5)	3
Others	8 (3)	3
Total	235	100

2.2　Design criteria

Most of the district regulations have the similar
design criteria of detention ponds. The following design
criteria are applicable to the large scale development of
housing lots.

(1) Estimation of peak discharge
i)　Rational formula is applicable to estimate the flood
    peak discharge.
ii) Time of concentration can be obtained by following
    equations.

Not developed;　　　$T = 0.83 \, L \, / \, I^{0 \cdot 6}$
Developed;　　　　　$T = 0.36 \, L \, / \, I^{0 \cdot 5}$

where T is the time of concentration (min), L is the
length of river (km), I is the slope of river.
iii) Coefficient of run-off is given as follows.

Not developed; depends on the geological and surface
               conditions (see Table 2)
Developed;　　　0.9 as a standard value.

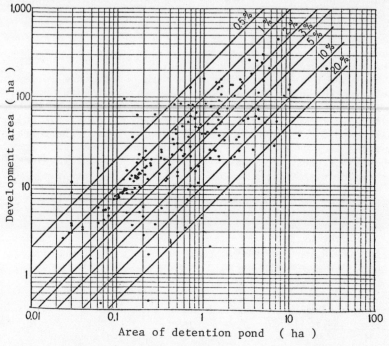

Fig. 1. Relation between development area and area of detention ponds

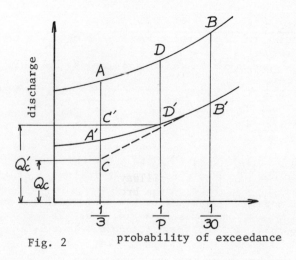

Fig. 2

iv) Design storm is obtained from duration-intensity curve of rainfall.

(2) Calculation of volume for flood control
    Principal concept of flood control by detention ponds is as follows.
    As for the flood smaller than 3-year flood, the peak discharge from developed area should be reduced to the present flow capacity in lower part of river. (The value A is reduced to A', in Fig. 2) When the present flow capacity is larger than A', the flood smaller than P-year flood should be controlled to the present level of flow capacity. (D → D') The peak discharge of P-year flood is obtained from the flow capacity in down stream. (Qc')
    As for the flood smaller than 30-year flood, the peak discharge of run-off from developed area should be reduced to the same level as before the land was reclaimed. (B → B')
    The out-flow from the catchment area is calculated from design storms before and after the land was reclaimed. Size of drain pipe, height of intake and outlet and the volume of detention pond are decided by the trial method to reduce the outflow discharge to the capacity of down stream.
    To control all floods smaller than 30-year flood, the volume of detention pond can be obtained by the following simple equation.

$$V = (r_i - \frac{r_o}{2}) \ T_i \ f \ A \ \frac{1}{360} \qquad (3)$$

where, V is a volume required for control (m^3), f is the coefficient of run-off when land was reclaimed, A is a catchment area (ha), r_o is the rainfall intensity equivalent to the flow capacity of down stream (mm/hr), r_i is a rainfall intensity at any duration T_i on the 30-year flood duration curve (mm/hr), T_i is a duration of rainfall (sec).
    This method is widely used in many local government regulations for its simplicity, and slightly larger volume is given by this method than that by trial one.

(3) Capacity for silt deposition
    The standard value of sediment run-off is 150 m^3/ha/y, and field data on the similar conditions might be considered on the design.

(4) Spillway capacity
    Discharge capacity of spillway should be greater than 1.44 times of the peak discharge brought by 100-year flood.

2.3 Measurement on the detention pond
    The big housing lots of 1316 hectars called KOHOKU New Town is now under development in the northern part of the City of Yokohama by the Japan Housing Co-operation. They have a plan to construct 25 tentative detention ponds

in this area, in which one housing lot with a detention pond
was selected to collect the hydrological data. The catchment
area of the test field is 110.5 hectares, of which 61.2 hectares was developed for housing lots. The detention pond was
arranged with an earth dam 13 m high at the lower end of
stream in that area.

The water level and discharge were measured at points A
and C in the stream, and the water level in the pond was taken
at B. (see Fig. 3) The discharge at point A, the outflow from
the undeveloped area, and the discharge at point C, the outflow from the developed area, are shown in Fig. 4.

The outflow from developed area has a good correspondence
to rainfall intensity. The coefficients of run-off are 0.41
and 0.46 at the first and second peak of outflow. These
values are about half that given by the design criteria, i.e.
0.9, which means the down stream area can be well protected
by the detention pond.

About 20 mm initial loss was observed in both areas.
This value is presumed to be comparatively large. The run-off
percentage from the reclaimed area was 0.49, in which the
initial loss was excluded, similar to the coefficient of peak
run-off.

The detention pond was not effective in keeping the outflow at the same level after the land was developed. However,
it is a most effective tool for reduction of peak run-off.

3.   OTHER CONTROL METHODS

3.1 Permeable pavement
A running test of permeable pavement is now being carried
out at 26 places on the sidewalks in Tokyo. The permeable
pavement consists of three layers; an asphalt surface layer
3 ∿ 4 cm thick having a high permeability, a roadbed filled up
with crusher-run 10 ∿ 20 cm thick having permeability and
storage capacity, and a sand filter 10 ∿ 20 cm thick at the
bottom for the protection of the roadbed.

The infiltration capacity of the test pavement decreased
gradually during a running test of two years. Table 3 shows
the change of permeability coefficient. The most recent values
are of the order of 10^{-3} cm/sec, and still pervious enough for
about 100 mm/hr rainfall intensity.

3.2 Run-off detention and land use
A survey of land use in Tokyo City was obtained by using
aerial photography. The results are shown in Table 4. The
land use is very sensitive to the run-off. The product of the
run-off coefficient f and area A is a good factor for evaluation of run-off. According to Table 4, a large part of the
run-off is yielded by roof and road.   As detention on
the roof is impossible for the usual Japanese

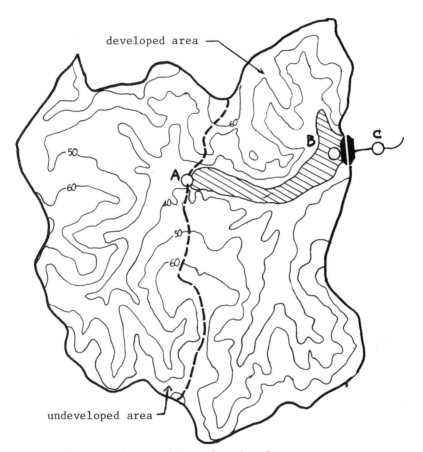

Fig. 3 Detention pond in a housing lot

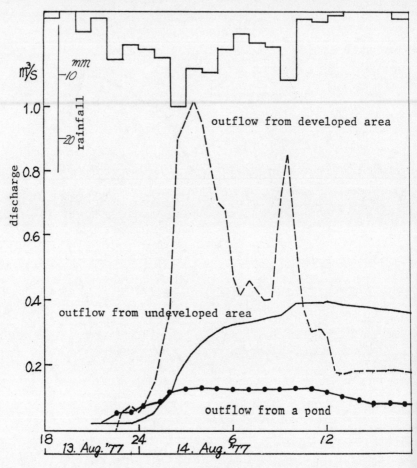

Fig. 4 Rainfall and out-flow observed in the test field

Table   2   Coefficient of run-off

Land condition	Coefficient of run-off
Steep mountain	0.75 ~ 0.90
Relief land & forest	0.50 ~ 0.75
Flat farm	o.45 ~ 0.60
Steep river	0.75 ~ 0.85
Mild river	0.45 ~ 0.75

Table   3   Results of permeability test of pervious pavement

Place	Asphalt mixture	Coefficient of permeability (cm/sec)		
		Mar. '74	Mar. '75	Mar. '76
U-rakucho	A	2.1×10^{-1}	9.1×10^{-3}	9.1×10^{-3}
	B	2.1×10^{-1}	4.8×10^{-3}	5.2×10^{-3}
	C	2.1×10^{-1}	7.4×10^{-3}	3.9×10^{-3}
	D	2.0×10^{-1}	6.9×10^{-3}	1.1×10^{-2}
Hatagaya	A	2.0×10^{-1}	2.0×10^{-1}	3.5×10^{-3}
	B	2.1×10^{-1}	1.9×10^{-1}	9.5×10^{-3}
	C	2.1×10^{-1}	2.0×10^{-1}	4.3×10^{-3}
	D	2.0×10^{-1}	4.8×10^{-2}	---
Kanamachi	A	2.1×10^{-1}	2.3×10^{-2}	3.9×10^{-3}
	B	2.2×10^{-1}	9.1×10^{-3}	5.2×10^{-3}
	C	2.1×10^{-1}	3.8×10^{-2}	4.3×10^{-3}
	D	2.1×10^{-1}	1.9×10^{-1}	3.9×10^{-3}

Table   4   Effect of land use on the run-off

Land use	Area ha	%	f	A X f	%
Road	9189	20.4	0.85	7810	26.6
Railroad	755	1.7	0.83	249	0.8
Open space (permeable)	12123	27.0	0.20	2425	8.3
Roof	19521	43.4	0.90	17569	59.8
water area	317	0.7	1.00	317	1.1
Park	2228	5.0	0.15	334	1.1
Open space (impermeable)	838	1.9	0.80	670	2.3
Total	44972	100.0		29374	100.0

house from the structural view point, the on-site detention should be considered on the garden and open space.

Re-construction of road into the permeable pavement is also an effective method to reduce the run-off from an urban district.

4   CONCLUSION

In Japan, the detention ponds are widely used with the large scale land development to prevent the flood disasters in down stream area.   The design criteria of detention pond was explained, and its effectiveness was shown by observed data.   Effectiveness of permeable pavement was also introduced with field data.

REFERENCES

1. On-site detention methods in urbanized area and their improvement ; Public Works Research Institute, No. 1174, 1977.
2. Design criteria of detention pond ; Japan Housing Co-operation, 1964.
3. Study on the permeable pavement ; University of Japan, 1976.

THE PREDICTION OF FLOOD MAGNITUDES ON UNGAUGED URBAN CATCH-
MENTS

Peter S Kelway

Senior Hydrologist, Northumbrian Water Authority

Synopsis

Techniques for predicting flood flows used in the Flood
Studies Report (FSR) and subsequently developed for urban
catchments by the Institute of Hydrology (IOH) are discussed.
These are augmented by techniques developed by Northumbrian ɛ
Water Authority for use in its area to predict the effects of
increasing urbanisation on developing catchments.

PREDICTION TECHNIQUES OF THE FLOOD STUDIES REPORT

The Flood Studies Report (NERC, 1975) contains two techniques
for predicting flood flows with specified return period on
ungauged catchments. Unfortunately, neither method adequately represents conditions on urbanised catchments, for the FSR
was essentially intended to deal with rural conditions.
Where urban terms appear in the various estimating equations
it is due to the necessity to recognise and allow for the
influence of an urban effect to reduce the error of estimate,
rather than a definite attempt to model the effect.

No claims were made in the FSR that either model could be
used to predict the effect of increasing urbanisation over a
particular catchment. Such a facility is of great importance
to engineers involved in Development Control investigations
and urban drainage design. Recent work by IOH (Packman,
1977) has led to modified forms being presented of both the
statistical and rainfall-runoff techniques.

The purpose of this paper is to question how far these modified methods can be expected to portray runoff behaviour on
urban catchments. Suggestions are made for improving some
aspects of the models.

The statistical approach

An estimating equation is given in the FSR to calculate the mean annual flood for each of a number of regions in the United Kingdom. Apart from South-east England, no urban term appears in the equation. To overcome this deficiency, IOH have introduced a factor by which the FSR estimate for the rural condition ("BESMAF") can be converted to an estimate for an urban catchment.

The growth factor

To facilitate the study of the increase in flood magnitude with changes in the degree of urbanisation (the "urban fraction") and return period, the concept of "growth factors" is introduced. This factor relates the flood flow under given conditions to a benchmark value, such as the mean annual flood for a specified urban fraction. The different rates of increase of flood magnitude with return period for varying degrees of urbanisation can be clearly seen using this dimensionless means of comparison.

To estimate flood flows for given return periods other than the mean annual, regional curves of growth factor against frequency are given in the FSR. The relevant factor is read from the curve at the required return period and multiplied by the mean annual flood.

Recent work has been carried out by IOH (Packman 1977) using the mean annual flood for a specified urban fraction as the benchmark. This work has suggested that the variation in growth factor with return period for varying urban fractions is considerable. Results from a number of gauged urban catchments suggest that the growth curves for different urban fractions do not maintain similar profiles with increasing return periods. They are coincident at a return period of about 10 years and gradients vary increasingly at larger returns. Relationships are not available for return periods in excess of 50 years and it is unclear whether results are truly representative of real conditions. Certainly no reliable indication can be given as to the likely relationship for long return periods.

The modified FSR statistical method depends on a knowledge of the variation of growth factor with both return period and urban fraction. It is therefore not applicable for return periods of greater than 50 years.

The rainfall-runoff approach

The fundamentals of the rainfall-runoff approach are adequately described in the FSR (Chapter 6), the model being

based on the Unit Hydrograph concept. Essentially, a profile
of the variation of rainfall with time during a storm is
"convolved" with a profile of the variation of runoff response to a unit of rainfall, producing a total hydrograph for
the catchment. The basic parameters of importance in the
context of urban runoff estimates are storm duration (D), the
percentage runoff (PR) and the time-to-peak (TP), the latter
being defined as the duration of time taken for water to flow
from the most extreme point in the catchment to the point at
which the estimate is required.

The basic model has been criticised elsewhere (Kelway, 1975).
At a seminar held at Birmingham University in March 1977
(Hamlin, O'Donnell, 1977) it was concluded that:

"The present FSR methods based upon the rainfall-runoff model:

(a)   should be limited to catchments where the urban percen-
       tage runoff did not exceed 25%;

(b)   ought not to be used to predict the effects of increas-
       ing urbanisation.

It is considered that these reservations are still valid,
despite recent slight modifications to the IOH model, as no
fundamental changes in philosophy have been introduced. The
model is recommended by IOH for use far outside the conditions suggested.

Comments on assumptions made in the modified IOH rainfall-
runoff model

1 The urban fraction (the FSR "URBAN" term) used in the
model is the "gray area" shown on the 1:63360 scale OS map
expressed as a percentage of the total catchment area. The
"urbanised" area so defined has an unknown proportion of
paved area. A given amount of paved area can in turn lead to
an uncertain percentage of runoff. Housing densities of as
low as 10 per hectare may be shown gray as well as factory
complexes with 70% imperviousness. A more sophisticated
urbanisation parameter may be required.

2 The soil classification index (the FSR "SOIL" term) is not
defined on the FSR maps for a number of areas in the United
Kingdom where the FSR may be used for predictions. Hence the
equation for rural PR may not be a reliable predictor, leading to unreliable estimates for the urban PR value. It is
therefore suggested that the soil index should be set at the
maximum of 0.5 in all areas of significant urbanisation, say
over 10%. This will overcome the problem of an index emas-
culated by the largely unknown effects of urbanisation on
soil characteristics, which leads to uncertainty in calculat-

ing the percentage of runoff. The maximum index value will maximise the urban effect, preventing possibly serious under-estimates of flow.

3 No floodplain storage parameter is present in the FSR estimating equations. The consequences in rural catchments are receiving attention by IOH at the present time. They have greater significance in the case of urban catchments on account of the likelihood of surcharging and blocking of urban water courses, particularly where culverting may exist over substantial lengths of channel. Storage effects may well cause under-estimates of TP for catchments where channel slopes are small, say less than 1%. A storage parameter could be derived to perform in such a way that it would modify the estimate of time-to-peak under critical conditions. The effect of this would be to allow more economic design, peak flow values being reduced to more realistic levels.

4 The location of urbanisation is not accounted for in the basic rainfall-runoff model. Development close to a problem area will produce a different effect than if development is far removed at the head of the catchment. If the rainfall-runoff model is to be used throughout the estimating pro-cedure, successful representation of catchment behaviour may only be possible by subdividing the catchment, calculating sub-catchment flows and then routing and summing the con-stituent flows.

5 The relationship between the return period of flow and the causative rainfall is to be taken as 1:1 in the modified model. However in the FSR much attention is given to the fact that rainfall of a given frequency does not necessarily yield a flood of the same frequency; indeed, there may be large differences between them. It is unclear why the return periods have been equated for the urban model.

Intuitively, one can assume that as urbanisation increases, rainfall and runoff frequencies become more directly related as other hydrological considerations diminish in importance. As the rarity of the storm increases, this relationship should intensify with increasing percentage of runoff. No such variation in the relationship is modelled by the modified IOH technique.

6 The reason for the adoption of the "50% Summer" rainfall profile is not adequately explained. In contrast to rural catchments, urbanised catchments will experience more events where significant responses to rainfall occur, due to the normally greater percentage of runoff. It would therefore be expected that a "peakier" profile than the 50% summer might be more reasonable, it being only marginally more severe than the standard 75% winter profile used in the unmodified

rainfall-runoff model.

Notwithstanding that, the dichotomy of profiles into summer and winter obscures the fundamental reason for adopting a particular profile, ie its peakedness. If standardised profiles are to be used, it would be preferable to refer to one set of profiles which relate to the whole population of rainfall events to be used for design, regardless of season.

7 The duration of the design storm D is related to mean annual rainfall and time-to-peak, as in the unmodified model. This relationship is intended to allow D to increase with elevation. However, storm duration is not directly related to mean annual rainfall and the equation for D tends to overestimate storm duration beyond values normally experienced. To maximise runoff peak, storm duration should be set equal to the time-to-peak.

8 Doubt has been raised (Bootman, 1977; Kelway, 1977a) as to the accuracy of rainfall amounts given for long return periods. There is no doubt that agreement up to a return period of, say, 20 years is good. However, the discrepancies often reach a factor of ten to one for Bilham estimates of 500 years. Bilham's equation will tend to suggest that heavy falls are more common over high ground than will the FSR method because it was based on data which ignored high-level stations where intense short-duration falls occur less often. However, in the North East two major storms in 1975 and 1976 produced intense falls over high ground and the very high FSR return periods for such storms must be considered suspect. At worst, Bilham estimates allow one an inbuilt "safety factor" when estimating falls of long return period for design purposes.

These limitations may help to explain why significant discrepancies often occur between flood predictions produced by the rainfall-runoff and the statistical methods. It is unfortunate that the model cannot facilitate the determination of the standard error of the estimate of a flood prediction. Instead, one has to learn by experience how reliable the model may be. In an area with no gauged urban catchments, it may be impossible to carry out an objective appraisal of the method. Therefore it becomes vital to obtain a number of estimates of urban flood flows, using different prediction techniques, before a definitive view can be obtained of their relative merits.

RELATIONSHIP OF MAGNITUDE OF FLOOD WITH RETURN PERIOD AND DEGREE OF URBANISATION

A computer model has been set up by the Northumbrian Water Authority to predict flows on ungauged rural and urbanised catchments, using the IOH rainfall-runoff technique. All necessary interpolations from the FSR maps and the various tables and curves are carried out by a number of subroutines. The procedures involved in determining the catchment parameters and convolving the rainfall profile and unit hydrograph co-ordinates are fully automated.

Relevant parameters were obtained for 65 catchments in North-east England where urbanisation was of significance to drainage engineers. The model was run for each catchment for a range of degrees of urbanisation and return periods.

Comparisons with urban mean annual flood benchmark

Figure 1 shows the increasing flood magnitude with return period for different degrees of urbanisation as predicted by the rainfall-runoff model. Growth factors are calculated by relating flood peak at a given return period to the mean annual flood *for the same urban fraction*. Each of the curves for the different sizes of urban fraction must therefore start at the point where the growth factor is unity at a return period of 2.33 years (ie relating to the mean annual flood).

When growth factors are calculated by relating peak flows for a T⁻year return period to the mean annual flow it is clear that the degree of urbanisation has little effect on the factor itself when the modified rainfall-runoff model is used. The family of "urban" curves depart markedly from the "rural" curve obtained from the original FSR model. This departure is due to the difference in the rainfall statistics between the two models; the calculation of percentage runoff is also different. The modified model is presumably intended to produce a reduction in growth factor as urbanisation increases and this effect does occur to a limited extent. The reduction from the rural condition to an urban fraction of 25% is much greater than the corresponding reduction from a fraction of 25% to 100% and this feature is unlikely to be experienced on a real catchment.

It is therefore possible that the change in growth factor for varying urban fractions is not realistic. There is little correspondence with the growth curves presented in the IOH modified statistical approach.

Comparisons with rural mean annual flood benchmark

Apart from the uncertainties already described, the use of the family of growth curves for prediction purposes has other shortcomings. In Figure 1 growth factors for T-year frequencies were related to mean annual flood flows for the same urban fraction. There are considerable uncertainties in the calculation of absolute mean annual peak flows for a given

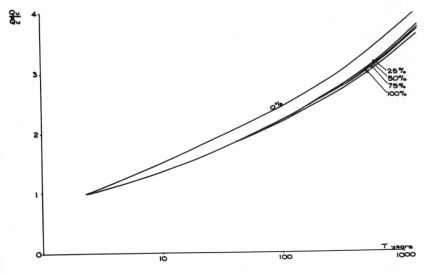

Figure 1.   Relationship between growth factor ($Q_u/\bar{Q}_u$) and return period (T) for different urban fractions. Q_u is the flood at a given return period and $\bar{Q}_u$ is the mean annual flood, both on the urban catchment.

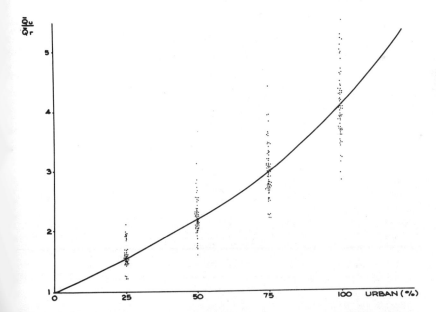

Figure 2.   Relationship between urban fraction (URBAN) and growth factor ($\bar{Q}_u/\bar{Q}_r$) expressing mean annual flood on the urbanised catchment ($\bar{Q}_u$) to that on the "rural" catchment ($\bar{Q}_r$).

urban fraction. The use of the curves for prediction purposes
would be highly error-prone as this unreliable mean annual
flood would be multiplied by an uncertain growth factor.

An alternative method of deriving a family of curves for prediction purposes is to relate the T-year flood for a specified
urban fraction to the more reliable T-year flood *for the rural
condition*. For the 65 sites used previously, Figure 2 shows
the curve for the mean annual flood for varying degrees of
urbanisation. The plotting points which are shown give an
indication of the considerable amount of scatter in the
growth factors for different catchments. This scatter is a
common feature of predictions obtained using the rainfall-
runoff model.

Figure 3 shows curves for the mean annual, 30-year and 1000-
year floods plotted on the same axes. The growth factor decreases with increasing return period as might be intuitively
supposed. The Possible Maximum Flood curve is plotted for
comparison. Values for the PMF condition were obtained by
running the rainfall-runoff model for special conditions
considered to represent a close approximation to the most
extreme event. The method used has been discussed in detail
elsewhere (Kelway, 1977b).

The family of curves produced can be used as a prediction
diagram. The rural estimate for a flood of a given return
period can be estimated by any one of a number of methods.
It is then simply multiplied by the relevant growth factor
obtained from Figure 3 for the return period and urban fraction for which the prediction is required.

As this prediction technique has been based on the rainfall-
runoff model, other methods were studied to determine whether
they perform in the same way.

An alternative approach to determine growth factors for urban conditions

As urban development proceeds, the percentage of runoff increases and the speed of flow also increases. The increased
rate of runoff may well be the most important factor in producing a greater hydrograph peak. This assumption was made
to develop a model to augment the IOH rainfall-runoff technique. The increased impermeability with urbanisation was
ignored, enabling the heightening of peak flow with increased
speed of runoff to be assessed in an absolute sense.

The IOH rainfall-runoff model was used to simulate conditions
on a test catchment. Storm duration, time-to-peak and storm
rainfall profile were varied in different permutations. A
family of curves such as in Figure 4 was derived for each

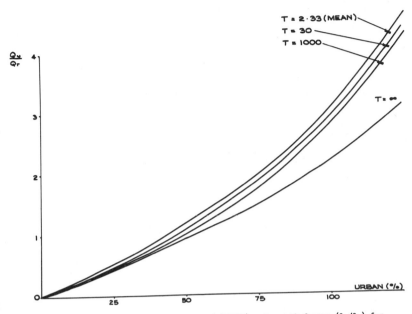

Figure 3.   Relationship between urban fraction (URBAN) and growth factor (Q_u/Q_r) for different return periods (T years). Q_u is the flood for a given urban fraction and Q_r is the flood on the "rural" catchment, both for the same return period.

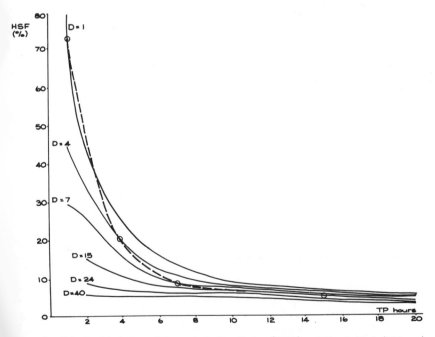

gure 4.   Relationship between Hydrograph Scaling Factor (HSF %) and Time-to-Peak (TP hours) r different storm durations (D hours).

profile, showing the relationship between time-to-peak of
flow with the peak flow itself. Each curve in the family
related to a particular storm duration. All peak flows were
related to the extreme flow condition, assumed to occur when
storm duration was 2 hours and time-to-peak was 1 hour; the
resulting percentage was termed the Hydrograph Scaling Factor (HSF). A given condition of time-to-peak, duration of
storm and rainfall profile produced a peak flow which could
be compared with conditions which were assumed to be the
extreme for the catchment.

When D and TP are equal, the flow will be maximised for that
particular catchment condition. For each rainfall profile,
a point was found on each of the family of curves where D and
TP were of equal value. The broken line on Figure 4 shows
the curve relating HSF to TP where D and TP are equal.
Co-ordinates of HSF against TP taken from this curve were
then transferred to a second graph. This procedure was repeated for each rainfall profile. Figure 5 shows the envelope of the various profile curves and also the modal curve,
denoted as the HSF curve. The small deviation in the curves
is remarkable, considering that over 50 different profiles
were used in the simulation.

Using the HSF curve

The HSF curve can be used to determine the effect on the
hydrograph peak of reducing the time-to-peak of flow.

If Q_o is the flow at the original TP value
$\quad$ Q_f is the flow at the final TP value
$\quad$ H_o is the HSF value from the curve at the original TP value
$\quad$ H_f is the HSF value from the curve at the final TP value

Then $Q_f = Q_o \times (H_f/H_o)$ $\qquad \cdots\cdots\cdots\cdots$ (1)

A subroutine was incorporated into the computer model to calculate the HSF values for different TP values. A series of
simulations was then carried out using the rainfall-runoff
model on the 65 catchments used previously. For a range of
degrees of urbanisation, the rainfall-runoff model was used
to determine the time-to-peak for flow. In each case, instead
of using the IOH technique to predict flow for the various
degrees of urbanisation, equation (1) was used. The value of
Q_o was the rural estimate and Q_f was the estimate for a given
degree of urbanisation.

Figure 6 shows the result of this simulation exercise which
was carried out for mean annual flood conditions. The scatter
is shown as in Figure 2 and it is noteworthy that the degree

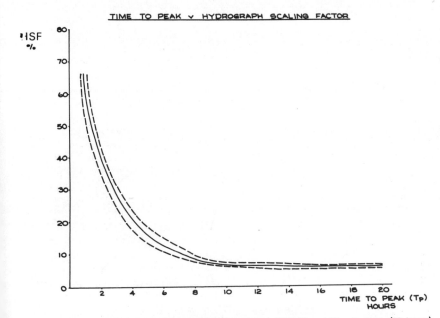

Figure 5. Relationship between Hydrograph Scaling Factor (HSF %) and Time-to-Peak (TP hours) in the condition where D and TP are equal. The broken lines indicate the envelope for all rainfall profiles studied. The solid line is the modal profile.

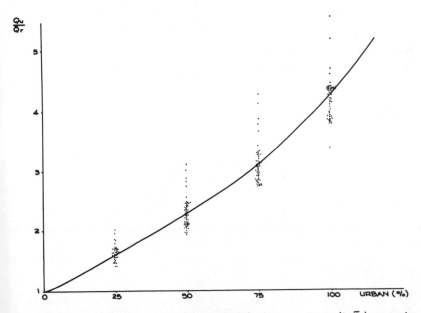

Figure 6. Relationship between urban fraction (URBAN) and growth factor $(Q_u/\overline{Q}_r)$ expressing the flood on the urban catchment (Q_u) to the mean annual "rural" flood $(\overline{Q}_r)$.

of dispersion has been considerably reduced by using the HSF curve for prediction.

The statistical model for predicting the effects of urbanisation

By operating the modified IOH statistical model on the 65 catchments used previously, a curve of the same form as Figures 2 and 6 was produced for the mean annual flood. However, the scatter was found to be very small, suggesting that the prediction technique was more stable.

Comparison of the different techniques for prediction

Figure 7 compares curves produced by the three different techniques. These show the relationship between the urban fraction and the growth factor of the urban mean annual flood over that for the rural condition. The HSF and IOH rainfall-runoff model curves are remarkably similar, while the statistical technique produces a curve which deviates significantly from them. The points plotted near the statistical curve were derived using the curves presented by Packman for an approximate assessment of the effect of urbanisation using the statistical method. Good correspondence is evident between the computer-based and approximate estimates.

It is impossible to say which of the methods is the most reliable without abundant sources of gauged urban catchment data. The standard error of estimate of the rainfall-runoff model cannot be directly determined on account of the model's nature. Even so, it is clear that the scatter obtained when using the model is considerable though it is unknown whether this represents real catchment variability or instability in the computation. While the scatter for the statistical model is less, the curve deviates considerably from the curves for the other two methods which, though both based on the same model, use quite different means to determine the growth factor.

Adoption of a prediction method

Bearing in mind the limitations of both the rainfall-runoff and statistical models and the lack of objective information for assessing the various techniques, difficulties will arise in the selection of a technique for prediction purposes.

It is important to realise that the use of the rainfall-runoff model will frequently produce results which deviate markedly from the curve in Figure 7 on account of the high degree of scatter which occurs using the technique. Two options are

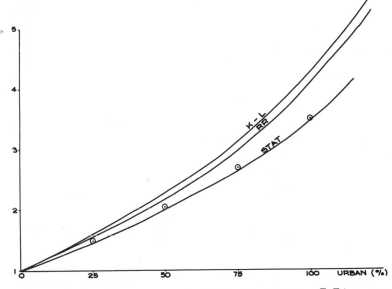

jure 7. Relationship between urban fraction (URBAN) and growth factor $(\bar{Q}_u/\bar{Q}_r)$ expressing an annual flood on the urbanised catchment $(\bar{Q}_u)$ to that on the "rural" catchment $(\bar{Q}_r)$ for e three prediction methods based on the FSR.

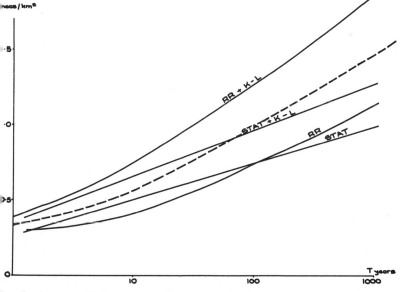

jure 8. Frequency diagram to assess flood conditions with varying return period on an anised catchment. Absolute discharge values (Q cumecs/sq km) are shown plotted against :urn period (T years).

open to the user. He can take the value obtained from Figur 7 and accept that the standard error is likely to be high, but indeterminate. Alternatively he can use the model to predict a value which may be in the right range for the catchment. However, the result is less safe as it is not known whether a low or a high estimate is likely to have been produced, rather than a value based on wide sample of catchment behaviour.

The policy of the Northumbrian Water Authority has been to calculate predictions using several different techniques based on the FSR. These values are then plotted together on a frequency diagram and a line of best fit is drawn through them. In this way, not only is a prediction based on severa techniques, but an envelope is obtained giving an indication of the likely confidence limits of the line of best fit.

An example of the estimation procedure adopted by the Authority is given below for the Ouseburn catchment at Three Mile Bridge in the Newcastle-upon-Tyne area. Computer-based estimates were obtained using each of the three models dis- cussed previously and refer specifically to the catchment. The codes used for the models are listed below.

STAT is the statistical technique of the FSR, modified for urban catchments.

RR   is the modified rainfall-runoff technique.

HSF   uses the HSF curve to predict the growth factor by assessing the effect on time-to-peak resulting from a change in urbanisation.

STAT + HSF takes the mean annual rural flood predicted by STAT and obtains its product with the HSF growth factor for the T-year flood on the urbanised catchment.

RR + HSF takes the mean annual rural flood predicted by RR and obtains its product with the HSF growth factor.

It has been found expedient to plot only the mean annual, 30-year and 1000-year urban estimates on a frequency diagram This reduces the amount of calculation to a minimum yet still allows adequate information for interpolation for other return periods.

The substantial variation in computer-based and prediction diagram estimates for growth factors and the predicted absolute flood flows is clearly seen in Tables 1 and 2.

The computer-based absolute estimates were plotted on the

frequency diagram in Figure 8. The line of "best fit" is shown by a broken curve. "Best" estimates of flood flow could then be obtained by reading flood flow values for any required return period.

The cost of prediction by this procedure is £1.50 for an average urban catchment on an IBM 370/168 computer at normal commercial rates. Manual effort is limited to the basic measurements from OS maps and interpretation of the results of the frequency diagram.

Table 1.   Growth factors for converting "rural" floods to values relating to the 25% urban fraction on the Ouseburn watercourse at Three Mile Bridge, Tyne and Wear.

Source	Mean Annual Flood		30-year Flood		1000-year Flood	
Method	Fig 7	Computer Model	Fig 3	Computer Model	Fig 3	Computer Model
HSF	1.61	1.92	-	1.92	-	1.92
RR	1.55	1.23	1.47	1.07	1.45	1.15
STAT	1.43	1.47	-	1.47	-	1.47

"Fig 3" and "Fig 7" relate to prediction diagrams shown previously.

Table 2.   Absolute values of flood flows for Three Mile Bridge on the Ouseburn watercourse.

		Rural Estimates		Estimates for 25% urban fraction			
		STAT	RR	STAT	RR	STAT+HRF	RR+HRF
Mean Annual Flood	PD	-	-	0.35	0.40	0.39	0.41
	Cmp	0.24	0.26	0.36	0.32	0.46	0.49
30-year Flood	PD	-	-	-	0.76	-	0.83
	Cmp	0.42	0.51	0.62	0.55	0.81	0.99
1000-year flood	PD	-	-	-	1.45	-	1.59
	Cmp	0.65	0.99	0.96	1.13	1.25	1.89

"PD" relates to the prediction diagrams of Figure 3 for the
30-year and 1000-year estimates (rainfall-runoff model only)
and Figure 7 for the mean annual flood estimate. "Cmp" relates to the computer-based estimates, these being plotted i
Figure 8. All flows are given in cumecs/sq km.

CONCLUSIONS

The two prediction methods put forward by the Institute of
Hydrology provide facilities to predict flood flows on ungauged urban catchments. However, the limitations and
uncertainties in the methods necessitate a close monitoring
of the operation of the models in a given region before they
can safely be used. In an area where no gauged data from
urban catchments is available, it will be advisable to use
more than one technique to obtain a "best" estimate of a
predicted value, with an associated indication of the
standard error of the value.

Some improvements in the rainfall-runoff model should be
possible and an extension of the applicability of the growth
curves for the statistical method will be welcomed. The HSF
technique may assist the prediction of effects of urbanisation.

ACKNOWLEDGEMENTS

The Author gratefully acknowledges the assistance of Mrs J
Wilkinson in the classification of rainfall profiles and the
derivation of the HSF prediction method. This work was
carried out as part of a thesis on rainfall analysis
techniques conducted under a Natural Environment Research
Council CASE studentship with Northumbrian Water Authority
to satisfy the requirements of a Ph D degree at the Universi
of Birmingham. Thanks are also due to the Director of
Planning and Scientific Services, Northumbrian Water Author
for permission to publish this paper.

REFERENCES

Bootman, A P (1977) Extreme Two Day Rainfall in Somerset.
Wessex Water Authority report, Bristol.

Hamlin, M J and O'Donnell, T (1977) The Flood Studies Report
- An Opportunity for Discussion. Report on a seminar at the
University of Birmingham.

Kelway, P S (1975) Contributions to Discussion at Flood
Studies Conference 7-8 May 1975. Inst Civ Eng, Conference
report: 12-14; 84-91.

Kelway, P S (1977a) Characteristics of Rainfall Conditions with Particular Reference to North-east England, J Inst WES, Lond 31, 4:251-268.

Kelway, P S (1977b) A Description of Procedures involved in using the Flood Studies Report and their incorporation into the Northumbrian Water Authority's Flood Estimation Scheme. Paper for Flood Hydrology Course, University of Newcastle-upon-Tyne.

Packman, J C (1977) The effects of Urbanisation on Flood Discharge Discussion and Recommended Procedures.   Report on Cranfield Conference, UK.

MEASURING THE EFFECTS OF URBANIZATION ON THE HYDROLOGIC
REGIMEN

Eugene J. Riordan[1], Neil S. Grigg[2], and Robert L. Hiller[3]

INTRODUCTION

Population growth, and individual mobility, have precipitated
a continued encroachment of agricultural land by residential
and commercial development in the United States. Urbanization,
or suburbanization, as this encroachment process has been
called, alters the hydrologic response of a watershed due to
the attendant increases in watershed impervious area, and in
stormwater removal efficiency of the natural drainage channels.
This alteration has resulted in significant adverse impacts on
the total hydrologic regimen of the watershed--two of the most
serious being the reduction in the adequacy of downstream
stormwater drainage facilities and the degradation of the
quality of stormwater runoff.

In an effort to reduce the flood damages incident to inadequate drainage facilities, local governments have undertaken
both structural flood control measures (channel improvements,
detention reservoirs, etc.) and non-structural flood control
measures (restrictive zoning of flood prone areas). However,
most local governments have been unable to establish satisfactory design levels for these flood control measures, and
have not been successful in apportioning the implementation
costs to the beneficiaries of these measures. These difficulties have promoted a research effort designed to develop
practical planning criteria (engineering, economic, legal,
and political) for an urban drainage program that will:

1. Graduate Research Assistant, Dept. of Civil Engineering,
    Colorado State University, Fort Collins, Colorado, USA.
2. Director, Water Resources Research Institute of the Univ-
    sity of North Carolina, Raleigh, North Carolina, USA.
3. Assistant Professor, Natural Resources Law, Dept. of Agr-
    cultural and Chemical Engineering, Colorado State Univer-
    sity, Fort Collins, Colorado, USA.

496

1. enable a local government to properly evaluate the
effects of various developments on the rainfall-runoff response of an urbanizing watershed, and

2. ensure the equitable recovery of <u>most</u> of the
expenses incurred by the local government in preventing flood
damages caused by those effects.

The writers have postponed criteria development for the
water quality degradation consequences of urbanization until
the problems related to managing the quantity aspects of
drainage are solved.

This paper reports on this research project, specifically
addressing the engineering criteria used to evaluate the
changes in watershed response. It summarizes previous work
reported in the literature on evaluating the changes in watershed response, discusses the viable approaches available for
evaluating this urbanization-induced consequence, and recom-
mends the approach that seems most practical for application
at the local level.

PAST WORK

Watershed urbanization has been shown to cause an increase in
the peak stormwater runoff rate, an increase in the volume of
stormwater runoff, and a decrease in the total runoff response
time for any particular storm. Since the mid 1950's a number
of studies have addressed the measurement of these particular
changes in basin runoff response. The results of these
studies--the measured effects of urbanization on peak stormwater runoff (Q_p), on the "time-to-peak" (t_p) and "lag time"
(t_L), and on the storm runoff volume (VOL)--are tabulated in
Appendix A. An attempt was made to express the results in
common units for easier comparison, however, interpolations of
the published expressions (equations and graphs) necessitated
by this manipulation were confined within the limitations of
those expressions.

Although the table of Appendix A is informative, the variability of the results suggests little, if any, help to a local
government that is attempting to identify the hydrologic consequences of a specific development project. The table
reveals that there is no universal hydrologic response formula
for estimating the effects of urbanization on stormwater
runoff hydrograph characteristics.

The listed works are not without merit, however. They all sub-
stantiate the notion of increasing Q_p with increasing urbanization (ranging from 1.8 to 8 times for the mean annual flood,
and 1.8 to 3.8 times for the 100-year flood), and with the
exception of one investigation (Doehring, 1975), they indicate

that the effects of urbanization on rare floods (100-year
flood event) are significantly less than on frequent floods
(mean annual flood event). In addition, each of the studies
presents a method for evaluating the effects of urbanization
and identifying the data requirements for that method. In the
absence of universal applicability of the results, the real
value of the studies becomes these methods and the data requirements they suggest.

TECHNIQUES FOR EVALUATING CHANGES IN HYDROLOGIC RESPONSE

Urban hydrology is recognized as a distinct field of hydrology
with its own set of predictive models and expressions for
representing urban watershed hydrologic response. Extensive
work has been done in developing usable equations and nomo-
graphs for evaluating peak stormwater runoff in urban areas
(Hicks, 1944; Tholin, 1959; Terstriep, 1969). These expressions were developed to answer the need for establishing
design criteria for urban drainage structures; and while they
are adequate for this purpose, they are generally not suitable
for evaluating the changes in the hydrologic response caused
by development in an urbanizing basin.

In addition, the predominant design criterion for construction
of these urban drainage structures--culverts, pipes, lined
channels, etc.--is the damage-free conveyance of the peak rate
of stormwater runoff from a design storm of some specified
frequency. This single design standard is not satisfactory:
it is not economical to construct drainage facilities for the
"ultimate" peak runoff rate of an urbanizing basin, nor is it
practical, in all cases, to continually enlarge (stage construction) drainage structures to accommodate the effects of
progressive urbanization. Detention or retention storage
facilities are reasonable alternatives that can effectively
accommodate this progressive urbanization and in some areas
the United States these facilities are developmental requirements. However, an additional stormwater response characteristic is required for the design of these detention facilities
the time distribution of stormwater runoff.

It becomes apparent that another set of predictive models and
expressions, or adaptations of the current ones, is required
for satisfactorily representing urbanizing watersheds. The
new models would predict the changes in peak runoff rate,
storm volume, and time distribution of runoff that can be
attributed to a particular development. In this way, proper
and equitable assignment of flood control benefits and lia-
bilities to that development can be assured. The available
techniques as described by the investigators listed in Appendix
A that would satisfy the requirements of this new set of "predictive models" can be classified into three categories:

1. Conceptual (lumped parameter) Models
2. Physically-based (distributed parameter) Models
3. Continuous Simulation Models

This categorization reflects the different amounts of data required by each technique as well as the philosophical approach embraced to model the rainfall-runoff process. For example, Conceptual Models are single transform functions that convert rainfall events into watershed runoff responses. The watershed runoff phenomenon is simply viewed as a "black-box" response ignoring all of the complex interdependent mechanisms of stormwater flow. Physically-based Models, on the other hand, do not ignore these mechanisms, but rather attempt to approximate the physical processes occurring within a watershed--interception, evapotranspiration, infiltration, overland flow, and channel flow-- that convert rainfall into stormwater runoff. Continuous Simulation Models are similar to the Physically-based ones but provide for continuous accounting of all of the water (subsurface as well as surface) within a particular watershed. Table 1 provides some detail concerning the data, technology, and expertise levels required for the three models.

ANALYSIS

The major drainage-related concerns of a local government are the reduction of flood damages, and the equitable recovery of the costs incurred by such reduction. In order to withstand judicial review, the cost assessment procedure adopted by a local government must be based on a sound engineering computation of the changes in hydrologic response caused by the assessed parcels. The choice of the engineering approach cannot rest solely on this legal criterion, however, but must also be based on the staff and resource capability of the local government. The approach selected must not demand personnel (both in numbers and in expertise) and financial resource requirements beyond the community's capability. A sophisticated technique for evaluating the changes in hydrologic response has little value if it cannot be supported. In addition, the effectiveness of the overall drainage program should be considered in the final selection of the engineering approach. The cost recovery objective may only be partially realized due to the following uncertainties of the program:

1. The flood control requirements of a drainage plan are based on normally uncertain projections of population growth and development trends within the planning area.

2. The implementation of any plan like a drainage program is quasi-political; approval of nonconforming developments must be accepted as a fact of life. This discretionary aspect significantly reduces the credibility of the master land use plans used for evaluating drainage requirements.

TABLE 1:   MODEL REQUIREMENTS

MODEL	DATA REQUIREMENTS					TECHNOLOGY REQ.			EXPERTISE REQUIREMENTS				
	REGIONAL RAINFALL-RUNOFF DATA (2-5YR)	WATERSHED PHYSIOGRAPHIC CHARACTERISTICS	EXTENDED RAINFALL PERIOD	INFILTRATION, EVAPOTRANSPIRATION, DETENTION STORAGE DATA	INITIAL PARAMETER ESTIMATION	CALCULATOR	COMPUTER FACILITIES	MODEL COMPUTER PROGRAM	UNIT HYDROGRAPH ANALYSIS	CONCEPTUAL MODELS	REGRESSION ANALYSIS	FLOOD FREQUENCY ANALYSIS	COMPUTER MODELING & MODEL FAMILIARITY
1. Conceptual Models	●	●	●			●	○		●	●	●	●	
2. Physically-based Models	●	●	●	●	●	○	●	●			○	●	●
3. Continuous Simulation Models	●	●	●	●	●	○	●	●			○	●	●

● Mandatory Requirements
○ Secondary Requirements

3. Actual construction of major flood control structures to accommodate the changes caused by urbanization generally occurs after a majority of the watershed is developed. Inflationary cost increases may exceed the interest accrued to drainage fees collected at the time of development approval.

The writers feel that the most inexpensive (in manpower as well as finances) engineering approach should be selected for the purpose of drainage planning and cost allocation. This translates into recommending that the adopted drainage program be based on a conceptual modeling approach perhaps similar to Espey's (Espey, 1969) or Stall's (Stall, 1970). In the cases where a local government has easy access to computer facilities and is able to support specially-trained personnel with computer modeling experience, the more appropriate approach would be physically-based models (Dempster, 1974) or some form of continuous hydrologic simulation (Lumb, 1976) if data availability permits. It should be noted that detailed design and operation of drainage facilities are quite another matter and are not being addressed here.

The writers would like to suggest that the engineering approach for the drainage program for a watershed be developed only once. Cost assessments for particular basins must be constant with time except for inflationary adjustments. It is unreasonable for local governments to penalize late (or early) developers because new data and techniques indicate the need for a revision of flood frequency information. If the community insists on changing the design and cost assessment rationale, they should be willing to absorb the associated revenue losses or refund the revenue gains of earlier assessments.

Finally, the writers would like to suggest that communities initiating data collection programs should carefully coordinate expenditures for data collection with their modeling needs. Certainly, detailed computer modeling exercises will not be feasible without an adequate data base. On the other hand, the conceptual modeling approach may yield satisfactory approximate results with fewer, less accurate, and less costly data requirements.

ACKNOWLEDGEMENTS

Financial support for this paper has been provided by the US Department of Interior, Office of Water Research and Technology through the Environmental Resources Center, Colorado State University.

APPENDIX A

LEGEND AND NOTES FOR: " LIST OF INVESTIGATIVE RESULTS"

$\underline{XR}$ $(A,B)_T$ = the ratio of the storm runoff characteristic $(\underline{X})$ for a watershed with A% of its area being impervious and B% of its area being sewered to the same storm runoff characteristic for the same watershed under rural conditions (approximately 0% impervious area and 0% sewered area) for the T year storm event (or for the watershed unit hydrograph when T = UH)

SYMBOL	STORM RUNOFF CHARACTERISTIC
X = Q	peak stormwater runoff rate
P	"time-to-peak" (time from the beginning of stormwater runoff to the peak stormwater runoff rate)
L	"lag time" (time from the centroid of excess rainfall to the centroid of direct stormwater runoff)
V	volume of direct stormwater runoff.

NOTES:

1. A blank (-) within the parentheses indicates that the investigator did not examine that parameter.

2. An asterisk (*) within the parentheses indicates that the storm runoff characteristic is relatively insensitive to the value of that parameter.

3. The investigators' qualitative description of the watersheds are listed when percentage values were not given and could not be estimated.

HP = high probability storm event   SF = single-family development   PD = partially-developed area
LP = low probability storm event   PUD = planned-unit development   PS = partially sewered area
RES = residential development   DEV = developed area   var = various

LIST OF INVESTIGATIVE RESULTS

YEAR	INVESTIGATOR	GEOGRAPHICAL AREA OF INVESTIGATION	AREA OF STUDY BASIN (mi)^2	EFFECTS OF URBANIZATION ON: Q_p	t_p & t_L	VOL
1955	BIGWOOD & THOMAS (2)	Connecticut	4.1(min) 1545(max)	$QR(RES)_{2.33}=3.5-5.5$	--	--
1961	CARTER (4)	Washington, D.C.	3.9(min) 546(max)	$QR(12,PS)_{2.33}=1.8$ $QR(12,100)_{2.33}=2.6$ $QR(100,100)_{2.33}=5.5$	$LR(*,PS)_{var}^{<.4}$ $LR(*,100)_{var}^{<.2}$	--
1961	WAANANEN (57)	Northern NJ, MI, PA, &VA	varies	$QR(DEV,-)_{HP}=3-4$	--	--
1961	WIITALA (59)	Detroit, MI	36.5 22.9	$QR(25,100)_{2.33}=2.3-2.7$	$LR(*,100)_{var}=.3$	$VR(25,100)_{var}\doteq 1$
1962	VAN SICKLE (54)	Houston, TX	38(min) 204(max)	$QR(DEV,100)_{UH}=2-5$	$PR(DEV,100)_{UH}\doteq .1$	--
1963	SAWYER (42)	Long Island, NY	31 10	$QR(DEV,-)_{2.33}>1$	--	--
1965	CRIPPEN (8,9)	Palo Alto, CA (Sharon Creek)	0.4	$QR(PD,-)_{UH}=1.4$	$PR(PD,-)_{UH}\doteq 1$ $LR(PD,-)_{UH}\doteq .7$	--
1965	JAMES (22)	Sacramento, CA (Morrison Creek)	72.7	$QR(30,30)_{2.33}=1.6$ $QR(30,30)_{100}=1.2$ $QR(100,100)_{2.33}=4.5$ $QR(100,100)_{100}=3.1$	$PR(100,100)_{var}^{<1}$	$VR(100,100)_{var}=$ 5.9-125
1965	ESPEY (17)	Austin, TX (Waller Creek)	4.13	$QR(27,50)_{UH}=1.5$ $QR(50,100)_{UH}=2.1$	$PR(27,50)_{UH}\doteq .5$ $PR(50,100)_{UH}\doteq .4$	$VR(25,-)_{var}\doteq 2$

LIST OF INVESTIGATIVE RESULTS

YEAR	INVESTIGATOR	GEOGRAPHICAL AREA OF INVESTIGATION	AREA OF STUDY BASIN (mi)	Q_p	EFFECTS OF URBANIZATION ON: t_p & t_L	VOL
1967	WILSON(60)	Jackson,MS	1(min) 10(max)	$QR(DEV,100)_{2.33}=4.5$ $QR(DEV,100)_{50}=3$	--	--
1968	ANDERSON(1)	Washington,D.C.	.0034(min) 570(max)	$QR(20,100)_{2.33}=3\text{-}4$ $QR(100,100)_{2.33}=6\text{-}7.7$ $QR(1\text{-}100,100)_{100}=2.4\text{-}3$	$LR(*,75)_{var}<.2$ $LR(*,100)_{var}\doteq.1$	--
1968	ESPEY(17)	Houston,TX	varies	$QR(50,50)_{UH}=3$	$PR(50,50)_{UH}\doteq.3$	--
1968	MARTENS(31)	Charlotte,NC & Central NC	.86(min) 865(max)	$QR(22,100)_{2.33}=2.4$ $QR(100,100)_{2.33}=4.7$ $QR(1\text{-}100)_{50}=1.9$	$LR(*,100)_{var}<.25$	--
1968	LEOPOLD(28)	Compilation of Results BY: CARTER, WITTALA, JAMES, ESPEY, ANDERSON, WILSON, MARTENS.		$QR(20,20)_{2.33}=1.5$ $QR(20,100)_{2.33}=2.5$ $QR(100,100)_{2.33}\doteq6$	--	--
1969	LULL & SOPPER(29)	Northeastern US,NH,MA,CT, NJ	4.5(min) 96.8(max)	$QR(DEV,-)_{var}>1$	--	--
1969	SARMA(40)	Indiana	.05(min) 19.3(max)	$QR(40,-)_{UH}=1.7\text{-}1.9$	--	--
1969	KINOSITA(26)	Tokyo,Japan (Syakuzii R.)	18.7	$QR(44,-)_{LP}=1.5$ $QR(100,-)_{LP}=2.5\text{-}4$	--	--
1969	RILEY(39)	Austin,TX (Waller Creek)	4.13	$QR(40,-)_{UH}>1.3$	$PR(40,-)_{UH}<.8$	--

LIST OF INVESTIGATIVE RESULTS

YEAR	INVESTIGATOR	GEOGRAPHICAL AREA OF INVESTIGATION	AREA OF STUDY BASIN $(mi)^2$	EFFECTS OF URBANIZATION ON: Q_p	t_p & t_L	VOL
1969	SEABURN(44)	Long Island,NY (East Meadow Brook)	31	$QR(28,65)_{UH}=2.5$	--	$VR(28,65)=1.1-4.6$
1970	FEDDES(18)	Bryan,TX	1.39 / 1.98	$QR(24,-)_{UH}\doteq2$	$PR(25,-)_{UH}=.5$ / $LR(25,-)_{var}=.6$	$VR(25,-)_{var}>1$
1970	STALL(46)	E. Cen. IL (Boneyard Creek & Kaskaskia R.)	3.58 / 12.3	$QR(75,100)_2\doteq8$ / $QR(75,100)_{50}\doteq4$	$PR(75,100)_{UH}\doteq.1$	--
1970	DA COSTA(10)	--	--	$QR(90,-)_{var}\ 3-12$	$PR(DEV,100)_{var}<1$	--
1971	REIMER(38)	San Diego,CA (Los Coches Creek)	.06(min) / 15(max)	$QR(DEV,50)_{100}=1.5-2.7$ / $QR(DEV,100)_{100}\doteq2$	--	--
1972	RAO(36)	Indiana & TX	.05(min) / 19.3(max)	$QR(40,-)_{UH}=1.9$	$LR(40,-)_{var}\doteq.6$ / $PR(40,-)_{UH}\doteq.4$	--
1973	JOHNSON(23)	Houston,TX	.05(min) / 358(max)	$QR(35,-)_2\doteq9$ / $QR(35,-)_{50}\doteq5$	--	--
1974	STANKOWSKI (47)	NJ	.6(min) / 779(max)	$QR(80,-)_2=3$ / $QR(80,-)_{100}=1.8$	--	--
1974	MCPHERSON (51)	Schwippe Valley Germany	19.5	$QR(DEV,-)_{var}=2$	--	--

LIST OF INVESTIGATIVE RESULTS

YEAR	INVESTIGATOR	GEOGRAPHICAL AREA OF INVESTIGATION	AREA OF STUDY BASIN $(mi)^2$	EFFECTS OF URBANIZATION ON:		
				Q_p	t_p & t_L	VOL
1974	DEMPSTER(14)	Dallas,TX	--	$QR(40,-)_2=1.35$ $QR(40,-)_{50}=1.16$ $QR(100,-)_{50}=1.36$	--	--
1974	DURBIN(16)	Santa Ana Valley,CA	3.7(min) 83.4(max)	$QR(DEV,-)_2=3-6$ $QR(DEV,-)_{100}\stackrel{\cdot}{=}1$	--	--
1975	BRAS(3)	Puerto Rico (Hypothetical Catchment)	.02	$QR(50,100)_{10}=1.3-2$ $QR(50,100)_{50}=1.1-1.2$	$PR(50,100)_{10}\stackrel{\cdot}{=}.65$ $PR(50,100)_{50}\stackrel{\cdot}{=}.8$	--
1975	DOEHRING(15)	Southeastern New England	34(min) 219(max)	$QR(-,-)_{2.33}=1-1.6$ $QR(-,-)_{100}=1.2-2.3$	--	--
1975	MCCUEN(32)	Baltimore,MD (Crayhaven Watershed)	.04	$QR(SF,-)_{2.25}\stackrel{\cdot}{=}2$ $QR(PUD,-)_{2-25}\stackrel{\cdot}{=}5$	--	--
1976	LAZARO(27)	Unity,MD	72.8 34.2	$QR(DEV,-)>1$	--	--
1976	CECH(5).	Texas Coastal Region(Houston, Galveston,Texas City)	varies	$QR(DEV,PS)_2=2-5$ $QR(DEV,PS)_{LP}=$ lesser effect	--	--

BIBLIOGRAPHY

1. Anderson, D. G. (1970) Effects of Urban Development on
    Floods in Northern Virginia. U.S. Geological Survey,
    Water Supply Paper 2001-C.

2. Bigwood, B. L., and M. P. Thomas. (1955) A Flood-Flow
    Formula for Connecticut. U. S. Geological Survey Cir-
    cular No. 365.

3. Bras, R. L. and F. E. Perkins. (1975) Effects of Urbani-
    zation on Catchment Response. ASCE Hydraul. Div., 101,
    HY 3:451-466.

4. Carter, R. W. (1961) Magnitude and Frequency of Floods in
    Suburban Areas. U. S. Geological Survey Paper 424-B.

5. Cech, I. and K. Assaf. (1976) Quantitative Assessment of
    Changes in Urban Runoff. ASCE Irrig. Drain. Div., 102,
    IR 1:119-125.

6. Chow, V. T. and B. C. Yen. (1976) Urban Stormwater Runoff:
    Determination of Volumes and Flowrates. U. S. Environ-
    mental Protection Agency Report No. EPA-600/2-76-116.

7. Crawford, N. H. (1971) Studies in the Application of
    Digital Simulation to Urban Hydrology. Hydrocomp Inter-
    national, Palo Alto, CA.

8. Crippen, J. R. (1965) Changes in Character of Unit Hydro-
    graphs, Sharon Creek, California, after Suburban Develop-
    ment. U. S. Geological Survey Prof. Paper 525-D, pp.
    D196-198.

9. _____. (1969) Hydrologic Effects of Suburban Development
    near Palo Alto, California. U. S. Geological Survey Open
    File Report.

10. Da Costa, P. C. C. (1970) Effect of Urbanization on Storm
    Water Peak Flows. ASCE, San. Engr. Div., 96, SA 2:187.

11. Dalrymple, T., W. B. Langbein, and M. A. Benson. (1952)
    Flood-Frequency Analysis. Manual of Hydrology, Part 3,
    Flood-Flow Techniques, U. S. Geological Survey Water
    Supply Paper 1543-A.

12. Dawdy, D. R. and J. M. Bergmann. (1973) Evaluation of
    Effects of Land-Use Changes on Streamflow. Paper presented
    at the Aug. 22-24, 1973 ASCE Irrig. Drain. Div. Specialty
    Conference, Agricultural and Urban Considerations in Irri-
    gation and Drainage, held at Ft. Collins, CO.

13. Dawdy, D. R., R. W. Lichty, and J. M. Bergmann. (1972) A
    Rainfall-Runoff Simulation Model for Estimation of Flood
    Peaks for Small Drainage Basins. U. S. Geological Survey
    Prof. Paper 506-B, pp. B1-28.

14. Dempster, G. R., Jr. (1974) Effects of Urbanization of
    Floods in Dallas, Texas Metropolitan Area. U. S. Geologi-
    cal Survey Water Resources Investigations No. 60-73.

15. Doehring, D. O., J. G. Fabos, and M. E. Smith (1975)
    Modeling the Dynamic Response of Flood Plains to Urbani-
    zation in Southeastern New England. Water Resources
    Research Center, University of Massachusetts at Amherst,
    Publication No. 53.

16. Durbin, T. J. (1974) Digital Simulation of the Effects of
    Urbanization on Runoff in the Upper Santa Ana Valley,
    California. U. S. Geological Survey Water Resources
    Investigations No. 41-43.

17. Espey, W. H., Jr., D. E. Winslow, and C. W. Morgan. (1969)
    Urban Effects on the Unit Hydrograph. Effects of Water-
    shed Changes on Streamflow, Proceedings, Water Resources
    Symposium No. 2, W. L. Moore and C. W. Morgan, eds.,
    University of Texas Press, Austin, TX, pp. 169-182.

18. Feddes, R. G., R. A. Clark, and R. C. Runnels. (1970) A
    Hydrometeorological Study Related to the Distribution of
    Precipitation and Runoff over Small Drainage Basins,
    Urban vs. Rural Areas. Texas A&M University, Water
    Resources Institute, Technical Report 28.

19. Grigg, N. S., et.al. (1975) Urban Drainage and Flood
    Control Projects: Economic, Legal, and Financial Aspects.
    Environmental Resources Center Completion Report Series
    No. 65, Colorado State University, Ft. Collins, CO.

20. Harris, E. E. and S. E. Rantz. (1964) Effect of Urban
    Growth on Streamflow Regimen of Permanente Creek, Santa
    Clara County, California. U. S. Geological Survey Water
    Supply Paper 1591-B.

21. Hicks, W. I. (1944) A Method of Computing Urban Runoff.
    Paper No. 2230, ASCE Transactions, 109, pp. 1217-1268.

22. James, L. D. (1965) Using a Digital Computer to Estimate
    the Effects of Urban Development on Flood Peaks. Water
    Resources Research, 1, 2:223.

23. Johnson, L. S. and D. M. Sayre. (1973) Effects of Urbani-
    zation on Floods in the Houston, Texas Metropolitan Area.
    U. S. Geological Survey Water Resources Investigations No.
    3-73.

24. Jones, D. E. (1967) Urban Hydrology--A Redirection. Civil Engineering, Aug. pp. 58-62.

25. _____. (1971) Where is Urban Hydrology Practice Today. ASCE Hydrul. Div., 97, HY 2:257-264.

26. Kinosita, T. and T. Sonda. (1969) Change of Runoff Due to Urbanization. International Assoc. of Scientific Hydrology, UNESCO/IASH, Publication No. 85, II.

27. Lazaro, T. R. (1976) Nonparametric Statistical Analysis of Annual Peak Flow Data from a Recently Urbanized Watershed. Water Resources Bulletin, 12, 1.

28. Leopold, L. B. (1968) Hydrology for Urban Land Planning-- A Guidebook on the Hydrologic Effects of Urban Land Use. U. S. Geological Survey Circular 554.

29. Lull, H. W. and W. E. Sopper. (1969) Hydrologic Effects from Urbanization of Forested Watersheds in the Northeast. Research Paper NE-146, U. S. Dept. of Agriculture, Forest Service.

30. Lumb, A. M. and L. D. James. (1976) Runoff Files for Flood Hydrograph Simulation. ASCE Hydraul. Div., 102, HY 10: 1515-1531.

31. Martens, L. A. (1968) Flood Inundation and Effects of Urbanization in Metropolitan Charlotte, North Carolina. U. S. Geological Survey Water Supply Paper 1591-C.

32. McCuen, R. H. and H. W. Piper. (1975) Hydrologic Impact of Planned Unit Developments. ASCE Urban Planning Dev. Div., 101, UP 1:93-102.

33. McPherson, M. B. and W. J. Schneider. (1974) Problems in Modeling Urban Watersheds. Water Resources Research, 10, 3:434-440.

34. Papadakis, C. and H. C. Pruel. (1972) University of Cincinatti Urban Runoff Model. ASCE Hydraul. Div., 98, HY 10: 1789-1804.

35. _____. (1973) Testing of Methods for Determination of Urban Runoff. ASCE Hydraul. Div., 99, HY 9:1319-1335.

36. Rao, R. A., J. W. Delleur, and P. B. S. Sarma. (1972) Conceptual Hydrologic Models for Urbanizing Basins. ASCE Hydraul. Div., 98, HY 7:1205.

37. Rao, R. G. S. and A. R. Rao. (1975) Analysis of the Effect of Urbanization on Runoff Characteristics by Nonlinear Rainfall-Runoff Models. Purdue University Water Resource Research Center Tech. Report No. 58, West Lafayette, IN.

38. Reimer, P. O. and J. B. Franzini. (1971) Urbanization's Drainage Consequences. ASCE, Urban Plan. Dev. Div., 97, UP 2:217.

39. Riley, J. P. and V. V. D. Narayana. (1969) Modeling the Runoff Characteristics of an Urban Watershed by Means of an Analog Computer. Effects of Watershed Change on Streamflow, Proceedings, Water Resources Symposium No. 2, W. L. Moore and C. W. Morgan, eds., University of Texas Press, Austin, TX, pp. 183-200.

40. Sarma, P. B. S., J. W. Delleur, and A. R. Rao. (1969) A Program in Urban Hydrology, Part II. An Evaluation of Rainfall-Runoff Models for Small Urbanized Watersheds and the Effect of Urbanization on Runoff. National Technical Information Service Tech. Report No. PB 189 043.

41. Savini, J. and J. C. Kammerer. (1961) Urban Growth and the Water Regimen. U. S. Geological Survey Water Supply Paper 1591-A.

42. Sawyer, R. M. (1963) Effect of Urbanization on Storm Discharge and Groundwater Recharge in Nassau County, New York. U. S. Geological Survey Prof. Paper 475-C, pp. C185-187.

43. Schneider, W. J. (1975) Aspects of Hydrological Effects of Urbanization. ASCE Hydraul. Div., 101, HY 5:449-468.

44. Seaburn, G. E. (1970) Effects of Urban Development on Direct Runoff to East Meadow Brook, Nassau County, Long Island, New York, U. S. Geological Survey Prof. Paper 627-B, pp. B1-14.

45. Snyder, F. F. (1958) Synthetic Flood Frequency. ASCE Hydraul. Div., 84, HY 5:1808,1-1808,22.

46. Stall, J. B., M. L. Terstriep, and F. A. Huff. (1970) Some Effects of Urbanization on Floods. Meeting Preprint 1130, ASCE National Water Resources Meeting, Memphis, TN.

47. Stankowski, S. J. (1974) Magnitude and Frequency of Floods in New Jersey with Effects of Urbanization. U. S. Geological Survey Special Report No. 28.

48. Task Force on Effect of Urban Development on Flood Discharges, Committee on Flood Control, Progress Report,
(1969) Effect of Urban Development on Flood Discharges--
Current Knowledge and Future Needs. ASCE Hydraul. Div.,
95, HY 2:287.

49. Terstriep, M. D. and J. B. Stall. (1969) Urban Runoff by
Road Research Laboratory Method. ASCE Hydraul. Div., 95,
HY 6:1809-1834.

50. Tholin, A. L. and C. J. Keifer. (1959) The Hydrology of
Urban Runoff. ASCE Sanitary Engr. Div., 85, SA 2:47-105.

51. UNESCO. (1975) Hydrological Effects of Urbanization, UNESCO
Press, Paris.

52. U. S. Dept. of Agriculture, Soil Conservation Service,
Indiana. (1973) Tech. Note (Engineering-2).

53. U. S. Dept. of Agriculture. (1975) Urban Hydrology for
Small Watersheds, Technical Release No. 55, Engr. Divi-
sion, U. S. Dept. of Agriculture, Soil Conservation
Service.

54. Van Sickle, D. (1962) The Effects of Urban Development
of Flood Runoff. Texas Engineer, 32, 12.

55. Viessman, W. (1966) The Hydrology of Small Impervious
Areas. Water Resources Research, 2, 3:405-412.

56. _____. (1966) The Hydrology of Small Impervious Areas.
Water Resources Research, 2, 3:405-412.

57. Waananen, A. O. (1961) Hydrologic Effects of Urban Growth--
Some Characteristics of Urban Runoff. U. S. Geological
Survey Prof. Paper 424-C.

58. Watt, W. E. and C. H. R. Kidd. (1975) QUURM--A Realistic
Urban Runoff Model. Journal of Hydrology, 27, DEC:225-
235.

59. Wiitala, S. W. (1961) Some Aspects of the Effect of Urban
and Suburban Development Upon Runoff. U. S. Geological
Survey Open File Report, Lansing, MI.

60. Wilson, K. V. (1967) A Preliminary Study of the Effect of
Urbanization on Floods in Jackson, Mississippi. U. S.
Geological Survey Prof. Paper 575-D, pp. D-259.

61. Wittenberg, H. (1975) A Model to Predict the Effects of
Urbanization on Watershed Response. Proceedings, National
Symposium on Urban Hydrology and Sediment Control, July 28-
31, 1975, UKY BU109, University of Kentucky, Lexington, KY.

URBAN STORM DRAINAGE

Ray K. Linsley

Hydrocomp Inc.

INTRODUCTION

The provision of urban storm drainage facilities is now
commonplace in most developed countries. Each new increment
of urbanization is required to provide facilities to drain its
area and connect with the existing system. Civil engineering
texts treat the determination of design flows as a simple
problem easily solved by simple handbook solutions, yet any
error in the determination of the design flow leads directly
to overdesign or underdesign of the system with a consequent
impact on costs. One wonders if owners, engineers, or
regulatory agencies ever seriously analyze how urban
drainage systems are designed.

Most urban drainage works are designed on the basis of
Mulvaney's formula (Mulvaney, 1851) or some modification
thereof. Such longevity of a methodology is relatively
uncommon in engineering, especially since Mulvaney did not
suggest the equation for urban drainage and, in his paper,
expressed considerable doubt as to its reliability. It
remained for others to take his procedure, ignore his
warnings, christen it the "rational method," and send it on to
posterity as the way to determine urban drainage flows.
Today, far more satisfactory methods are available but the old
method persists. I intend to discuss some of the
misconceptions which contribute to this persistence and to
show that much better procedures are available for use.

URBAN DRAINAGE AND FLOOD CONTROL

The problems of urban drainage and flood control have much in
common. Both tasks involve the disposal of excess water in
such a way that damage by flooding is minimized. The solution
in both cases is often the same--enhancing the conveyance of
the channel system which carries the water away. Why have

512

modern methods of flood control planning not been more widely adopted in urban storm drainage? The answer seems to lie in the persistence of several myths about urban drainage.

The low cost of drainage
A common myth is that the cost of urban drainage does not justify an elaborate hydrologic design. This is patently false. Rawls (1969) found that the average cost of storm drainage per hectare was about $650 (1963 dollars). Adjusted to 1977 dollars this is about $1750 per hectare. Thus, the drainage cost for one hectare of land is roughly equal to two man weeks of effort or more.

Few people really recognize the magnitude of the costs of urban drainage. Total historical costs for storm drainage in the United States is estimated at about $40 billion (U.S. National Water Commision, 1973). For the same period total investments in flood control and irrigation were about $28 billion each. Storm drainage costs over U.S. history exceed the cost of major flood control works! The estimated cost of the Chicago Deep Tunnel Plan for storm drainage is $2.5 billion. The most spectacular of main river projects rarely cost as much. As a diversion it may be noted that cost of highway drainage has been about 25 percent of total highway costs--an annual investment of a billion dollars or more in the United States. Surely it is impossible to justify a simple approach to drainage design in the face of such figures.

Lack of field data
A second excuse for use of simplified hydrologic methods in drainage design is the lack of field data. No one can deny that there is a lack of data. Indeed, there are almost no data. This point raises the question as to why only a few people feel that urban hydrologic data are important. Possibly by the circular argument that no data are required when using the rational method? Because the cost of urban drainage does not justify the cost of data collection? In other situations the absence of data has tended to provide the impulse for significant data collection programs, and for "major" projects, significant study effort has always been deemed necessary.

The rational method really is adequate
A myth which has persisted for many decades is that the rational method really does give adequate designs. Many extensive storm drainage systems have been built and seem to function successfully. What better evidence of adequacy could be desired? The rational method tends to overdesign because of built-in assumptions in the procedure and because of an inherent tendency among users to "be safe." Thus, the system which seems to work without flooding may indeed be

adequate—in fact, overadequate, and hence, too costly. Cost
is a proper matter of concern to cities which can reach the
state of financial bankruptcy—witness New York City. Without
field data it is impossible to determine whether a drainage
design performs according to the designer's intent, and in any
case, few designers ever go back to check. The assumption of
adequacy seems to be based on no more than the fact that an
actual design has not been compared with a cost effective
design.

What better method is there?
Of all the beliefs regarding the rational method the query
regarding a better method has the most foundation. Until the
mid-1940's, there really was no better method. The unit
hydrograph concept (Sherman, 1932) was of doubtful value for
the urban scene because of the lack of data from which to
derive unit hydrographs. The concept of synthetic unit
hydrographs (Snyder, 1938) might have been helpful. The study
of overland flow by Izzard (1946) was definitely a
contribution to improved analysis. Routing methods which
began to find extensive use about 1940 (McCarthy, 1938;
Rutter, Graves, and Snyder) also could have contributed to a
better solution of the urban drainage problem. The
computations would, however, have been tedious, and it was not
until the digital computer became reasonably available that it
could be said that a practical alternative solution was in
hand.

WHY IS A BETTER DESIGN METHOD REQUIRED

Several factors establish the need for a better method for
dealing with urban storm drainage. Perhaps the most important
reason is the tremendous investment in urban drainage
facilities. That such large investments may be poorly
designed should be a sobering thought for hydrologists.

Cost is not the only factor involved. The effect of
installing storm drainage facilities in urban areas is to
transmit higher flood peaks to areas downstream. Urbanization
increases runoff volume by increasing impervious area. The
drainage system which is normally much more hydraulically
efficient than the natural drainage channels of the area
accelerates the flow from the area. The effect is usually
greater during the more frequent floods. Rare floods may be
altered only slightly, if at all. The increased peaks may
have undesirable consequences for riparian lands downstream of
the urban area. Hence, urban drainage studies should consider
the possibility of reducing outflow peaks by use of storage.

We have recently become aware of the fact that the storm
runoff from urban areas transports substantial amounts of
pollution to the streams, lakes, and estuaries into which it

is discharged. Concentrations of pollutants may exceed those
in the effluent from secondary treatment plants, and total
annual quantity of pollutants may exceed that from the
treatment plant. Effective management of water quality
requires careful assessment of the pollutants in urban runoff
and of their impact on the stream. A simple, steady-state
Streeter-Phelps solution is no longer adequate. It cannot be
assumed that the critical water quality condition occurs
during a hot summer day when streamflow is at or near minimum.
Urban runoff comes only with storms! Hydrology has become
basic to the environmental engineering task.

These three factors—cost, flood augmentation, and pollution
problems—require that our analysis of urban hydrology be as
accurate as data and techniques permit. Indeed, if it is
considered necessary to treat all urban runoff before it is
discharged to the receiving waters the cost of water quality
control will exceed all other aspects of water resource
management.

WHAT ARE THE REQUIREMENTS OF AN ADEQUATE URBAN HYDROLOGIC STUDY

It cannot be said that all solutions which have been
programmed for computer are significant replacements for the
older methods. Many of the programs which are now available
are only slightly improved over the rational method. It seems
to this writer that the study for planning a storm drain
system requires the following characteristics:

(1)   It should correctly assess the probability
       distribution of flow at all critical points within
       the proposed system. To accomplish this the
       synthetic storm derived from rainfall-intensity-
       frequency analysis must be abandoned in favor of a
       simulation technique which uses a rainfall data
       series of substantial length (many years) so that
       the statistical properties of natural rainfall are
       properly represented.

(2)   The rainfall data series should be transformed to a
       runoff series using a model which has been
       demonstrated through calibration to be physically
       representative of the processes which actually occur
       in nature.

(3)   The runoff series should be converted to flows by
       routing the runoff through a first trial layout of
       the proposed system so that a flow-probability
       relation can be computed for each point within the
       system where such information is required.

(4) The streamflow series should be augmented by
     simulation of the washoff of pollutants into the
     storm drain system and of the concentration of these
     pollutants in the flow at each point of discharge
     from the system. If reservoir storage is to be used
     internally in the system, data may also be needed at
     the proposed reservoir sites.

(5) The effect of the proposed urban drainage system on
     downstream flows should be determined from the
     analysis, and corrective measures planned if
     required.

The above requirements are solely with regard to the
hydrologic procedures. An extensive socio-economic evaluation
is required to test various alternative configurations and to
establish the proper level of protection to be offered by the
system. This is all part of the planning process in which the
hydrologic analysis is only a portion, albeit an extremely
important portion.

The need for a computer-based simulation model for urban
drainage design has been recognized, and a dozen or more such
models are in use. It does not follow, however, that a model
which has been programmed for solution on a digital (or
analog) computer is necessarily effective. Many of the
models, typified by the Storm Water Management Model (SWMM)
(Environmental Protection Agency, 1971) are merely routing
models. The rainfall of a single storm event--either
synthetic or actual--is input. Runoff is computed by a simple
runoff coefficient or a loss rate computation, thus retaining
some of the principle sources of error in the rational formula
approach.

The models which conform to the specifications cited above are
primarily those continuous deterministic models which are
descendants of the original Stanford Watershed Model. Such a
one is Hydrocomp Simulation Programming (HSP). Figure 1
illustrates the agreement achieved during calibration of this
model on the Querecual River near Barcelona, Venezuela. An
assertion that a model does not need calibration is merely a
restatement of the concept that anyone can estimate the runoff
coefficient of the rational equation; any model can be run
without calibration by estimating the parameters. If the
opportunity for calibration exists, it should be seized in
order to eliminate the need to rely totally on judgment.
Using the calibrated model, 10 years of precipitation data
were input to the model and flow frequency curves were
computed for several points within a proposed drainage system
for Barcelona (Figure 2). This particular example was for
demonstration--normally a longer synthetic flow record would
be produced.

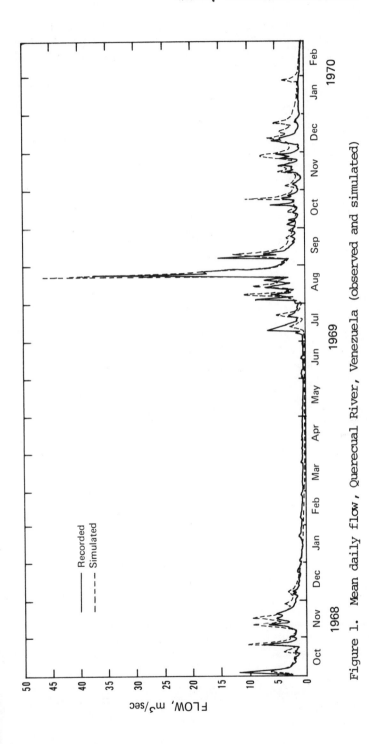

Figure 1.   Mean daily flow, Querecual River, Venezuela (observed and simulated)

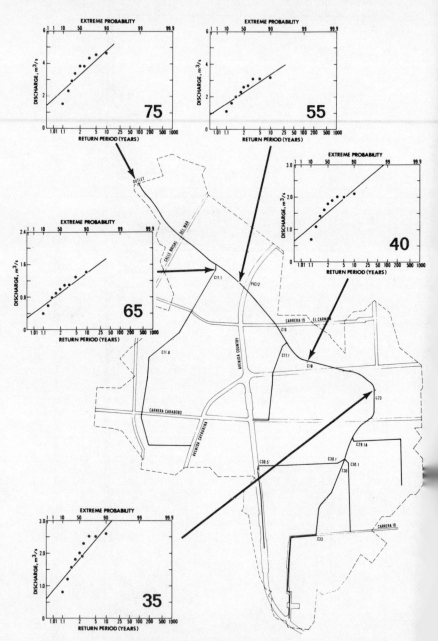

Figure 2.   Flood frequencies in proposed storm drains,
Barcelona, Venezuela

The question of how detailed the routing system should be may need more investigation. Clearly where the terrain is very flat, a careful accounting of head losses in manholes, bends, and junctions is required in order to minimize slope and hence excavation and the number of relift stations required. Where slopes are reasonably steep, all of the fine detail can probably be ignored with no serious error. Since the upstream laterals are usually sized to meet a minimum diameter requirement, detailed routing of overland flow and flow in the small laterals seems hardly necessary. As one moves downstream in the system, pipe sizes become larger and cost per unit length increases rapidly. Reliable routing in these major conduits is necessary to assure the use of the minimum satisfactory size. Even so, since the capacity of standard precast pipe increases between 15 and 30 percent between successive sizes, routing need not necessarily be ultra-detailed. This consideration is important since the routing computations are the most time-consuming part of the simulation, and an excessively detailed routing system coupled with a continuous model would require very large amounts of computer time. Surely a correct definition of frequency (which requires continuous simulation) is often more important than a detailed routing computation.

Once the necessary frequency curves are defined as in Figure 2, it is a relatively simple matter to select the flow for the specified probability level, and calculate the pipe size required to contain it. The use of a systematic modeling program assures a "balanced" system, i.e., all parts are designed with the same procedure and are consistent. No part of the system is overdesigned while other portions are underdesigned. Many tests have shown (Figure 3) that continuous simulation can reliably reproduce flood frequency curves.

When storage is proposed as a means of reducing peak flows, continuous simulation becomes imperative. The effect of a reservoir on flood peaks depends more on the shape and volume of the inflow hydrograph than on the peak flow. As alternative reservoir configurations are tested, the position of individual floods in the frequency series will change. A high, sharp peak with little volume may be greatly reduced, while a moderate but sustained peak with large volume may be modified only slightly and move upward in the series. To correctly define the frequency curve of reservoir outflow, a substantial number of flood events must be tested.

Frequency curves showing the percent of time that each water quality factor is at various values can also be produced in the simulation. Figure 4 is an example of such a curve for dissolved oxygen. In fact, Figure 4 represents the results of three simulations assuming different levels of treatment at an

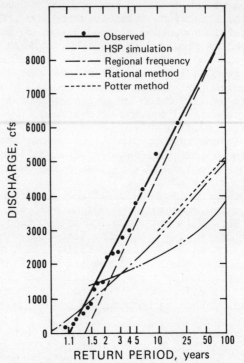

Figure 3. Flood frequency,
Rapid Creek, Iowa

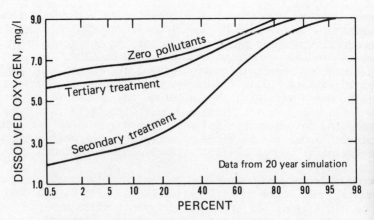

Figure 4. Dissolved oxygen curves, South Platte
River, Colorado

upstream treatment plant. Only through this type of
comparison can one truly evaluate the socio-economic benefits
(or losses) by changing treatment methods. Substantial
research is still needed to establish quantitative
relationships between quality parameters and benefits or
damages. For estimating the washoff of pollutants extensive
sampling of the water quality in storm drains is required to
determine the nature and quantity of pollutants in the
particular city. The usual sampling of water quality in
streams on a monthly interval is far from adequate.
Calibration of a quality simulation model requires a
substantial period of time—several years if possible—with
daily samples together with several weeks distributed
throughout the year during which three to six samples are
taken daily to define diurnal fluctuations. If an automatic
sampler is used, the manual sampling may not be necessary.
Analyses of these samples should be quite complete so that
information is available on all important quality factors,
Quality control decisions should not be based solely on
dissolved oxygen.

FURTHER RESEARCH

The Stanford Watershed Model and its derivatives have been in
use for 11 years and have been applied to several hundred
watersheds in all parts of the world. No major defects have
been detected, and as far as I know, no one has found a way to
significantly improve on the algorithms of SWM. Some
improvements will no doubt evolve, but extensive trials will be
needed to verify them. It seems, however, that we may assume
that we have a fairly effective system for hydrologic
simulation.

Water quality simulation is a rather different problem. Many
models exist, all using substantially similar algorithms.
Extensive studies have been performed with these models, but
because of the very serious shortage of data on water quality
variations in natural rivers and lakes, it has been almost
impossible to test any of the algorithms in depth. Until
better data are available, it will be impossible to really
evaluate the adequacy of present quality models and determine
whether further algorithm development is necessary. It is
likely that the answer will be "yes."

Benefit-cost analysis has been systematically applied to many
types of water resources development but rarely to urban
problems (or highway drainage). The arbitrary adoption of
design return periods may be a token acknowledgment of the
problem but more is needed. In short, the emphasis of
research on urban water problems might best be directed to the

structure of planning within which the hydrologic results are used than to hydrologic procedures themselves.

I hope that we will not be long in recognizing that hydrology is an all inclusive science. "Urban hydrology" is not significantly different from "rural hydrology." There is no need to identify "forest hydrology" or "agricultural hydrology" as specialties. All of these fields must rely on the same basic principles since the hydrologic processes are essentially the same on forests, agricultural land, and urban land. The end product of analysis is different in each case but the methods of analysis can be the same.

Some of the delay in adoption of new techniques for urban applications stems from those who say that "urban problems are different" or that "there has been little or no research on urban hydrology." These statements imply that procedures developed using data from rural watersheds are not applicable in the urban scene. Why not continue with the rational method until a good urban model is developed? If we can forget the specialities and work as hydrologists, improvements in all aspects of hydrology will come much faster.

REFERENCES

Izzard, C.F. (1946) Hydraulics of Runoff from Developed Surfaces. Proc. High. Res. Board, 26:129-150.

McCarthy, G.T. (1938) The Unit Hydrograph and Flood Routing. North Atlantic Div., U.S. Corps of Engineers.

Rutter, E.J., Q.B. Graves, and F.F. Snyder (1938) Flood Routing. Trans. Am. Soc. Civil Eng., 104:275-294.

Sherman, L.K. (1932) Streamflow from Rainfall by the Unit Hydrograph Method. Eng. News Rec., 108:501-505.

Snyder, F.F. (1938) Synthetic Unit Hydrographs. Trans. Am. Geophys. Union, 19, 1:447-454.

U.S. Environmental Protection Agency (1971) Storm Water Management Model. Water Pollution Control Research Series, Washington, D.C.

U.S. National Water Commission (1973) Water Policies for the Future. Washington, D.C. p. 506.

SEDIMENT MANAGEMENT CONCEPTS IN URBAN STORM WATER SYSTEM DESIGN

Harold P. Guy

Hydrologist, U.S. Geological Survey, Reston, Virginia

ABSTRACT

Storm drainage systems can be designed which will greatly
reduce peak rates of runoff and the amount of sediment and
pollutants normally transported from urbanizing and urban
areas to receiving water bodies. Reduction in peak flow
rates reduces the potential for serious channel enlargement
and additional sediment problems downstream from the development. Optimum design can be achieved through good land-use
planning that is well coordinated with natural drainage.
This in turn, will make it possible to minimize excavation
and soil exposure during construction, and provide a maximum
of individual and (or) community onsite storm water detention storage. The resulting storm drainage system would
usually have a lower initial cost and result in a more
esthetically pleasing neighborhood than generally exists with
conventional designs; but, may cause loss of convenience and
be more costly to maintain.

INTRODUCTION

Sediment management problems in storm water system design
correlate with many aspects of storm flow management. A few
decades ago the interest in storm water system design was
mainly that of controlling floods and providing maximum
convenience to the user. The ideal design required that the
system be capable of moving all, or nearly all, inflows of
water and sediment from the area in question without delay.
Also, the design too frequently addressed only the problems
at the site and ignored effects on other parts of the drainage basin, especially those downstream (Poertner, 1974).
The result of these inadequate designs created the need for
extensive downstream remedial measures to (1) prevent flood
damage, (2) prevent erosion and deposition problems, and
(3) provide storage and treatment to remove pollutants. Where
remedial measures could not be applied, problems with flooding,

erosion, deposition, and pollution reduced property values or even caused property to be abandoned.

This paper assumes the premise that good planning and design of storm water systems should be aimed at solving sediment problems associated with storm water management. That is, if these sediment problems are adequately handled, then most of the flow and pollution problems would be minimized. To accomplish this, it is necessary to go beyond our conventions of merely providing for storm water detention. It is necessary to make use of comprehensive planning techniques to ensure that heavy sediment loads during the relatively short-term construction period can be minimized and accommodated and that long-term runoff-generated pollutants, including sediment, can be managed.

It is important to remember that a design to eliminate all inconvenience and minor property damage is fundamentally unreasonable and almost certainly infeasible. A major storm should be expected to cause some erosion, damage to lawns and other vegetation, and damage to unwisely located structures, but flooding and sediment damage to buildings and essential facilities should be prevented. It is also unreasonable, infeasible, and perhaps undesirable to prevent some sediment and pollutants from moving through the system. The objective should be to minimize inconvenience to the user and to prevent the movement of harmful loads of sediment and pollutants to the neighborhood downstream.

A DEFINITION OF THE STORM WATER SYSTEM

A storm water system is composed of both natural and man-made elements that harmoniously absorb precipitation and move and store runoff and sediment. The system must accomplish this in a manner that will (a) prevent or minimize damage to both onsite and offsite property or loss of life after infrequent or unusual storms, and (b) eliminate or minimize inconvenience or disruption of activity both on and off the sites after more frequently occurring, less intense storms. This dual purpose usually requires the use of a "blue-green" development concept which can accommodate a good storage system as well as a "majo and a "minor" flow system. Jones (1967) and the National Association of Home Builders (1974) describe the minor system the pipe or channel to carry the runoff for the convenience factor and the major system the seldom used but essential floodway. Storage must be planned not only as a part of both the minor and major systems, but also upstream from these components; in other words, the storage system must usually be installed and maintained on both the public and private secto Driscoll (1977, written communication) has noted the importan of pertinent time and space scales relative to the many facet of water quality in storm water systems. See figures 1 and 2

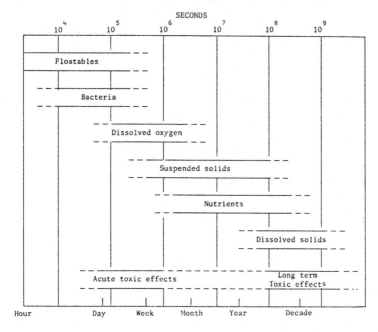

Figure 1.--Relative time scales for some aspects of water quality
(modified from Driscoll, 1977).

DESIGN PRINCIPLES

Priniciples for sound sediment management in storm water systems
go further than considering the sizing of storm water detention
basins. The principles identify and apply "natural" engineering
techniques to preserve and enhance desirable physical features
of the site. More specifically, they recognize the potential for
a heavy unmanageable sediment input to the system during the
development phase of the area, the potential for increased flooding and downstream erosion because of increased imperviousness,
and the potential for movement and accumulation of pollutants
from the developed area. The design must also recognize the need
for cost effectiveness, including maintenance, convenience to the
property users, the impact on water resources, and, above all,
pertinent aspects of safety. Bintzler and Scheib (1975) illustrate many of these principles in a paper titled, "Urban Development by Resources Planning", where the development plan is
derived from a melding of the natural resources inventory and
the objectives to be attained in the development.

Planning
The potential for unmanageable sediment input to the drainage
system both during construction and post construction of a
development can be handled to a considerable extent during the
planning phase of the project. That is, effective land utilization is needed that provides for green belts and cluster development. This, in turn, will make it possible to minimize the

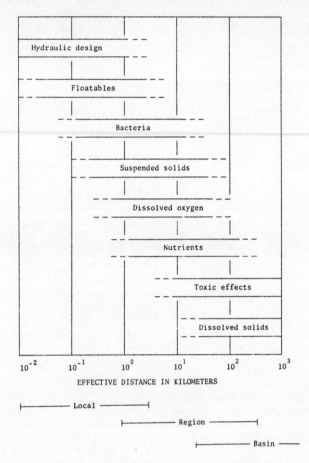

Figure 2.--Relative space scales for some aspects of water
          quality (modified from Driscoll, 1977).

amount of land which will be disturbed at a given time and to
minimize street length, storm drainage, and other utilities.
(Note: Excavation for streets, storm drainage, and other utili
ties is a serious environmental impact, and therefore the volu
of earthwork needed for installation must be minimized and thu
the need for careful planning of terrain use). The design of
lot and street patterns and grades must be coordinated with
components of natural drainage and ponds or lakes to complemen
the green belts, a part of the provision to reduce runoff and
sediment movement (National Association of Home Builders, 197!
The general design or pattern of land utilization may also
affect the relative amounts of private sector and public sect
storm water storage in the drainage system -- maintenance of
public sector system is considered somewhat more reliable tha
that of the private sector. An example of good land-use plan
ing and drainage layout has been described by Lindley (1976)

development of a 3.1 km² area in northern Illinois.

During construction

Given the best planning possible, good management during construction is needed which will reduce erosion of soils from the
construction area by minimizing the area exposed. Where land
may be exposed for more than a few weeks, temporary vegetative
or non-vegetative stabilization must be used. This in turn will
minimize the concentration of overland runoff and sediment movement. The U.S. Environmental Protection Agency (1973, p. 68)
shows that the expected relative effectiveness of growing vegetation in reducing soil loss to range from 90 percent for annual
ryegrass to 99 percent for grass sod, and that the effectiveness
of some mulches ranges from 90 percent for 3400 kg/ha of wood
cellulose fiber to 98 percent for 4500 kg/ha of hay or straw.

Where temporary land stabilization is impractical to use in controlling sediment, then ponding of excess flows (usually near
the site boundary) may need to be an integral part of the construction process. Such ponding may be either the offchannel or
inchannel types. Offchannel ponds are excavated basins exclusive
of any natural stream and which are located to receive upland
sediment laden flows. Inchannel ponds are usually formed by construction of a dam across the "valley", with a "hardened" outflow
system, to form storage above the level of the original channels.
In some situations, inchannel ponds can be formed by localized
excavation in the original channel to form a greatly enlarged
flow section (Kuo, 1976). Regardless of the type of pond used,
it must be installed at the beginning of the construction period
and periodically cleaned and otherwise maintained for effectiveness. This ponding device can sometimes be designed so that it
will be a permanent feature of the landscape; that is, a part of
an esthetically pleasing "blue-green" environment.

During construction in areas having complicated water and sediment management systems, it is sometimes necessary to use a
variety of flow and sediment handling systems, other than ponds,
such as temporary or permanent diversion terraces and channels,
level spreaders, and down drains. These devices can be used with
varying intensity and variety to meet desired objectives. Their
main function is to contain overland flow concentrated from small
areas in a manner which will reduce erosion, and also to induce
deposition of sediment which has been inadvertently entrained.
These devices generally require more frequent maintenance than
the ponds. They may also be expected to fail rather easily by
overtopping from high intensity storms or from plugging by sediment deposition. Such failure may result in more damage from
sediment than would have occurred without their installation.

In designing sediment control systems it is necessary to remember
that the physical flow-control elements have limited trap efficiency to retain sediment. Such efficiency depends greatly on
the particle size of the sediment, which usually has a wide

range, from sand (0.062 mm to 4 mm) to clay (<0.002 mm). The
sand is relatively easy to trap and thus it should be so trapped
if it may accumulate in the channels downstream. Chen (1975)
has published graphs showing the expected trap efficiency of
ponding devices.

Fine sediments on the other hand are difficult to trap by physical
control methods, and therefore the major emphasis must be to prevent their erosion--hence the need for good planning methods and
stabilization of construction areas to prevent erosion. The fine
sediments, because of their ease of suspension, do not affect the
physical character of channels as coarse sediments do; but they
seriously affect the appearance of and light transmission through
the flow. It takes only a few grams of colloidal material to
affect the apperance or character of a large body of water.

In addition to the problems with maintenance, possible failure,
and limited effectiveness, the installation of these nonvegetati
sediment control devices, including ponds, is often another element in construction which may adversely impact the environment.
It must also be remembered that such devices, if temporary in
nature, must be removed in the cleanup operation.

Mainly because of the relatively large supply of fresh sediments
available, impact from movement of pollutants during constructi
is normally considered minimal. Occasionally, however, an exte
sive fuel spill may occur or someone may improperly dispose of
toxic substances. The latter is frequently the reason that san
tary and storm drainage systems are closed during construction.

Post construction
Significant sediment aspects likely to be encountered after co
struction is complete and when the landscape becomes stabilized
have been noted in the literature (Peavy, 1976) (American Publ
Works Association, 1969) (Guy, 1975). These include (1) the
movement of pollutants which are adsorbed on the sediment and
(2) the potential for serious channel enlargement downstream
because of increased runoff intensity and volume, thus causing
problems even further downstream (Fox, 1976) (Robinson, 1976).
The first aspect must be dealt with as a long-term continuing
threat by preventing the movement of sediment and these pollut
ants from the area. The second is of a more intermediate dura
tion, that is, the channel enlargement process may complete it
self in a decade. In some instances the extra flow may need
be accommodated by enlarging or riprapping the offsite channe

Just as good planning and erosion control by surface stabiliz
tion are the most practical means of sediment management during construction, good planning and detention storage are the
most practical means of sediment management during post const
tion. Some useful principles to accomplish good planning and
onsite storage goals include:

1. On a large scale, contiguous communities must collaborate to plan drainage that extends across jurisdictional boundaries.

2. Individual developments must include drainage systems compatible with the community system.

3. The system should include both major and minor storm water drainage components for both flow and sediment management which will provide cost-effective property protection and convenience.

4. The development of impervious surfaces should be minimized even at the cost of some increased inconvenience and maintenance. This can greatly decrease the cost of drainage installation and improve ground-water recharge. For example, do not use a curb-and-gutter system if it can be avoided, reduce the width of paved streets compatible with the forthcoming use of smaller vehicles, or use pervious paving where conditions permit (Diniz, 1976).

5. Use small-area ponding wherever practical (roof tops, lawns, parking lots, and green areas) to compensate for the necessary development of impervious surfaces. This can greatly reduce the first cost of the small-area drainage system.

6. Use larger-area ponding where terrain permits. These ponds may be more cost effective and more practical to maintain than small-area storage but require more careful design for safety, appearance, recreational use, potential ground-water recharge, and maintenance provisions.

7. The use of enclosed components of the drainage system should be minimized wherever existing natural systems can accommodate runoff and sediment and (or) wherever the local public will act responsibly toward open channels.

8. Erosion of soils should be minimized by appropriate design and maintenance provisions to prevent downstream sediment problems.

9. Trade offs between construction, amortization, operation, and maintenance costs must be used to provide a minimal initial present value.

As implied in these principles, it is self-evident that good sediment management will also help minimize pollution of water resources downstream from a given developed area. Models do not aid in the specific design of sediment control systems according to a review by Jennings and Mattraw (1976) of several urban-flow and water-quality models.

The characteristics of urban strom water are reported by Field (1975) as shown in table 1. Sartor and Boyd (1972), show that storm water pollution loads, while occurring for relatively short periods to time, may range in strength from the equivalent of super-strong to very diluted sanitary sewage. In fact, torm water may contain dangerous levels of toxic materials not often found in sanitary sewage. The effect of elapsed time n street loading of contaminants is indicated in figure 3

TABLE 1.   Characteristics of urban storm water
(modified from Field, 1975)

(Data in milligrams per liter, except where noted)

	Selected data
BOD$_5$	1-700
TSS	2-11,300
Total solids	450-14,600
Organic N	0.1-16
NH$_3$N	0.1-2.5
Total PO$_4$	0.1-125
Chlorides	2-25,000
Oils	0-110
Phenols	0-0.2
Lead	0-1.9
Total coli	200-146 x 10^7/L

for different urban uses. These curves help to explain some of
the variability in pollution loading of urban-area storm water
runoff.

The data on storm water pollution loads clearly show that urban
runoff is a major problem, and therefore serious consideration
must be given to reduction in movement of pollutants. Otherwis
heavy costs for storing and treating storm water for removal of
such pollutants may be necessary. As noted by Lynam and others
(1977), Chicago's rock-tunnel plan is scheduled at $2,700,000,(
to relieve flooding and provide treatment of storm water runof
in the older part of the city. A system of landscaped swales

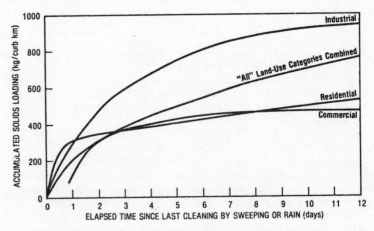

Figure 3.--Accumulated solids loading on streets for differer
land-use categories (from Sartor and Boyd, 1972).

grass drains, instead of a curb-and-gutter system in areas of
light traffic and on-site detention storage, wherever practical,
would help to minimize direct runoff and help to trap the contaminants. In many instances, good planning of land use and
proper storm drainage design should accomplish a 10-fold decrease
in sediment and other pollutant movement from a developed area
over that encountered by conventional planning and design.

JURISDICTIONAL RESPONSIBILITIES

In order to implement these principles of good drainage design,
it would be necessary for many jurisdictions to change their
regulations. Jones (1967) in a study of different codes dealing
with drainage in residential developments found many unnecessary
and poorly conceived requirements. These include the provision
that the first storm water inlets must be within 91 m of the
drainage divide, all lots must drain from front to back, all
lots must drain from back to front, roof drains must be piped
to the roadway curb, all storm drains must be sufficiently deep
to permit gravity outfall from all footing drains, the minimum
street grade shall be 0.5 percent, and curbside flow capacity
must accommodate a storm with a recurrence interval of 50 years.

Urban drainage requirements are needed which will recognize and
capitalize upon the natural environmental conditions and provide
for cost-effective improvements to accommodate the natural
forces necessary for reasonable utilization of the area, both
on-site and off-site. Jurisdictions must also provide assurance
that facilities will be maintained as needed because a poorly
maintained ineffective drainage system can be a damaging and
costly liability to the community.

Jurisdictional responsibilities can be modified in a wide
variety of ways. Some jurisdictions may be content with changing a few outmoded codes, but others work to attain broader
goals. In Maryland, for example, the Director of Water Resources
Administration of the Department of Natural Resources has asked
local governments, under the authority of the Natural Resources
Law, Title 8, Subtitle 9 of the State of Maryland, to make storm-
water-management policy, or an approved equal program, a requirement of an "Acceptable Operating Sediment Control Program" by
January 1, 1979. Maryland's sediment control program has been
working since 1970. This action to bring the storm water management into the ongoing sediment control program conforms with
a prior implication in this paper that good sediment management
will axiomatically yield most of the desired storm water management goals.

CONCLUSIONS

It is not necessary to have precise data and manuals of concise
design criteria for the land-use planner and the designer of

urban storm drainage systems to effect a substantial reduction
in the movement of sediments and other pollutants from urban
areas over that attained by more conventional designs. The more
desirable design can usually be accomplished with a considerable
reduction in environmental impact and reduction in first cost.
A more esthetically pleasing landscape can also be effected for
the user, though he may find some loss of convenience and increase in local maintenance costs. Considering the mandate of
Public Law 92-500 in the United States, which requires that procedures and methods be set forth to control nonpoint sources of
pollution, the general methods prescribed should also reduce
long-term community costs.

The principles of good sediment management in urban storm drainage systems are, in the simplest terms, (1) plan for cluster
type development which has generally been shown to result in
efficient land-use and minimum excavation for underground storm
drains and other utilities, (2) minimize bare soil exposure to
intense rains during construction through good construction timing and judicious use of vegetative and other soil stabilization
methods, (3) use a minimum of street and other connected impervious areas consistent with needed convenience and reasonable
maintenance costs, and (4) provide for detention and spillway
storage on small areas and in community ponds or lakes to reduce
peak flows and trap sediment and pollutants. These principles
minimize the potential for sediment to move into the system as
well as afford the best possibility that it can be removed from
the flow near its source if it does become entrained. These
principles also address the need to reduce the movement of
storm drainage pollutants that may harm down-stream waters.

Though many of these concepts are already being applied in storm
drainage system design, they are generally used on a fragmented
basis because of political pressures or because the planner and
designer are not yet "comfortable" with the whole set of principles. Part of the problem is that good planning and design
are too often treated as a mechanical process. As Barker (1973
has noted, there is too often "a desire to crank up long-dormant
mathematical models, develop new modeling techniques, and atte
to collect a quantity of heretofore nonexistent data that will
prove that, yes, we do have a very complex and diverse problem
on our hands." Though the problem is complex and diverse, considerable progress can be made if it is attacked on a multidis-
ciplinary basis.   For example, the use of "natural drainage
flow techniques" requires knowledge of vegetative flow retard-
ance, channel infiltration capacities on a seasonal basis, and
social acceptance of a "wet" channel at a specific location.
Jurisdictional responsibilities should also be multidisciplin
and provide for proper enabling regulations and system mainte
ance.

REFERENCES

American Public Works Assoc., 1969, Water Pollution Aspects of Urban Runoff: U.S. Dept. of the Interior, Washington, D.C.

Baker, P. E., 1977, Areawide waste treatment management of nonpoint pollution: Presented to Amer. Soc. of Civil Engineers spring convention, Dallas, Texas, Apr. 25-29, 1977, 16 p.

Bintzler, R. C. and Scheib, W. L., 1975, Urban development by resource planning: Proc. of Natl. Symp. on Urban Hydrology and Sediment Control, Univ. of Kentucky BU 109, p. 17-22.

Chen, C. N., 1975, Design of sediment retention basins: Proc. of Nat. Symp. on Urban Hydrology and Sediment Control, Univ. of Kentucky BU 109, p. 285-298.

Diniz, E. V., 1976, Quantifying the effects of porous pavements on urban hydrology: Proc. on Nat. Symp. on Urban Hydrology, Hydraulics, and Sediment Control, Univ. of Kentucky BU 111, p. 63-70.

Field, Richard, 1975, Coping with urban runoff in the United States: Water Research, v. 9, p. 499-505.

Fox, H. L., 1976, Channel alteration in an urbanizing watershed: A case history in Maryland: Proc. of Nat. Symp. on Urban Hydrology, Hydraulics, and Sediment Control, Univ. of Kentucky BU 111, p. 105-113.

Guy, H. P., 1975, an overview of non-point water pollution from the urban-suburban arena: Proc. Southeastern Regional Symposium on Non-point Sources of Water Polution, Virginia Polytechnic Institute and Univ., Blacksburg, Va., p. 45-66.

Jennings, M. E. and Mattraw, H. C., 1976, Comparison of the predictive accuracy of models of urban flow and water-quality processes: Proc. of Nat. Symp. on Urban Hydrology, Hydraulics, and Sediment Control, Univ. of Kentucky BU 111, p. 239-243.

Jones, D.E., Jr., 1967, Urban hydrology--a redirection: Amer. Soc. of Civil Engineers, Civil Engineering v. 37, no. 8, p. 58-62.

Kuo, C. Y., 1976, Sediment routing in an instream settling basin: Proc. of Nat. Symp. on Urban Hydrology, Hydraulics, and Sediment Control, Univ. of Kentucky BU 11, p. 143-150.

Lindley, R. W., 1976, A case study: Green trails development, Lisle, Illinois: Proc. of Nat. Symposium on Urban Hydrology, Hydraulics, and Sediment Control, Univ. of Kentucky BU 111, p. 245-255.

Lynam, B. T., and others, 1977, Managing storm runoff in the Chicago-land area, in Urban runoff quality measurement and analysis: Amer. Soc. of Civil Engineers Fall Convention and Exhibit, Preprint 3091, p. 1-34.

National Assoc. of Home Builders, 1974, Land developement manual: Washington, D. C.

National Association of Home Builders, 1975, Residential storm water management; objectives, principles and design considerations: The Urban Land Institute, National Assoc. of Home Builders, and Amer. Soc. of Civil Engineers, Washington, D. C., 64 p.

Peavy, H. S., 1976, Sedimentation from an established urban watershed: Proc. of Nat. Symp. on Urban Hydrology, Hydraulics, and Sediment Control, Univ. of Kentucky BU 111, p. 163-169.

Poertner, H. G., 1974, Practices in detention of urban storm water runoff: Amer. Public Works Assoc. Special Report No. 43, Chicago, Illinois.

Robinson, A. M., 1976, The effects of urbanization on stream channel morphology: Proc. of Nat. Symp. on Urban Hydrology, Hydraulics, and Sediment Control, Univ. of Kentucky BU 111, p. 115-127.

Sartor, J. D. and Boyd, G. B., 1972, Water pollution aspects of street surface contaminants: U.S. Environmental Protection Agency, EPA-R2-081, 236 p.

U.S. Environmental Protection Agency, 1973, Processes, procedures, and methods to control pollution resulting from all construction activity: EPA 430/9-73-007, Supt. of Documents, U.S. Govt. Printing Office, Washington, D. C.

URBAN DRAINAGE: THE EFFECTS OF SEDIMENT ON PERFORMANCE AND
DESIGN CRITERIA.

P. Ackers

Hydraulics Consultant, Binnie and Partners, London.

INTRODUCTION

Although comprehensive information about the concentration
and grading of sediment in sewers and storm drains is lacking,
it is well known that silt, sand and coarser solids gain
access to drainage systems in appreciable quantity. If they
are not transported through the system by the flow, a
maintenance exercise is required to extract them if their
presence is not to seriously impair hydraulic performance.

Various criteria have been proposed for self-cleansing
conditions, frequently expressed as a limiting velocity (BSI,
1968) but sometimes as a limiting shear stress at the
perimeter of the flow section. In fact the old "Maguire rule"
for laying a 12 inch pipe at 1 in 120, a 24 inch at 1 in 240
etc. is a shear stress criterion, as average shear stress is
proportional to the product of slope and hydraulic mean depth.
Recent information suggests, however, that neither a constant
value of limiting velocity nor of shear stress adequately
represents the competence of a pipe flowing partly full to
transport sediment.

This problem has recently been highlighted in the drainage
system of a capital city overseas, which has suffered from
serious sedimentation, long lengths of main sewer now
operating surcharged with an appreciable reduction of capacity
through deposition. The available information on the transport capacity of drainage systems was therefore reviewed, in
the hope of understanding this particular system better and
hence to draw up suitable criteria for the design of a
replacement or supplementary system.

AVAILABLE INFORMATION

May (1975) reviewed the subject, but found that different

535

authorities had produced a range of criteria that left
considerable uncertainty over their application and relevance
to practical situations. Some researchers had been concerned
with the initial motion condition in smooth pipes i.e. that
state of flow that could just displace individual particles
resting on a smooth solid invert. This zero transport
condition is not an appropriate criterion in practice,
however, wherever an appreciable concentration of solids has
to be moved through the pipeline.

In the early 1950's,an important programme of research was
undertaken at the University of Iowa. This was summarised
by Laursen (1956) and included work by Craven and Ambrose.
Craven (1953) examined transport under full-bore conditions,
using 50 mm dia. and 140 mm pipes and uniform sands in the
range 0.25 to 1.62 mm. He considered the maximum transport
rate with a given hydraulic gradient with no deposition in
the pipe; and also the transport function with a mobile sediment bed.

Ambrose (1953) extended the research to cover free-surface
flow with a sediment bed, but discovered that this mode of
operation is essentially unstable: he had to generate a
uniform sediment deposit under full-bore conditions, the
performance then being examined under free-surface conditions.

Laursen's summary identified several features that are
crucial to the problem. He recognised that for full-bore
flow, the joint application of a friction equation and
sediment transport equation defined both the discharge
capacity and the transport capacity. One form of sediment
function expressed the limiting condition of maximum transport
without deposition, and hence if the sediment supply was
less than this the pipeline would operate without a sediment
bed, whilst if the supply was greater deposition would occur.
A different function was derived to describe transport in the
latter case. It follows that if a given pipe is supplied with
known water and sediment fluxes, then with full-bore flow the
hydraulics are defined: only one solution of the resistance
and transport functions exists, yielding the necessary depth
of deposit (which might in some cases be zero) and hydraulic
gradient.

The Iowa researchers also recognised both empirically and
theoretically that free-surface flow with a sediment bed was
inherently unstable. If a clean pipe operating at a prescribed gradient and flow rate with a free-surface was
supplied with sediment exceeding its no-deposit capacity,
sedimentation would begin to occur. This would increase the
roughness (especially if dunes occurred) and reduce the
velocity and the transport capacity. This would accelerate
deposition (consistent with the surplus rate of supply) until

finally the pipeline ran full-bore. If it could generate
an increased hydraulic gradient through surcharge, the gradient and depth of deposition might then achieve values
simultaneously satisfying both the resistance and transport
equations.

Laursen recognised that if there had been well established
sediment transport formulae a full theoretical description
would emerge by combining it with a suitable friction
equation. The uncertainties surrounding the available
equations instead led the Iowa team to empirical relationships for their particular range of experiments and sediment
sizes.

Important research has been undertaken recently at the
University of Newcastle-upon-Tyne (Novak and Nalluri, 1975)
and a programme is also under way at Wallingford. The
Newcastle programme has covered the initial motion condition
of discrete particles on a smooth bed; of patterns of
grouped particles; and of sediment transport at the limit of
no-deposit.

The present paper is, however, based not on new experimental
data but on the application of established formulae for
sediment transport and resistance to provide a theoretical
framework to describe the performance of sediment transporting
urban drainage systems. The main interest lies in the
conclusions about performance, which confirm several of the
Iowa findings. At this stage, however, the results are not
sufficiently confirmed to form a basis of general design,
because there is uncertainty about some of the assumptions
and coefficient values.

THEORY

The Colebrook and White equation (1939) for resistance is
used, in the form:

$$V = - \sqrt{(32gRS)} \, \log \left[\frac{k_s}{14.8R} + \frac{1.255\nu}{R\sqrt{(32gRS)}} \right] \qquad (1)$$

where

V = average velocity (Q/A)
g = gravitational acceleration
R = hydraulic mean depth (A/P)
S = hydraulic gradient, or invert gradient
    for uniform free-surface flow
k_S = equivalent sand roughness
ν = kinematic viscosity of the fluid
A = cross sectional area of flow section
Q = volumetric discharge of fluid
P = wetted perimeter of flow section

Where there is a sediment bed, the roughness k_S is taken as the weighted average value for the pipe walls and bed:

$$k_s = \frac{P_w k_w + P_b k_b}{P_w + P_b} \tag{2}$$

where the suffices w and b refer to the wall and bed respectively.

The viscosity ν depends on temperature, and k_w and k_b have to be assessed from the character of the pipe walls (construction and sliming) and sediment bed (plane or duned, and sediment size).

The Ackers and White equation (1973) for sediment transport has been used, although it was derived for open channels with an active bed, rather than closed conduit flow. This equation has a sound theoretical basis and has been confirmed for a wide range of conditions,(White, Milli and Crabbe, 1975). It may therefore be more reliable when scaled to field conditions than other equations based on small scale laboratory research that may be subject to scale effects. Minor adjustments to the basic formula are required to take account of a non-rectangular flow section. The basic equation is:

$$G_{gr} = C \left(\frac{F_{gr}}{A_c} - 1 \right)^m \tag{3}$$

where G_{gr} = a transport function, $\left(\frac{Xd}{sD} \right) \left(\frac{v_*}{V} \right)^n$

F_{gr} = a mobility number, $\dfrac{V^{1-n} \, v_*^{\,n}}{\sqrt{g(s-1)D} \left\{ \sqrt{32} \log \frac{10d}{D} \right\}^{1-n}}$

C = empirical coefficient
A_c = initial motion condition (empirical)
m = empirical exponent

Also v_* = shear velocity, $\sqrt{(gRS)}$
X = sediment mass flux as ratio of fluid mass flux (equivalent to a concentration by weight)
D = particle diameter (the 35 percentile value is appropriate with graded sediments)
d = equivalent flow depth
s = relative density of solids to fluid.

The coefficients A, C, m and n are all functions of a form of particle-size Reynolds number.

$$D_{gr} = \left\{ \frac{g \ (s-1)}{\nu^2} \right\}^{\frac{1}{3}} D \qquad (4)$$

The empirical coefficient values are given by:

For coarse material $(D_{gr} \geqslant 60)$

$$n = 0 \qquad (5)$$
$$A_c = 0.17 \qquad (6)$$
$$m = 1.50 \qquad (7)$$
$$C = 0.025 \qquad (8)$$

For transitional sizes $(1 < D_{gr} < 60)$

$$n = 1.00 - 0.56 \log D_{gr} \qquad (9)$$

$$A_c = \frac{0.23}{\sqrt{D_{gr}}} + 0.14 \qquad (10)$$

$$m = \frac{9.66}{D_{gr}} + 1.34 \qquad (11)$$

$$\log C = 2.86 \log D_{gr} - (\log D_{gr})^2 - 3.53 \qquad (12)$$

For the non-rectangular section of flow in a pipe, the equivalent flow depth d is defined by the flow area divided by an effective width for sediment transport. This is assumed to be the pipe diameter if the water surface is above 0.5D; and the actual water surface width if below that elevation. These definitions also apply when there is a sediment bed but for pipes more than half-full of sediment the effective width is taken to be the bed width.

A computer program was prepared to solve these friction and sediment equations simultaneously. Input consists of basic data (viscosity ν , sediment relative density s, sediment size D, pipe roughness k_w, sediment bed roughness k_b, pipe diameter D_0); and flow conditions (proportional depth of water, proportional depth of sediment, hydraulic gradient, S). Output consists of velocity V, discharge Q, shear velocity v_*, shear stress $\tau = v_*^2$, sediment concentration X. Extracts from the tabulated data for a particular set of conditions have been plotted to illustrate system performance; they are not intended to represent a general practical case however.

RESULTS

To illustrate the performance of a pipe carrying sediment, the following case was chosen: d = 1.5 m; D = 0.2 mm;

k_W = 0.6 mm; k_b = 6 mm; s = 2.65; S = 0.5 x 10^{-3}. The
proportional depths of sediment calculated were zero, 0.1,
0.25 and 0.5; and proportional water surface levels of 0.25,
0.5, 0.75 and full bore. Surcharged operation was included
with hydraulic gradients up to 10^{-2}. These selected results
are given in figs 1 and 2 (part-full and full-bore respectively) in the form of sediment concentration against discharg

Fig 1 shows two families of curves, one linking points with
equal sediment depth but varying water depth, the other
linking points of equal water surface level but varying
degrees of sedimentation. Fig 2 has three families of lines;
equal sediment depth; equal hydraulic gradient; and equal
total sediment transport rate. The implications of these
diagrams are best illustrated by an example, marked a-g.

Consider a discharge of 0.8 m^3/s of clean water where the
concentration is initially below 50 mg/l (point a, fig.1).
The transporting capacity is high enough for the invert to
remain free of deposit and point c shows that the hydraulic
resistance law determines a depth of flow of a little under
half-full.

Now add an appreciable sediment load of 2000 mg/l, point b.
This exceeds the actual transporting capacity of the clean
invert pipe (point c) of 660 mg/l (antilog $\overline{4}$.84) so
deposition occurs progressively. When the pipe is one tenth
full of sediment, point d shows it to be operating at a
proportional depth of about 0.6, with transport capacity
430 mg/l i.e. less than when the invert was clean. Deposition then progresses through point e (sediment depth 0.25,
water level about 0.7d, transport 400 mg/l) to full bore
flow flow at f. Without surcharge, the transport capacity
has dropped to 200 mg/l and the pipe is about 0.35 full of
sediment.

Point f on fig 1 is also marked on fig 2 for full-bore flow.
Assuming sufficient depth is available for unlimited
surcharge beyond the invert gradient of 0.5×10^{-3}, the depth
of deposition will continue to increase to point g, where
the sediment depth is a little over the half diameter mark
and (by interpolation) the hydraulic gradient has increased
three fold to about 1.5 x 10^{-3}.

This example for a steady discharge illustates the inevit-
ability of sedimentation and pipe surcharge if the long term
average sediment concentration exceeds the capacity of the
clear pipe-line. It accords with the Iowa researchers'
findings that operation part-full with a sediment bed is not
stable, because an increased depth of deposit reduces the
transport capacity.

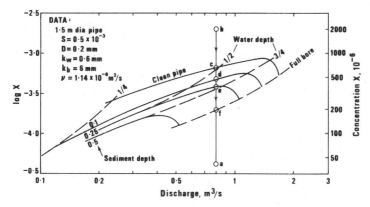

FIG.1   **TRANSPORT IN SEDIMENTED PIPE:
FREE SURFACE AND NO SURCHARGE**

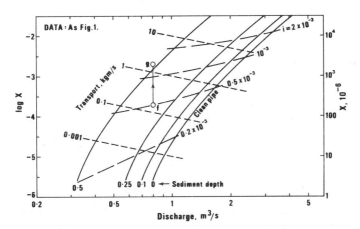

FIG.2   **TRANSPORT IN SEDIMENTED PIPE:
FULL BORE WITH SURCHARGE**

TENTATIVE DESIGN CONCLUSIONS

Measurements from a particular overseas city drainage system suggest that the values of k_W and k_b used in the above exampl were not realistic for that case; values of k_W = 3 mm and k_b = 30 mm would be more appropriate. The combination of Colebrook and White with Ackers and White can in theory be used for any chosen conditions, and these roughness values were applied to the cases of no deposition and 0.1 proport-ional deposit, with a sediment size of 0.4 mm (based on actua samples). It is then possible to plot, as in fig 3, a design chart in terms of required gradient against pipe-full discharge, with lines indicating pipe diameter and sediment concentration. Entering fig 3 with a known pipe full discharge and sediment concentration determines the <u>maximum</u> pipe diameter and <u>minimum</u> gradient acceptable.

These overall results can also be converted to simple exponential formulae that provide a close approximation to the full solution.

$$D_o = C_1 Q_o^{0.43} \tag{13}$$

$$V_o = C_2 D_o^{0.325} \tag{14}$$

where C_1 and C_2 are tabulated below (D_o in metres, V_o in m/s, Q_o in m^3/s).

TABLE 1. Coefficients in tentative design criteria.

$10^6 X$ (mg/l)	Clean pipe C_1	C_2	0.1 sediment bed C_1	C_2
100	1.36	0.62	1.46	0.56
200	1.26	0.74	1.36	0.65
500	1.16	0.90	1.23	0.83

It must be stressed that these can not be regarded as definitive equations for design. Further investigation is required to confirm, or establish, a general theory applicabl to the full range of conditions likely to be found in sewers and storm drains. It is interesting, however, that the concept of a single value for a "self cleansing" velocity is not upheld by this theoretical analysis. It increases by 50 percent for a 5-fold increase in sediment concentration; and by 25 percent for each doubling of pipe diameter.

COMPARISON WITH PREVIOUS RESEARCH

Because methods of presentation differ, it is not easy to

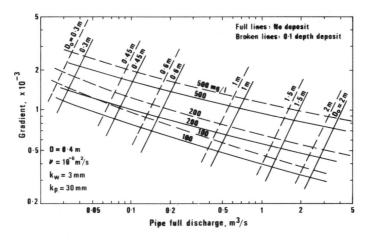

FIG. 3   **TENTATIVE DESIGN CRITERIA FOR PIPES TRANSPORTING SEDIMENT**

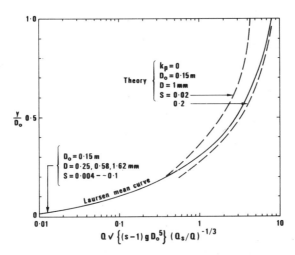

FIG. 4   **COMPARISON OF PRESENT THEORY WITH LAURSEN**

compare this theory with previous research, for example
that at Iowa and Newcastle, although some overall comparisons
have been possible.

Fig 4 shows the results of applying the present theory to
the Iowa pipe and sediment sizes. Laursen (1956) plotted

$$Q \left\{ (s-1)gD_o{}^5 \right\}^{-\frac{1}{2}} (Q_S/Q)^{\frac{1}{3}} = \text{function } (y/D_o) \qquad (15)$$

where Q_S/Q is the limiting volumetric sediment transport
concentration for deposition to be avoided and y/D_o is the
proportional depth of flow. The scatter of experimental data
was represented by a single curve. The present theory yields
different curves depending the assumed values of pipe roughness and slope. Two sample curves are shown in comparison
with Laursen's average curve, confirming the general form
of the overall function with quite close agreement for D =
1 mm, $i \simeq 0.2$ with a smooth pipe and no deposition (Craven and
Ambrose used some very steep gradients). A similar comparison was also made with Laursen's curve for full-bore flow
with a sediment bed, agreeing with the trend of the data.

Abdelrahman,(1977) used the Novak and Nalluri equation for
established transport (equ. 28 of N and N, 1976, rather than
equ. 25, which gives appreciably different results) with the
Manning (Crimp and Bruges) equation for friction loss to
establish design curves on a similar framework to fig. 3.
Using the same equation to calculate transport rates for
hydraulic conditions taken from fig 3 yields the following
reasonably close comparison

TABLE 2. Comparison of present theory with Novak and Nalluri.

Pipe diameter (m)	0.6			1.0			2.0		
Gradient (10^{-5})	138	90	70	103	66	50	74	45	32
Discharge (l/s)	208	167	147	700	550	480	3620	2830	2400
Sediment concentration, 10^{-6}									
A and W/C and W	500	200	100	500	200	100	500	200	100
N and N (equ. 28)	880	470	320	680	350	230	550	260	150

CONCLUSION

The combination of a friction law and a sediment transport
law defines the performance of a sewer or storm-drain carrying
sediment. Particular equations have been tested in this way
and demonstrate that certain observed features can be predicted theoretically. The method has been used to give
tentative design criteria for steady flow, but these are
subject to revision in the light of research results and

field information, which should improve not only the understanding of the physical process but also provide more reliable coefficient values. Comparison of the theoretical calculations with laboratory transport data has been very limited but appears promising. It is hoped that discussion of the relevance of available theoretical and empirical methods to practical design will be stimulated by the paper.

ACKNOWLEDGEMENTS

The writer thanks his colleagues G. Thompson and J. Loveless for computer programming and supplementary calculation; and also acknowledges the help provided by discussions at HRS Wallingford and Newcastle University.

REFERENCES

Ackers, P and White, W.R. (1973), Sediment transport: new approach and analysis. Proc. ASCE, 99, HY11, 2041-2060.

Abdelraham (1977) Incipient sediment motion on fixed smooth beds and critical assessment of the sewers hydraulic design formulae, unpublished report, Dept. Civ. Eng'g, University of Newcastle upon Tyne.

Ambrose, H.H. (1953), The transportation of sand in pipes, II, free-surface flow, Proc. 5th Hydraulics Conf. Bulletin 34, Studies in Engineering, State Univ., Iowa, USA, 77-88.

B.S.I. (968), Sewerage, Code of Practice CP2005, British Standards Instn, London.

Colebrook, C.F. (1939), Turbulent flow in pipes, with particular reference to the transition region between the smooth and rough pipe laws, J. Instn. Civ. Engrs, II, 133.

Craven, J.P. (1953), The transportation of sand in pipes, I., full-pipe flow. Proc 5th Hydraulics Conf, Bulletin 34, Studies in Engineering, State Univ. Iowa, U.S.A. 67-76.

Laursen, E.M. (1956), The hydraulics of a storm-drain system for sediment transporting flow, Bull. No. 5, Iowa Highway Research Board.

May, R.W.P. (1975), Deposition of grit in pipes: literature survey, INT 139, Hydraulics Research Station, Wallingford.

Novak, P. and Nalluri, C. (1975), Sediment transport in smooth fixed bed channels. Proc. ASCE, 101, HY9, 1139-1154.

White, W.R. Milli, H., and Grabbe, A.D. (1975), Sediment transport theories: a review, Proc. Instn. Civ. Engrs, Part 2, 59, 265-292.

A SUSPENDED SOLIDS MODEL FOR STORM WATER RUNOFF

R K Price, Hydraulics Research Station, Wallingford, UK
G Mance, Water Research Centre, Stevenage, UK

ABSTRACT

A model to predict the runoff of suspended solids from an urban storm water catchment has been developed as an extension of the water quantity model which forms the basis of a new design and simulation method. The suspended solids model assumes separate daily deposition rates on paved and roof surfaces and a constant input per rainfall event. Time-dependent erosion, transport and runoff of material in suspension on each surface are modelled to give the input to the sewer system in which the material is transported to the outfall. The model is applied to observed events on a catchment at Stevenage and results indicate that the model provides a realistic description of runoff.

INTRODUCTION

Storm runoff from urban areas can adversely effect water quality in receiving streams. Considerable research, particularly in the USA, has gone into statistical analysis of water quality data to quantify pollution loads for more effective water resource planning. With the advent of powerful computers the deterministic modelling of storm water runoff quality has become increasingly attractive, and models such as the SWMM (Metcalf and Eddy Inc et al 1971) and STORM (Roesner et al 1973) have been developed. These models enable the research engineer to simulate the behaviour of pollutants in storm sewer systems and provide the planner with objective guidance in decision making.

Both SWMM and STORM assume that the runoff of a particular pollutant from a given area is directly related to the instantaneous discharge from that area and the amount of pollutant available. Whereas this assumption gives a reasonable indication of water quality it can lead to considerable errors. Consequently there is value in studying a more detailed prediction of runoff from a reasonably small urban area to see if the prediction can be refined. The purpose of this paper is to describe an experimental numerical model which will give further insight into the detailed runoff of pollutants, and in particular suspended solids, from an urban area.

A difficulty in developing such a model has been the lack of laboratory studies on certain phenomena such as the pick-up and/or erosion of particulate solids from a surface under rainfall. Consequently a number of calibration parameters have had to be included within the model. The evaluation of these parameters has therefore depended on the

546

availability of good quality data from an urban stormwater sewer. Fortunately such data have been collected from the Shephall catchment in Stevenage New Town since 1975. In addition this catchment has been used in the development of the quantity model which forms the basis of a new design method for storm sewers; see Price and Kidd (1978). Therefore, the simulation model for suspended solids has been conceived as a further development of the quantity model on which the new design method is based. A particular feature of the quantity model is the separation of the above and below ground phases in urban runoff. The following description of the suspended solids model highlights this distinction.

SURFACE RUNOFF MODEL

Consider a storm sewer network consisting of a convergent tree network of pipes each with an impermeable contributing area of the order of 500 m^2. Suppose that each contributing area has a particulate solids loading of $\overline{m}_s$ per unit area. This mass, together with a particulate solids loading of $\overline{m}_r$ per unit area from rainfall, is available for erosion, deposition and transport into the pipe system.

The runoff of suspended solids is directly dependent on the runoff of rainwater. A version of a particular rainfall/runoff model developed by Kidd (1978), and described by Price and Kidd (1978), is used to predict suspended solids runoff. This model is based on the non-linear reservoir equation describing the runoff, q (in mm/hr) from a unit area, giving

$$\frac{dS}{dt} = i - q \qquad \qquad ....(1)$$

$$S = Kq^n. \qquad \qquad ....(2)$$

Here S is the storage, i is the rainfall intensity, K is the storage constant and n is the non-linearity parameter. It is assumed that no flow occurs before the rainfall has filled a depression storage, h, which is uniformly distributed over the area. The total runoff, Q, from the area is given by

$$Q = Axq \qquad \qquad ....(3)$$

where A is the area and Ax is the proportion of A contributing to the runoff. However, in developing the suspended solids model it was recognised that there would be a need to define a realistic mean velocity for the overland flow. Consequently the runoff was interpreted as coming off a conceptual rectangular area with flow in the direction of the side of length $A^{\frac{1}{2}}x$. Equations (1), (2) and (3) now become

$$\frac{dS'}{dt} = A^{\frac{1}{2}}xi - q' \qquad \qquad ....(4)$$

$$S' = K'A^{\frac{1}{2}}xq'^n \qquad \text{and} \qquad ....(5)$$

$$Q = A^{\frac{1}{2}}q' \qquad \qquad ....(6)$$

where $K' = KA^{n/2}x^n$ as suggested by Price and Kidd (1978) and $q' = A^{\frac{1}{2}}xq$.

Consider next the erosion of particulate solids from an impermeable surface. There are two important phenomena to be modelled: the disturbance of particles by the impact of raindrops on the surface and the entrainment of particles due to the shear stress generated by the flow of water over the surface. These phenomena can be modelled by assuming that the rate at which particles are lifted into suspension by the raindrops is directly proportional to some power of the rainfall intensity, and the entrainment of particles is proportional to the excess of the shear stress, τ, over some critical stress, τ_{ce}. Therefore, the rate of erosion of particulate solids from the surface can be defined by $a_i i^{\gamma} + a_e \lambda_e (\tau - \tau_{ce})$ where $\lambda_e = 1$ for $\tau \geqslant \tau_{ce}$ and $\lambda_e = 0$ for $\tau < \tau_{ce}$, and a_i and a_e are constant parameters. It is understood that no erosion of particulate solids can occur if the total deposit has been removed.

There will also be a deposition of particles. The rate of deposition will depend on a variety of factors such as the turbulence intensity, the shear stress and the concentration. For simplicity it is assumed that the rate of deposition is proportional to the excess of some critical shear stress, τ_{cd}, over the actual shear stress, giving $a_d \lambda_d (\tau_{cd} - \tau)$ where $\lambda_d = 0$ for $\tau \geqslant \tau_{cd}$ and $\lambda_d = 1$ for $\tau < \tau_{cd}$, and a_d is a constant parameter.

Consider now a strip of surface of unit width with flow of water along the strip. From Equation (5) the effective instantaneous mean depth of water is $Kq'^n + h$, where h is the storage depth as defined above. If M is the instantaneous mass of suspended solids per unit area then the rate of decrease of mass due to transport off the strip is $- Mq/(Kq'^n + h)$. Consequently conservation of mass gives the equation

$$\frac{dM}{dt} = a_i i^{\gamma} + a_e' \lambda_e (\tau - \tau_{ce}) + a_d \lambda_d (\tau_{cd} - \tau) - \frac{dm_r}{dt} - \frac{Mq'}{Kq'^n + h} \qquad ....(7)$$

where $- dm_r/dt$ is the rate of input per unit area of particulate solids by the rain. Here i is in mm/hr, M is in g/m^2, h is in mm and t is in hrs.

If the rate of input of particulate solids by the rain is assumed to be proportional to the remaining mass in the air and the rainfall intensity, then

$$\frac{dm_r}{dt} = - \frac{m_r i}{t_r i_r} \qquad ....(8)$$

where $t_r i_r$ is some fixed depth of rainfall. Equation (8) gives

$$m_r = \bar{m}_r \exp \left\{ - \int_o^t \frac{idt}{t_r i_r} \right\} \qquad ....(9)$$

where $\bar{m}_r$ is the initial mass of particulate solids in the atmosphere per unit area of surface.

It remains to define τ for flow over the surface. For very small flow depths over a rough irregular surface τ can be defined in terms of a mean velocity gradient. The mean velocity is $q'/(Kq'^n + h)$, so

$$\tau = \frac{kq'}{(Kq'^n + h)^2} \qquad ....(10$$

Rewriting Equation (7):

$$\frac{dM}{dt} = a_i i^\gamma + a_e \lambda_e \left[\frac{kq'}{(Kq'^n+h)^2} - \tau_{ce}\right] + a_d \lambda_d \left[\tau_{cd} - \frac{kq'}{(Kq'^n+h)^2}\right]$$

$$+ \frac{m_r i}{t_r i_r} \exp\left\{-\int_o^t \frac{idt}{t_r i_r}\right\} - \frac{Mq'}{Kq'^n+h} \qquad ....(11)$$

and the rate of mass/runoff from an area is given by $A^{\frac{1}{2}}x \, Mq'/(Kq'^n+h)$.

Equations (4), (5) and (6) are solved as described by Price and Kidd (1978) with the solution for q' used in Equation (11). This equation is first solved in integral form to find M; the rate of runoff of mass is then evaluated. To reduce computing time the runoff of flow, and therefore mass, is calculated for standard paved and roof areas per gully. The runoff from an actual contributing area for an individual pipe is then found by adding the instantaneous runoffs from the paved and roof areas deduced as appropriate multiples of the runoffs from the standard areas.

PIPE ROUTING MODEL

The transport of suspended solids in a pipe is primarily a function of the mean flow velocity with variations from the mean velocity across a section leading to a longitudinal diffusion of the solids. In the present model the flow is calculated using a one step variable parameter version of the routing method. This method uses an explicit finite difference solution of the equation

$$\frac{\partial}{\partial t} \int^Q \frac{dQ}{c(Q)} + \frac{\partial Q}{\partial x} = 0 \qquad ....(12)$$

such that the truncation error for the finite difference scheme is $[a(Q)/c(Q)] \partial^2 Q/\partial x^2$, where this term is small compared with the terms in Equation (12); see Price (1977). Here $c(Q)$ is defined by

$$c(\bar{Q}) = \frac{1}{B} \frac{d\bar{Q}}{dy} \qquad ....(13)$$

where $\bar{Q}$ is defined using the Colebrook resistance equation:

$$\bar{Q} = A(32gRS)^{\frac{1}{2}} \log_{10}\left(\frac{14.8R}{k_s}\right) \qquad ....(14)$$

A(y) is the cross-sectional area of flow for depth y, B(y) is the surface width, R(y) is the hydraulic radius, g is the acceleration due to gravity, s is the bottom slope and k_s is the boundary roughness length. Similarly $a(Q)$ is defined by

$$a(\bar{Q}) = \frac{\bar{Q}}{2\bar{B}s} \qquad ....(15)$$

If the dispersion of suspended solids in pipe flow is ignored the equation describing the convection of the solids is

$$\frac{\partial}{\partial t}(CA) + \frac{\partial}{\partial x}(CQ) = 0 \qquad \qquad ....(16)$$

where C is the concentration of the solids. However, to ensure conservation of solids it is better to work with the rate of mass flow, $\dot{m}$, rather than C. Rewriting Equation (16) gives

$$\frac{\partial}{\partial t}\left(\frac{\dot{m}}{V}\right) + \frac{\partial \dot{m}}{\partial x} = 0 \qquad \qquad ....(17)$$

where V is the mean velocity of flow. Equation (17) is solved using an explicit finite difference scheme with conditions to ensure that $\dot{m} \geq 0$ and $V > 0$. It is assumed that there is no storage of solids at manhole junctions.

MODEL APPLICATION

The model has been applied to the experimental catchment at Stevenage. This catchment is residential and has an area of 142.55 ha. The associated storm sewer system drains to a single outfall where the depth of flow is recorded continuously by an Arkon recorder. This depth is converted to discharge using a rating curve which has been calibrated by dilution gauging up to 0.24 proportional depth. A flow activated sampling system has been installed at the point of depth measurement and this has enabled the variation in time of the concentration of suspended solids to be estimated. A description of the measuring equipment and the interpretation of the results is given by Mance and Harman (1978).

Initial runs with the model showed that the mass runoff from contributing areas was too late with high concentrations on the recession of the runoff hydrograph. Two main reasons were deduced for this deficiency in the model: over-simplification of the rainfall/ runoff relationship and the non-uniform distribution of the particulate solids. Pending further research, the rainfall/runoff model had to be retained as it was. However, the non-uniform distribution of particulate solids was included by adjusting the concentration in the runoff from a contributing area by an arbitrary factor. This factor can be interpreted in terms of the proportional length, ℓ_s, of solids deposited on the conceptual strip in the rainfall/runoff model. So the rate of decrease of mass due to transport off the strip becomes $- Mq'/\ell_s(Kq'^n+h)$.

Further runs with the model demonstrated that with $\ell_s = 0.15$ the erosion of particulate solids by the shear stress of overland flow alone produced peak concentrations and mass flow rates at the outfall which lagged behind observed peaks. However, when the erosion of particulate solids was made dependent on the impact of raindrops on the surface, peak concentrations occurred at about the right time. Consequently, within the limitations of the present model the erosion of particulate solids can be regarded as being due primarily to the impact of raindrops. There still remains, however, the possibility that a better quantity model for surface runoff with a more realistic distribution of particulate solids carried by wind and deposited during previous storm events could show that erosion by the shear stress is more important than indicated by the present model. This dilemma needs to be investigated further, preferably under laboratory conditions.

Adjustment of the deposition parameters showed that this phenomenon is not important but it does provide a useful refinement when calibrating the model.

Two events with good quality data were available during the preparation of this paper, namely those of 4.3.75 and 23.6.75. The first event was more interesting in that it included several runoff peaks. Consequently, it was selected for calibrating the model, while the second event was reserved for the verification stage.

In both events the loading of particulate solids was determined from the observed amount of total solids washed off and distributed between the roof and paved areas in the proportions indicated by results obtained in the field. A number of runs of the model were then carried out for the first event, changing the parameters until the best fit was obtained. From the discussion above the most significant parameters were a_i and γ (see Equation (11)) and the final values for these parameters were $a_i = 0.4$ and $\gamma = 1.5$. The values of all the relevant parameters are summarised in Table 1.

TABLE 1 Parametric values

Standard paved area	*Standard roof area*
$A = 300.0$ m^2	35.0 m^2
$x = 0.204$	0.679
$K = 0.12$	0.07
$h = 1.2$ mm	0.4 mm
$n = 0.6$	0.6

$$a_i = 0.4$$
$$\gamma = 1.5$$
$$ka_e = 2.0 \text{ gm}$$
$$\tau_{ce}/k = 2.5 \text{ /hr}$$
$$ka_d = 2.5 \text{ gm}$$
$$\tau_{cd}/k = 2.0 \text{ /hr}$$
$$t_r i_r = 1.5 \text{ mm}$$

The observed and predicted hydrographs and pollutographs for the first event are shown in Fig 1. Note that the difference in timing of the last peak is due primarily to timing errors in the gauge records.

A study of the recorded and predicted hydrographs and pollutographs and the corresponding predicted rainfall/runoff hydrographs for the standard paved and roof areas (see Fig 2) shows that the first small peak in the outfall hydrograph is due almost entirely to runoff from the roofs. In addition, nearly all of the particulate solids are washed off the roofs during the first 40 min. The runoff of suspended solids from the paved areas reflects the rainfall intensity, though the last peak at 150 min is limited because of the small amount of particulate solids remaining for erosion.

Finally, the model was applied to the second event as shown in Fig 3. This has a single runoff peak which is over-predicted by the model. The peak of the mass flow rate of suspended solids is similarly over-predicted, but the errors are small and acceptable.

CONCLUSIONS

A runoff quality model for suspended solids has been developed and calibrated. Initial simulations indicate a very good agreement between predicted and observed mass flows. Roof runoff has a more immediate impact on quality than road runoff, although the latter contributes the greater pollutant load. Within the limitations of the model the

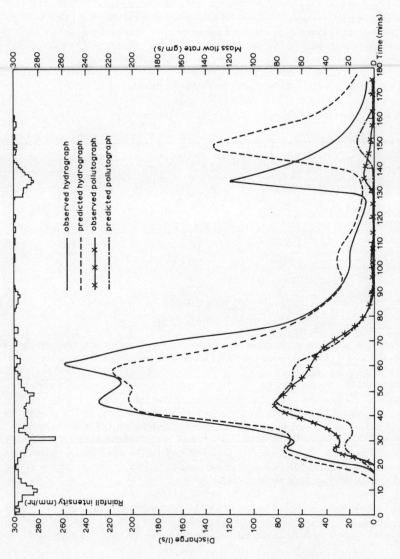

Figure 1. Observed and predicted hydrographs and pollutographs for event 4/2/75

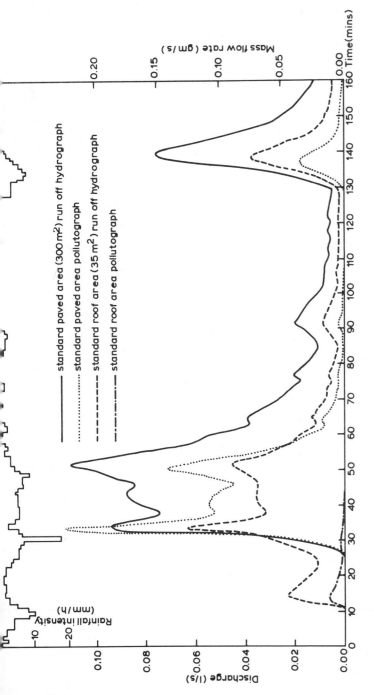

Figure 2. Observed and predicted run off hydrographs from standard paved and roof areas for event 4/3/75

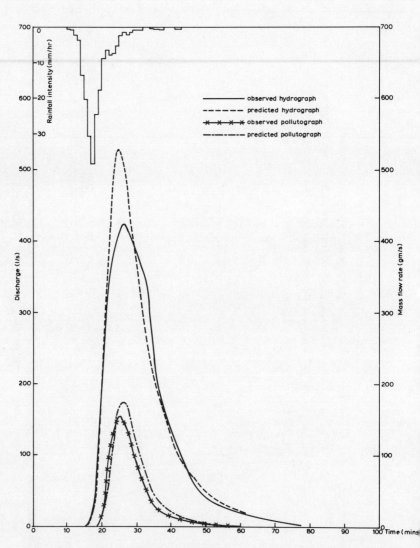

Figure 3. Observed and predicted hydrographs and pollutographs for event 23/6/75

impact of raindrops is more significant than the shear stress due to overland flow in eroding particulate solids. There is a need for further testing and refining of these conclusions both in the laboratory and in the field. In particular laboratory research should help to define more closely the functional form for the erosion of solids by the impact of raindrops.

The computer program for the model was developed on an ICL 1904S computer at the Hydraulics Research Station. The program uses 54K of store and takes about 15 min computing time to simulate the two events above.

ACKNOWLEDGEMENTS

The work described in this paper forms part of the research programmes into urban drainage at the Hydraulics Research Station and the Water Research Centre. The paper is published with the permission of the Director of Hydraulics Research and the Director of the Water Research Centre.

REFERENCES

Kidd, C. H. R. (1978) A calibrated model for the simulation of the inlet hydrograph for fully sewered catchments. International Conference on Urban Storm Drainage, Southampton.

Mance, G. and Harman, M. M. I. (1978) The quality of urban storm water runoff. International Conference on Urban Storm Drainage, Southampton.

Metcalf and Eddy Inc et al (1971) Environmental Protection Agency Storm Water Management Model, Vols I, II, III and IV. Report Nos 11024 DOC07/71, 08/71, 09/71 and 10/71, Environmental Protection Agency, Washington.

Price, R. K. (1977) FLOUT − a river catchment flood model. Hydraulics Research Station Report, IT 168.

Price, R. K. and Kidd, C. H. R. (1978) A design and flow simulation method for storm sewers. International Conference on Urban Storm Drainage, Southampton.

Roesner, L. A., Monser, J. R. and Evenson, D. E. (May 1973) Computer program documentation for the stream quality model QUAL-II (STORM). Prepared for the Environmental Protection Agency, Systems Development Branch, Water Resource Engineers, Walnut Creek, California, 84p.

SEDIMENT CONTROL FROM URBAN CONSTRUCTION AREAS IN THE U.S.

B.J. Barfield, T.Y. Kao and C.T. Haan

University of Kentucky, Lexington, Kentucky

ABSTRACT

The problems caused by eroded sediment in urban areas are different from those in rural areas. Urban residents normally perceive the sediment problem as essentially a nuisance although damages do occur which result in economic loss to government as well as individuals.

The main procedure for making predictions of the magnitude of eroded sediment under varying control techniques is the Universal Soil Loss Equation along with a sediment delivery ratio. Simulation models have been developed but the complexity of the inputs required limits their applicability. The Universal Soil Loss Equation is discussed in detail.

The main system used for trapping eroded sediment is the sediment detention reservoir. A new DEPOSITS model, which has potential for rational design of detention reservoirs, is described. Other control techniques are discussed.

INTRODUCTION

Sediment can be eroded when soil is exposed to rainfall energy and flowing water regardless of whether the location is urban or rural since the same physical laws apply. The problems generated by the eroded sediment, however, are somewhat different. In a rural setting, unchecked sediment movement from an agricultural field results in a potential loss of fertility and productivity from the field. In addition, damages are done downstream. At an urban construction site, the loss of sediment from the construction area has essentially no economic impact on the site itself, but considerable offsite damage can occur. In the rural setting, therefore, there is an inherrent long term benefit to controlling erosion whereas at a development site, erosion control is essentially a liability to the developer while the sediment produced may be a liability to the community as a whole. Another difference between the sediment problems in urban and rural areas lies in its duration. In rural areas, sediment production is a perennial problem due to tillage operations which occur repeatedly throughout the year on an annual basis. At an urban construction site, land disturbance normally lasts only for a period of a few months after which the surface is rapidly stabilized by either permanent vegetation or paving.

The techniques of analysis and management of soil erosion from rural sites are well developed and tested. Unfortunately, not all of the techniques are directly applicable to urban sites. In recognition of this difference, sediment control was included as a major topic in a series of four symposiums held at the University of Kentucky starting in 1974. (See Haan et al., 1977 for details of the symposiums.) The presentations have included research papers dealing with new concepts as well as papers on application and case studies illustrating actual practices used by consultants and government agencies to analyze and solve sediment problems. This paper is an overview of the techniques presented at the symposiums as well as other analysis and control techniques used in the United States. It is not intended to be an exhaustive literature citation, but represents the author's biases as to the more important techniques.

THE URBAN SEDIMENTATION PROBLEM

Over 1,000,000 acres in the U.S. are denuded each year in construction (Meyer, 1974) yielding quantities of eroded sediment much higher than that from rural areas. Estimates range from as high as 1000 times as much (EPA, 1976) to around 100 times. Comparative data are shown in Figure 1. The eroded sediment causes relatively little damage to the site itself, however, it can muddy streets increasing street cleaning costs, clog storm sewers, and fill reservoirs and streams. For a developer to allow unchecked erosion from a construction site represents a serious abuse of the rights of those who must suffer the consequences.

Coincident with the increased sediment production during construction, there is an increase in runoff in areas where construction is completed. The combination of these two factors causes a change in the morphology of unimproved drainage channels which convey urban storm drainage. Diniz and Moore (1974) summarize results from Guy (1970) showing the effects of changing land use on both channel stability and sediment yield. (See Table 1) During periods

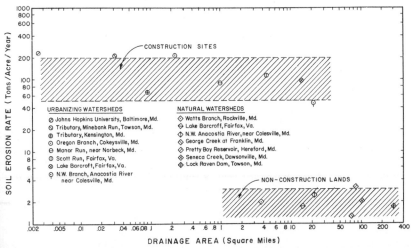

Figure 1. Soil Erosion Rates from Construction and Non-Construction Lands (Chen, 1974).

Table 1
Effect of Land Use Sequence on Relative Sediment Yield
and Channel Stability From Diniz and Moore (1974)

Land Use	Sediment Yield	Channel Stability
A. Natural forest or grassland	Low	Relatively stable with some bank erosion
B. Heavily grazed area	Low to moderate	Somewhat less stable than A
C. Cropping	Moderate to heavy	Some aggradation and increased bank erosion
D. Retirement of land from cropping	Low to moderate	Increasing stability
E. Urban construction	Very heavy	Rapid aggradation and some bank erosion
F. Stabilization	Moderate	Degradation and severe bank erosion
G. Stable urban	Low to moderate	Relatively stable

of channel instability, stream bank erosion adds considerably to sediment yield.

The increased runoff from urbanization is a problem which must be managed indefinitely whereas the sedimentation problem is relatively short lived. Consequently there is a tendency to ignore the sediment problem and emphasize control of urban stormwater. For example, Ward et al. (1977) point out that most detention structures built to control runoff during construction are designed for flood control and not sediment control.

In addition to the short run aspect of the urban sediment problem as compared to the urban runoff problem, the nature of the damage resulting from eroded sediment is different from that of runoff. The damage from eroded sediment which the urban individual directly attributes to soil erosion is essentially nuisance type disturbance whereas increased runoff can result in property damage which the landowner directly attributes to increased runoff. Although eroded sediment clogs streams and contributes to flooding, the affected individuals do not normally make the direct connection. As a result of these differences, most cities have an ordinance dictating flood control measures associated with urbanization whereas fewer have ordinances dictating sediment control measures.

THE EROSION PROCESS

Soil erosion by water involves detachment of the soil, transport downslope, and subsequent deposition (Meyer, 1974). Detachment is caused by raindrop impact and the shear forces of flowing water. Detached soil comes from rill and interrill areas. Rills are essentially small channels through which the runoff is conveyed.

Interrill erosion results primarily from the impact of raindrops while rill erosion is caused both by raindrop energy and by the tractive force of the flowing water.

A complex interrelationship exists between the components of the processes that cause detachment, transport, and deposition. Meyer (1974) divides the soil erosion process into four subprocesses; 1) detachment by rainfall, 2) detachment by runoff, 3) transport by runoff, and 4) transport by rainfall. The interrelationship between processes are shown schematically in Figure 2.

Based on the Meyer model of erosion, it can be seen that the amount of sediment transported can never be greater than the transport capability of the conveying channel. As is quite often the case, the amount of sediment eroded from a site exceeds the capacity of the conveyance channels to transport materials resulting in deposition in the channels. Therefore, the sediment yield from a watershed is frequently less than the sum of the erosion from each of the sub-watersheds. This makes the problem of predicting sediment yield more complicated than simply predicting the amount of soil eroded over the entire watershed. The methods used in the U.S. to predict soil erosion and sediment deposition will be discussed in subsequent sections.

MODELING THE SEDIMENTATION PROCESS

Soil Erosion
Soil Erosion as used in this report refers to the soil moved from a given site or field which can be represented by a given set of physiographic and topographic features. Sediment yield from a watershed is the sum of the soil erosion from the entire watershed minus that deposited on field boundaries. The two methods used to analyze soil erosion are the Universal Soil Loss Equation which is basically a regression model and simulation. The Universal Soil Loss Equation is widely used in predicting soil erosion in urban areas and will be discussed more thoroughly than simulation.

The Universal Soil Loss Equation. The Universal Soil Loss Equation was developed by Wischmeier and Smith (1965) to be used as an aid in planning conservation

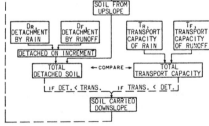

SOIL EROSION PROCESS

Figure 2. Schematic of the procedure used to simulate the process of soil erosion by water. Four sub-processes are evaluated on each successive slope-length segment, and soil movement is routed downslope as illustrated (Meyer, 1974).

measures for farmland. The basic equation is

$$A = RKLSCP \tag{1}$$

where A is the average soil loss in tons/acre for the time interval represented by the factor R.

R is a factor which gives the erosive power of rainfall and runoff. In the U.S. it is normally given as the product of the rainfall energy and the maximum thirty minute intensity of rainfall divided by 100 (EI_{30} index), and has the units of ft tons/acre or joules/m^2. Values have been tabulated for average annual, individual storm, and probability levels for 181 stations in the United States (Wischmeier and Smith, 1965). An example is given in Table 2. Average annual values have also been distributed according to the percentage occurring in an individual month. Examples for Kentucky are shown in Table 3. Limitations of the EI_{30} index have been discussed by Hudson (1971). He reports good results in Africa when using an index which is the kinetic energy of rainfall for intensities greater than 25 mm/hr (1 in./hr).

K is the soil erodability factor which is a function of soil type and antecedent moisture conditions. Values have been determined experimentally for many sites around the world. Wischmeier et al. (1971) propose a nomograph for estimating K values as shown in Figure 3. Gilley et al. (1976) report troubles with the use of Figure 3 on highly disturbed soils such as strip mines.

LS is the length-slope factor which adjusts for the fact that soil erosion increases as slopes and lengths increase. Values for LS used in the U.S. are shown in Figure 4. The solid lines represent the range of experimental data and the dashed lines are projections. Care should be taken when using the dashed lines, particularly on long steep slopes. The slope length is the length from the point of origin of overland flow to a point where the flow enters a channel or where the slope decreases such that deposition occurs. If the slope is non-uniform, corrections should be made as indicated by Foster and Wischmeier (1974).

C is the cover factor which relates the effectiveness of soil cover in dissipating rainfall energy. Values of C have been summarized for many cropping sequences (Wischmeier and Smith, 1965) used in agriculture and for mulches used for stabilizing urban construction sites. The C value for mulches and for grasses is given in Table 4 and Figure 5. C values for tall weeds and for forested areas can be estimated from procedures outlined by Wischmeier (1975).

P is the management practice factor which relates in agricultural lands to such things as contour plowing, terracing, and strip cropping. With the exception of terracing, the practice factor should be set equal to 1.0 for urban construction sites. Chen (1974) proposed that P be set equal to one and that the C value be called the control-practice factor given by

$$C = C_s \cdot C_r \cdot C_t \cdot C_e \cdot C_o \tag{2}$$

where C_s = the control factor due to surface stabilizing or protecting treatments such as seeding, mulching, and netting

Table 2
Probability Rainfall Factors for Various Kentucky Locations*

Rainfall Factors

Location	Average Annual	Annual High One Year in		Single Storm High One Storm in	
		5 Years	20 Years	5 Years	20 Years
Lexington	175	248	340	80	151
Louisville	175	221	286	59	85
Middlesboro	150	197	248	52	73
Cairo, Ill.	250	349	518	101	173
Evansville, Ind.	200	263	362	55	86
Cincinnati, Ohio	175	211	299	48	69
Nashville, Tenn.	200	262	339	68	99
Huntington, W. Va.	150	173	233	49	89

* From Wischmeier and Smith, 1965.

Table 3
Rainfall Erosion Index Distribution Table by Months

	Western Kentucky* (R = 200; 250)	Eastern Kentucky* (R = 150; 175)
January	4 percent	3 percent
February	5 percent	4 percent
March	8 percent	6 percent
April	9 percent	6 percent
May	11 percent	8 percent
June	13 percent	17 percent
July	14 percent	20 percent
August	13 percent	16 percent
September	8 percent	9 percent
October	6 percent	4 percent
November	5 percent	3 percent
December	4 percent	4 percent
	100 percent	100 percent

* From Wischmeier and Smith, 1965.

Table 4
Control Factor, C_s for Vegetative Covers**

Type of Cover	Estimated Value of C_s
None (Denuded Condition)	1.0
Temporary seeding (90% stand):	
First month	0.6*
Second month	0.4*
Remainder of first year:	
Annual ryegrass	0.1
Perennial ryegrass	0.03 - 0.05
Small grains	0.05
Millet or sudangrass	0.05
Field bromegrass	0.03
Permanent grasses (90% stand)	0.01
Grass sod (laid immediately)	0.01

*   When seeded ground is mulched, use the lower value of C_s for mulch
       (Figure 4) or for grass (Table 4).

**   after (Chen, 1974)

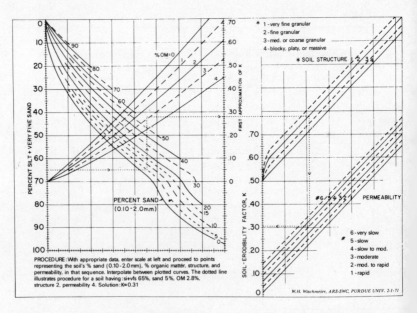

Figure 3. Nomograph for Erodibility Factor (Wischmeier et al., 1971).

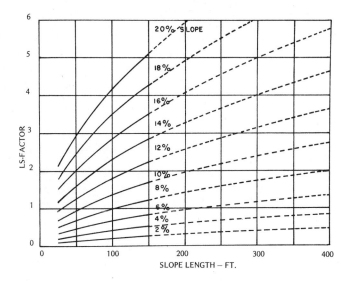

igure 4. Topographic effect graph used to determine LS-factor values for different combinations of slope steepness and slope length (from Wischmeier and Smith, 1965).

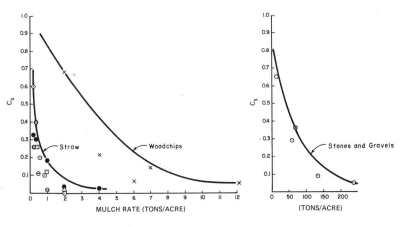

Figure 5. Control Factor, C_s, of Surface Mulching (after Chen, 1974).

C_r = the control factor due to runoff-reduction practices such as diversion berms, interceptor dikes, benches and terraces, sodded ditches, level spreader and sectional downdrains

C_t = the control factor due to sediment trapping measures such as the sediment basins or basins with chemical flocculants

C_e = the control factor due to restricting the spatial and/or temporal exposure of the denuded site to the rainfall and runoff erosion

C_o = the control factors associated with other practices that are not included herein.

He further presents procedures for determining each of the factors.

The Universal Soil Loss Equation has been used under many different circumstances with varying degrees of accuracy. A good summary of these uses is given in a book published as a tribute to Wischmeier (Soil Cons. Soc. of Amer., 1976).

Simulation. Simulation of soil erosion using overland flow transport equations is a concept that has arisen in recent years. Numerous models have been presented with varying degrees of sophistication and complexity. All of the models must address the following issues:

(1) Detachment of sediment by both flow and rainfall energy.
(2) Transport of sediment by runoff.
(3) Deposition of sediment when the amount available exceeds the transport capacity.

Sediment detachment by rainfall can be modeled as a function of rainfall energy (Bubenzer and Jones, 1971; Wischmeier and Smith, 1965), intensity of rainfall (Meyer and Wischmeier, 1969; Moldenhauer and Long, 1964) or depth of rainfall (Negev, 1967). Sediment detachment by flowing water has been modeled as a function of the bed shear stress in excess of the critical shear for initiating motion (Foster and Meyer, 1975). As shown in Figure 2, the sediment transported downslope at any point is the lesser of either the transport capacity or the total amount of detached material to be transported. If the amount of soil available for transport exceeds the transport capacity, then deposition occurs. Modeling of deposition is an important attribute of any erosion model since it allows one to extend the Universal Soil Loss Equation or other erosion models to predict sediment yield.

Although the simulation of erosion by basic erosion mechanics has led to a greater understanding of the erosion processes, the models require input data that are not routinely available. Hence, the Universal Soil Loss Equation is still the basic equation most often used for analyzing soil erosion.

Sediment Yield

Sediment yield is simply the amount of sediment eroded minus the amount deposited in grass, field boundaries, conveyance channels, etc. Neibling and Foster (1977) divide sediment yield predictive techniques into three categories

(1) Methods based on statistical analysis.

(2) Modified versions of the Universal Soil Loss Equation with a delivery ratio.

(3) Simulation models based on erosion mechanics and sediment transport relationships.

Since urbanization is a dynamic process, it is not feasible to collect a data base for statistical analysis, therefore methods two and three are the techniques which are useful for predicting urban sediment yield.

Delivery Ratios. The sediment delivery ratio for a watershed is simply the ratio of soil eroded to that which exits from the watershed. Using this concept, sediment yield is then

$$SY = \text{Erosion} \times DR \tag{3}$$

where SY is the sediment yield and DR is the delivery ratio. Boysen (1974) and others have proposed that sediment yield in urban areas can be predicted by using the Universal Soil Loss Equation to predict erosion and then assuming a delivery ratio of one. This would be acceptable if one is concerned with long term sediment yields from areas where all conveyance channels were paved and all areas on which erosion occurs were adjacent to conveyance channels. In this case the infrequent heavy rains would tend to clean out streets, gutters, etc. and the long term delivery ratio would be equal to erosion. If one is interested in short term predictions or in modeling erosion where the runoff from construction sites must flow through an area where permanent deposition can occur, (i.e. grass strips, woodlands, etc.) then the delivery ratio will be less than one. Unfortunately, there are no well evaluated relationships available for predicting delivery ratios under these conditions.

Neibling and Foster (1977) propose a method for predicting the average annual sediment transport capacity through natural vegetation using factors available from the Universal Soil Loss Equation as well as factors derived from limited amounts of data. The analysis is as yet untested, but serves as the best estimate available to date. They assume that the transport capacity is proportional to flow energy and slope. They further assume that flow energy can be related to the EI_{30} index. Their final equation is

$$T_c = r \cdot R \cdot T \cdot L^2 \cdot S^B \cdot C_{ro} \cdot C_T \tag{4}$$

where T_c is the transport capacity for the period used in determining R, r is a proportionality constant relating EI_{30} units to Σ (runoff volume X peak discharge) for the period used in determing R, R is the rainfall factor in EI_{30} units in the Universal Soil Loss Equation, T is the transportability factor of the soil particles relating the ability of a particle to be transported, C_{ro} is the runoff reduction factor relating the effects of vegetation on runoff, B is a slope exponent which is a function of particle diameter and specific gravity, and C_T is a relative transport factor indicative of surface roughness and its influence on effective shear stress to transport soil particles. Graphical procedures are presented for determining these factors under a limited number of conditions.

Research at the University of Kentucky has also been oriented toward developing procedures for predicting deposition in vegetal filters (Kao et al.,

1975; Tollner et al., 1976; Barfield et al., 1977). The procedures are well developed for artificial media but are as yet untested on real grasses. Demonstration plots are planned for 1978 to evaluate the procedures under real conditions.

Simulation Models. Several simulation models have been developed using basic erosion mechanics and overland flow transport relationships to predict sediment yield (Simons et al., 1977; Williams and Berndt, 1976; David and Beer, 1975; Curtis, 1976; Smith, 1977; Li et al., 1977). All of these models describe the sedimentation process in great detail and provide insights into the sedimentation process. The inputs required are extensive and not normally available, which makes the models difficult to use as a predictive tool.

Reservoir Sedimentation

Sediment detention reservoirs are the most commonly used structures for trapping sediment eroded from a construction site. The trapping of sediment by a detention structure depends on the following factors (Ward et al., 1977):

(1) inflow sediment graph
(2) inflow hydrograph
(3) hydraulic characteristics of the basin
(4) basin geometry
(5) particle size distribution of the sediment
(6) settling characteristics of the basin

The most commonly used method for predicting the trapping efficiency of a reservoir is that of Brune (1953) relating trapping efficiency to the ratio of reservoir capacity to annual inflow. Other less well used relationships which are based on steady state flow are discussed by Ward et al. (1977). The problem with the relationship of Brune is that varying particle sizes are not accounted for. Since the fall velocity of all of the particles obviously affects the potential for trapping, this is a serious deficiency.

Ward et al. (1977) developed a non-steady state model known as DEPOSITS which accounts for variations of flow through the reservoir, particle fall velocity, reservoir geometry, and the type of outlet structure. The assumption was made that the particles settled at the terminal velocity and that a plug of flow maintains its identity. Account is also taken of the loss in storage due to deposition. Verification of the DEPOSITS model has been limited due to the limited experimental information available; however, the agreement between predicted and observed trapping has been excellent for those data analyzed.

Changes in Stream Morphology

As mentioned in a previous section, changes occur in sediment load and runoff volumes and rates as urbanization takes place. These changes result in altered channel morphology for natural drainage channels. Fox (1976) and Robinson (1976) give descriptions of the changing morphology and Apmann (1974) presents equations which can be used to predict the amount of channel erosion or aggradation.

METHODS OF REDUCING SOIL EROSION AND LIMITING SEDIMENT YIELD

A good summary of techniques for reducing soil erosion and sediment yield is given by the U.S. Environmental Protection Agency (1972, 1976). These techniques may be categorized as:

(1) Procedures to stabilize the soil surface.
(2) Procedures to stabilize drainage channels.
(3) Procedures to prevent sediment from entering drainageways.
(4) Procedures to trap sediment which has entered drainageways.
(5) Management schemes to limit the time that a denuded area is exposed to erosive forces.

Stabilizing the Soil Surface

The soil surface can be stabilized with straw mulches, woodchips, artificial mulches, chemical substances, and with vegetation. As shown in Figure 5, a straw mulch rate of 2 tons/acre reduces erosion by 90 percent. The advent of hydromulchers has made mulching much easier. Straw mulching is frequently used with a chemical binder such as asphalt on steep slopes.

Artificial mulches and chemical stabilizers are used in critical areas due to the high cost per unit area. Vegetation is the best permanent solution, but should normally be protected with a mulch to prevent erosion during early development.

Terraces or diversions are also used to stabilize the soil surface. The function of the diversions is to break up the slope length resulting in a reduction in erosion. The use of diversions can reduce erosion by up to 50 percent depending on the vertical interval between adjacent diversions.

Procedures to Stabilize Drainage Channels

Drainage channels can be stabilized to prevent erosion by stabilizing the channel sides and by stabilizing the channel bottom. Channel sides can be stabilized with vegetation or with rock or concrete. Channel bottoms can be stabilized with vegetation if flows are intermittent or with rock or concrete for continuous flows. Channel grades can be reduced by constructing drop structures.

Procedures to Prevent Sediment from Entering Drainageways

The best place to trap sediment is onsite. If this cannot be done by stabilizing the soil surface, then a trap can frequently be installed between the denuded site and a drainageway. Vegetative filters are frequently used. The width may vary from a few feet on each side around a storm inlet to a few hundred feet wide next to a major channel. Laboratory and field studies indicate that vegetative filters can trap up to 100 percent of incoming sediment.

Straw filters are finding increasing use. These filters are constructed of hay bales placed around the entrance to drainageways. Rock filters are also used in a similar manner. In recent years, fiber glass filters have also been used.

Procedures to Trap Sediment Which has Entered Drainageways

Sediment trap structures vary in sophistication from simple straw bale check dams in an intermittent drain to large reservoirs. Temporary check dams are constructed of straw bales, rock, gabions, and wood. These check dams reduce the flow rate and allow the larger particles to be deposited. Little information is available on their trap efficiency. Reservoirs, if properly designed and constructed, can trap up to 100 percent of incoming sediment (Ward et al., 1977).

Management Schemes to Limit Time of Exposure

By proper timing of construction operations, it is possible to limit exposure of denuded sites to those seasons where the probability of highly erosive rainfall is low (see Table 3). Chen (1974) illustrates a simple procedure for estimating the effect of timing of operations on erosion. Chen's procedure is a modification of that proposed by Wischmeier and Smith (1965).

Cost Effectiveness of Operations

Chen (1974) proposed methods of evaluating the cost effectiveness of various erosion techniques. Examples are shown in Figures 6 and 7 for 1974 prices in Maryland. By using these type curves and an allowable soil loss in the Universal Soil Loss Equation, it is possible to select the combinations of practices which will control sediment production to within allowable limits as shown in Figure 8. Chen (1974) shows how this can be translated into an average concentration of sediment in the runoff.

LEGAL CONSIDERATIONS

As governmental agencies propose legislation to limit eroded sediment, it is important for consideration to be given to methods of enforcement as well as the technology available to meet the standards. One type of control is known as performance standards. Under this type legislation, the engineer is allowed to use any type design available to meet the standards. An example would be the

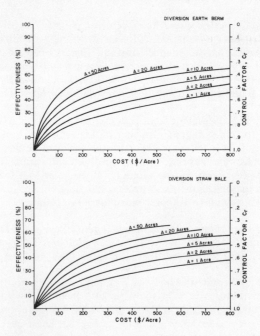

Figure 6. Typical Cost Effectiveness Curves of Diversion Practices
(From Chen, 1974)

U.S. EPA water quality standards for suspended solids which allow 30 mg/ liter average and 100 mg/liter maximum suspended solids. In order for the law to be properly enforced, an inspector must take runoff samples during storm events with a high return period. Since this is difficult to accomplish, the law is difficult to enforce.

An alternative to performance standards is known as a design standard. In this case a given design is required and the runoff can have any concentration of suspended solids as long as the system is properly installed. The design standard is easy to enforce but limits the ingenuity of the engineer. It also eliminates pressure on the developer to properly maintain the system. Both performance and design standards are used in the U.S.

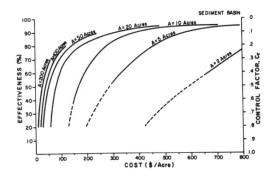

Figure 7. Typical Cost Effectiveness Curves of Basin Trapping Practice (From Chen, 1974)

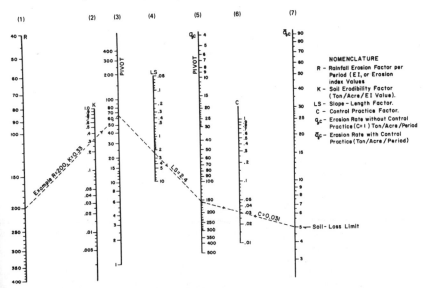

Figure 8. Nomograph for On-Site Erosion Control Planning (After Chen, 1974)

SUMMARY

The problems caused by eroded sediment in urban areas are different from those in rural areas. Urban residents normally perceive the sediment problem as essentially a nuisance although damages do occur which result in economic loss to government as well as individuals.

The main procedure for making predictions of the magnitude of eroded sediment under varying control techniques is the Universal Soil Loss Equation along with a sediment delivery ratio. Simulation models have been developed but the complexity of the inputs required limits their applicability.

The main system used for trapping eroded sediment is the sediment detention reservoir. A new DEPOSITS model, which has potential for rational design of detention reservoirs, is described. Other control techniques are discussed.

REFERENCES

Apmann, R.P. (1974) The Influence of Urbanization on Stream Channel Behavior. Proceedings 1974 National Symposium on Urban Rainfall and Runoff and Sediment Control. UK BU 106, College of Engineering, University of Kentucky, Lexington, Ky. pp. 183-189.

Barfield, B.J., E.W. Tollner and J.C. Hayes. (1977) Prediction of Sediment Transport in Grassed Media. ASAE Paper No. 77-2023. American Society of Agricultural Engineers, St. Joseph, Michigan.

Boysen, S.M. (1974) Predicting Sediment Yield in Urban Areas. Proceedings 1974 National Symposium on Urban Rainfall and Runoff and Sediment Control. UK BU 106, College of Engineering, University of Kentucky, Lexington, Ky. pp. 199-204.

Bubenzer, G.D. and B. Jones. (1971) Drop Size and Impact Velocity Effects on the Detachment of Soils Under Simulated Rainfall. Transactions of American Society of Agricultural Engineers, 14(4):625-628.

Brune, G.M. (1953) Trap Efficiency of Reservoirs. Transactions of American Geophysical Union, 34(3):407-418.

Chen, C.N. (1974) Evaluation and Control of Soil Erosion in Urbanizing Watersheds. Proceedings 1974 National Symposium on Urban Rainfall and Runoff and Sediment Control. UK BU 106, College of Engineering, University of Kentucky, Lexington, Ky. pp. 15-23.

Curtis, D.C. (1976) A Deterministic Urban Storm Water and Sediment Discharge Model. In Proceedings of National Symposium on Urban Hydrology, Hydraulics, and Sediment Control. UK BU 111, College of Engineering, University of Kentucky, pp. 151-162.

David, W.P. and C.E. Beer. (1975) Simulation of Soil Erosion − Part I. Development of a Mathematical Erosion Model. Transactions of American Society of Agricultural Engineers, 18(1):126-129, 133.

Diniz, E.V. and W. Moore. (1974) Changes in Sedimentation Characteristics of an Urbanizing Watershed, Dallas and Collin Counties, Texas. Proceedings 1974 National Symposium on Urban Rainfall and Runoff and Sediment Control. UK BU 106, College of Engineering, University of Kentucky, Lexington, Ky. pp. 15-23.

Foster, G.R. and L.D. Meyer. (1975) Mathematical Simulation of Upland Erosion by Fundamental Erosion Mechanics. USDA Publication ARS-S-40, U.S. Dept. of Agric., Washington, D.C. pp. 190-207.

Foster, G.R. and W. Wischmeier. (1974) Evaluating Irregular Slopes for Soil Loss Predictions. Transactions of American Society of Agricultural Engineers 17(2):305-309.

Fox, H. (1976) Channel Alteration in an Urbanizing Watershed: A Case History in Maryland. Proceedings National Symposium on Urban Hydrology, Hydraulics, and Sediment Control. UK BU 111, College of Engineering, University of Kentucky, Lexington, Ky. pp. 105-115.

Gilley, T.E., G. Gee and D. Bauer. (1976) Particle Size Distribution of Eroded Spoil Material. Farm Research, Nov-Dec, 1976:35-36.

Guy, H.P. (1970) Sediment Problems in Urban Areas. Circular 610-E, U.S. Geological Survey, Washington, D.C.

Haan, C.T., B.J. Barfield and T.Y. Kao. (1978) Urban Stormwater Management in the U.S. Proceedings International Conference on Urban Storm Drainage, University of Southampton, United Kingdom.

Hudson, N. (1971) Soil Conservation. Cornell University Press, Ithaca, N.Y.

Kao, D.T.Y., B.J. Barfield and A.E. Lyons, Jr. (1975) On-Site Sediment Filtration Using Grass Strips. Proceedings National Symposium on Urban Hydrology, Hydraulics, and Sediment Control. UK BU 109, College of Engineering, University of Kentucky, Lexington, Ky. pp. 73-82.

Li, R.M., R. Simons and L. Shiao. (1977) Mathematical Modeling of On Site Soil Erosion. Proceedings International Symposium on Urban Hydrology, Hydraulics, and Sediment Control. UK BU 114, College of Engineering, University of Kentucky, Lexington, Ky.

Meyer, L.D. (1974) Overview of the Urban Erosion and Sedimentation Process. Proceedings National Symposium on Urban Rainfall and Runoff and Sediment Control. UK BU 106, College of Engineering, University of Kentucky, Lexington, Ky. pp. 15-23.

Meyer, L.D. and W. Wischmeier. (1969) Mathematical Simulation of the Process of Soil Erosion by Water. Transactions of American Society of Agricultural Engineers, 12(6):754-758.

Moldenhauer, W.C. and D. Long. (1964) Influence of Rainfall Energy on Soil Loss and Infiltration Rates – I. Effect over a Range of Textures. Soil Science Society of America Proceedings, 28(6):813-817.

Negev, M.A. (1967) A Sediment Model on a Digital Computer. Department of Civil Engineering, Standford University, Standford, California. Technical Report No. 76.

Neibling, W. and G.R. Foster. (1977) Estimating Deposition and Sediment Yield from Overland Flow Processes. Proceedings International Symposium on Urban Hydrology, Hydraulics, and Sediment Control. UK BU 114, College of Engineering, University of Kentucky, Lexington, Ky.

Robinson, A.M. (1976) The Effects of Urbanization on Stream Channel Morphology. Proceedings National Symposium on Urban Hydrology, Hydraulics, and Sediment Control. UK BU 111, College of Engineering, University of Kentucky, Lexington, Ky. pp. 105-115.

Robinson, C.R. (1974) Erosion Control During Pipeline Construction. Proceedings National Symposium on Urban Rainfall and Runoff and Sediment Control. UK BU 106, College of Engineering, University of Kentucky, Lexington, Ky. pp. 15-23.

Simons, D.B., A.J. Reese, R.M. Li and T.J. Ward. (1977) A Simple Method for Estimating Sediment Yield in Soil Erosion: Prediction and Control. Proceedings of a National Conference on Soil Erosion. Soil Conservation Society of America, Ankeny, Iowa. pp. 234-242.

Smith, R.E. (1977) Field Test of a Distributed Watershed Erosion Sedimentation Model In Soil Erosion: Prediction and Control. Proceedings of a National Conference on Soil Erosion, Soil Conservation Society of America, Ankeny, Iowa. pp. 201-210.

Soil Conservation Society of America. (1976) Soil Erosion: Prediction and Control. Soil Conservation Society of America, Ankeny, Iowa.

Tollner, E.W., B.J. Barfield, C.T. Haan and T.Y. Kao. (1976) Suspended Sediment Filtration Capacity of Simulated Vegetation. Transactions of the American Society of Agricultural Engineers, 19(4):678-682.

U.S. Environmental Protection Agency. (1972) Guideline for Erosion and Sediment Control Planning and Implementation. Report EPA-R2-72-015, U.S. EPA, Washington, D.C.

U.S. Environmental Protection Agency. (1976) Erosion and Sediment Control, Surface Mining in the Eastern U.S. Publication EPA-625/3-76-006. U.S. EPA, Cincinnati, Ohio.

Ward, A.D., C.T. Haan and B.J. Barfield. (1977) The Performance of Sediment Detention Reservoirs. Proceedings International Symposium on Urban Hydrology, Hydraulics, and Sediment Control. UK BU 114, College of Engineering, University of Kentucky, Lexington, Ky.

Ward, A.D., C.T. Haan and B.J. Barfield. (1977) Simulation of the Sedimentology of Sediment Detention Basins. Research Report No. 103, University of Kentucky Water Resources Institute, Lexington, Ky.

Williams, J.R. and H.D. Berndt. (1976) Sediment Yield Prediction Based on Watershed Hydrology. ASAE Paper No. 76-2535, American Society of Agricultural Engineers, St. Joseph, Michigan.

Wischmeier, W.H. (1975) Estimating the Soil Loss Equation Cover and Management Factor for Undisturbed Areas. USDA-ARS Publication ARS-S-40. U.S. Dept. of Agriculture, Washington, D.C. pp. 118-124.

Wischmeier, W.H., C. Johnson and B. Cross. (1971) A Soil Erodibility Nomograph for Farmland and Construction Sites. Journal Soil and Water Conservation 26(5):189-193.

Wischmeier, W.H. and D.D. Smith. (1965) Predicting Rainfall Erosion Losses from Cropland East of the Rocky Mountains. Agr. Handbook No. 282. U.S. Dept. of Agriculture, Washington, D.C.

POLLUTION IN STORM RUNOFF AND COMBINED SEWER OVERFLOWS

Oddvar Lindholm and Peter Balmér

Norwegian Institute for Water Research, Oslo

INTRODUCTION

Concern about storm water pollution is not new. In England, as
early as 1893, Wardle states "... the first storm washings contain quantities of putrescible organic matter ... they are very
foul and often contain as much as the sewage itself."

Although some early investigations indicated considerable levels
of polluting matter in storm water runoff, it was not until
the late 60-ies that serious concern arised about the polluting
effects of storm water runoff. Since then numerous reports on
pollutional levels in storm water runoff have appeared. As can
be seen from the compilations of data made by Lager and Smith
(1974), the scattering of reported data is considerable.

Few reports have provided data on the annual discharge of
pollutants in storm runoff and in combined sewer overflows.
There is also a general lack of knowledge on the correlation
between catchment parameters and the level of pollution in
urban runoff. Likewise, few attempts have been made to compare
annual discharge of pollution in storm runoff in separate sewer
systems to those in combined sewer systems.

In Norway a project was initiated in 1974 with the objectives
of estimating the level of annual discharge of pollution in
storm runoff, and to investigate in what way the pollutional
levels were influenced by:

- The sewerage system
- Different degrees of urbanization
- Time from the previous rainfall
- Time from start of the rain event
- The runoff intensity.

The precipitation, the runoff and concentrations were measured in three combined sewer systems and in four separate sewer systems, and approximately 10 000 chemical analyses were performed.

CATCHMENT DESCRIPTION

The seven catchments were located in the cities of Oslo, Sandefjord and Trondheim. The degree of urbanization, expressed by the percentage of impervious surfaces, varies between 10 percent and 97 percent.

In Table 1 the most important characteristics of the catchments are presented.

Table 1.   Field characteristics.

Location	Sewerage system	Area ha	Impervious surfaces percent	Population density persons/ha	Average slope percent	Field descriptions
Oslo 1	Combined	219	69	342	2.8	Residential (flats) and commercial
Sandefjord	"	380	12	25	2.5	Residential (low housing)
Trondheim 1	"	21	37	93	1.1	Residential (flats) and commercial
Oslo 2	Separate	10	97	-	5.3	City centre commercial
Oslo 3	"	37	43	155	4.1	Suburban residential (flats)
Oslo 4	"	37	33	123	9.3	Suburban residential (mixed flats and low housing)
Trondheim 2	"	20	18	30	5.3	Suburban residential (low housing)

The combined sewerage systems of the Oslo 1 and Sandefjord catchments have a large culvert as main sewer. Deposits were observed to accumulate during dry weather flows and to be flushed out during wet weather. It is believed that there is a dry weather depositing also in the Trondheim 1 catchment because many of the sewers are not self-cleaning (velocity < 0.6 m/s) during dry weather flow. In the Trondheim 2 catchment some construction work occured during the monitoring period. In the Oslo 4 catchment there was also some construction activity, but to a smaller extent than in the Trondheim 2 catchment.

METHODS

Rain gauging
All Oslo catchment areas were provided with recording pluviographs of the tipping bucket type (Plumatic, Kongsberg Våpen-
fabrik A/S), that recorded every 0.2 mm of precipitation. The
pluviograph for the Oslo 1 catchment was located one km from
the centre of the catchment area. The pluviographs in the
other catchments were located within the catchments.
In the Sandefjord and Trondheim catchments rain was gauged
with continuously recording float pluviographs (Fuess),
located within the catchments.

Flow measurements
In all the catchments with separate sewers, flow was recorded
by V-notch weirs. The flow recording stations were also part
of a research programme on urban hydrology and are described
in detail by the Norwegian Water Resources and Electricity
Board (1974). In the Oslo 1 catchment flow was calculated
from the liquid level in the sewer. In the Sandefjord catchment flow was recorded in a Parshall flume downstream an overflow. Flow through the flume and flow diverted at the overflow weir were calculated from the liquid level.
In the Trondheim 1 catchment a flume was constructed in the
sewer and was calibrated at several levels with a current
metre.

Sampling
All sampling was performed manually. A person was always on
duty for each catchment and hurried to the sampling station
when he understood that a rain was approaching the field.
In some rain events runoff had started before the sampler had
arrived. Samples were extracted by dipping a one litre plastic
bottle in the waste stream. Between 5 and 20 samples were
taken during each rain event, depending on the characteristics
of the rain. The number of samples extracted and the time
interval between samples were decided by the sampler.
Interval between samples was as short as 2 min. during rapidly
changing conditions, and as much as 2 hours during extended
low intensity rains. The total number of samples during a
rain event was usually between 5 and 20.

Analytical methods
The 7 day biochemical oxygen demand (BOD_7) was determined by
the dilution method. Chemical oxygen demand (COD) was determined according to Standard Methods (1971). Total phosphorus
(tot-P) and total nitrogen (tot-N) were determined with automated versions of the Norwegian Standard procedure (1974)
(1975). Suspended solids (SS) were determined according to
Norwegian Standard (1973). Lead, copper and zink were determined with atomic adsorption with a Perkin Elmer 306 instrument.
At low concentrations a Perkin Elmer 300 SG with graphite fur-
nace (HGA 472) was used.

Calculations

Pollution from urban runoff is in this paper defined as the
additional material transport that occurs in a storm sewer or
in a combined sewer during a rain event. Dry weather flows
and concentrations were measured 5-10 times in each catchment.
Both dry weather flow and mass transport were small compared
to the wet weather conditions. In the calculations dry weather
flow and mass transport are thus assumed constant.

The calculations have been performed as follows:

Q_n = Discharge during wet weather at the time t_n, m^3/s

C_n = Concentration of a component during wet weather at
t_n, g/m^3

Q_d = Discharge during dry weather, m^3/s

C_d = Concentration of a component during dry weather, g/m^3

t_m = Time at end of runoff from the rain event, s.

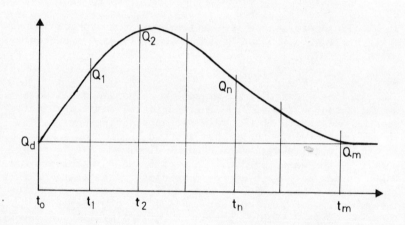

Mass transport of a component during a rain event due to precipitation is M_r g:

$$M_r = \left(\sum_{n=1}^{m} \tfrac{1}{2}(Q_n \cdot C_n + Q_{n-1} \cdot C_{n-1})(t_n - t_{n-1}) \right) -$$
$$- Q_d \cdot C_d \cdot (t_m - t_o).$$

The total mass transport of a component due to precipitation of all the measured rain events is M_T g and the corresponding storm runoff volume V_T m^3.
The overall mean concentration, C_T g/m^3, for all measured rain events for a component is hence:

$$C_T = \frac{M_T}{V_T} \; .$$

The calculated annual discharge per unit area, M_y kg/km^2,year of a component due to precipitation is found by the formulae:

$$M_y = C_T \cdot P \cdot \phi \cdot 10^3, \qquad \text{where}$$

P = annual precipitation, m
ϕ = average runoff coefficient, dimensionless.

From the definitions above it follows that for combined sewered catchments deposits in sewers that are scoured during wet weather flow are included in the urban runoff pollution. This is not equal to the pollution load on the receiving water as this is influenced by factors as overflow settings etc.

The average runoff coefficients are estimated on basis of recorded precipitation and runoff and the percentage of impervious surfaces.

It is also assumed that snowmelt that represents a minor part of the total runoff, has the same concentrations as storm runoff due to rain.
Data used in the calculations are shown in Table 2.

Table 2.   Data used in calculations.

Location	Average runoff coefficient	Annual precipi- tation	Number of rain events in material	Number of samples in material
		m		
Oslo 1	0.35	0.74	18	142
Sandefjord	0.17	0.75	9	88
Trondheim 1	0.40	0.86	13	99
Oslo 2	0.60	0.74	6	51
Oslo 3	0.20	0.79	9	52
Oslo 4	0.25	0.75	11	69
Trondheim 2	0.20	0.86	14	82

RESULTS

The average concentrations in storm runoff are shown in Table 3
and the annual discharges are shown in Table 4.
For Oslo 2 there are two sets of data; the first set includes
a very large rain event (corresponding to a 10-year return
period), and the last set does not include this rain event.

Table 3.   Average concentrations in storm runoff.

	Location	BOD_7 g O/m^3	COD g O/m^3	SS g/m^3	VSS g/m^3	Tot-P g P/m^3	Tot-N g N/m^3	Lead g Pb/m^3	Zink g Zn/m^3	Copper g Cu/m^3
Combined sewered catchment	Oslo 1	200	530	721	188	2.4	8.2	0.45	1.07	0.17
	Sandefjord	103	268	424	168	4.0	14.4	0.08	0.64	0.11
	Trondheim 1	-	352	510	193	3.0	-	-	-	-
Separate sewered catchment	Oslo 2	-	244	1038	189	1.2	4.2	0.82	1.73	0.52
	Oslo 2	-	160	303	75	0.6	3.2	0.41	0.57	0.19
	Oslo 3	-	73	367	46	0.5	4.9	0.10	0.17	0.04
	Oslo 4	-	63	86	33	0.8	5.9	0.05	0.32	0.13
	Trondheim 2	-	74	929	72	0.3	2.3	0.07	0.10	0.03

Table 4.   Annual discharge of pollutants in storm runoff.

	Location	Annual discharge kg/km^2, year								
		BOD_7	COD	SS	VSS	Tot-P	Tot-N	Lead	Zink	Copper
Combined sewered catchment	Oslo 1	518	1373	1867	487	6.2	21.2	1.2	2.8	0.44
	Sandefjord	131	340	537	213	5.1	18.2	0.1	0.8	0.14
	Trondheim 1	-	1210	1755	665	10.3	-	-	-	-
Separate sewered catchment	Oslo 2	-	1083	4609	839	5.3	18.6	3.6	7.7	2.3
	Oslo 2	-	710	1345	333	2.5	14.2	1.8	2.5	0.84
	Oslo 3	-	108	543	68	0.7	7.3	0.15	0.25	0.06
	Oslo 4	-	117	159	61	1.6	10.9	0.1	0.59	0.24
	Trondheim 2	-	127	1600	123	0.5	4.0	0.12	0.20	0.05

DISCUSSION

When the impact of storm runoff pollution is to be evaluated,
discharges during rain events as well as annual discharges
have to be considered. As runoff occurs only 5-10 percent of
a year, the loadings during rain events may be very high.
This is also due to the fact that a few large rain events are
responsible for a main part of the annual discharge.   To
illustrate this point it may be noted that in the most intensive 1/4 hour in a single rain event, the load of organic

matter (COD) from the 10 ha Oslo 2 catchment corresponded to
that of untreated domestic sewage from 160 000 persons.

From Table 4 it may also be noted that a single rain event that
incidently is included in the measurements, increases the calculated annual load 2-3 times. All calculation of annual discharges will thus suffer from serious uncertainties as long as
it is not verified that the captured rain events are representative.

In Table 5 the annual storm runoff pollutant mass transport
due to rain is compared to that of the domestic sewage from
the same catchment.

Table 5.   Annual pollutant mass transport due to storm runoff
           as a percentage of untreated domestic sewage from
           the same catchment.

		Location	Storm runoff/sewage, %			
			COD	SS	Tot-P	Lead
Combined sewered catchment		Oslo 1	8	26	2	81
		Sandefjord	56	180	28	143
		Trondheim 1	36	108	23	–
Separate sewered catchment		Oslo 3	2	17	0.5	56
		Oslo 4	2	4	1	29
		Trondheim 2	10	209	1.5	182

The relative annual load from storm runoff is very variable
from catchment to catchment and from constituent to constituent.
The extremes are represented by 0.5% P in storm runoff compared
to untreated sewage in Oslo 3, and 209% suspended solids compared to untreated sewage from Trondheim 2. If the wastewater
is treated, the relative load of storm runoff increases and may
be significant also on an annual basis.

From table 1 and 5 it can be seen that the pollutional load
from domestic sewage decreases more rapidly than pollution
from the storm runoff, when the population density decreases.
This implies that the load from storm runoff becomes relatively
more important as the population density decreases. However,
it is still believed that storm runoff from large communities
is the largest threat to water resources.

In the figures 1-5 the annual discharge of five different constituents in storm runoff are shown versus the percentage of
impervious surfaces in the catchments. For all constituents
there is an obvious correlation between these factors. The
exceptions for suspended solids can be explained by construction work in the catchment.

For all constituents except lead it is a clear tendency for the
combined sewered catchments to have more pollution in storm
runoff than the separate sewered catchments. This is probably
due to the flushing of pipe deposits in combined sewers during
wet weather periods. A rough estimate from the diagrams
indicates that 50 percent of storm runoff pollution in combined
sewered areas originates from pipe deposits and 50 percent from
surfaces in the catchments. The relation between pollution
from pipe deposits and surfaces will of course vary considerably as the self-cleaning capacity of the pipes is influenced
by many parameters as slope, diameter, etc.

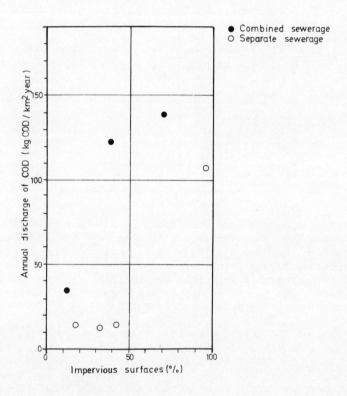

Figure 1. Annual discharge of COD vs. percentage of impervio
           surfaces in the catchment.

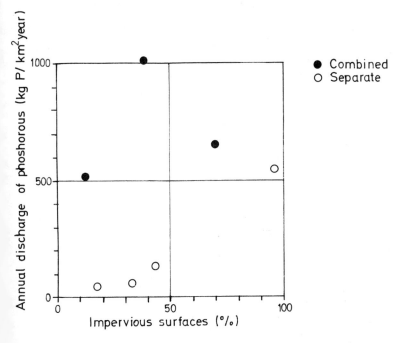

Figure 2.   Annual discharge of total phosphorus vs. percentage
of impervious surfaces in the catchment.

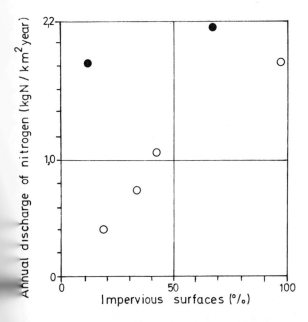

gure 3.   Annual discharge of total nitrogen vs. percentage
of impervious surfaces in the catchment.

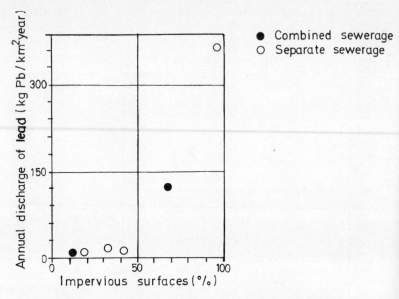

Figure 4.   Annual discharge of lead vs. percentage of
            impervious surfaces in the catchment.

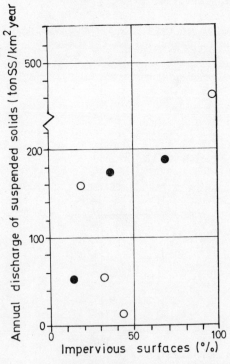

Figure 5.   Annual discharge of suspended solids vs. percentag
            of impervious surfaces in the catchment.

SUMMARY

Storm runoff pollution has been monitored in seven Norwegian urban catchments. Three of the catchments were served by combined sewers and four by separate sewers. The annual mass transport of pollutants varies considerably from one catchment to another. However, there seems to be a correlation between pollution load and percentage of impervious surfaces in the catchment. The increased pollutant mass transport caused by rain in catchments served by combined sewers, seems to be about twice the mass transport in catchments served by separate sewers.

REFERENCES

Lager, J.A. and Smith, W.C. (1974) Urban Storm Water Management - An Assessment. Environmental Protection Technology Series EPA-670/2-74-040.

Norwegian Standard NS 4733 (1973).

Norwegian Standard NS 4725 (1974).

Norwegian Standard NS 4743 (1975).

Norwegian Water Resources and Electricity Board (1974) Urbaniseringens innvirkning på avrenningsforhold i små nedbørfelt. Prosjektrapport nr. 1: Introduksjon av måleprogram og målefelter.

Wardle, T. (1893) On Sewage Treatment and Disposal. John Heywood, London.

AN ASSESSMENT OF THE IMPORTANCE OF ROADSIDE GULLY POTS IN
DETERMINING THE QUALITY OF STORMWATER RUNOFF.

I J Fletcher, C J Pratt and G E P Elliott*
Department of Civil and Structural Engineering and *Department
of Physical Sciences, Trent Polytechnic, Nottingham, England.

ABSTRACT

Roadside gully pots on an urban estate have been monitored
for a period of one year to obtain measurements of various
pollution parameters. The results of an analysis of these
parameters for the first year of study is presented and
initial conclusions drawn on the relative importance of the
different sources of pollution and of the effect of rainfall,
dry periods, seasonal variations and human activity upon the
quality of the stored water within the gully pots.

Examination of the stormwater runoff, (using an automatically
triggered sampling machine) indicates that the quality of the
'first flush' may be substantially attributed to the discharge
of gully pot liquors.

Three different forms of pollutant run off curve have been
identified dependent upon the origin and nature of the
material in question. Equating the levels of pollutants in
gully pots to those found in the stormwater runoff has
provided an indication of the principal origins of the
pollutant.

The preliminary results of a laboratory simulation of the
removal of solids from a gully pot are presented.

INTRODUCTION

Gully pots are placed along the roadside to act as inlet
points for rainwater and stormwater runoff to the sewerage
system. Their principal purpose is to remove solid material
transported by the runoff which may subsequently cause sewer
blockage. They also act as a water seal to prevent the release
of bad odours from the sewer.

Although in recent years the nature and polluting character of
stormwater runoff has been the attention of several studies

586

For any given storm the degree of pollution is determined by the following factors:-

1   Intensity and duration of rainfall;

2   Length of the preceding dry period, which controls the build-up of pollutants in gutters and the quality of stored water;

3   Seasonal variations that occur in the rainfall pattern, temperature(which affects the degradation of organic matter), leaf fall, use of salt as a de-icer; and

4   Effectiveness of Council cleansing procedures.

The present study is being undertaken to gain an insight into the relative importance of the above listed factors in influencing the quality of stored water in roadside gully pots. The significance of this water in determining the degree of pollution of stormwater runoff is also being considered.

CATCHMENT CHARACTERISTICS, SAMPLING AND ANALYSIS

The catchment selected for study is at the Clifton Grove estate on the south side of Nottingham, England. Clifton Grove is a small urban estate covering an area of some 10.6 hectares, of which approximately 5.5 hectares are impervious, and which consists totally of middle income private housing. Built between 1973 and 1976 the development is spaciously arranged with areas of open land designated "amenity areas" inter-spersing areas of housing arranged around cul-de-sacs. In accordance with modern practice the estate is sewered on a separate system with the stormwater runoff draining directly into the River Trent.

The Clifton Grove catchment consists of seven subcatchments defined by the "arms" of the sewers. Each subcatchment being divided into microcatchments that are drained by individual gully pots.

The whole system contains 108 gully pots receiving runoff from the surface of roadways and paved areas. Each pot normally holds about 95 litres of water resulting in a total volume of stored water of 10,240 litres, there being no other storage of water in the system. Four gully pots were selected for detailed examination of pollution parameters over an extended period and the pots chosen were situated in micro-catchments which illustrated different physical features. The following catchment characteristics were considered:-

1   Impervious area.

2   Slope.

3   Number of houses (important in determining the degree of human activity in the catchment).

(1,3,4,5,6,7 and 9) the role of gully pots in removing sediments and concentrating polluting matter is not known to have
formed the core of any major study. As early as 1900 Folwell
(2) noted the accumulation with subsequent putrefaction of
organic matter and drew attention to the inefficiency of
cleansing cycle. The U.S. Dept of the Interior (9) produced a
general report on the pollution of storm runoff which included
a minor section on gully pots. The report indicated BOD values
from 35 to 225mg/l in the supernatant liquid and concluded
that "catch basins (gully pots) may be one of the most
important single sources of pollution from stormwater flows".
Other works including Sartor et al. (5) and Tucker (6) have
come to similar conclusions. Tucker noted the anaerobic
breakdown of organic material and reported wide variations in
stored water quality between gully pots at the same and
different locations, with values of BOD up to 350mg/l and the
COD up to 965mg/l.

Pollution of stormwater
The pollution of stormwater results from the contamination of
rainwater through contact with various substances from the
time of origin in the atmosphere until the moment of its
discharge into a receiving body of water. This quality variation may be brought about by such a wide variety of different
events that the complex changes which result are difficult to
isolate. However, in broad outline pollution of stormwater
runoff may be attributed to the following principal sources:

1   Open land contributes dust and soil particles which
     accumulate in roadside gutters and it acts as a source of
     nutrients and micro-organisms. In heavy storms runoff can
     occur directly off open land resulting in high solid
     loads.

2   Vegetation and animal activity leads to the deposit of
     grasscuttings, leaves, faecal matter, etc.

3   Roof and road surfaces provide collecting areas for
     pollutants and are themselves sources of inorganic solids,
     cement, sand, eroded road material and salt.

4   Motor Traffic acts as a source of oil, exhaust gases,
     petrol, particles from car bodies and tyres.

5   Factories are a sources of dust and gaseous emissions in
     the atmosphere that subsequently settle or are washed
     down in rainfall.

6   Human activity within a catchment leads to the accumu-
     lation of litter, detergents from car washing and garden
     fertilizers.

The relative importance of these sources of pollution at any
given location will be governed by the type of land use ie
urban, industrial or commercial, and by individual catchment
characteristics such as building density, slope, soil type etc

Presence of open land which might contribute to the runoff entering the gully pot.

Situation on the estate. (cul-de-sac feeder road).

Table 1 outlines the nature of gully pot catchments chosen.

Table 1

Gully Pot	Slope	Imp. Area.	No. of Houses	Open Land	Situation
1	0.044	140m^2	0	No	Subsidiary Rd.
6	0.038	680m^2	4	Yes	Cul-de-sac
7	0.040	513m^2	2.5	Possibly	Feeder Road
8	0.047	270m^2	2	Little	Cul-de-sac

Sampling of the gully pot liquors was initially undertaken weekly but subsequently this was reduced to fortnightly. In addition, other gully pots were sampled during dry periods and randomly selected samples of another 25 were taken on one day to assess the variation of pollution parameters and test the significance of a single sample in representing the quality of the stored water. All samples were drawn from a depth of 110mm below the surface of the water using a 1 litre bottle attached to a rod.

Sampling of the total runoff was achieved with a Rock and Taylor Multipurpose sampling machine which could receive a maximum of 48 x 500ml samples. The machine was triggered by a float switch that simultaneously activated a constant head/continuous injection doser upstream of the sampler. The doser discharged lithium chloride at a constant rate of 1 ml/sec and the lithium chloride surface water mixture was sampled downstream. The concentration of lithium in the stream enabled the flow rate in the sewer to be calculated. Samples were drawn from the stormwater flow over a period of 100 sec, there being a cycle time of 270 sec between samples.

RESULTS AND COMMENTS

The results of the analysis to date are presented in Tables 2 & 3. Figures 2 to 4 and 6 to 8 show graphically the temporal variation of some of the pollution parameters.

(a) Seasonal fluctuations and dry periods
     Seasonal variation may be clearly seen in the results. Of all the parameters investigated ammonium appeared to behave in the most predictable fashion (Fig.3). It retained consistently low values throughout the winter months, high values only being associated with summer dry

Table 2
Variation and mean values of parameter concentrations for the whole observation period. (All concentrations in mg/l).

	SS	DS	BOD	COD	DO	NH$_4^+$	NO$_3^-$	CL$^-$	pH	Ca^{2+}
Maximum	213	17,475	46	163	11.2	9.13	132	9800	12.94	272
Minimum	0.2	35	0.9	8.7	0	0.05	0.8	1.8	6.30	4
Mean	21.4	397	6.8	39.1	6.1	0.96	7.8	142.3	8.35	30.7
Standard Deviation	20.6	1,114	8.3	23.4	3.2	1.49	14.2	578	0.74	30.2

Table 3
Variation of parameter concentration for a randomly selected sample of 25 gully pots, samples being collected on the same day. (All concentrations in mg/l).

	SS	DS	BOD	COD	NH$_4^+$	NO$_3^-$	CL$^-$	pH	Ca^{2+}	ZN^{2+}
Maximum	252.8	292.5	41.6	428.6	4.62	28	36	8.38	35.8	0.570
Minimum	2.8	62.5	0.25	0	0.21	2.5	9.5	7.55	17.6	0
Mean	36.0	143.2	13.7	67.3	1.39	9.6	18.2	7.87	25.7	0.090
Standard Deviation	61.8	52.6	11.4	111.2	1.03	6.3	6.5	0.20	3.7	0.132
95% Confidence Limits	64.5%	13.5%	31.3%	60.4%	26.9%	25.4%	13%	0.91%	5.3%	52.4%

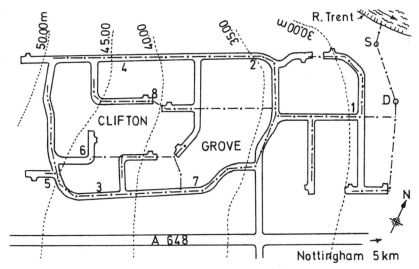

D   Manhole containing dosing equipment.
S   "   "   sampling   "

Figure 1. Plan of Clifton Grove Estate, Nottingham,
showing main storm sewer system.

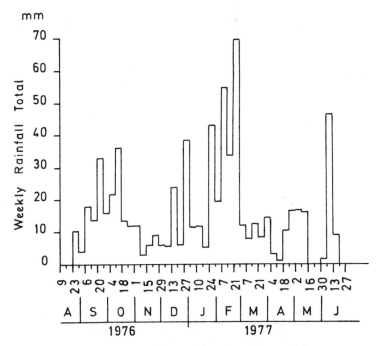

Figure 2. Weekly rainfall totals for observation
period, for Meteorological Office raingauge 116959:
weeks beginning Monday 09.00 G.M.T.

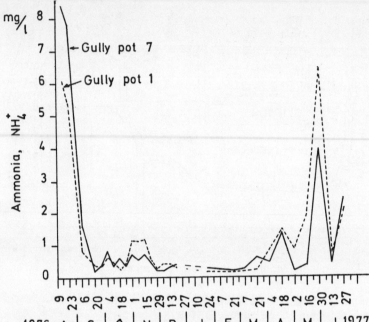

Figure 3. Temporal variations in nitrate concentration in gully pot liquors 1 and 7.

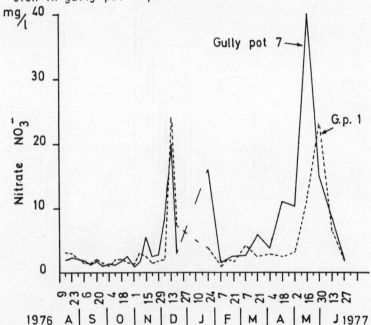

Figure 4. Temporal variations in nitrate concentration in gully pot liquors 1 and 7.

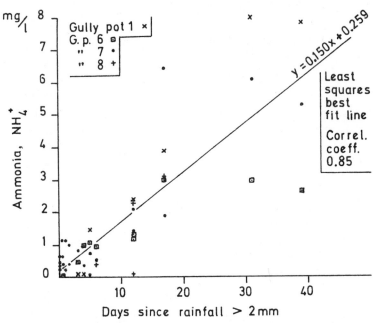

Figure 5. Correlation of ammonia concentration with number of days since rainfall greater than 2mm for gully pots 1,6,7 and 8.

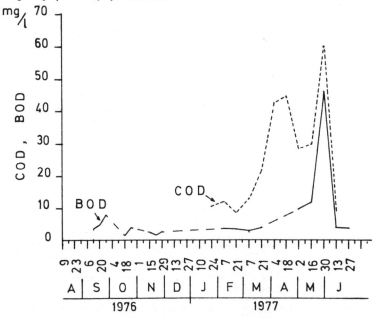

Figure 6. Temporal variations in B.O.D. and C.O.D. in gully pot liquor 1.

periods. Nitrate also showed high values during the summer (Fig.4) and was possibly the source of the high ammonium values through the action of anaerobic bacteria. It is interesting to note that in the summer of 1977 nitrate values did not decrease through this conversion mechanism, possibly due to the fresh input of nitrate from car-wash liquors.

However, in the summer of 1976 when car washing was prohibited due to the water shortage low nitrate levels were recorded. High nitrate values were also recorded during periods of winter road salting, the input being thought attributable to the impure rock salt employed. This influx of nitrate did not significantly affect ammonium values due principally to the inhibition of bacterial action by the low temperatures. The build up of ammonium over a dry period is illustrated by Fig.5.

Values of BOD and COD (Fig.6) showed a similar but more variable pattern in comparison to that observed for ammonia, low values occurred during winter months and high peaks were coincident with summer dry periods. The inhibition of bacterial action is reflected in the BOD values observed during the winter months. BOD remained virtually constant and winter dry spells did not cause any significant build up in oxygen demand.

Dissolved solids (Fig.7) mirror in a general fashion the changes already mentioned for specific parameters. The increase in pollutants and dissolved material over dry periods may be attributed to the following events:-

1   Input from sweeping and wind blown material e.g. grasscuttings and organic debris providing fresh material for 2).

2   Anaerobic digestion of organic matter to soluble compounds by bacterial action.

3   Input of soluble material from human activity i.e. car washing.

4   Evaporation increasing effective concentration (minor).

In winter the major event to influence dissolved solids concentration was road salting, (Fig.8). This was not only the cause of sharp increases in the values of dissolved solids, sodium and chloride but also increases were observed for nitrate, potassium and calcium. These parallel increases being thought attributable to the impure rock salt employed.

Autumn leaf fall was considered to have no significant effect upon this catchment. Tucker (6) had reported clogging of gully pots by leaves with subsequent high increases in suspended solids, BOD and COD. However, due

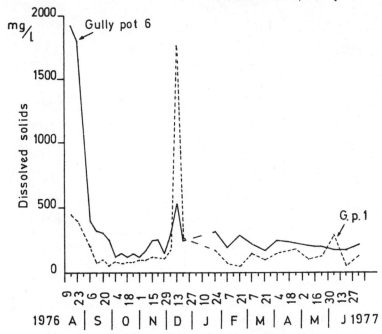

Figure 7. Temporal variations in dissolved solids
concentration in gully pot liquors 1 and 6.

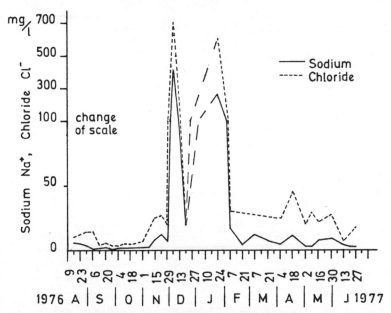

Figure 8. Temporal variations in sodium and
chloride concentration in gully pot liquor 1.

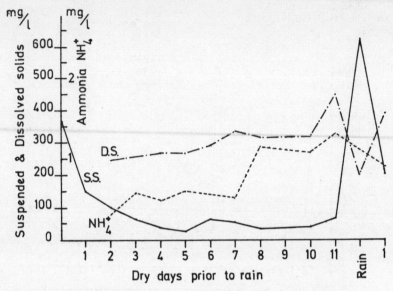

Figure 9. Variation over a dry period of suspended solids, dissolved solids and ammonia concentrations in gully pot liquor 4.

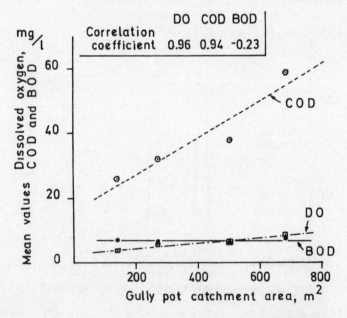

Figure 10. Correlation of mean values of dissolved oxygen, B.O.D. and C.O.D. with gully pot catchment area.

to regular street sweeping and the presence of only a few deciduous trees of any size on the estate no changes in quality from this cause were identified.

A significant parameter in specifying the quality of stored water is the dissolved oxygen level. Dissolved oxygen was observed to fall to zero after a period of 4 to 7 days dry weather. This resulted in septic conditions in the gully pot followed by the development of anaerobic digestion with subsequent increases in the values of BOD, COD and dissolved solids. During hot weather the production of hydrogen sulphide was evident and the gully pots developed a pungent smell.

Day to day examination of gully pot liquor during a dry period (Fig.9) showed a decrease in the amount of suspended solids following an inverse exponential form. The period of this decrease ranged from a few hours to five days dependent upon the type of material input (fine dusts associated with building sand had the slowest settling rate). General increases in dissolved solids were also observed. Observations during a rainstorm showed a sharp increase in suspended solids due to wash off from road surfaces and the stirring of bottom sediments. The levels of dissolved solids decreased to the same value as those of the incoming rainwater and then adjusted to a new steady state dependent upon the nature of the soluble material input.

(b)   Human activity and catchment size

The nature and level of human activity in the micro-catchment of a gully pot is thought to be a significant factor in determining the chemical composition of the pot liquor. Unusually high values of parameters that are not mirrored in measurements from other gully pots may often be ascribed to human activity in the catchment. Some of the observations which may be attributed to human influence are listed below:-

1   Oil washed from road surfaces and in particular driveways was often present in the gully pots as a thin film on the surface of the liquor.

2   Frothiness due to car-wash liquids was commonly observed. Extreme cases of this were associated with abnormal nitrate levels, and during dry periods this may have had an influence upon ammonia levels.

3   In the summer months grasscuttings were found in abundance in gully pots and gutters and their presence could contribute to BOD levels during dry periods.

4   pH values up to 12.9 were recorded, probably due to the use of garden fertilisers and other agricultural chemicals which were blown or washed into the gully

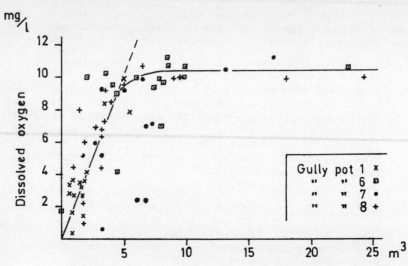

Figure 11. Variation of dissolved oxygen concentration with runoff through gully pots 1,6,7 and 8 for week prior to analysis.

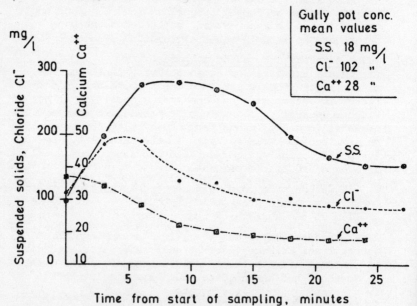

Figure 12. Pollutant runoff curves corresponding to storm water runoff (6 Jan 1976) illustrating the removal of different materials from the Clifton Grove Catchment. (Note mean values of pollutant concentrations in observed gully pots prior to storm).

pots.

5   Salting of driveways was found to add large quantities
    of dissolved material above and beyond that due to
    council road saltings.

The level of human activity might be expected to correlate
well with catchment area in an urban estate i.e.the larger
the catchment area the greater the likelihood of human
activity affecting the gully pot. Fig.10, a plot of mean
values of COD and BOD against catchment area, shows that
values of COD correlate well with catchment area i.e.
human activity, whilst the values of BOD appear to be
independent of area. This suggests that the polluting
input due to human activity is principally not organic in
nature. Dissolved oxygen (Fig.10) showed a dependence
upon catchment area, however, this was less a result of
the human influence but was directly related to the
volume of runoff through the gully pot. A graph of total
runoff through the gully pot for the week prior to
analysis plotted against values of dissolved oxygen is
given in Fig. 1 . The graph indicates that to keep a
gully pot fully aerated a runoff volume of $6m^3$ per week
is necessary. However, to keep dissolved oxygen at a
reasonable level e.g. above 6mg/l, and assuming a typical
weekly summer rainfall of 6mm a catchment area of $500m^2$
would be needed. The majority of gully pot catchment
areas at Clifton Grove are smaller than this, e.g. gully
pot 1, which is consequently rarely fully aerated. It
would seem that to reduce pollution of gully pots by
increasing the dissolved oxygen level, larger catchment
areas should be employed.

(c)   Council cleansing procedures
The normal council cleansing procedure was to have the
estate swept by hand fortnightly and the gully pots
emptied on average once every 4 months. However, during
the period of this study the gullys were emptied only
once. The frequency of street sweeping appeared to be
sufficient to keep the gutters and the estate generally
clean and tidy. Despite the fact that pots were only
emptied once, no exceptional depths of settled material
were noted and a frequency of cleansing every 4 to 6
months appeared to be adequate as far as solid material
was concerned. However, when the gully pot liquors were
at their most polluting i.e. during summer dry periods,
such a frequency was inadequate to maintain the pots in
a wholesome state.

(d)   Storm runoff
Three principal types of pollutant runoff curve have
been identified (Fig.12). The type of curve followed is
dependent upon the nature of the material being removed

and its origin. For example a parameter whose principal
origin is in gully liquors will follow a curve of type 3
in figure 12, this type of removal is typical of ammonia,
calcium and potassium. Pollutants whose prime origin is on
the road surface or atmosphere will follow curves of
types 1 or 2. The position of the peak of these curves is
determined by the removal coefficient (i.e. ease of
mobilisation) of the material. For example, suspended
solids have a low removal coefficient and consequently
peak concentrations are observed later in the storm.
Soluble matter such as salt possesses high mobility
giving rise to an earlier peak concentration. Comparison
of levels of parameters as found in stormwater runoff
compared to parameter levels found in gully pots confirms
this argument (Fig.12). During winter months nitrate
values were also found to be higher in runoff than in
gully pots for the storms analysed to date, examination of
roof runoff revealed its principal origins to be in rainfall.

The 108 gully pots on the Clifton Grove estate contain a
total volume of some 10,240 litres. Comparing this volume
to the storm runoff volume of 1,045,000 litres for the
catchment over a 1 hr M5 rainfall of 19mm, reveals that
gully liquors contribute some 1% of the total runoff.
However some 40% of daily rainfall at Clifton Grove is
equal to or less than 1mm resulting in a runoff volume of
which gully liquors may contribute upwards of 20% of the
total runoff. Bearing in mind that the greater part of the
original gully liquors are discharged in the earlier
stages of runoff the significance of gully pot quality in
determining the day to day polluting nature of the 'first
flush' of storm runoff can be seen.

(e)   Laboratory simulation
Initial results from a laboratory simulation of the
removal of settled material from gully pots have indicated
that for all but the lowest flows complete mixing of the
storm water inflow and the pot liquor is established
rapidly and prevails during runoff. The removal of solids
by the disturbance of the bottom sediments is dependent
upon:-

1   rate of flow of water into the gully pot;

2   size of the gully pot;

3   depth of sediment within the pot; and

4   mass of sediment available for release.

The typical removal of sediment is illustrated by Figure
13. The rising limb, when material is being released from
the bottom sediments, can be defined by an equation of the
form:

$$c = A(1 - e^{Bt})$$

where c = concentration of suspended solids in the outflow

  t = time

  A and B are constants dependent upon flow rate.

And the falling limb, after all available material has been released by an equation of the form:-

$$c = Ae^{-Bt}$$

The mass of sediment resuspended and released in runoff is relatively small compared to the mass remaining in the bottom of the gully pot. This indicates the general efficiency of gully pots in trapping heavier particulate matter. However, the finer sediment washed over is of greater significance in determining the polluting character of storm runoff.

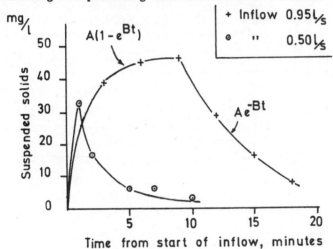

Figure 13. Laboratory simulation of the removal of solid material from a gully pot under steady flow conditions. (Initial depth of material in gully pot was 50 mm).

CONCLUSIONS

There are a large number of variables which determine the quality of the stored water in any one gully pot or the effect of any one source of pollution. As a result it is difficult to isolate factors and express their effect in a quantitative manner from such a short study. However, the following general conclusions may be drawn from the work to date:-

1 The most important climatic influence affecting quality of gully pot liquors is summer dry periods. These dry periods result in sharp increases in the values of BOD, COD, nitrate, ammonia and dissolved solids.

2   The loss of dissolved oxygen and the consequent anaerobic conditions is possibly the major cause of pollution.

3   Over dry periods human activity in the vicinity of a gully pot causes a significant deterioration of the pot liquor quality.

4   Human activity leads to variation in quality between gully pots.

5   COD values for the liquor appear to be dependent upon human activity/catchment area. BOD appears to be unaffected.

6   Dissolved oxygen levels are related to catchment area. The introduction of wider spacing between gully pots would increase average dissolved oxygen levels and hence reduce pollution of runoff due to gully liquors.

7   Comparison of pollutant runoff curves and values of certain parameters in gully pots provides an indication of the principal origins of pollutants.

8   Gully pots are efficient in the removal of heavier particulates from stormwater runoff.

REFERENCES

1   Bryan, E. H., (1972) Quality of stormwater drainage from urban land, Wat. Res. Bull, 8, 3:578.

2   Folwell, A.P., (1900) Sewerage.

3   Pravoshinsky, N.A., and Gatillo, P.D., (1969) Determination of the pollutional effect of surface runoff. Proc. 4th Int. Conf. on Wat. Pollut. Res, Prague, 187-195.

4   Preul, H.C., (1976) Urban Runoff Characteristics. U.S. EPA 600/2 - 76 - 217.

5   Sartor, J.D., Boyd, G.B., and Agardy, F.J., (1969) Water pollution aspects of street surface contaminants. J. Wat. Pollut. Cont. Fed, 46, 3 : 458 467.

6   Tucker, C.G.J., (1974) Pollution loads in runoff from urban areas. MSc thesis, Trent Polytechnic, Nottingham, England.

7   Weibel, S.R., Anderson, R.J., and Woodward R.L., (1964). Urban land runoff as a factor in stream pollution. J.Wat. Pollut. Cont. Fed, 36 914 - 924.

8   Wullschleger, R.E., Zanoni, A.E., Hansen, C.A., (1976). Methology for the study of urban storm generated pollution. U.S. EPA 600/2 - 76 - 145.

9   U.S. Dept. of the Interior (1969) Water pollution aspect of urban runoff.

THE QUALITY OF URBAN STORM-WATER RUN-OFF

by G. Mance and M. M. I. Harman

Water Research Centre, Stevenage Laboratory

INTRODUCTION

The quality of urban storm-water run-off has received little attention in the United Kingdom since Wilkinson (1956) demonstrated that storm-water run-off from a separately sewered housing estate was of much poorer quality than the original rainwater.

The Technical Committee on Storm Overflows and the Disposal of Storm Sewage (1970) discussed the problem of pollution resulting from the discharge of storm sewage from combined sewer systems. They favoured the adoption of separate sewer systems for new developments, but acknowledged that highly polluting flows could occur when predominantly industrial areas were so drained.

The extent of the pollution originating from storm-water run-off was indicated by the Trent study (Lester, 1975), when it was observed that the water quality of the River Tame, Staffordshire, deteriorated rapidly during storm flows arising in the Birmingham area. A considerable discrepancy was found between the observed pollutant loads in the river and the loads predicted from all known sources, this excess load, particularly of heavy metals, being attributed to the run-off of accumulated deposits from urban surfaces.

Much of the world's literature regarding the composition of urban run-off water was reviewed in 1975 by Loehr. His paper high-lighted the wide variation in the quality of urban run-off, with reported mean and maximum BOD values ranging from 12 to 160 mg/l and 18 to 7700 mg/l, respectively. An initial attempt to assess the factors that influence the composition of urban surface run-off in Halifax, Nova Scotia (Waller, 1972) illustrated the importance of local climate, topography, and street cleaning practices.

All the published information indicates that urban storm-water run-off can be a significant source of pollution to the receiving stream.

The Working Party on Storm Sewage (Scotland) (Nicholl and McGillivray, 1977) recently recommended a return to the construction of combined sewer systems in Scotland, as even with the provision of overflow storage tanks this would result in significant financial savings. The poor quality of urban storm-water run-off influenced their decision as did the problem of wrong connections between surface and foul sewers. However, apart from Wilkinson's work (1956) on suspended solids and BOD, little direct information on the quality and quantity of run-off from separately sewered systems in the United Kingdom existed in 1974, when a study of urban run-off from Stevenage New Town was initiated by the Water Research Centre under contract to the Department of the Environment. In 1975 a study of the suspended solids load of storm run-off from a small area of Nottingham was completed (Tucker, 1975) and confirmed Wilkinson's observation of a first flush effect. More recently this first flush has been reported from a study of urban run-off in Hendon (Ellis, 1976), where the maximum concentration of suspended solids observed in any individual storm was directly related to the total discharge. This paper reports some of the findings of the Stevenage catchment work.

DESCRIPTION OF THE STUDY AREA

Stevenage New Town is situated in Hertfordshire, 30 miles north of London, and covers an area of 2500 ha with a population of 69 000 (Stevenage Development Corporation, 1971). The town has separate sewer systems for foul and storm waters, surface drainage being discharged without treatment to the Stevenage Brook.

Within the New Town the run-off from a well defined sub-catchment of mature housing (completed in 1959) has been studied in detail. The Shephall sub-catchment, which has an area of 142.55 ha, of which 32.8 ha are impervious, drains to a single point of discharge, and houses a population of 9500 at a density of 3.6 persons per dwelling, with one motor vehicle for every 5 persons.

Storm run-off arising from this catchment is sampled from the final sewer 200 m above its outfall to the Stevenage Brook. The depth of discharge in the storm sewer is recorded continuously and converted to rates of discharge using the method given in Hydraulics Research Paper No. 4 (Ackers, 1963). The stage discharge relationship used was confirmed by lithium dilution gauging. A flow-actuated sampling system, providing discrete samples at pre-set time intervals and installed at the point of flow measurement, allowed the mass flow of pollutants emitted during individual storms to be calculated.

Throughout the study the aerial deposition of pollutants was measured by analysis of rainwater samples collected by a continuously exposed polythene funnel and bottle. The quality of run-off from roofs of houses was monitored by flow-weighted sampling of the water discharged from a house-roof downpipe, using a tipping-trough sampler designed at WRC. Grab samples of water from four road-drain catchpits were also taken at least once a week.

During this study the daily mass flow of pollutants in the surface drainage from the whole of Stevenage New Town was measured over a two-year period. Two-hourly samples collected at a gauge located 1 km below the New Town were bulked in proportion to flow each day.

RUN-OFF QUALITY

The water-quality data collected for 80 run-off events have been summarized in Table 1 in which the flow-weighted mean and the maximum concentrations observed are listed for all constituents. Four snow-melt run-off events were sampled and these have been summarized separately. Snow-melt is of poorer quality than rainfall run-off for a number of possible reasons: heavy applications of road salt during snowfall, less efficient utilization of petrol by motor vehicles, and contact between snow and car bodywork. In Stevenage, road-salting operations occur when either severe ground frost or snow is predicted. When snowfall does not occur subsequent rainfall run-off is heavily contaminated and the high maxima of suspended and soluble solids and chloride were caused by the application of road salt to the catchment. The maximum concentrations of ammonia and heavy metals were associated with prolonged dry weather in the summer months.

Two estimates of the flow-weighted mean concentrations have been presented. The first is the value calculated from all the observations obtained and the second is the arithmetic mean of the flow-weighted mean concentrations calculated for each run-off event. The latter procedure yields a different result as all storms are treated equally irrespective of the magnitude of the discharge, which can change by a factor of 100 or more.

The flow-weighted mean and the maximum concentrations recorded in each run-off event are log-normally distributed and are linearly correlated, but both are independent of the mean and maximum flow rates and of the total discharge observed, so that simple prediction of the concentrations of pollutants using hydrological data only is not very successful.

Table 1. Summary of the quality of rainfall and snow-melt run-off from the Shephall catchment

	Rainwater run-off			Snow-melt run-off	
	Flow-weighted mean concentration	Arithmetic mean of flow-weighted mean concentrations	Maximum concentration recorded	Flow-weighted mean concentration	Maximum concentration recorded
			mg/l		
Suspended solids	112	133	1386	263	1893
Soluble solids	258	390	2342	880	4544
Total solids	364	523	2623	1144	4904
Chloride	49	64	1008	413	2724
Soluble organic carbon	9.7	11.2	30.6	11.0	13.3
Nitrate-N	1.7	2.5	6.9	4.1	9.0
Total inorganic-N	2.0	3.5	7.6	4.4	9.3
			µg/l		
Nitrite-N	50	182	4000	90	170
Ammoniacal-N	283	789	6000	180	230
Copper	28	23	350	50	270
Manganese	110	160	1700	149	1680
Lead	205	254	3700	423	4000
Zinc	271	322	1630	355	1930
Total heavy metals	613	812	5880	968	7880

Concentrations of heavy metals are linearly related to those of suspended solids within each run-off event, but the nature of this relationship changes between events, with the gradient of the regression lines ranging from 0° to 180°. Regression analysis of the data demonstrated that the concentrations of all constituents were linearly related to the rate of discharge for any individual run-off event, but variation between storms was very high. This variability was not reduced by treating rising and recession limbs of hydrographs separately.

First-flush effect

A first-flush effect was observed in this study but was not a constant feature of all run-off evants as suggested in earlier studies (Wilkinson, 1956; Tucker, 1975; Ellis, 1976). The frequency distribution of the time of occurrence of the peak concentration of suspended solids in relation to the time of occurrence of peak flow is shown in Fig. 1. Similar distributions were observed for the other constituents, but within individual run-off events the peak concentrations did not occur simultaneously for all constituents. The percentage occurrence of peak flow rate and concentrations in various time intervals from the start of discharge are given in Table 2. With the exception of ammonia and the heavy metals, peak concentrations occurred within the first 40 minutes of flow in 80 per cent of the events studied.

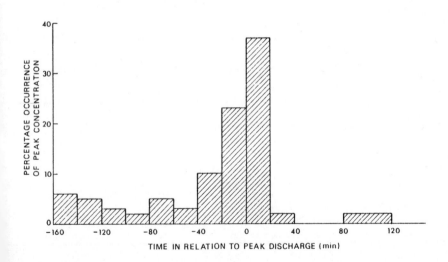

Figure 1.   Frequency distribution of peak concentration in relation to time of peak discharge.

Table 2. The percentage occurrence of maximum flow and
concentrations in time intervals from the start of
discharge. The range of cumulative percentage
distribution of mass flows is also given

	Time since start of discharge (min)				
	0-20	20-40	40-60	60-100	>100
Flow	25	32	6	11	26
Suspended solids	57	25	5	10	3
Soluble solids	70	17	1	5	7
Total solids	78	15	3	2	2
Chloride	81	7	3	2	7
Soluble organic carbon	46	42	8	-	4
Nitrate-N	82	4	5	4	5
Total inorganic-N	77	5	14	4	-
Nitrite-N	73	22	5	-	-
Ammoniacal-N	38	6	22	17	17
Copper	50	28	13	5	4
Manganese	42	37	14	1	6
Lead	48	32	4	9	7
Zinc	33	29	17	17	4
Total heavy metals	46	31	9	8	6
Range in cumulative mass discharged per time interval (as percentage of total)	29-38	53-65	69-84	80-100	88-112

It has been suggested that the polluting effect of stormwater discharges may be reduced significantly by treatment of
the first flush. As can be seen from the last line of Table 2,
this would reduce the total mass of pollutants discharged by
between 69 and 80 per cent if the first hour of discharge
were treated. But even after this time has elapsed the
quality of the discharge could still be polluting in character
(Table 3); for example, the mean concentration of suspended
solids is far in excess of the current standard for effluent
from sewage-treatment works.

Influence of the duration of the antecedent dry period
Wilkinson (1956) observed that the quality of urban run-off
decreased with an increase in the length of the antecedent
dry period. Multiple regression analysis of the data for the
mass of pollutants discharged by individual run-off events
indicates that, except for chloride and ammonia, between 60
and 90 per cent of the variance can be explained by the total
volume discharged, the length of the antecedent dry period,
and the magnitude of the previous run-off event. However,
the antecedent dry period and the magnitude of the previous
event only accounted for a maximum of 6 per cent of the

variance explained. The characteristics of the current rainfall run-off event are the dominant factor in determining the
mass of pollutants discharged and, as observed by Waller
(1972), antecedent dry period has little statistical effect on
the quality of run-off.

Table 3. Variation of the flow-weighted mean concentration
with time from the start of discharge. Summary of all
run-off events studied

	Time since start of discharge (min)				
	0-20	20-40	40-60	60-100	>100
	concentration (mg/1)				
Suspended solids	143	118	96	90	81
Soluble solids	288	278	249	224	201
Total solids	426	391	339	308	272
Chloride	53.7	54.3	50.4	43.3	34.9
Soluble organic carbon	9.7	9.8	8.9	8.3	7.3
Nitrate-N	1.78	1.71	1.65	1.59	1.55
Total inorganic-N	2.21	2.07	1.94	1.83	1.77
	concentration (μg/1)				
Nitrite-N	66	58	42	35	31
Ammoniacal-N	367	309	250	209	190
Copper	31	30	27	25	24
Manganese	133	121	95	87	85
Lead	224	224	193	180	173
Zinc	290	286	261	249	239
Total heavy metals	677	661	576	540	521

SOURCES OF POLLUTANTS

The rate of aerial deposition of pollutants on the catchment
was constant when measured over a period of 8 or more days,
but for shorter periods of measurement variation was high.
The observed rates of deposition were within the ranges
reported for a national survey in 1975 (Cawse), for example
the reported range for lead was between 46.6 and 183.3 μg/m^2 d.
Using the measured rate of aerial deposition on the catchment,
together with knowledge of the antecedent dry period and mass
discharged by the run-off events studied, it is possible to
estimate the proportion of the mass of pollutants in urban
run-off that originates from aerial deposition (Table 4).
Results for soluble organic carbon have been omitted because
samples were contaminated by leaching from the polythene
collection funnel. The estimate of ammonia exceeds the output

by a large factor, but a similar result has been obtained in
recent work in Sweden (Malmquist, personal communication).
This may represent undetected contamination of the samples by
bird faeces. Phosphorus analyses, however, did not confirm
contamination from this source, and the rate of aerial
deposition of ammonia obtained in this study is low when
compared with the range of values (1.4-6.6 mg/m^2 d) reported
for six UK sites by Cawse (1974).

Table 4. Rates of aerial deposition and the percentage
of the mass of pollutants discharged that is
derived from aerial deposition

	Rate of aerial deposition	Percentage contribution to run-off
	mg/m^2 d	
Suspended solids	26.1	23
Soluble solids	77.1	30
Total solids	103.0	29
Chloride	14.6	28
Nitrate-N	1.2	68
Ammoniacal-N	1.2	417
	µg/m^2 d	
Nitrite-N	25.7	50
Copper	51.2	96
Manganese	51.4	49
Lead	127.6	54
Zinc	159.6	38

Another source of material for removal by rainfall run-off
will be that eroded from the catchment surface by weathering
and human activity. Independent assessment of the quality of
run-off from road surfaces is difficult, but the quality of
water draining from house roofs is more readily determined.
Flow-related sampling of the water discharged by downpipe
from a house roof showed that weathering of roofing materials
was of little importance as a source of pollutants in this
particular study. However, in Sweden where copper and zinc
are common constituents of materials used for roofing,
guttering, and building fixtures, corrosion leads to a significant increase in run-off of copper and zinc (Malmquist and
Svensson, 1977).

In recent years it has been demonstrated that road-run-off,
particularly from major trunk roads and motorways, is an
important source of heavy metals and organic compounds
(Hedley and Lockley, 1975; Laxen and Harrison, 1977). However

in the Shephall catchment the traffic density is light, with a maximum number of vehicles using the main artery road of 9500 per day. While motor vehicles represent a significant source of metals, which are emitted in exhaust fumes and released by corrosion, the suspended solids, BOD, and organic compounds are derived from other sources. The deposition of animal faeces, litter, grass cuttings, and leaves in autumn, with their subsequent decay, must contribute to the pollutant load of run-off water, as will erosion of the road surfaces, wear of car tyres, and road-salting operations.

An indirect estimate of the rate of accumulation of material on paved and road areas can be obtained using the measured rates of aerial deposition and weathering of house roofs and the rate of accumulation on the whole impervious catchment, deduced from the mass exported by each run-off event. The values of these estimates are given for particulate solids in Table 5. The input from roads and paved areas dominates, and there is potential for a considerable improvement in run-off quality with increased frequency and efficiency of street-cleaning operations.

Table 5. Estimates for the rate of accumulation of particulate matter on the catchment

Average rate of accumulation on impervious area $(mg/m^2 d)$	Rate of accumulation on roads and paved areas $(mg/m^2 d)$	Rate of aerial deposition $(mg/m^2 d)$	Rate of decay of roofing materials $(mg/m^2 d)$	Input per rainfall event (mg/m^2)
104.5	148.8	26.1	16.7	55.5

Two earlier studies have indicated that road drain catchpits might be sources of polluted water (Waller, 1972; Tucker, 1975) and this has been confirmed. In the area under study there are 656 catchpits containing an estimated 100 m^3 of poor-quality water (Table 6), which is a significant volume in comparison with the mean volume of 363 m^3 of run-off water discharged in the events studied. Even after flushing by the heaviest rains, the water retained in a catchpit becomes anoxic within 24 hours with a consequent deterioration in water quality. Thick bacterial scums are commonly observed at the water surface, and it is possible that the removal of this scum by the initial run-off may represent a significant proportion of the first flush of suspended solids and BOD. During a period of dry weather, especially in summer, there are significant changes in water quality. Ammonia concentrations increase as oxidized nitrogen is reduced by bacteria (Fig. 2). At the same time the total concentration of heavy

metals increases, possibly indicating bacterial mobilization
of the metals from the bottom sediments.

Table 6. Mean and maximum concentrations in
water in catchpits of road drains

	Concentration (mg/l)	
	Mean	Maximum
Suspended Solids	70.7	1032.0
Soluble solids	282.0	5860.0
Total solids	252.7	5964.0
Chloride	76.4	1829.0
Soluble organic carbon	28.9	255.0
Ammoniacal-N	0.7	4.5
Nitrite-N	0.08	0.34
Nitrate-N	0.8	3.0
Total copper	0.02	0.04
Total manganese	0.20	0.62
Total lead	0.17	0.60
Total zinc	0.11	0.25

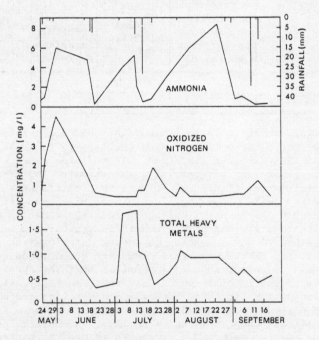

Figure 2. Change in concentrations of heavy metals, ammonia,
and oxidized nitrogen in road-drain catchpit water
in relation to rainfall.

Figure 3 records the mean change in the concentration of
heavy metals in a catchpit in relation to the occurrence of
rainfall. This change in concentration is dependent on the
concentration before the rainfall occurred. If the initial
concentration was high then a decrease in concentration
always occurred. High rainfall normally had the same effect,
except when the initial concentration was low, in which case
an increase in concentration occurred. This increase may
possibly be due to the catchpits having submerged outlets, so
that, at low rates of run-off, when vertical mixing is poor,
the poorer quality run-off water displaces the cleaner catchpit water, before it is itself discharged. Also, during dry
periods the catchpit water evaporates and this provides a
small volume of storage, which at low rainfalls would restrict
or even prevent discharge from the catchpit and this may also
lead to increases in the concentration of pollutants. This
pattern of change in concentration in response to rainfall
was observed for all constituents, with the exception of
oxidized nitrogen, for which rainfall caused an increase in
concentration.

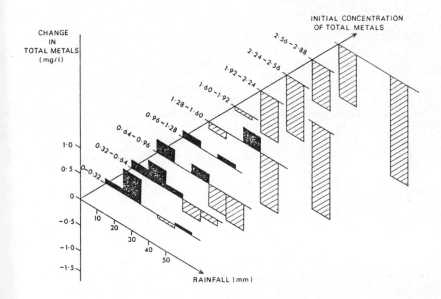

Figure 3.   Mean change in concentration of heavy metals in
            catchpit water in relation to rainfall and initial
            concentration.

Examination of data for grab samples of the quality of catchpit water before and after rainfall events indicates that the
effects of this reservoir on the mass flow of particulate
matter through the system are variable. At times catchpits

may trap an estimated mass of particulate solids equivalent to
50 per cent of the total mass discharged whilst for other
events as much as 20 per cent of the mass discharged may have
originated from the catchpit waters.

POLLUTED RUN-OFF FROM STEVENAGE NEW TOWN

Annual yields of material per unit area of impervious catchment
have been calculated for both the Shephall sub-catchment and
the Stevenage New Town (Table 7) and show good agreement. The
Shephall sub-catchment is a residential area with two small
neighbourhood shopping centres. The New Town has a large
industrial area, an unknown proportion of which is connected
to the foul sewerage system to avoid contamination of surface
water by pollutants spilled in such areas as loading bays,
etc. The difference in annual yield of soluble solids from
the two catchments is attributed to the presence of run-off
from the industrial area in the New Town drainage.

Table 7. Mean yield from the Shephall sub-catchment and
estimates of the annual yield from the Shephall
sub-catchment and Stevenage New Town

	Shephall		Stevenage
	Yield per hour of run off	Annual yield	Annual yield
	mg/m^2 h	kg/ha year	
Suspended solids	78.4	148	99
Soluble solids	154	168	596
Total solids	248	318	695
Chloride	26.4	28	76
Soluble organic carbon	6.8	5.7	11
Nitrate-N	0.9	1.2	3.0
	$\mu g/m^2$ h	g/ha year	
Nitrite-N	91.5	171	265
Ammoniacal-N	105.7	966	864
Copper	20.7	16	18
Manganese	103.5	41	30
Lead	176.5	41	55
Zinc	173.0	66	59

The mean yields of pollutants per unit area of Shephall per
hour of rainfall have been calculated (Table 4) and when
these are used to estimate the masses exported from the
larger catchment, for any given discharge, the level of

agreement is good. For example, for one autumn storm with a discharge of 159 x 10^3 m^3 lasting 16 hours, it was predicted that 11.5 tonnes of suspended solids were exported compared with 13.9 tonnes observed, 22.4 tonnes of soluble solids were predicted compared with 24.6 tonnes observed, and 15.5 kg of ammoniacal nitrogen were predicted compared with 19 kg observed. Further examination of the relationship between the masses of suspended solids discharged by the same storm events from the sub-catchment and the total catchment demonstrates that they are linearly related in proportion to the areas of the two catchments.

CONCLUSIONS

1. Run-off from a purely residential area is of poor quality and carries a significant pollutant load; particularly noteworthy are the mean concentration of suspended solids and the maximum concentrations observed for ammoniacal nitrogen, chloride, and heavy metals.

2. Snow-melt run-off is of much poorer quality than rainfall run-off because of road-salting operations, less efficient performance of the internal combustion engine in the cold at low speeds, and contamination of the snow by direct contact with the underside of vehicles.

3. A 'first flush' (i.e. occurrence of the peak concentration within the first 40 minutes of discharge) was observed in 80 per cent of the events studied.

4. The flow-weighted mean concentrations of pollutants in the run-off decreased with the increasing duration of discharge but had not decreased by more than 40 per cent even after 100 minutes of discharge.

5. The length of the antecedent dry period and the magnitude of the previous run-off event have little effect on the quantity of pollutants discharged.

6. Most of the particulate solids in run-off originate on roads and paved areas.

7. Roof run-off makes a negligible contribution to the pollutant load of run-off and has a diluting effect on road run-off.

8. Road drain catchpits represent a significant reservoir of poor quality water, the behaviour of which is difficult to predict.

9. Flushing of road drain catchpits during rainfall can contribute significantly to the pollutant load of run-off.

10. The annual yield of material per unit area from the
residential sub-catchment is similar to that calculated per
unit area for the whole New Town, which includes a large
industrial area.

ACKNOWLEDGEMENTS

Much of the work reported was funded by the Department of the
Environment. This paper is published with the Department's
approval and by permission of the Director, Water Research
Centre.

REFERENCES

Ackers, P. (1963) Tables for the Hydraulic Design of Storm-
drains, Sewers and Pipe-lines. Hydraulics Research Paper
No. 4. H.M. Stationery Office, London.

Cawse, P. A. (1974) A survey of atmospheric trace elements
in the UK (1972-73). A.E.R.E. Report R7669. H.M. Stationery
Office, London.

Cawse, P. A. (1975) A survey of atmospheric elements in the
UK: Results for 1974. A.E.R.E. Report R8038. H.M. Stationery
Office, London.

Ellis, J. B. (1976) Sediments and water quality of urban
stormwater. Water Services, 730-734.

Hedley, G., and Lockley, J. C. (1975) Quality of water discharged from an urban motorway. Water Pollution Control,
74(6), 659-674.

Laxen, D. P. H., and Harrison, R. M. (1977) The highway as
a source of water pollution: An appraisal with the heavy
metal lead. Water Research, 11, 1-11.

Lester, W. F. (1975) 'Pollutant River: River Trent,
England'. In 'River Ecology' (ed. B. A. Whitton). Studies
in Ecology Volume 2, Blackwell Scientific Publications,
Oxford, 489-513.

Loehr, R. C. (1975) Non-point pollution sources and control,
269-299. In 'Water Pollution Control in Low Density Areas'.
Proceedings of a Rural Environmental Engineering Conference
1975 (ed. W. J. Jewell, and R. Swan). Publ. for Univ. of
Vermont by University Press of New England.

Malmquist, P- A., and Svensson, G. (1977) Urban stormwater
pollutant sources. In 'Effects of Urbanization and Industrial
ization on the Hydrological Regime and on Water Quality
(Proceedings of the Amsterdam Symposium, October 1977), 31-38:
IAHS Publ. No. 123.

Ministry of Housing and Local Government. (1970) Final
Report of the Technical Committee on Storm Overflows and the
Disposal of Storm Sewage. H.M. Stationery Office, London.

Nicholl, E. H., and McGillivray, R. (1977) Report of the
Working Party on Storm Sewage (Scotland): A Review.
Presented at the Annual Conference (Brighton, September 1977)
of the Institute of Water Pollution Control.

Stevenage Development Corporation. (1971) Stevenage Household Survey, March 1971. 63pp.

Tucker, C. G. J. (1975) Pollution loads in runoff from
urban areas. M.Sc. Thesis, Loughborough College of Technology.

Waller, D. H. (1972) Factors that influence variations in
the composition of urban surface runoff. Water Pollution
Research in Canada, 7, 68-95.

Wilkinson, R. (1956) The quality of rainfall run-off from
a housing estate. Instn Publ. Hlth Engrs J., 55, 70-78.

EVALUATING EFFECTIVENESS OF URBAN DRAINAGE PROJECTS

By Neil S. Grigg[1]

ABSTRACT

In the drive to increase the effectiveness of urban public investments, drainage and flood control often receives little notice. It is possible to apply simple but effective multiple objective planning tools to the analysis of these projects. Techniques currently in use in Denver, Colorado, and Los Angeles, California, are cited. These techniques can be used for: water planning; programming and budgeting; prioritizing and investment timing; cost allocation; and *post facto* evaluation.

EVALUATING EFFECTIVENESS OF UDFC PROJECTS

Urban drainage and flood control UDFC projects are desirable and in some cases essential components of the infrastructure of cities. Wealthy developed countries exhibit stronger demand for the amenity aspects of drainage, but even the poorest human settlements need a basic level of drainage for the purposes of sanitation, protection from flood damage, and to allow convenient living during and after periods of rainfall. This paper has as its objective to define effectiveness criteria for UDFC projects and to articulate the state-of-the-art in evaluating such effectiveness.

CASE FOR EVALUATION

In a day where demands on the public purse are very pressing, there is a natural tendency to examine carefully the effectiveness of every investment. Symptoms of this in the USA include the "sunset laws" and the "zero-based" budgeting techniques currently being used in the Federal Government. Public works managers will increasingly need to defend their proposals and programs with hard facts and logical demonstrations of effectiveness. Where they are able, public works managers should, in fact, have trained socio-economic analysts on their staffs. In all cases, managers should develop such capabilities themselves; certainly, the skill to develop and present the facts.

---

[1]Director, Water Resources Research Institute of The University of North Carolina, NCSU, Raleigh, NC.

UDFC has always been a difficult program component to justify. Once a pipe or channel is built, it is not a visible and obvious improvement to the infrastructure. A park, library, or paved road has much more presence. To present UDFC in its best light, then, the public works manager must be prepared to evaluate effectiveness.

In fact, evaluation really needs to be carried out at five different steps of the implementation process:

1. Master Planning - Selecting the best alternatives
2. Programming and Budgeting - To get projects implemented
3. Prioritizing and Investment Timing - To set the sequence of implementation
4. Cost Allocation - To see that beneficiaries pay appropriate costs
5. *Post facto* Evaluation - To set the stage for further action and improve the management process

Evaluation of UDFC projects is thus seen to really be a subset of the water resources planning and management process.

ELEMENTS OF EVALUATION

In an earlier paper Grigg (1975) traced the elements of evaluating UDFC projects. Such evaluation must take into account benefits which take into account protective, economic, and amenity objectives which are fulfilled by projects. Table 1 summarizes the usual operational objectives to be met by UDFC.

Table 1. Objectives of UDFC and Measures of Performance

Objectives	Performance Measure
Protective	
To minimize property damage from all types of flooding	Average Annual Property Damage
To eliminate loss of life due to flooding	Expected Loss of Lives
To alleviate health hazards from water hazards caused by unsanitary conditions	Absence of Hazards
To reduce traffic accident hazards due to street flooding	Presence (Absence) of Hazards
Other Economic	
To enhance neighborhood land values by improving the urban environment	Measured Land Values

(Continued)

Table 1 (Continued)

Objectives	Performance Measure
	Other Economic
To reduce street maintenance costs by prevention of run-off damage	Expected Maintenance Budget
To reduce liability of property owners and land developers associated with runoff-producing land development	Presence (Absence) of Potential Liability
	Amenity
To improve the visual and aesthetic impact of the urban environment	Scale of Aesthetic Value
To provide recreational opportunities where possible	Quantity of Recreational Opportunities
To make urban life more convenient by the reduction of delays and other inconveniences associated with drainage problems	Travel Time, Cleaning Bills, etc.

In evaluating UDFC projects, a multi-objective framework must be used. Such a framework must be practical enough for application by public agencies. One of the problems of many elaborate multi-objective evaluation procedures is complexity. A large regional flood control district in Denver, Colorado, (The Urban Drainage and Flood Control District, UDFCD) has recently begun testing a practical procedure based on multi-objective techniques but stressing simplicity (Grigg, *et al*, 1977). The procedure utilizes a process such as that shown in Figure 1.

The UDFCD has listed the specific objectives shown in Table 2 as those which appear to warrant most consideration in the Denver area. These are framed more operationally than those in Table 1.

Table 2. Objectives of Projects in UDFCD

1. To reduce flood damage and maintenance requirements to public and private property and facilities.
2. To enhance the value of land and other property in neighborhood services by UDFC facilities.
3. To improve the visual impact and aesthetic quality of the urban landscape.
4. To improve the quality of water resources.

(Continued)

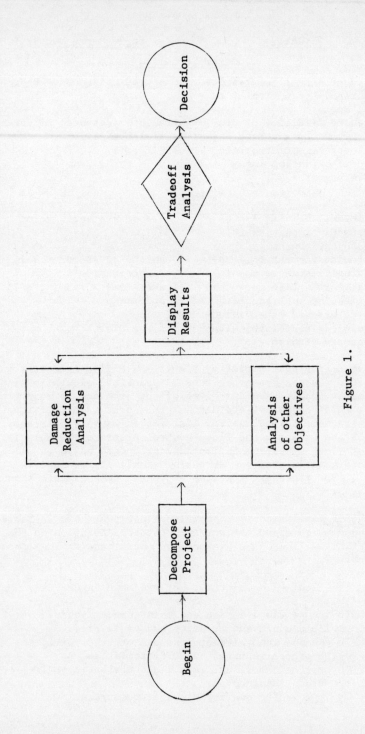

Figure 1.

Table 2 (Continued)

5. To preserve the ecological and environmental values of natural floodplain areas.
6. To reduce threat to life from floods.
7. To reduce UDFC-related health hazards.
8. To enhance recreational opportunities.
9. To reduce public inconvenience.
10. To reduce traffic hazards from flood waters.
11. To enhance emergency vehicle movement.
12. To make effective use of public funds.

The UDFCD methodology must be applied under the normally difficult constraints found in urban areas of:

1. Budget constraints
2. Environmental constraints
3. Public acceptability
4. Changing priorities
5. Development goals
6. Growth management

The application of the methodology used can be seen from the matrix display given in Table 3. Such a display represents, in reality, the state-of-the-art of practical, multiobjective evaluation studies. The aim is to put as much clear, concise information before the decision maker as possible.

Another very interesting evaluation process is currently being used by the Los Angeles County Flood Control District (LAFCD). Their process was developed to meet the following five criteria: (Vance, 1977)

1. It must have a reliable, objective quantitative method for evaluating flooding problems and assessing the relative merit of project candidates.
2. It must include adequate public participation so that all community concerns are discovered and dealt with before the project enters the design and right-of-way acquisition processes.
3. It must recognize the practical and legal importance of environmental impact considerations, striving to optimize environmental protection and project performance.

(Continued)

4. It must consider enough projects so that the
    normal project mortality rate is accommodated
    with enough survivors to sustain the desired
    construction rate.
5. It must be efficient and economical.

The LAFCD uses a procedure known as the "funnel screen." All
possible projects are screened for economy, community support,
engineering feasibility, and environmental acceptance. If the
final screen is passed, an environmental impact report (EIR)
is prepared and used as a project document to obtain internal
approval. Only then does the project enter the design, budgeting and right-of-way acquisition process.

The four screens described above are applied systematically.
The economic screen uses an innovative approach called the
"Beneval" equation. "Beneval" is an acronym for "Benefit
Evaluation of Urban Storm Drains." An equation has been de-
veloped to measure the contribution of a series of independent
factors on project value. The resulting equation is:

$$V = 319Q - 766,080S - 78L +$$
$$32P + 2.7T + 0.07AV - 44,000 \tag{1}$$

where:

    V is the project value in dollars
    Q is the design Q in cfs
    S is the slope of the flow path in feet/foot
    L is the length of the flow path in feet
    P is the population along the flow path
    T is the average daily traffic count through the
      flooded area
    AV is the assessed valuation along the flow path

The Beneval equation was derived by analysis of projects receiving voter approval during past years. In this sense, it
seeks to measure "market signals" better than techniques such
as benefit-cost analysis.

WATER QUALITY PROBLEMS

Pollution from stormwater runoff is a serious problem. Nevertheless, it is difficult to integrate water quality management
considerations into the multiple objective UDFC planning process at the present time. Perhaps the reason is that we are
still looking for effective solution strategies for this
problem. Hopefully, such strategies will be found soon and
projects can be optimized with water quality included.

Table 3. Matrix Display of Benefits of Alternatives in Decision Unit 103-105

Alternative / Objective	NON-STRUCTURAL MEASURES ONLY		10-YEAR HARD CHANNEL AND SELECTED NONSTRUCTURAL MEASURES		10-YEAR SOFT CHANNEL AND SELECTED NONSTRUCTURAL MEASURES	
	+	−	+	−	+	−
1. Damage Reduction	Insignificant reduction of property damage to existing property		Reduces Annual Damages $75,000		Reduces Annual Damages $75,000	
2. Effective Use of Public Funds		Annual cost including maintenance $15,000		Annual cost including maintenance $52,000		Annual cost including maintenance $41,000
3. Improve Visual Impact	No change to existing visual impact		Positive impression achieved by careful landscaping	Concrete channel appearance will be objectionable to some citizens	Very significant visual improvement by soft channel design	
4. Enhance Recreation	No change in recreation opportunities		No change	Presence of lined channel prevents future option of developing trails	Opportunity exists to integrate bike/hiking trail in with channel design	
	Net benefits Indeterminate B/C Ratio Indeterminate Investment Required = $120,000		Net Benefits = $23,000/year B/C Ratio = 1.44 Investment Required = $500,000		Net Benefits = $34,000/year B/C Ratio = 1.83 Investment Required = $360,000	

CONCLUSION

Evaluating the effectiveness of UDFC projects is a necessary
capability for all public works departments. Although the
techniques currently available are rather simple, they can be
used to improve the decision-making process. They represent
a logical extension into multi-objective analysis for the
drainage engineer and public works manager alike.

REFERENCES

Grigg, Neil S., *et al*, (1975), "Urban Drainage and Flood Control Projects: Economic, Legal and Financial Aspects," Environmental Resources Center Report No. 65, Colo. State University, July 1975.

Grigg, Neil S., Tucker, L. S. and Urbonas, B. (1977), "A
Practical Methodology for the Evaluation of Urban Drainage and
Flood Control Projects," submitted to the Journal of Water Re-
sources Planning and Management, ASCE.

Vance, Harold A. (1977), "Planning for Success," Paper given
at International Symposium on Urban Hydrology, Hydraulics and
Sediment Control, University of Kentucky, July 18-21, 1977.

ACKNOWLEDGEMENTS

Financial support for this paper has been provided by the US
Department of Interior, Office of Water Research and Technology,
through the Environmental Resources Center, Colorado State
University.

A PROCEDURE FOR CALCULATING THE COST OF STORM WATER SEWER CONSTRUCTION

D. M. Farrar, Transport and Road Research Laboratory, Crowthorne, UK
P. J. Colyer, Hydraulics Research Station, Wallingford, UK

INTRODUCTION

The Hydraulics Research Station (HRS) and the Transport and Road Research Laboratory (TRRL) are engaged on research into the design, construction and maintenance of sewer pipelines; HRS is concerned with hydraulic design (Price, 1978) and TRRL with structural design (Boden et al, 1977). To make engineering decisions on a rational basis, it is essential to have a detailed knowledge of the materials and operations involved in construction, and their costs.

Information on costs can be obtained either from the study of successful tenders, or from site studies of construction operations. The Water Research Centre (1977) has used the former approach of examining tender costs. This approach has the advantage that the actual total cost to the client is obtained (apart from the effect of claims and variations). It is probably, therefore, the most suitable approach for broad considerations of national or regional planning. The disadvantage is that the breakdown of costs in the tender documents is not very detailed, is not presented in a standardised form in the United Kingdom, and does not necessarily represent the real cost distribution.

The alternative approach of using site studies can be used to build up a detailed picture of the resources and operations involved. Farrar (1977) has used this approach to develop a procedure for calculating the cost of laying rigid sewer pipes with a granular bedding. The procedure is a simple one, requiring a very limited data base. Bramwell (1974) has independently examined the resources required for pipeline and manhole construction using an extensive data base and large computer program. The disadvantage of the resource cost approach is that the results may differ considerably from actual tender costs, depending on such factors as the extent of works ancillary to the actual pipeline, the contractor's other commitments, and his estimate of the state of the economy.

This paper describes a construction cost model which has been incorporated into a set of programs for the design and simulation of storm sewer systems. The procedure used by Farrar was selected as a simpler approach requiring less information on the contractor's methods of operation and giving an output expressed in resource costs. Its use is discussed and illustrated by reference to a real drainage scheme.

627

DATA

A full description of the data used is given elsewhere (Farrar, 1977). This section of the paper describes how the data was obtained, and the next section how it is used in the present cost model.

Timing of elements

Observations were made at a number of sewerage contracts, in which the materials, plant and labour elements employed to construct rigid pipes with flexible O-ring joints were recorded and timed. The full sequence of operations, where required, is: breakout of carriageway, excavation, trench support, pipe bedding, pipe-laying and testing, backfilling and compaction, re-instatement of carriageway, and removal of surplus soil. The rate of trench excavation was found to be very dependent on the ease of operation of the excavator, and ranged from 0.04 m^3/min for close sheeting in unstable ground to 1 m^3/min for trenches with battered sides in open field. It was not possible to obtain reliable times for excavation in rock or some other difficult ground conditions.

Efficiency of operations

It was not possible to make sufficiently detailed site studies to determine the variability in the times of the operations, or in the time lost in non-productive work. An indication of the overall performance of the pipe-laying gang was obtained by comparing the time that should have been taken if all the operations described in the previous paragraph had been carried out with no interruptions or variations with the time actually taken to construct a length of pipeline. The ratio of these times was found to be in the range 35-70 per cent.

Cost of materials, plant and labour

The cost of supplying pipes to site in bulk is taken as 20 per cent less than manufacturers' list price for clay or Class 'M' concrete. The cost of contractor's plant, materials and labour is based principally upon Davis et al (1976); these costs are necessarily approximate and will vary with locality and circumstances.

CALCULATION OF COSTS

Input

The information available on site conditions and the contractor's mode of operation is often minimal at the design stage. Site conditions do, however, have a substantial influence on cost. The input required for each cost calculation has been minimised as far as possible, but this minimum is essential for a realistic estimate. The input required is summarised in the table below.

The definition of site conditions is required to establish the need for excavation and re-instatement of carriageway, and the feasibility of using trenches with battered sides. The definition of ground conditions is required to establish the rate of excavation and the need for trench sheeting.

TABLE 1   Data requirements

Input	*Form*
Internal diameter of pipe	(m)
Average depth to invert	(m)
Maximum depth to invert	(m)
Length of run	(m)
Site conditions	1. Road
	2. Pavement
	3. Unpaved urban (eg gardens, verges)
	4. Open field
Ground conditions	1. Less than 1 service per 10 m run (ie rural)
	2. 1 service per 2-10 m run (ie suburban)
	3. More than 1 service per 2 m run (ie dense urban)
	4. Unstable (rural)[†]

† Requiring continuous support but not any specialised geotechnical process.

Assumptions

The cost of storm sewer construction is calculated using the times of operations and costs obtained as described in the previous section. In addition, the following assumptions are made:

(a)   The utilisation of plant is 50 per cent, and of labour 75 per cent (ie costs are multiplied by 1/0.5 for plant and 1/0.75 for labour).

(b)   Pipe-laying runs are sufficiently long for the cost of bringing plant on to site to be negligible.

(c)   An allowance is made for the cost of ancillary plant (eg traffic lights). This would not be sufficient for a site where extensive de-watering was required.

(d)   A crane is employed to lift pipes greater than 750 mm internal diameter.

Output

The cost model can be used to give a breakdown of total cost subdivided into the cost of each operation, and also subdivided into materials, plant and labour. The calculation is readily performed on a computer. The majority of the calculations were carried out on an ICL 1904S (128K words) but calculations have also been carried out on a mini-computer.

The structural design of the sewer is not considered in detail in the present model (Farrar, 1977). A single design is assumed which is structurally satisfactory for depths of cover between 1 m and 3 m, which are the depths employed in the majority of storm sewers. Alternative designs will be required in some circumstances for depths outside this range.

Costs are given at January 1975 prices; no attempt has yet been made to up-date these prices since to date they have only been used for purposes of cost comparison and optimisation.

APPLICATION

Storm drainage systems have traditionally been designed to convey without surcharge the discharges produced by rainstorms of arbitrarily selected return periods. The relative costs and performance of alternative designs have not normally been investigated. The Hydraulics Research Station has developed a design procedure which incorporates such evaluation, in order to place the design of storm drainage on a more sensible economic basis. The procedure makes use of the construction cost program in two ways:

(a)   To optimise depth, gradient and diameter for minimum construction cost (Price, 1978).

(b)   To compare construction cost with performance of designs based on different rainfall return periods.

Some applications of the cost program to engineering design are now described.

The cost of overdesign

It is sometimes claimed that accurate hydraulic design of pipe diameters is not important because the costs of the pipes themselves are insignificant in relation to total construction costs. The program was used to investigate this claim. The pipe cost as a proportion of total construction cost is shown in Figure 1 for a range of typical depths. Site condition 1 and ground condition 2 were selected to give costs in typical suburban conditions. To assist the comparison of engineering alternatives, depths of cover have been plotted rather than the depths to invert used by the program. The relationship between cost and diameter is of course discrete, but continuous lines have been drawn in the figures to clarify trends. Figure 1 shows that while the costs of smaller pipes normally account for less than 20% of total construction costs, the cost of large pipes may account for over 50%.

This might suggest that accurate hydraulic design is not required for the smaller pipe sizes. However, Figure 2 illustrates the cost of overdesign as the percentage increase in construction cost over the construction cost of a pipe one diameter increment (75 or 150 mm) smaller. Results for a standard cover depth of 2 m are illustrated, but similar results were obtained for other depths. The following conclusions may be drawn from Figure 2:

(a)   For pipes of 300 mm diameter and less, the percentage cost increases are relatively small.

(b)   From 375 mm up to 1200 mm diameter, the percentage cost increase tends to decrease with increasing diameter.

(c)   Very large increases occur at specific points in the diameter range. These are related to the following assumptions in the cost program:

— pipes over 300 mm diameter require increased trench widths.

— pipes over 750 mm diameter require craneage.

— trenches of 3 m depth or greater require closed sheeting and the use of a tracked excavator. (For pipes over 975 mm diameter in the present example.)

— above 1200 mm, pipes are available only in 150 mm increments.

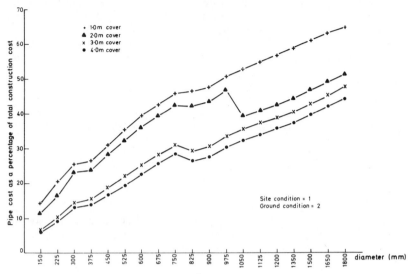

Figure 1. Pipe cost as a proportion of total construction cost

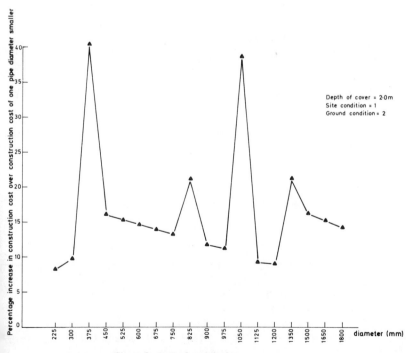

Figure 2. Cost of overdesign

It is concluded that specific construction techniques have a significant influence on the cost of the total range of sewer pipe sizes, and that detailed hydraulic design is justified at least for pipes of 300 mm and above.

The results shown in Figure 2 were checked using real pipe network data for a storm drainage catchment in south-west England (Hydraulics Research Station, 1976). The typical tree-shaped network contained 112 pipes between 150 mm and 1125 mm diameter, mostly at gradients steeper than 1 in 100. By changing all the designed pipe diameters (above a specified minimum of 150 mm) by one or two increments and calculating the total construction cost of the network it was found that:

Reduction by one increment produced a cost reduction of 12%.
Reduction by two increments produced a cost reduction of 19%.
Increase by one increment produced a cost increase of 16%.
Increase by two increments produced a cost increase of 36%.

These figures conform with the results in Figure 2. The percentage increases were larger than the corresponding reductions because of the number of pipes affected by the requirement to maintain a minimum diameter. In the existing system 12 of the 112 pipes were already at the assumed minimum diameter (150 mm), and after one reduction the total rose to 36.

Effect of construction depth
The effect of construction depth on total costs (again for suburban conditions) is shown in Figure 3, in which a pipe cover of 2 m has been taken as the base condition. The figure shows the same discontinuities as in Figure 2. A depth reduction from 2 m cover to 1 m cover introduces construction cost savings of 10 to 25%. Depth increases to 3 m or 4 m produce cost increases of about 10 to 70% and 15 to 90% respectively.(There would be an additional cost increase of about 5% for larger pipes because a stronger class of pipe would be required at 4 m cover). Such large percentage changes show that the selection of pipe depth is an important factor in the design of sewerage schemes.

The results in Figure 3 were also checked by comparison with the real pipe network (Hydraulics Research Station, 1976). Since most of the network had not been constructe it was possible to make the idealised assumption of a uniform depth of cover throughou the network. Taking a pipe cover of 2 m as the base case, a reduction to 1 m introduce a cost reduction of about 15%. An increase in depth of cover to 4 m caused a cost increase of 58%. Again these figures conform with the results in Figure 3.

Effect of site condition and ground condition
These conditions are usually outside the designer's control, but it is desirable to know their significance in relation to other factors over which the designer does exert some control. The sensitivity of the cost model to changes in these conditions was, therefore, examined using the real pipe network.

The construction cost of the scheme was calculated using pipe diameters designed for th catchment and typical depths, varying both the site condition and ground condition indices (as defined in Table 1) from 1 to 4. The results of the 16 tests are shown in Figure 4, from which the following conclusions may be drawn:

(a)   There is little or no cost difference between ground conditions 1 and 2, nor between ground conditions 3 and 4, nor between site conditions 3 and 4.

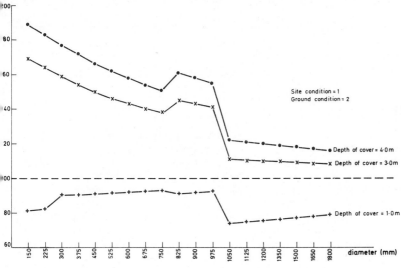

Figure 3. Effect of depth on construction cost

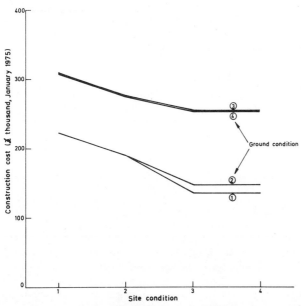

Figure 4. Effect of site condition and ground condition on
construction cost of a real network

(b)   Ground conditions 3 and 4 are between 40% (at site condition 1) and 80% (at site condition 4) more expensive than ground conditions 1 and 2.

(c)   Site condition 2 is 10-15% cheaper than site condition 1, and site conditions 3 and 4 are 15-35% cheaper than site condition 1.

The largest single factor in this group is, therefore, whether the ground condition is of type 1-2 or 3-4.

CONCLUSIONS

An investigation into sewer construction practice has yielded a flexible cost model which is being used in conjunction with hydraulic design calculations to permit a more logical cost-based approach to storm drainage design. The model is based on resource costs, which may not represent the total costs of a drainage scheme. The effects of pipe diameter, depth and site and ground conditions on cost have been evaluated. It is concluded that detailed hydraulic design is justified for all pipe diameters above 300 mm; that considerable savings may be achieved through the minimisation of pipe depth; and that construction in densely developed urban conditions or in unstable ground causes particularly high construction costs.

ACKNOWLEDGEMENT

The work described in this paper formed part of the research programmes of the Transport and Road Research Laboratory and the Hydraulics Research Station, and is published with the permission of the Directors of Research.

REFERENCES

Boden, J. B., Farrar, D. and Young, O. C. (1977) Standards of site practice – implications for innovation in the design and construction of buried pipelines. Conf. on Opportunities for Innovation in Sewerage, Reading.

Bramwell, D. M. (1974) Computer aided systems in civil engineering using drainage as the prime data base. Ph.D. Thesis, Dept. of Civil Engng., University of Aston, Birmingham.

Davis, Belfield and Everest, Eds. (1976) Spon's architects' and builders' price book. 101st Edition, London (E. and F. N. Spon Ltd.).

Farrar, D. M. (1977) A procedure for calculating the cost of laying rigid sewer pipes. Department of the Environment, Department of Transport, TRRL Supplementary Report SR333: Crowthorne (Transport and Road Research Laboratory).

Hydraulics Research Station (1976) Pool Valley Storm Drainage Scheme, Report No EX 757.

Price, R. K. (1978) A design and simulation method for storm sewers. International Conference on Urban Storm Drainage, Southampton.

Price, R. K. (1978) Design of storm sewers for minimum construction cost. International Conference on Urban Storm Drainage, Southampton.

Water Research Centre (1977) Water and sewage cost data. Technical Report TR 61.

CROWN COPYRIGHT 1977.

DESIGN OF STORM SEWERS FOR MINIMUM CONSTRUCTION COST

R. K. Price, Hydraulics Research Station, Wallingford, U.K.

ABSTRACT

A comparison is made of two design methods for urban storm sewers by which pipe diameters and depths are determined for a minimum construction cost. The first method is based on an extension of the discrete differential dynamic programming approach adopted by Mays and Yen (1975) whereas the second method uses non-linear programming and regards the pipe diameters as being continuous variables. In both methods the construction cost is minimised for constraints imposed on the flow and depth of each pipe. The construction cost is determined from a routine developed by the Transport and Road Research Laboratory. The methods can be applied to a sewer system of any size, though for large systems a new pipe-by-pipe procedure has been adopted. A description is given of the application of both methods to the redesign of an existing sewer system, together with a discussion of their relative merits and disadvantages.

INTRODUCTION

The large capital investment in sewerage and the need for the renovation and renewal of aging pipe systems is forcing planning authorities to exercise a more stringent financial control over new sewerage projects. Whereas it is possible that a more efficient use of resources can be made in each of the four main areas of planning, design, construction and maintenance, this paper considers methods available to the design engineer so that he can minimise the construction cost of a new sewerage system.

Arguably a complete philosophy for the design of a sewerage system has the objective of producing a system which minimises both the construction and damage costs. Currently the quantification of possible damage costs at a particular location is very uncertain compared with corresponding construction costs. In addition modelling techniques which can provide an accurate prediction of surface flooding of specified return period in an urban catchment are only beginning to be developed. Consequently it is unlikely that a satisfactory design method using the complete approach can be developed in the immediate future. However, if damage costs are ignored there are a number of attractive possibilities for designing sewer systems such that the construction costs alone are minimised. The two main possibilities are the calculation of pipe slopes and diameters for a given layout and the optimisation of the plan layout.

636

One of the most advanced methods for optimal plan layout is that of Mays (1976) in that it can design a system for time-dependent flows, but the penalty common to all such methods is a very large computing time and their solutions are generally sub-optimal. There are also serious doubts as to the feasibility of plan optimisation in the UK except in a few specialised cases, for example in very flat catchments where constraints are few. Consequently more research effort has been put into deriving efficient techniques for calculating pipe slopes and diameters when the layout is given. The more notable methods developed to date are due to Danjani and Hasif (1974) who employ linear programming, Merrit and Bogan (1973), Argaman et al (1973), Walsh and Brown (1973) and Mays and Yen (1975) who use dynamic programming techniques, and Gupta et al (1976) and Lemieux et al (1976) who adopt non-linear programming. Each method appears to have its advantages and disadvantages. The main criticism levelled at methods using linear and dynamic programming is that their solutions can be strongly sub-optimal (see Lemieux et al, 1976), whereas the non-linear programming methods are suspect when the objective cost function has large discontinuities. Both of the non-linear programming methods produced by Gupta et al (1976) and Lemieux et al (1976) were produced for continuous convex cost functions. However, a realistic cost function applicable to UK conditions has discontinuities due to the various techniques of laying pipes at different depths; see Farrar and Colyer (1978). It is possible that a cost function with such a structure would make it difficult for the non-linear programming techniques to converge to an optimal solution in certain cases.

To explore further the relative advantages and disadvantages of the dynamic and non-linear programming techniques two new methods were derived for the design of pipe depths and diameters in a storm sewer network; the first is an extension of the discrete differential dynamic programming (DDDP) method developed by Mays and Yen (1975) and the second uses a quasi-Newton algorithm developed by Gill et al (1976). To overcome the considerable cost involved in computing an optimal solution simultaneously for all pipes in a large system both methods adopt a pipe-by-pipe approach to design.

PIPE DEPTH-DIAMETER OPTIMISATION

The objective of the optimisation is to minimise the total construction cost (including cost of pipe, trenching, laying, resurfacing etc) of a sewer system according to certain constraints. The construction cost for an individual pipe and its upstream manhole is determined from the cost function produced by the Transport and Road Research Laboratory and reported by Farrar and Colyer (1978). The constraints include

(i)   the depth of cover for each pipe is greater than some minimum value;

(ii)   the depth of cover for each pipe is less than some maximum value;

(iii)   the diameter of each pipe is adequate for the pipe to convey a given discharge; and

(iv)   the pipe-full flow velocity in each pipe is greater than some minimum value to prevent deposition of sediment.

The current versions of the methods described in this paper test for constraints (i) and (ii) only at the upstream and downstream ends of each pipe. Constraint (ii) is not required for structural reasons as the cost function is applicable for any depth. However, this constraint is included so that certain conditions, such as a specified outfall depth, are achieved.

To ensure savings in computer time the design discharges, Q, at each manhole are determined from the Rational formula

$$Q = CAi \quad\quad\quad ....(1)$$

where CA is the total contributing impervious area and i is the rainfall intensity. However, the final design of the pipes is done using an accurate simulation method; see Price and Kidd (1978). This procedure is described in more detail below.

DISCRETE DIFFERENTIAL DYNAMIC PROGRAMMING (DDDP) METHOD

The DDDP method has been described by Mays and Yen (1975) and Yen et al (1976) and the method used in this paper is similar.

Consider a convergent tree sewer network consisting of a number of pipes each of which is connected to an upstream and a downstream manhole. The set of inflow (upstream) invert elevations, $\{S_n\}$, of the set of stages $\{n\}$ (pipes and manholes) are related to the set of outflow (downstream) invert elevations $\{\tilde{S}_n\}$, by

$$S_{n+1} = \tilde{S}_n$$
$$S_n \geqslant \tilde{S}_n + \epsilon L_n \quad\quad\quad ....(2)$$
$$\tilde{S}_n = S_n - D_n$$

where ϵ is zero for manhole stages and equals the minimum permissible slope for pipe stages, L_n is the length of the pipe at the nth stage, and D_n is the drop in elevation across the nth stage. If the return (or cost) of stage n is $r_n(S_n,D_n)$ the recursive relations used in minimising the total cost are

$$f_n(\tilde{S}_n) \begin{cases} = \displaystyle\min_{D_n} \; [r_n(S_n,D_n)] & n = 1 \\ = \displaystyle\min_{D_n} \; [r_n(S_n,D_n) + f_{n-1}(\tilde{S}_{n-1})] & n = 2, ..., N \end{cases} \quad ....(3)$$

where $f_n(\tilde{S}_n)$ is the minimum cost of the system down to stage n for n = 1, ..., N.

The DDDP method first assumes a set of inflow and outflow invert elevations (or states) called the trial trajectory. A number of other states are then defined in the neighbourhood of the trial trajectory forming a corridor; see Fig 1. The conventional dynamic programming approach is then applied to states within the corridor according to the relationships (2) and (3) and the constraints on pipe diameter and depth of cover outlined above. The least cost trajectory is sought and then adopted as the improved trajectory to form a new corridor. Mays and Yen refer to the process of forming a corridor and finding the least cost trajectory as an iteration. With each iteration the width of the corridor is reduced until the relative change in cost between successive least cost trajectories is less than some permissible error in cost or the width of the corridor is less than a specified value.

All tests of the method reported below use a 5 lattice point corridor with state increment Δs halved at each iteration, provided the absolute relative change of the minimum cost of the last and previous iterations is less than a specified value. This follows the conclusions reached by Mays and Yen.

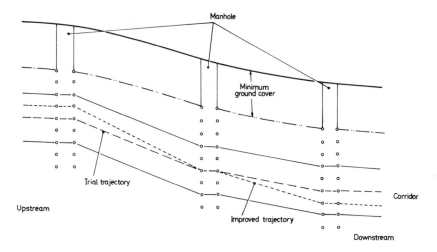

Figure 1. Sketch of stage – corridor representation (after Yen et al 1976)

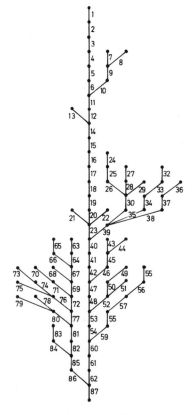

Figure 2. Schematic diagram of the pipe system for St Marks Rd., Derby

NON-LINEAR PROGRAMMING (NLP) METHOD

A survey of various numerical methods for constrained optimisation is given by Murray (1976). The second model used in this paper is based on an algorithm developed by Gill et al (1976) and described in Murray's survey.

Consider the vector, x, of all pipe gradients in a network and assume that the depth, d_1, of the manhole upstream of the first pipe in the main branch is prescribed. Suppose also that invert levels of incoming pipes and the outgoing pipe are aligned at each manhole. If peak discharges are known at all manholes the diameter (treated as a continuous variable) for each pipe can be determined from the upstream discharge and the pipe gradient. Consequently the total construction cost, F, can be defined in terms of x and d_1. The objective now is to minimise $F(x)$ subject to the constraints $c(x)$, including limits on depth of cover and flow velocity in each pipe and the possible requirement that prescribed invert levels for manholes at the top of branches other than the main branch are achieved.

The problem is first transformed to canonical form by including additional slack variables to convert the inequality non-linear constraints to equality constraints. Then the augmented Lagrangian function

$$L(x, \underline{\lambda}, \rho) = F(x) - \underline{\lambda}^T c + \rho c^T c \qquad \qquad ....(4)$$

if formed, where ρ is a penalty parameter and $\underline{\lambda}$ is the vector of Lagrange multipliers defined by

$$\underline{\lambda}(x) = (A^T(x)A(x))^T \, A^T(x)g(x) \qquad \qquad ....(5)$$

Here $A(x)$ is the Jacobian of $c(x)$ and $g(x)$ is the gradient vector of $F(x)$. Given estimated of $\underline{\lambda}$ and ρ, L is minimised using a quasi-Newton algorithm subject to the bounds on the original and slack variables. The solution is then used to generate better estimates for $\underline{\lambda}$ and ρ, and the process is repeated until a sufficiently accurate solution has been achieved. The optimisation procedure is carried out in this model using the FORTRAN subroutine SALQDR from the National Physical Laboratory Algorithms Library.

Several versions of the model have been investigated. Earlier versions concentrated on pipe depths as the independent variables. However, this led to problems with pipe gradients becoming too small during the optimisation. Therefore, the current version of the model uses pipe gradients as the independent variables with the manhole depth at the top of the main branch defined as an additional variable when it is not specified independently.

A general disadvantage of the NLP model is that predicted pipe diameters have eventually to be adjusted to the nearest commercial sizes. If this is done after the derivation of the solution, such as is recommended by Lemieux et al (1976), the spare capacity available in some pipes would be wasted. A method of partially overcoming this difficulty is explained below.

DESIGN USING AN ACCURATE SIMULATION MODEL

A criticism of both of the above methods is that pipe diameters are designed using the Rational method. This method is generally accepted as being inaccurate compared with methods using hydrograph simulation, such as the TRRL method (Watkins, 1962). The replacement of the Rational method by, for example, the TRRL method in the optimising methods above would greatly increase computing time. However, the pipe diameters can be designed using an accurate simulation method in the following manner.

Consider a pipe network with N isonodal levels. Here the nth isonodal level consists of those manholes which are $(n-1)$ manholes from the outfall (regarded as the root manhole). Suppose that the design peak discharges at all manholes on the first isonodal level have been calculated using an accurate hydrograph simulation model for the sub-catchments feeding the manholes. These calculated discharges can replace discharges predicted using the Rational method. However the design discharges for the remaining manholes are still to be calculated using the Rational method. This raises the difficulty that the discharges calculated by the Rational method may have a different return period from those adopted for the manholes on the first isonodal line. Consequently the return period used in the Rational method is adjusted to the return period, R, required for the Rational method to give the largest peak discharge for the manholes on the first isonodal line as predicted by the more accurate simulation method.

Generally the design rainfall intensity, i, for the Rational method can be defined by

$$i = \frac{a_1 R^{a_2}}{(a_3 + t_c)^{a_4}} \qquad \qquad ....(6)$$

where t_c is the time of concentration and a_1, a_2, a_3 and a_4 are parameters for a specified location. Knowing the contributing area, A, to a manhole, the design discharge, Q, and t_c, the corresponding return period is given by

$$R = [\frac{Q}{A} \frac{(a_3 + t_c)^{a_4}}{a_1}]^{1/a_2} \qquad \qquad ....(7)$$

This value for R is used in the optimising methods.

Once an optimal solution to the pipe network has been found the simulation method can be used to route the design hydrographs through the pipes between the first two isonodal levels to define new design discharges at the manholes on the second isonodal level. The procedure can now be repeated for the reduced pipe network with $(N-1)$ isonodal levels and so on until all the pipes have been designed for discharges calculated using the simulation method.

The particular simulation method used in the two design methods is that developed by Price and Kidd (1978).

DESIGN OF LARGE NETWORKS

The optimal design of large networks according to the constraints is generally precluded by the limit on computer storage and the cost of running the programs. However, if the designer is content with a sub-optimal solution which is close to the optimal, the procedure for designing with the hydrograph simulation method suggests a modified solution as follows.

Suppose that the N isonodal level pipe system is part of a larger network. Once the diameters of the pipes between the first two isonodal lines have been finally designed a new pipe network with N isonodal lines 2, 3, ..., N, N+1 (referred to the larger network) can be defined and the procedure repeated. If a new branch with more than N isonodal levels is introduced results from the original network are stored while pipes in the branch are designed.

An advantage of this approach is that at the beginning of each optimisation most of the pipes will, in general, be close to an optimal design, thus reducing the number of iterations required. A disadvantage of the approach is that constraints on invert levels at the bottom of a large system may be difficult to achieve. This difficulty can probably be overcome by a careful initial definition of individual minimum and maximum depths of cover for each pipe.

Finally the above step-by-step method of producing the final design for the pipes down a large system enables the non-linear programming method to be adapted for discrete pipe diameters in the following manner. Having calculated the diameters to convey the specified discharges at the manholes on the first isonodal level of an N isonodal level system, the diameters of the pipes between the first two isonodal levels are adjusted to the nearest commercial sizes which will convey the design discharges. The optimisation for the N isonodal system is then repeated keeping the diameters of the top pipes fixed and insisting that their slopes should always be such that the design discharges can be conveyed in the pipes. The design of the larger network proceeds in the same way as described above.

APPLICATION OF THE METHODS

To assess the viability of the two methods for practical design they were both applied to the redesign of the sewer system for the St Mark's Road catchment in Derby. This catchment was used in the calibration of the hydrograph simulation model as reported by Price and Kidd (1978).

The catchment drains an area of 10.4 ha, has 87 pipes (see Fig 2) and is effectively flat over much of its extent such that runoff is controlled primarily by the artificial camber on roads and other paved areas. However, the ground slope at the top end of the catchment is steeper and the change in ground slope has an important effect on the design of pipe gradients and depths downstream. Design rainfall profiles of different durations and return periods were obtained from the Meteorological Office and the design rainfall intensity, i, for use with the Rational method was defined by

$$i = \frac{12.1 \ R^{0.25}}{(0.063 + t_c)^{0.75}} \qquad \qquad ....(8)$$

where R is in years, t_c in hours and i is in mm/hr; see Yen (1975).

In each run of the computer programs for the two methods the existing pipe gradients and manhole depths were used to define the initial conditions. It was assumed that the invert levels of all incoming pipes and the outgoing pipe to a manhole were the same. Of course, in practice, pipes from a side branch generally have their soffit levels at or

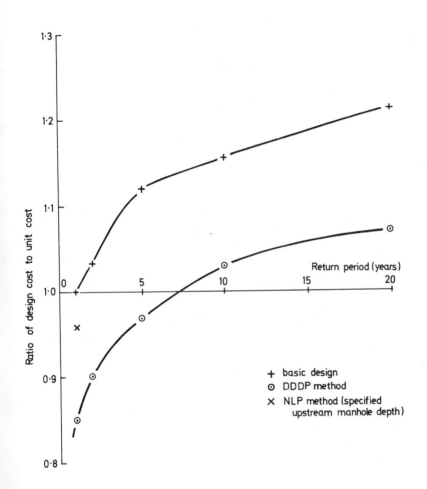

Figure 3. Relative costs of designs using basic DDDP and NLP methods

above the soffit level of the outgoing pipe to prevent unnecessary backwater effects in the side branch. Consequently the costs of the designed systems described below would have been reduced further if this less restrictive constraint was adopted.

No account was taken of constraints imposed on the original designer due to other services. These are partly taken into account by the cost routine, but any restriction on the alignment of the pipes other than the ground slope would increase the costs of the designed systems. Also pipe diameters were allowed to decrease downstream. Again the non-decreasing diameter constraint would increase the construction costs.

Finally, the invert level of the outfall manhole was assumed to be arbitrary. This assumption avoids the problem of a specified outfall depth which is encountered in many practical problems. However, from the results below for the Derby catchment this would only be a problem for the larger return periods. For typical design return periods of one or two years the existing outfall depth could be achieved without difficulty.

To have an objective comparison of any savings in construction cost using an optimised design the Derby system was also redesigned (for existing pipe gradients) using the new design method (referred to below as the basic method) which incorporates the hydrograph simulation model (Price and Kidd, 1978). As for the optimising methods, pipe diameters were allowed to decrease downstream. Designs were obtained for rainfall events with return periods of 1, 2, 5, 10 and 20 years. The cost of the 1 year design was £69 655 and this was used as the unit cost to compare all other designs; see Fig 3.

Consider the DDP method first. Initial tests of the model successively increasing the size of the isonodal subsystem showed that a 7-isonodal subsystem produced acceptable results. The model was then applied to the Derby catchment for the five return periods above. From Fig 3 there is a reduction of between 11 and 15% in the cost of the basic design for each return period. Table 1 shows that for the 1 year design storm pipe diameters calculated by the DDDP are generally smaller along the main branch than those predicted by the basic method. A study of the DDDP design reveals that there is a trade off between pipe diameter and slope such that the discharge capacity of each pipe is closer to the design discharge.

Experience with the NLP method has confirmed that a non-linear programming technique using derivatives of the cost function and the constraint functions produces reasonably consistent results even though the cost function has a number of discontinuities. The development of the method has not yet been completed though results using a 5 level isonodal subsystem for the 1 year event with specified values of d_1 are shown in Fig 3 and Table 1.

CONCLUSIONS

A preliminary comparison of the two methods indicates that the DDDP method may be preferable for practical design work. This is because

(i)   results using the DDDP method are more consistent than for the NLP method;

(ii)   computer storage requirements based on current computer programs are less for the DDDP method (47K) than for the NLP method (75K); and

(iii)   the DDDP method takes approximately 400 seconds computing time on an ICL 1904S computer compared with more than 900 seconds for the NLP method when

TABLE 1   Redesigns of Derby system for 1 year event (main branch)

Pipe number	Upstream ground level (m)	BASIC DESIGN			DDDP DESIGN			NLP DESIGN		
		Upstream manhole invert (m)	Pipe diameter (mm)	Cost (£)	Upstream manhole invert (m)	Pipe diameter (mm)	Cost (£)	Upstream manhole invert (m)	Pipe diameter (mm)	Cost (£)
1	58.60	56.60	150	1211	57.95	150	864	56.60	150	1128
2	55.10	53.10	150	1061	54.44	150	758	53.53	150	868
3	51.90	50.01	150	1149	51.24	150	827	51.01	150	857
4	49.30	47.15	225	1394	48.57	225	1037	48.49	225	1057
5	47.80	45.70	300	818	46.75	225	576	46.59	225	602
6	47.22	45.60	300	622	46.49	225	444	46.16	225	475
11	46.90	45.47	375	1118	46.09	300	847	45.89	300	877
12	46.90	45.31	450	877	45.89	300	609	45.69	300	634
14	47.00	45.28	375	744	45.75	300	466	45.49	300	636
15	46.20	45.19	375	1571	45.39	375	1499	45.28	375	1531
16	46.60	45.05	375	297	45.25	300	250	45.07	375	285
17	46.10	45.02	450	933	45.20	300	644	45.03	300	665
18	46.25	44.99	450	373	44.99	300	295	44.84	375	336
19	45.82	44.97	375	1612	44.86	375	1672	44.80	375	1686
20	45.87	44.72	525	691	44.67	375	556	44.62	375	563
23	46.00	44.77	450	1176	44.61	375	1066	44.49	300	978
40	45.90	44.68	525	3084	44.42	450	2824	44.16	450	2903
41	45.90	44.47	525	474	44.16	450	417	43.95	450	476
42	45.75	44.44	375	2123	44.11	525	2470	43.90	375	2261
47	45.70	43.88	375	708	43.97	450	793	43.14	450	873
48	45.70	43.72	375	342	43.91	525	399	43.09	375	388
53	45.70	43.61	375	2308	43.89	525	2059	43.01	450	2238
54	45.73	42.65	450	1169	43.80	525	838	42.74	375	1077
60	46.00	42.51	450	3449	43.78	375	2362	42.53	450	3435
61	45.90	42.09	450	2849	43.28	375	2291	42.37	375	2510
62	46.14	41.90	450	1769	42.90	300	1552	42.05	375	1631
87	46.30	41.70	450	1789	42.21	300	1346	41.71	300	1564
(88)		41.50			41.33			40.93		

optimising the network of 87 pipes as shown in Fig 2.

However more research is still needed to confirm these conclusions.

Experience with the DDDP method in the redesign of an existing catchment shows that there may be a number of opportunities to make significant savings in the construction cost of storm sewer systems. By a careful arrangement of the way in which catchment data are prepared for input to the computer program it should be possible to make the DDDP method almost as easy to use as the existing TRRL method. The value of the optimising method would be that it would avoid the subjectivity of the iterative procedure in which pipe gradients (and depths) are first specified, pipe diameters are designed, adjustments are made to the original gradients to satisfy the velocity constraint or to reduce the size of downstream pipes, pipe diameters are redesigned, and so on until the designer is satisfied.

Finally it should be remembered that a system designed with, say, the TRRL method generally has a number of pipes with significant spare capacity over the design discharges due to the increments in commercial pipe sizes. It is possible that this spare capacity contributes to the large safety factor available in most storm sewers before surface flooding occurs; see Bettess et al (1978). Research is underway to explore the variation in this safety factor for systems designed using standard hydrograph methods, such as the TRRL method, and the optimising methods above. Additional research is being done into the incorporation of storage tanks and pumping stations in an optimised design.

ACKNOWLEDGEMENTS

The work described in this paper forms part of the research programme into urban drainage at the Hydraulics Research Station and is published with the permission of the Director of Hydraulics Research.

REFERENCES

Argaman, Y., Shamir, U. and Spivak, E. (1973) Design of optimal sewerage systems. ASCE, J. Environ. Eng. Div., 99 (EE5), pp 703-716.

Bettess, R., Pickfield, R. and Price, R. K. (1978) A surcharging model for storm sewer systems. International Conference on Urban Storm Drainage, Southampton.

Danjani, J. S. and Hasit, Y. (1974) Capital cost minimization of drainage networks. A.S.C.E., J. Environ. Eng. Div., 100 (EE2), pp 325-337.

Farrar, D. M. and Colyer, P. J. (1978) A procedure for calculating the cost of storm water sewer construction. International Conference on Urban Storm Drainage, Southampton.

Gill, P. E., Murray, W., Picken, S. M., Wright, M. H. and Long, E. M. R. (1976) SUBROUTINE SALQDR — a quasi-Newton algorithm to find the minimum of a function of N variables subject to non-linear constraints when first derivatives are available. National Physical Laboratory Algorithms Library.

Gupta, J. M., Agarwal, S. K. and Khanna, P. (1976) Optimal design of waste water collection systems. A.S.C.E., J. Environ. Eng. Div., 102 (EE5), pp 1029-1041.

Lemieux, P. F., Zech, Y. and Delarue R. (1976) Design of a stormwater sewer by non-linear programming – I. Con. J. Civ. Eng., 3, pp 83-89.

Mays, L. W. (1976) Optimal layout and design of storm sewer systems. Ph.D. Thesis, Dept. of Civil Eng., University of Illinois at Urbana-Champagne, Ill.

Mays, L. W. and Yen, B. C. (1975) Optimal cost design of branched sewer systems, Water Res. Res., 11, No 1, pp 37-47.

Merrit, L. B. and Bogan, R. H. (1973) Computer-based optimal design of sewer systems. A.S.C.E., J. Environ. Eng. Div., 99 (EE1), pp 35-53.

Murray, W. M. (1976) Methods for constrained optimisation. In Optimisation in Action, Edited by L. C. W. Dixon, Academic Press, pp 217-51.

Price, R. K. and Kidd, C. H. R. (1978) A design and simulation method for storm sewers. International Conference on Urban Storm Drainage, Southampton.

Walsh, S. and Brown, L. C. (1973) Least cost method for sewer design. A.S.C.E., J. Environ. Eng. Div. 99 (EE3), pp 333-345.

Watkins, L. H. (1962) The design of urban sewer systems, HMSO, RRL Technical Paper 55.

Yen, B. C. (1975) Risk based design of storm sewers. Hydraulics Research Station Report IT 141.

Yen, B. C., Wenzel Jr., H. G., Mays, L. W. and Tang, W. H. (1976) Advanced method-ologies for design of storm sewer systems. Water Resources Centre Report No 112, University of Illinois.

THE BENEFITS OF URBAN STORM DRAINAGE: COMPUTER MODELLING AND
STANDARDISED ASSESSMENT TECHNIQUES

John B. Chatterton Severn/Trent Water Authority

Edmund C. Penning-Rowsell Middlesex Polytechnic

INTRODUCTION

When a decision to install or up-rate an urban storm drainage
system involves a choice of design standards, the decision
may then include a trade-off between residual flood damage to
houses, other property and traffic disruption costs, against
the increased cost of the scheme. For example, a system
designed for the 10-year rainfall event may not protect all
the property in the area from flooding. To give full protection however, perhaps with a scheme designed for the 25-year
event, might substantially increase the cost. This increased
cost may not be justifiable in relation to the residual flood
damage averted. To establish whether increasing design
standards to reduce or eliminate residual flood damage is
justified the engineer requires information on flood damage to
the properties at risk.

The investigation of flood damages, and more particularly
flood damages averted, already plays an essential part in the
economic analysis of urban flood alleviation schemes undertake
by the regional water authorities. Indeed such analysis is
required by the Ministry of Agriculture, Fisheries and Food in
support of applications for Government grants in England and
Wales to demonstrate that the total costs incurred by the
community in alleviating flooding is matched by the benefits
to be gained by the works. However, the data and guidelines
on potential flood damage to property at risk have until
recently been minimal.

In 1967 the Institution of Civil Engineers recommended the
systematic collection of flood damage data and its application
to benefit assessments. More recently, Section 24/5 of the
1973 Water Act requires the new water authorities to undertake
'surveys of their areas in relation to their drainage
(including flood protection) functions'. The guidance notes

(MAFF, 1974) clarified the Act and stressed the importance of
cost-benefit procedures in future appraisals of flood alleviation and warning schemes. Similarly, the National Water
Council's Standing Committee on Sewers and Water Mains,
through its Working Party on the Hydraulic Design of Storm
Sewers, has pointed to the need for more attention to assessing the benefits of storm drainage systems (National Water
Council, 1977). The tide of opinion in this field is also
turning in favour of design based on cost-effectiveness rather
than solely on hydraulic considerations.

STANDARD DEPTH/DAMAGE DATA

Techniques for estimating flood magnitudes and their recurrence intervals are well advanced (NERC, 1975) but it is
essential to complement these techniques and data with standard
nationally applicable flood damage data before accurate benefit
analyses are possible. To this end, the Middlesex Polytechnic
Flood Hazard Research Project in collaboration with the
regional water authorities and the Ministry of Agriculture,
Fisheries and Food has been investigating methods of assessing
and standardising the benefits to the community of protecting
urban land from flooding (Penning-Rowsell and Chatterton, 1977).
In addition to providing standard damage data a flexible
computer model has been developed to standardise benefit calculations and enable rapid assessments of the benefits of
alternative design standards.

There are two basic approaches to the assessment of flood
damages: firstly, the collection of information on damages as
they occur, and secondly a theoretically based or 'synthetic'
approach which involves costing likely potential damage from
flood events with a range of magnitudes and durations drawing
on the accumulation of sporadic information and experience on
flood damage from various flood events. The latter approach
was developed by White (1945) in the USA and is widely
employed there by the Corps of Engineers and other government
agencies (Grigg and Helveg, 1975; Day and Lee, 1976). White
introduced the concept of the depth/damage curve to relate
potential damage in a property to depth of flooding. In
reality, damage is a function of not only depth, but flood
duration, flood water velocity, effluent content etc. - the
hydrological variables - as well as the size and function of
the property, quality and quantity of contents etc. - the
socio-economic variables. This second approach was taken in
the Middlesex research project to provide synthetic, nationally applicable depth/damage information and replace the
ad hoc procedures currently in use, for example, estimating
potential damage as 10 percent of house value or an arbitrary
multiple of the property rateable value.

It is customary to distinguish between tangible and intangible
flood damages based on whether or not monetary damages can be

assigned to the consequences of flooding. Our research has
attempted to systematise the calculation of direct and indirect
tangible damages for any sequence of flood events within a
clearly defined flood prone area – the benefit area – whilst
at the same time appreciating that reliance should not be
placed on tangible economic benefits alone. Many other
factors need careful consideration when evaluating the potential for flood protection. These include important amenity
and conservation issues, personal anxiety and political
feeling (Penning-Rowsell and Chatterton, 1976). If the public
decides that flood alleviation is needed despite an unfavourable economic analysis then this is their choice; the ultimate
responsibility for public expenditure lies with the public.

The land use of a flood prone area profoundly affects the
likely flood damage characteristics. Houses are affected
differently from offices and shops, which in turn suffer
different kinds of damage from those experienced in industrial
premises. Nine land use types have been identified and guidelines produced for assessing the effect of different depths
of flooding on each:

1. Residential dwellings
2. Agricultural buildings
3. Non-build-up land (non agricultural)
4. Non-domestic residential
5. Retail trading and related services
6. Professional and office
7. Public buildings and community services
8. Manufacturing and extractive industries
9. Public utilities and transportation

Alternative methods of obtaining flood damage information for
all property types have been derived. Basically these comprise
either site surveys or the use of standard depth/damage
information. The choice of method depends on the size of the
benefit area under investigation or the accuracy of the assessment. A hierarchical land use classification has been developed to allow flexibility in benefit assessments. For generalised surveys the user can record each property as the
'average' dwelling or the 'average' shop and select the
appropriate depth/damage data set. For more detailed assessments it is necessary to record, for example, each semi-
detached house, and note the social class of the occupant as
an indicator of the quality of the contents. In the retail
sector the user can separate a butcher's from a grocer's shop
and select the data appropriate to each of these specific
properties. At the highest level of detail the user can
conduct site surveys using standard check lists. By applying
standard susceptibility data to the inventory items found in
each property depth/damage data specific to individual
buildings can be created.

Standard depth/damage data for residential property was
compiled from three basic sources: typical ground floor plans
(needed for the likely dimensions of each property), the
standard susceptibility data applied to the building fabric
details of these properties and the inventory susceptibility
data applied to the inventory for each property.   It was
impractical to mount large scale surveys to discover the
ownership and quality of furniture and household durables
within different types of dwellings so secondary sources were
used to derive this information.   For instance, the variation
in household durable ownership for the whole of Britain was
derived from a market research sample of 35,000.   Similarly,
damage to retail properties was amalgamated from information
from the estates departments of 41 companies representing
nearly 8,000 shops whilst the National Census of Distribution
was a major data source on average retail stock values and
shop sizes.   Thus using the cumulative experience of loss
adjusters, chartered surveyors and structural engineers
enabled the collection of full and detailed information on the
susceptibility of individual properties to flood damages.   The
range of standard depth/damage information thus produced is
illustrated for the residential sector in Figure 1.

The assessment of potential flood damage to industrial
premises probably presents the most difficult problems in
cost-benefit analysis of urban storm drainage schemes owing to
the lack of any comprehensive past damage information and the
great diversity in size, intensity and function of industrial
premises.   A questionnaire has therefore been developed to
provide information on the susceptibility of premises to
physical damage and the likely magnitude of disruption to
production.   However, if questionnaire response is incomplete
estimates can be made either from similar types of premises
successfully surveyed within the benefit area or from the 180
manufacturing and related properties, all with flood plain
locations, surveyed during the course of the research (Penning-
Rowsell and Chatterton, 1977, Appendix 5.1).

It is not possible to standardise the relationship between
direct and indirect damage.   Too many site-specific variables
affect the disruption of roads and other communication
systems to use fixed coefficients relating resulting indirect
losses to physical damages, which in certain circumstances
may be negligible.   Standard data has been derived for
commercial and industrial premises measuring indirect damage
in terms of loss of trading profit.   This can be significant,
and indeed industrial flood damages can produce a multiplier
effect when severe disruption of production can affect a
local community's economic welfare and possibly even disturb
regional and national economies.

COMPUTER MODEL

In calculating the costs and benefits of urban storm drainage
schemes it is obviously desirable for water authorities and
consulting engineers to follow a standard evaluation
procedure to ensure a consistent allocation of resources.
The production of standard depth/damage relationships and
guidelines for the estimation of other benefits provides the
basic potential flood damage data. When applied to the land
use mix within the benefit area, and related to the flood
stages for different magnitude events, the annual benefits of
protecting against floods of a range of recurrence intervals
can be calculated. A standard evaluation procedure for urban
flood protection schemes is illustrated in Figure 2. It
should be emphasised that this evaluation procedure examines
the cost-effectiveness of a range of alternative schemes
rather than establishing a single benefit figure for some
arbitrary design standard. Using the computer model benefit
assessments may be as detailed or generalised as required.
However, level of generality of the standard depth/damage data
sets selected should correspond with the accuracy of the land
use, topographic and hydrological/hydraulic data inputs.
The major components of the model are given in Figure 3.

Standard depth/damage data sets are assembled using the PREASS
computer model and stored as sub-files for selection in
program DDASS. Thus it is possible to run successive benefit
assessments using data for different flood durations,
different flood warning lead times or upper and lower
confidence limits of the standard data sets. Certain data is
specific to each assessment. Broadly this can be divided into:

1. Land use data for the benefit area;
2. Topographic and location data relating land use data to
    Ordnance Datum and location within the benefit area.
    This enables each property to be related to the approp-
    riate flood stages;
3. Hydrological/hydraulic data applicable to the catchment(s)
    draining into the area, thus enabling flood stages to be
    estimated for specific flood events.

The amalgamation of these with the standard damage data,
suitably updated for inflation, and other non-standard data
on traffic flows, public utilities, emergency services and
intangible effects within program ESTDAM enables frequency/
damage relationships and ultimately discounted (i.e. present
value of) annual benefits to be calculated as an estimate of
the capital sum it would be worthwhile to invest on storm
drainage works.

ESTDAM is flexible so that the three basic data inputs - land
use, standard depth/damage and hydrological data - can be
manipulated to suit the specific requirements of the benefit

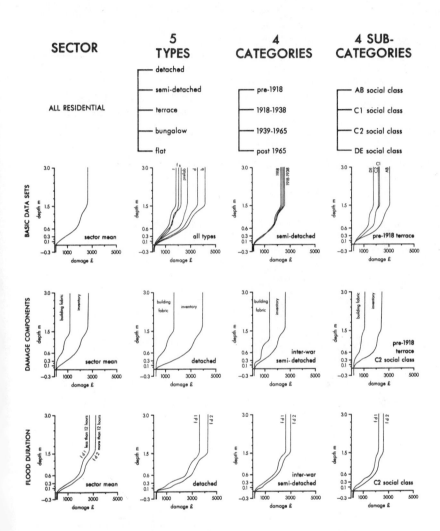

Fig.1 (page1) Examples of standard residential depth/damage
information

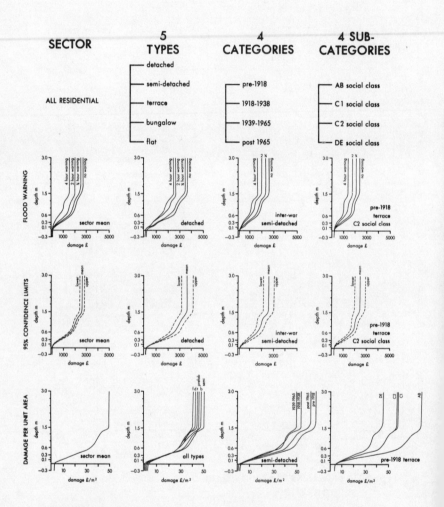

Fig. 1 (page 2)

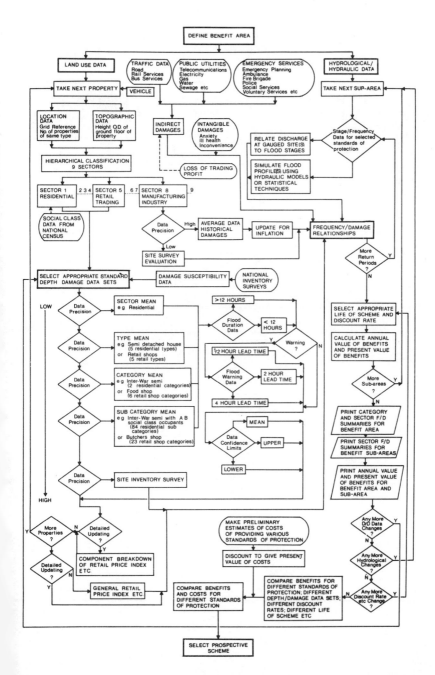

Fig.2 Standard evaluation procedure for urban flood protection scheme

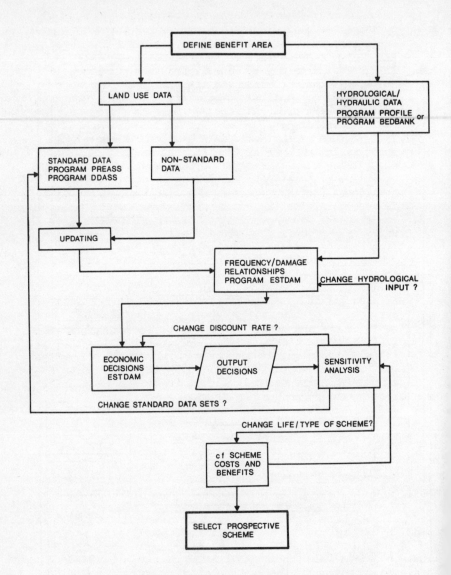

Fig. 3 Major components of urban benefit computer model

assessment:

1. Land use of the benefit area may be coded either simply by sector (e.g. residential property) or by type (e.g. semi-detached house) or category (e.g. inter-war semi) according to the detail of the benefit analysis;
2. The benefit area may be sub-divided into sub-areas to any specification, including the amalgamation or sub-division of pre-determined sub-areas, so that the parts of the benefit area contributing most to the benefits may be identified;
3. The standard depth/damage data available for computer analysis is for 2 flood durations (less than 12 hours and greater than 12 hours), 4 flood warning lead times (residential data only) and for upper and lower confidence limits. The user may select whichever data is appropriate to the scheme within DDASS;
4. The standard depth/damage data can be changed or up-dated to any desired specification;
5. If no standard depth/damage data is available for a specific property the appropriate sector or category average, if available, can be applied;
6. The depth/damage data produced for each property can be related either to selected flood profiles actually surveyed or statistically derived (using PROFILE), or to river bed and bank or embankment data (using BEDBANK), to produce stage/frequency damage data;
7. Indirect and other damages (e.g. road traffic disruption, loss of trading profit etc.) can be added to the direct damages prior to the calculation of average annual benefit and subsequent discounting;
8. Damages are converted to discounted benefits for any standard of protection scheme, for any length of scheme life, for any discount rate;
9. An overall up-dating option is included to allow for inflation and other price changes;
10. Suppression of any of the printout is possible (sub-area, area and frequency/damage totals, discounted benefit tables etc.).

SENSITIVITY ANALYSIS

The accuracy of the benefit estimation for urban storm drainage is dependent on many factors. These include the accuracy of the land use survey of the benefit area, the appropriateness of the standard depth/damage data and the reliability of the meteorological and hydrological data used to determine the return periods of the various flood extents and depths. However, one of the main advantages of computer analysis is the opportunity it creates for rapid sensitivity analysis by re-running the model with alternative assumptions and data, allowing the present value of benefits to be compared with each successive analysis.

Some of the variables which could have significant influence on the present value of benefits are given below:

1. Flood extent simulations and their assigned return periods;
2. Level of aggregation of land use data;
3. Selection of depth/damage information;
4. Selection of discount rate and scheme life.

Thus an incorrect simulation of flood extent could lead to incorrect inclusion or exclusion of properties within the benefit area. An underestimation of the assigned return period for a particular flood stage may increase the calculated benefits substantially. Adjustment of data or control cards in ESTDAM allows the effect of each of these variables on calculated benefits to be explored. Equally important are the decisions made when selecting the depth/damage data sub-files within DDASS; use of long flood duration data or upper condifence level data will increase calculated benefits.

The Ashton Vale floods

The effect of selecting different variables and data sets on the stability of calculated annual benefits is illustrated in a theoretical benefit study within Ashton Vale, Bristol. Following repeated flooding in Ashton Vale, most severe in 1968, a series of culverts, tunnels and trapezoidal channels have been constructed by Wessex Water Authority to contain storm discharges with a recurrence interval of one in one hundred years.

The benefit model described above was used to provide an estimate of part of the benefits to be derived from such a scheme, and in particular the benefits of protecting against floods up to the magnitude of the 1968 event estimated to have a 64-year recurrence interval.

In complex urban areas where many streams are culverted and the sewerage outfalls are an integral part of the drainage system, flood discharges of similar magnitudes can have quite different depth/extent relationships. It is difficult to predict the extent and depth of a flood of any specific frequency, yet this is crucial to accurate benefit assessment. A flood of the 1968 magnitude is unlikely to cover exactly the same area again even assuming the tidal situation in the river Avon is repeated. However, reliable 1968 flood depths in 262 dwellings are available and to overcome some of the problems of local hydraulic anomalies a generalised flood profile was computed using trend surface analysis (O'Leary, Lippert and Spitz, 1966) to represent a flood comparable to the 1968 event. This simulated profile was stepped down, differentially within sub-areas or reaches exhibiting similar hydraulic characteristics, to correspond with flood extents for known lesser floods.

The ESTDAM benefit assessment model was used to extract and

Table 1: Present value of benefits using alternative sets of
parameters

Present value of benefits (£)	Percent of 'standard' parameter benefits	Flood duration	Flood warning	Depth/damage data	Land use data	Evacuation of vehicles	Flood extent	Other benefits	Return periods	Life of scheme (years)	Discount rate (%)
		(Parameter changes)									
669000	216.9								2		
456600	148.1	2			S	N				50	9
446300	144.7			U	S	N				50	9
413000	133.9	2			S	N				50	
403600	130.9			U	S	N				50	
375800	121.8	2			S						
367400	119.1			U	S						
356900	115.7	2								50	
355000	115.1				S						
353500	114.6	2									9
353400	114.3						E				
347200	112.6			U						50	
344000	111.5			U							9
342700	111.1			L	S						
336900	109.2									50	
335400	108.8	2	4		S						
333700	108.2										9
326700	105.9	2									
325600	105.6			L						50	
322800	104.7		4		S						
322500	104.7			L							9
317900	103.1			U							
315600	102.3	2	4							50	
312700	101.4	2	4								9
310600	100.7					N					
308400	100.0	1	N	M	F	Y	S	68	1	25	10
303500	98.4		4							50	
300700	97.5		4								9
298100	96.6			L							
289000	93.7	2	4								
282300	91.5	2									12
280000	90.8		4			N					
277900	90.1		4								
274700	89.1			U							12
266500	86.4										12
257500	83.5			L							12
249700	81.0	2	4								12
240100	77.9		4								12
161200	52.3								3		
105100	34.1						L				
92800	30.1		4				L				

Notes:

<u>Flood duration</u> 1   Less than 12 hours
             2   Greater than 12 hours

<u>Flood warning</u>   N – No warning
                        4 – 4 hours effective lead time

<u>Depth/damage data</u>   U – Upper confidence limit (95% level) of
                            selected data set
                        M – Mean of selected data set
                        L – Lower confidence limit (95% level) of
                            selected data set

<u>Land use data</u>   S – Sector averages used as calculated from
                            depth/damage data files
                        F – Full land use code given and appropriate
                            depth/damage data selected

<u>Evacuation of vehicles</u> N – Vehicles assumed not moved before flooding
                        Y – Vehicles assumed moved (each residential
                            property is assigned 0.7 x vehicle in
                            accordance with national car ownership
                            statistics)

<u>Flood extent</u>   S – 'Standard' extent as estimated from trend
                            surface analysis and adjusted to simulate
                            actual flood extents corresponding to
                            return period 1
                      L – Profile lowered by –0.3m in all sub-areas

<u>Return periods</u>   1 – 12, 25, 40 and 64 years as estimated for
                            the 4 flood extents used
                      2 – 5, 15, 30, 64 years
                      3 – 20, 30, 50, 64 years

<u>Other benefits</u>   68– Indirect etc damages added to the highest
                            magnitude flood as estimated from the
                            1968 event (£288000) and updated for
                            inflation
                      E – a cruder estimation adding smaller amounts
                            to the lesser magnitude floods and increasing
                            the indirect damages to £500,000 for highest
                            magnitude flood

<u>'Standard' parameters</u> The benefits as calculated assuming protection
against all events upto and including the 1968
flood in Ashton Vale. The events assume no
warning and rapid dispersal of flood waters
after tide levels in the Avon are reduced.

sum the correct depth/damage data for each property within each
land use sector to produce expected damage for each of three
flood events. After addition of non-standard data (indirect
damages) the stage/damage relationships were converted to
discounted annual benefits using a 10 percent treasury
discount rate and a scheme life of 25 years. Mean values of
depth/damage data assuming no flood warning and less than 12
hours flooding were used as input data to DDASS. Using the
benefit assumptions made above as a 'standard' the data sets
and control parameters have been altered to illustrate the
sensitivity of both economic and hydrological decisions.

Table 1 summarises the results of repeated benefit calculations
using selected data sets and different economic and hydrological parameters and Table 2 indicates the single parameters
which contribute to increasing or decreasing the benefits.
Assuming that the treasury discount rate is constant at 10
percent the quality of land use data and type of depth/damage
data selected are most significant in benefit calculations.
In Ashton Vale, standard data relating to long duration
flooding or the upper confidence limit of the mean data,
assuming no provision for likely damage reducing actions as
a result of no flood warning, when applied to sector average
land use data for a proposed scheme with a 50 year life,
increases calculated benefits by more than 30 percent.
Certain parameters when taken together, for example long
duration data assuming damage reducing actions have not been
taken, cancel each other out and approximate the benefits
attributed to the 'standard' parameters. Altering the extent
and depth of the flood profile has a profound effect on
reducing annual benefits and stresses the importance of
accurate hydraulic data. The accurate calculation of indirect
benefits for each flood stage significantly enhances benefits.

It must, however, be stressed that long duration, confidence
limit and warning data were not used for the industrial and
much of the retail related services (warehouses etc.) as data
specific to these properties was collected during site surveys.
The true effect of each of these parameters is masked, since
a substantial proportion of the benefits is derived from
these sectors (67 percent in the 'standard' parameter
calculation). Table 2 shows that for the residential sector
alone event damage for the 1968-type flood is increased by
27 percent using long duration data, 12 percent using upper
confidence limit data and decreased to 49 percent of 'standard'
parameter damage using data assuming an effective 4-hour
warning lead time.

The effect of land use data precision The sector average land
use data simulates a benefit assessment recording each
residential property as the 'average' dwelling, each retail
property as the 'average' shop, etc. Average depth/damage data

Table 2   The effect of specific parameters on changing
the present value of benefits

Parameter change	Percent of 'standard' present value of benefits	Percent of 'standard 64-year event damages (Residential Sector)
1 Sector average land use data	115.1	107.5
2 (Increase Indirect Benefits)	(114.3)	
3 50 year scheme life	109.2	
4 9 percent discount rate	108.2	
5 Long flood duration	105.9	127.2
6 Upper confidence limit data	103.1	112.0
7 Evacuation of vehicles	100.7	103.9
8 4-hour flood warning lead time	98.4	49.0
9 Lower confidence limit data	96.6	88.0
10 12 percent discount rate	86.4	
11 Reduction of flood extent/depth	34.1	

Increase Annual Benefits

. . . .

Decrease Annual Benefits

for all industrial properties in Ashton Vale was applied to
each factory or warehouse at risk. Table 1 shows that this
approach increases likely future benefits by about 15 percent.
This is partly because the residential sector average overestimates likely damages in this area of largely poorer
quality interwar semi-detached housing, yet the sector average
is compiled from a complete cross-section of UK properties.
The difference between the generalised benefits for the 64-year
flood in the residential sector is some £65,000 or an increase
of 7.5 percent. Thus in a poor area this sector average
approach will raise calculated benefits and in a richer area
it will depress them. More important is an increase of
industrial benefits where a difference of £430,000 is found
using the gneralised approach, an increase of 66 percent.
This discrepancy arises from the sector average per square
metre of floor space being biased towards the small intensively used workshops with high damage potential. When this
average is applied to the large factories their damage
potential is grossly exaggerated. This points to the need for
individual site surveys in industrial premises.

The effect of return period on discounted benefit calculations
The figures in Table 1 include calculations only up to the
64-year event - the Wessex Water Authority estimate of the
1968 flood - and therefore the annual benefits shown may be
lower than in reality for a scheme since the contributions of
the larger floods to this total are not included.

If the return periods of the lesser magnitude floods are
altered from the best estimates the effect on the annual
benefit and subsequent discounted benefit totals is very
profound. For example, if the largest flood is maintained as
the 64-year event but the return periods of the lesser events
altered from 12, 25 and 40 years to 5, 15 and 30 years the
discounted benefits increase to £669,000 or over double the
figure from the best estimate return periods. Again,
increasing the return periods of the same lesser floods to
20, 35 and 50 years reduces the discounted benefits to only
£161,200. Clearly the calculated economic viability of a
storm drainage scheme may depend critically on the probabilit-
ies assigned to quite minor events which generally receive
somewhat scant attention from the design engineer. Perhaps
fortunately the return period estimates for the smaller
events, given limited rainfall or flow data, are likely to
be more accurate than those of the larger events.

Changing in isolation the return period estimate of the
largest event under consideration should be undertaken with
some caution. Comparisons between benefit tables can be
distorted when using different flood series. As a general
rule, when comparing benefit assessments with different
assumptions care should be taken to make sure that the largest

flood is given a constant return period.

CONCLUSION

When designing storm drainage or flood protection schemes it is preferable to optimise the cost/benefit ratio by comparing expected benefits for alternative design standards. The computer model discussed above facilitates this process by enabling the decision-maker to select the discounted benefits most appropriate to the design flood conditions and the proposed scheme. The model is now interactive so that the data sets and the control options fixing the economic and hydrologic assumptions for each assessment may be readily changed.

ACKNOWLEDGEMENTS

The authors would like to acknowledge the financial assistance of the Natural Environment Research Council and Middlesex Polytechnic and also thank Alison Shepherd for drawing the diagrams. The opinions expressed in the paper are those of the authors.

REFERENCES

Day, H. J. and Lee K. K. (1976) Flood damage reduction potential of river forecast. Journal of the Water Resources Planning and Management Division, ASCE 102 WR1: 77-78

Grigg N. S. and Helveg, O. J. (1975) State-of-the-art of estimating flood damage in urban areas. Water Resources Bulletin 11 (2): 379-90

O'Leary, M., Lippert, R. H. and Spitz, O. T. (1966) Fortran IV and map program for computation and plotting of Trend Surfaces for degree 1 through 6. Computer Contribution 3, State Geological Survey, University of Kansas

Ministry of Agriculture, Fisheries and Food (1974) Guidance notes for water authorities. Memorandum, Water Act 1973 Section 24. MAFF, London

Natural Environment Research Council (1975) Flood Studies Report, Vols I-V. NERC, London

National Water Council (1977) Standing Committee on Sewers and Water Mains, working party on the hydraulic design of storm sewers. 2nd progress report (mid 1975 - mid 1977). NWC, London

Penning-Rowsell, E. C. and Chatterton, J. B. (1976) Constraints on environmental planning: the example of flood alleviation. Area 8(2): 133-8

Penning-Rowsell, E.C. and Chatterton, J. B. (1977) The benefits of flood alleviation: a manual of assessment techniques. Saxon House, Farnborough, England

White, G. F. (1945) Human adjustment to floods. University of Chicago, Department of Geography Research Paper 93

HYDROLOGICAL REGIME OF A RURAL URBAN CATCHMENT IN CZECHOSLOVAKIA.

Jaroslav BALEK

Sen. Res. Of., Stavebni Geologie, Gorkého 7, Prague 1, Czechoslovakia.

ABSTRACT

As a part of an extensive project concerned with the hydrogeo-chemical mass balance in the agricultural region of Czech-Moravian Hills, a complex hydrological balance is studied in relation to the activity of man in the agriculture and cattle management. Results clearly indicate different runoff pattern of the rural urban catchment when compared with the other non-urbanized areas.

Water balance results of the rural urban catchment of Pojbuky in the Upper Želivka basin have been compared with the hydro-logical regime of the agricultural catchment Vočadlo, the drainaged field Samšin and forested catchment Hartvikov, all allocated in the vicinity of Pojbuky drainage area.

Table 1: Catchment area.

Catchment	Drainage km^2	Forest km^2	Field km^2	Pasture km^2	Urban km^2	Altitude m.a.s.l.
Pojbuky	2.04	.25	.75	.62	.42	659
Vočadlo	0.59	.14	.42	.03	–	570
Samšin	0.06	–	.06	–	–	493
Hartvikov	0.98	.98	–	–	–	672

The urban agricultural area of Pojbuky is inhabited by 130 people, the local cooperative owns 255 cattle, 655 pigs and 700 other animals. The other catchments are almost uninhabited and no cattle management practiced. Such a selection of catch-ments was made with the aim to obtain basic information on the influence of man on the hydrological regime under various agricultural conditions. Continuous analysis of water quality in streams, springs, wells and precipitation, supplies additional information on the fluctuation of water quality and its fluctuation according with changing agricultural pattern.

The latter results will be presented at another occasion
(Balek et al., 1978).

The hydrological water balance was calculated partly by standard methods; for selected periods of a year there was applied a
model based on the Guelph approach (1975). The model was used
mainly for the separation of three basic components of the
hydrograph. Three component separation was found as essential
for successful simulation of the geochemical mass balance.

Basic water balance results are presented in Tab. 2 for relatively dry hydrological year 1976 and in Tab 3 for relatively
wet year 1977.

Table 2: Water balance in 1976

Catchment	Precipitation mm	Total runoff mm	Evapotranspired mm
Pojbuky	616	304.6	311.4
Vočadlo	594	111.4	482.6
Samšin	537	118.8	418.2
Hartvikov	803	92.2	710.8

Table 3: Water balance in 1977

Catchment	Precipitation mm	Total runoff mm	Evapotranspired mm
Pojbuky	944	566.1	377.9
Vočadlo	979	250.2	728.8
Samšin	784	215.5	568.5
Hartvikov	920	128.7	791.3

No significant difference in the groundwater level and soil
moisture was recorded at the beginning and end of the balance
intervals.

A difference between the rural urban catchment and the other
three is clearly visible on Tables 4 and 5, with the values of
annual surface runoff, subsurface flow and groundwater flow,
again for dry and wet year.

Table 4: Hydrograph separation, dry year 1976

Catchment	Total runoff mm	Surface runoff mm	Subsurface flow mm	Groundwater flow mm
Pojbuky	304.6	123.7	104.9	76.0
Vočadlo	111.4	19.4	55.8	36.2
Samšin	118.8	-	81.0	37.8
Hartvikov	92.2	9.0	22.4	60.8

Table 5: Hydrograph separation, wet year 1977

Catchment	Total runoff mm	Surface runoff mm	Subsurface flow mm	Groundwater flow mm
Pojbuky	566.1	186.4	230.3	149.4
Vočadlo	250.2	77.9	119.8	52.5
Samšin	215.5	66.5	129.7	19.3
Hartvikov	128.7	18.6	46.8	63.3

The results from dry and wet year, even if preliminary, clearly indicate an increase of surface runoff as dependent on the urbanisation of the rural area. Also other runoff components in the rural urban areas are higher in the comparison with the other nonurbanized catchments, very likely owing to the reduced evapotranspiration rate from the urban part of the catchment. This has been partly proved by the direct transpirational measurement in the forested catchment by using the method as described by Balek, Pavlik (1977).

Also the mass balance of the basic elements indicated profound differences in the fluxes of the elements through urban, agricultural and forested catchments. Regarding the results obtained in 1976/77, continuous measurement of the electric conductance in streams, springs, wells and soil water will be provided in 1978 with the aim to obtain an additional information on more accurate separation of the hydrograph components, accordingly with the method described by Balek et al. (1978).

References

1. Balek, J., 1975. Bratislava Symp. on math. mod. IAHS-AISH Publ. No. 115.
2. Balek, J., Pavlik, O., 1977. Sap stream velocity as an indicator of the transpirational process. Journal of Hydrol. 34 (1977), 193.
3. Balek, J., Moldan, B., Pačes, T., Skořepa, J., 1978. Hydrological and geochemical mass balance in small forested and agricultural basins Symp. on Water Quality Modelling, Austria 1978.

THE EFFECT OF STORM DRAIN OUTFLOW ON THE ECOSYSTEM OF THE ALSTER LAKE AND ITS CANALS IN THE CENTRE OF HAMBURG

Hubert Caspers
Institut für Hydrobiologie und Fischereiwissenschaft der Universität Hamburg

Supported by the Deutsche Forschungsgemeinschaft.

Water bodies located in cities represent especially interesting objects of study for limnologists, because their ecosystems are are simpler than those of natural lakes, in which multifaceted evolutionary processes have taken place over the centuries. Interrelationships among the biotic components of the community in such urban lakes are relatively simple and a causality can often be established for their trophic levels and evolutionary history. The Alster in the center of Hamburg is an example of such a water body. This shallow lake is fed through a number of canals, into which storm drains empty (Fig. 1).

After an extensive fire in the central part of the city in 1842, a drainage canal system was designed by the English engineer W. Lindley. In this regard, Hamburg was a progressive city for its time. For financial reasons, however, the drainage system was constructed to allow both rainwater and domestic sewage to flow through the same ducts into the Elbe. As a precaution, storm-overflows were provided to drain excess water from heavy rainstorms into the Alster and the canals within the city. Thus, we have an ecosystem that receives sudden periodic inputs of raw sewage. This is brought about by the melting of snow in the winter, but most often by thunderstorms in summer. The effects have been investigated for many years, and the results of these studies have been used as evidence for the need of a separate sewerage system.

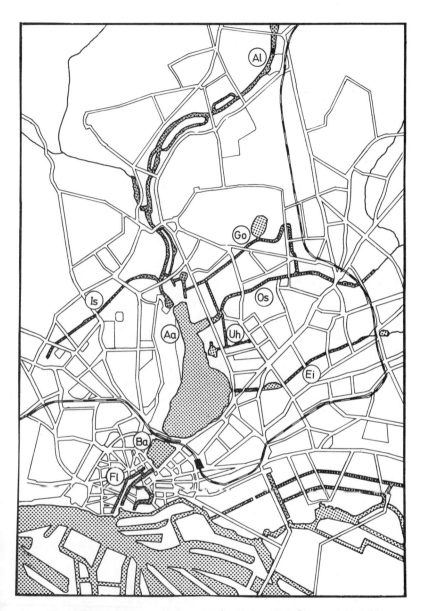

Fig. 1. The Alster Lake and its canal system in
        Hamburg
        Al = Alsterlauf   Uh = Uhlenhorster Kanal
        Aa = Außenalster   Ei = Eilbekkanal
        Ba = Binnenalster   Fl = Fleete = Outlet of
        Is = Isebekkanal   the Alster into
        Go = Goldbekkanal   the Elbe
        Os = Osterbekkanal

In the last decades a total destruction of the ecosystem has been observed occasionally, especially
in certain restricted locations within the canal
net. These catastrophies were first noted as mass
mortalities of fishes. In the days that followed,
entire sections of the canals became anaerobic.
Both the zooplankton and phytoplankton died off extensively, and the benthic fauna was largely decimated. This phenomenon has repeatedly led to pro-
tests by the citizenry.

Recent investigations, however, have revealed that
regeneration occurs rather quickly. The benthic
community, however, restores itself to normal relatively slowly after the oxygen supply has been re-
stored to the water: It is incorrect to speak in
terms of a "total destruction of the biological
community". Even during the anaerobic phase, some
elements of the phytoplankton survive in sufficient
numbers to rapidly repopulate the area, returning
the species diversity and biomass to a condition
normal for the trophic state of the water body.

The benthic fauna of the city's water bodies is subject to relatively large fluctuations in species
diversity and biomass. Zones with firm sediment show
a greater variety of species than muddy regions,
which are limited to only a few elements, chiefly
tubificid worms and chironomid larvae. Firmer zones
of benthos can support massive populations of the
Dreissensia, a freshwater mussel. The water bodies
of the city give the impression of being generally
similar to natural ponds and small lakes in Northern
Germany. A strong influx of wastewater bring about
a reduction in the total plankton biomass of about
65 %. The regeneration of the phytoplankton depends
upon the sedimentation of the detritus, so that illumination can again reach to a depth of about 2 m
in the canals. The availability of light is the de-
ciding factor. With the return of the phytoplankton
the oxygen content in the water is rapidly improved. A massive development of phytoplankton species is accompanied by temporary periods of oxygen
supersaturation, and by pronounced fluctuations in
the 24 hour oxygen curves (Fig. 2). After a discharge of sewage, an immediate rise in BOD_1 is detected. The oxygen consumption shows a 100 % increase after just a few hours, and this condition
lasts for several days (Fig. 3).

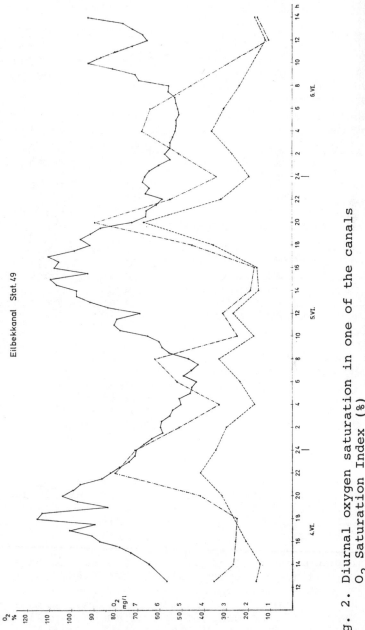

Fig. 2. Diurnal oxygen saturation in one of the canals
O₂ Saturation Index (%) ─────────
O₂ Consumption (% - 12 hr) ─.─.─.─.─.
O₂ Consumption (mg/l - 12 hr) ·········

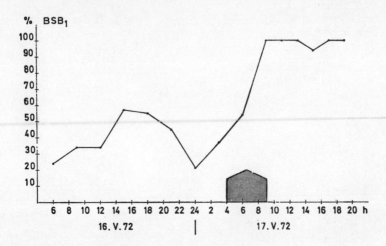

Fig. 3. Change in oxygen consumption after a sewage
    discharge (shaded arrow)

The various "blooms" of phytoplankton are related
to the fact that the waste water brings with it an
enrichment of nutrients into the aquatic ecosystem.
In summer, most of these nutrients are incorporated
in phytoplankton biomass (Fig. 4).

The regeneration of the phytoplankton brings with
it a reordering of location in which the biomass is
concentrated within the community. Small phytoplankton elements displaying a high reproductive rate
comprise a relatively large amount of the total biomass. Gradually the system fluctuations dampen out
until the community structure again displays the
diversity that existed before the water received
the discharge of wastes.

The phytoplankton in the highly enriched city water
bodies displays an amazingly high species diversity.
A considerable proportion of the species consists
of those designated species, as indicators for the
saprobic system.

The high concentrations of dissolved nutrients responsible for the reproductive rate of the phytoplankton generally remain relatively constant, provided the decomposition processes in the benthos
keep pace with the demands of the ecosystem. It is
important to remember that the water bodies in the
city have a very high trophic level. This is not

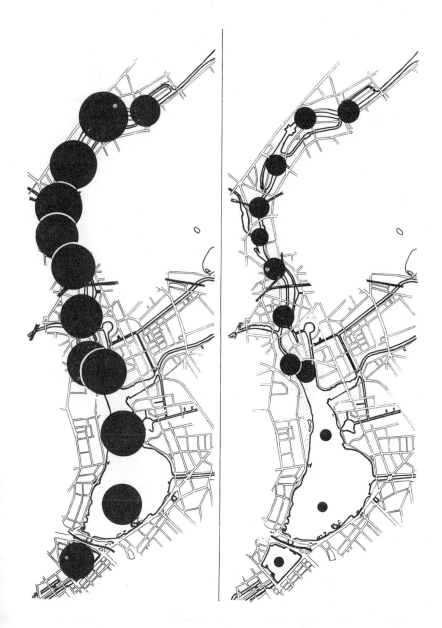

Fig. 4. Comparison of relative winter (left) and
summer (right) nitrate content

related, however, to a reduction of the species diversity, which would normally be expected in waters
subject to strong organic enrichment. In summer,
the limnological terminology applied to these water
bodies is ß-mesosaprobic.

A reduction in the populations of organisms is
clearly seen in the benthos. Here, not only a reduction in the biota to only a few ecological elements
takes place, but the biomass is also greatly decreased. The cause of this is the anaerobic conditions brought about by the repeated sedimentation
of detritus. The trees standing along the bank drop
massive numbers of leaves into the water in autumn,
the cellulose of which is only very slowly decom-
posed. The leaves form a layer over the finer detritus and partially, or at times almost fully, limit
the access of the upper mud layer to oxygen. The
evolution of methane and hydrogen sulfide observed
in the canals demonstrates that the decomposition
of organic materials is essentially limited to anaerobic bacteria.

A certain improvement in the situation can be seen
in the littoral zones of the Alster, when submerged
plants develop over large areas. Wherever these
plant associations survive, an enrichment in the
benthic elements is immediately perceived. A vast
array of aquatic animals live among the plants. Auf-
wuchs also shows a strong development. Unfortunately, this strip of littoral flora is decimated by
dredging operations by the city authorities, so in
this region a regeneration of the aquatic ecosystem
is prevented.

The result of the hypertrophic state in such city
water bodies is that the entire production is limited to plankton.

The reduction of solar illumination in autumn causes
the cessation of the massive phytoplankton development. The dominance passes to zooplankters which
depend upon detritus as their food sources. At this
time peak populations of rotifers and copepods are
recorded. The species diversity of the zooplankton
is greater in summer, because many specialized fee-
ders are represented, which are able to consume
particular phytoplankton groups. Some of the chains
are clearly discernible. It is an unstable system,
however, and the disappearance of particular phytoplankton elements from the community leads to the

elimination of the corresponding zooplankters that depend upon them for nourishment.

In winter, the water bodies of the city are unequivocally characterized as   -mesosaprobic, even in periods when no wastewater inflow occurs. The hypertrophic conditions remain.

As described, the dynamics of the trophic system in this highly productive shallow water body are significantly influenced by the allochthonous detritus input. This provides the best means to characterize the ecosystem of the city's water bodies. It can be concluded that in spite of the strong influence of wastewater outflow, the habitat still supports a living ecosystem. The time during which the entire water system displays anaerobic conditions is confined to only a few days. It is amazing that recovery and regeneration of the plankton begins in only 2 to 3 days, which leads to a return to a normal condition in terms of both community structure and biomass. The production bringing this about is confined to the plankton, both phytoplankton and zooplankton, while the benthos is limited to minimum numbers of species and biomass corresponding to the large load of detritus imposed upon it. The ecosystem described is therefore unbalanced. It typifies a residual type of a normal shallow water body. Nevertheless it is a habitat.

An improvement in the ecosystem through the expansion of the benthic community can only be expected when the outflow of rainwater is fully eliminated, or at least limited to a small quantity. The limnological investigation of such water bodies in cities reveals interrelationships of community and trophic factors, which allow for relatively clear interpretations of cause and effect relationships concerning the development of the biological community and production of its biomass. The evaluation of the results for practical considerations of water resource management provides motivation for such studies.

References

Caspers, H. (1953) Die Bodentierwelt und Biologie des Hamburger Alsterbeckens und der Stadtkanäle (Quantitative Untersuchungen zur Kennzeichnung eines Stadtsees). Mitt. d.Hamb.Zool.Mus.u.Inst. 52, 9-60

Caspers, H. (1954a) Der Lebensraum der Hamburger
        Alster. Biologische Untersuchungen zur
        Kennzeichnung eines Stadtsees. Städte-
        hygiene 5, 125-127

-        (1954b) Die Biologie von Elbe und Alster
        GWF 95, 638-643 (gleicher Titel: Des-
        infekt. u. Gesundheitswesen 46, 150-151)

-        (1976) Hydrobiologische Entwicklung und
        biocoenotische Struktur der Hamburger
        Stadtgewässer. Sonderband Alster, Mitt.
        Geol.Paläont.Inst.Univ.Hamburg, 267-354

-    & Karbe, H. (1967) Vorschläge für eine sa-
        probielle Typisierung der Gewässer. Int.
        Revue ges.Hydrobiol. 52, 145-162

-    & Mann, H. (1961) Bodenfauna und Fischbe-
        stand in der Hamburger Alster. Ein quan-
        titativ-ökologischer Vergleich in einem
        Stadtgewässer. Naturw.Ver.i.Hamburg,N.F.
        V, 89-110

-    & Penzhorn, H. (1976) Trophie und Saprobität
        Gütebeurteilung der Hamburger Stadtge-
        wässer. Sonderband Alster, Mitt.Geol.
        Paläont.Inst.Univ.Hamburg, 413-431

-    & Schulz, H. (1960) Studien zur Wertung der
        Saprobiensysteme. Erfahrungen an einem
        Stadtkanal Hamburgs. Int.Revue ges.Hy-
        drobiol. 45, 535-565

-        (1962) Weitere Unterlagen zur Prüfung
        der Saprobiensysteme. Ibid. 47, 100-117

Penzhorn, H. (1976) Populationsdynamik und trophi-
        sche Struktur des Planktons Hamburger
        Stadtgewässer. Sonderband Alster, Mitt.
        Geol.Paläont.Inst.Univ.Hamburg, 355-411

POTENTIAL FOR USING STORM RUNOFF WARNINGS IN THE OPERATION OF
PUMPED SEWERS IN COASTAL TOWNS

J.A. Cole and G.P. Evans

Water Research Centre, Medmenham Laboratory, P.O. Box 16,
Henley Road, Medmenham, Marlow, Bucks, SL7 2HD.

1. A QUESTIONNAIRE AND ITS OBJECTIVES

In connection with a survey of the potential benefits of a
national weather radar network, in collaboration with the Central
Water Planning Unit and the Meteorological Office, the Water
Research Centre sent a special enquiry as in Table 1 to
engineers dealing with the drainage works of coastal towns in
England and Wales. Stormwater runoff has to be pumped in low-
lying coastal areas. Our aim was to discover what use might
be made in future of storm runoff warnings:

> (a) to save costs of operating the pumps, and

> (b) in more efficient deployment of the work force

given that a 3 hour warning of heavy rainfall due to affect a
particular urban area could be provided by a weather radar.
Benefits besides (a) and (b) can be foreseen, e.g. in alerting
the police and the public to flood risks due to flows exceeding
sewer capacity and in preparing sewage works for large inflows
of storm-diluted sewage, but these were judged to be too
complicated for the questionnaire approach.

2. DEPLOYMENT OF THE WORK FORCE

The following emergency tasks could be set in action, with
improved effectiveness, if a 3 hour warning of heavy rainfall
were available to the urban drainage engineer:

> (i)   alerting personnel working in sewers

> (ii) unblocking of pump-house screens

> (iii) unblocking of road gullies and culverts

679

(iv)   start-up of diesel alternator sets.

Item (i) is a vital safety matter, of which every sewage worker is well aware. Of course, it is particularly relevant for larger systems or when major reconstruction work is taking place. Existing procedures recall men from sewers at the first hint of rain, but with radar-based warnings, tasks which might otherwise have begun can be postponed to avoid interruption.

Unblocking operations at screens are very necessary as a huge amount of screenings is liable to occur during heavy storm flows, especially in the shallow gradient trunk sewers of coastal towns. Where comminutors are installed this problem rarely arises. Surface blockages are best avoided by preventive maintenance, but some hasty checking of known trouble spots would be worth doing if given an warning.

Item (iv) is a good precaution, in case of electrical mains failure associated with thunderstorm activity. It is generally only necessary to warm up the diesel set (leaving it running on low power). unless the intention is to use it as a substitute for mains for economic reasons, as will be discussed in Section 3.

3.   COSTS OF PUMPING SEWAGE

Storm runoff in urban area is a relatively rapid phenomenon compared to that from rural areas, and typically flow rates in main sewers reach peak values within 30 minutes of the rainfall maxima. In order to achieve maximum effectiveness in storing runoff, predrainage of coastal sewers and wet wells is necessary: usually pumps are controlled automatically to achieve this result, but circumstances may call for manual override, to bring in extra pumps in advance of forecast rainfall. Table 2 instances some coastal towns, for which sewage pumping data are presented.

Cost savings can be expected from the following actions:

1.   avoidance of maximum demand charges

2.   allowing gravity drainage, if feasible, above normal pumping levels

3.   allowing trunk sewers to partly fill at end of storm, so reducing pumping lift

4.   reducing pump-switching operations.

Action 1 entails prejudging the hydrograph of storm runoff, then fitting in the best pumping schedule.

Figure 1 a) indicates a hypothetical forecast of flow entering storage in the large diameter sewer system. Three different pumping rates are assumed possible and two pumping strategies are shown. The effect of these two strategies on the amount in store is shown in Figure 1 b), where it is seen that both strategies satisfy the criterion that overflow should not take place. The storage limit S_{max} in the present example corresponds to the gravity overflow h_{max} as explained in Figures 1 c) and 1 d).

In some situations S_{max} is the limit above which surcharge of the sewers, flooding of streets and inundation of the pumping station could occur. It may be possible to avoid penal pumping charges at rate P_3, by keeping to rate P_2 and continuing for longer. In the latter case critical conditions are reached sooner and would clearly have been tolerable only with a reliable flow-forecast.

The ideal pumping strategy is a matter for operational research and would include the option to allow a substantial degree of storage within the sewer and gravity overflow where this is possible (action 2 and 3).

Diesel power is able to substitute for some of the peak demand load, so may be considered as an additional means of reducing the maximum demand charge. This has to be weighed on the relative operating costs only, since diesel units are reckoned as permanent standby installations whose other costs have to be met in any case.

In respect of action 4, Japanese workers (1) have reported that they have sought to minimise pump switching operations, in order to reduce their maintenance costs: the algorithm to achieve this minimisation is not detailed by these authors, but clearly a search procedure would be able to select useful rules of this character, using forecasts of rainfall and runoff response.

CONCLUDING REMARKS AND ACKNOWLEDGEMENTS

The presentation of these brief findings from a questionnaire aims to stimulate discussion on the interaction of urban hydrological forecasting and the management and operation of coastal sewage pumping stations.

The authors are grateful to the various Water Authorities and their agencies who have contributed replies amplified in many cases by discussion. Acknowledgement is given to the Director of the Water Research Centre for permission to present this paper.

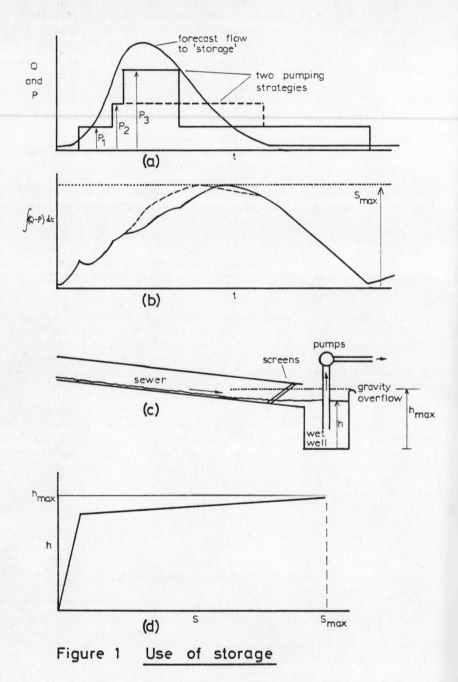

Figure 1   Use of storage

REFERENCE

1. Shuji Kato et al., 1977 Optimal Control of Sewage and
    Rain - Pumps Based on Influent Flow Prediction. Hitachi
    Review, $\underline{26}$,(7), pp. 235-240.

Table 1

Questionnaire

Operation of coastal sewage-pumping stations

1. Name of authority or agency.

   For each major pumping station :

2. Are the pumps installed at a sewage works or at a pumping station elsewhere on the sewer line?

3. Are the pumps manned on a 24 hour basis?

4. What is the installed pump capacity?

5. What is the head through which sewage is lifted?

6. What is the dry weather flow?

7. What is the average flow?

8. What is the power drawn by the pumps at average flow and how much does this cost?

9. What is the peak power drawn by the pumps and how much does this cost?

10. Is there a peak demand tariff in operation for charging purposes?

11. Have there been any cases of surcharge/flooding? Please give dates of events.

12. If so have these been caused by inadequate pump capacity or by failure of screen raking devices?

13. Could a 3 hour advance warning of heavy rainfall have assisted by the earlier start-up of pumps? (Thereby resulting in the reduction of peak demand charges).

14. Could a 3 hour advance warning of heavy rainfall have assisted by the more frequent raking of screens?

15. Could a 3 hour advance warning of heavy rainfall have assisted by the earlier opening of penstocks which lead to sea outfalls.

16. Could a 3 hour advance warning of heavy rainfall be of value in any other way?

17. Could you give an estimate of time available for storage in the sewers and wet well given average flow as in 7 above

Table 2 Some examples of coastal sewage pumping stations

	$\bar{Q}$ m³/day	Installed pumping capacity ÷ $\bar{Q}$	Peak Pump Power kW	Price of power p/kWh	Pump manning	Available storage time at $\bar{Q}$
Weston-super-Mare	14 000	47	900 (1/3 mains 2/3 diesel alternators)	4.2 (flat-rate)	call out	
Tilbury, Essex	11 400	6.3	165	1.638 (but max. demand tariff applies)	24 hours/day	1 hr. 20 mins
Hull, East	81 000	27	410	0.48 (but max. demand tariff applies approx. £1.2/kW)	24 hours/day	>4 hrs.
Hull, West	81 000	33	300	1.05 (but max. demand tariff applies approx. £1.2/kW)	24 hours/day	>4 hrs
Fleetwood (Wyre Borough Council)	16 000	25	450	3.097 (flat-rate)	visited on alternate days	
Anchorsholme (Blackpool Borough Council)	25 100	58	1850	3.079 (flat-rate)	24 hours/day	12 hrs.
Manchester Square (Blackpool Borough Council)	60 000	55	3350	3.079 (flat-rate)	24 hours/day	12 hrs.
Fylde Borough Council	20 000	25	910	2.2	No but automatic alarm system covers	12 hrs.
Sheppards Row, Exeter	3 050	17	180	1.865 (but max. demand tariff applies approx. £0.75/kW per summer month £5/kW per winter month)	Unmanned	4 hrs.

FLOOD SIMULATION IN PARTLY URBANISED CATCHMENTS

J. C. Packman
Institute of Hydrology, Wallingford, UK.

INTRODUCTION

It has been recognised for some while that urbanisation can
have a quite dramatic effect on the flood response of a catchment. The conversion of pervious to impervious surfaces
causes a reduction in rainfall losses and thus an increase in
the percentage of storm rainfall yielding direct runoff. The
alterations and improvements to the drainage system cause a
reduction in travel times and thus an increase in rapidity of
response. The combination of increased storm runoff and a
more rapid response leads to more runoff in less time - an
increase in flood potential.

Besides increasing flood potential, urbanisation also brings
the need for better flood control. There exists, therefore,
the need for a design flood estimation procedure capable of
accounting for urbanisation. However, a study of literature
(Packman 1977a) has indicated that accurate estimation of
the effect of urbanisation is difficult, and several studies
have yielded quite different results. It is apparent that
urbanisation cannot be satisfactorily accounted for by a
simple factor (such as the proportion of the catchment paved).
The author (Packman, 1977a,b) has outlined several other
factors that can affect the degree of modification of the
rainfall-runoff process, the most significant of which were:
(i) the typical percentage runoff from significant storm
events for the catchment in its natural condition; (ii) the
severity of the storm event; (iii) the location of the urban
development within the catchment; and (iv) the degree of
improvement made to the drainage system, and how this inter-
acts with the increased impervious area. Some recommendations
have been made (Packman, 1977b; IH, 1978) for the estimation
of urbanised design floods taking account of the first two
factors, and in this paper the present stage of development
of a method taking account of the third factor is described.

The fourth factor is at present considered only implicitly, being catered for within the index used to describe the extent of impervious area (URBAN - the proportion of the catchment under urban devopment). While this is not ideal, URBAN is easily obtained and is probably all that would be available at the planning stage. Furthermore, the statistical dependence between impervious area and improvements to the drainage system would complicate the development of meaningful regression equations involving each factor explicitly.

LOCATION OF URBAN DEVELOPMENT IN PARTLY URBANISED CATCHMENTS

Location of urban development has a twofold effect on catchment response. Firstly, it affects the relative scale of response from different parts of the catchment. Urban development in the previously dry, non-contributing areas of the catchment will increase percentage runoff more than urban development in the previously wet, contributing areas. Secondly, it affects the phasing of response from different parts of the catchment. Urban development upstream may cause the urban response to coincide with and reinforce the slower rural response from downstream, while urban development downstream may cause the urban response to pass before the rural response has arrived. In order to try to model these effects, experiments have been carried out with a subcatchment model, FLOUT (Price, 1977).

FLOUT

FLOUT is a simple planning model in which the catchment can be broken down into first and second order channel segments with discrete and distributed lateral inflows. Subcatchment routing uses the unit hydrograph method and channel routing uses a variable parameter Muskingham-Cunge method. It is recommended that individual subcatchments should contain at least 10% of the total catchment area. The model allows different time periods for subcatchment and channel routing; the recommended time period for subcatchment routing is less-than-or-equal-to $T_p/5$ where T_p is the time to peak of the unit hydrograph, and for channel routing is L/2c where L is channel length and c is maximum wave speed. To fit the model to ungauged catchments, estimates of subcatchment time to peak, percentage runoff and baseflow are required, and also estimates of channel wave speed and attenuation parameter. For UK conditions, estimates can be made using the Flood Studies Report (N.E.R.C., 1975), wave speeds being estimated from the difference in time to peak at upstream and downstream ends of the reach, and values for attenuation parameter being estimated from channel breadth and slope. Full details of the model and fitting methods can be found in Price, 1977. The model was fitted to the Silk Stream catchment in North London.

SILK STREAM CATCHMENT

The Silk Stream catchment has an area of 30.27 km^2. The period
of record is from 1928 to 1944, during which time the impervious
area grew from 14% to 21% (URBAN area 40% to 56%), estimates
based on linear interpolation between values taken from
Ordinance Survey 1:10560 maps dated 1912 and 1935. The new
urban area was however located towards the outfall of the
catchment, the upper areas remaining, even now, undeveloped.
A plan of the catchment showing the extent of the urban area
in 1912 and 1935 is given in fig 1. Also shown is the chosen
subcatchment scheme.

Flow data were obtained from a structure which has now been
demolished and it is thus impossible to be certain of the
quality. There were thought to be bypassing problems in the
later record. Rainfall data were obtained from an autographic
gauge at Stanmore for events before 1931 and from an autographic
gauge at Edgeware after 1931. Only one gauge was used for
each event so no idea of areal distribution is possible. Also,
relative timing errors between rainfall and runoff are indeterminate. In spite of these reservations, the data are thought
to be good enough for testing the ability of FLOUT to simulate
observed flood discharges.

UNIT HYDROGRAPH ANALYSIS

Nineteen events were selected from the record and subjected to
unit hydrograph analysis according to the procedure of the
Flood Studies Report (N.E.R.C., 1975) - excepting that since
SMD data were not available, the antecedent catchment wetness
index, CWI, for each event was set to a value corresponding to
zero SMD and zero 5 day API. Since most events were winter
events, it was not felt this departure from the recommended
procedure would affect the results significantly. A summary
of event data is given in table 1. The annual maximum flow
in that year is included in table 1 to give some idea of
severity of event.

While there is some indication of an increase in peak discharge
with urbanisation, the derived unit hydrographs fall into two
groups, events 1, 2, 3, 4, 11, and 13 exhibiting a slower more
attenuated response, and the remainder exhibiting a much
flashier response. The slower response tends to be associated
with higher percentage runoff (Event 10 is probably a summer
thunderstorm event not covering the whole of the catchment and
virtually missed by the raingauge). This suggests the flashy
response derives mainly from the downstream urban area, while
the longer response occurs when the upstream undeveloped area
is contributing significantly. Either response type is
capable of producing the maximum flood in any year.

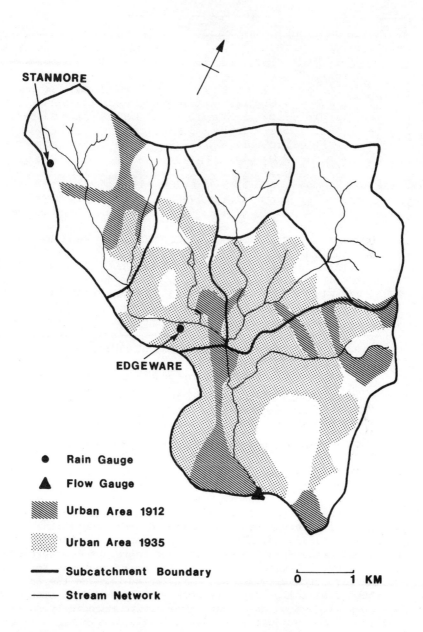

STANMORE

EDGEWARE

●   Rain Gauge

▲   Flow Gauge

    Urban Area 1912

    Urban Area 1935

━━ Subcatchment Boundary

── Stream Network

0   1 KM

Fig.1 SILK STREAM CATCHMENT.

TABLE 1 : EVENT DATA

No	Date	D	P	Q_p	AMQ	PR	q_p	T_p
1	16/11/29	9.5	23.6	3.63⎤		16	56	3.0
2	27/11/29	14.5	24.0	4.57⎬	5.89	30	34	5.0
3	6/12/29	4.0	12.5	5.89⎦		41	39	3.25
4	11/12/30	6.0	11.5	2.97	2.97	31	31	5.0
5	23/10/32	5.5	17.1	3.63	3.63	11	84	1.5
6	15/11/33	6.0	14.4	2.99	2.99	11	100	1.8
7	7/7/36	2.0	18.3	4.83	4.83	7	128	1.7
8	21/5/37	7.5	17.1	5.55⎤		15	108	1.8
9	13/8/37	2.0	13.5	6.38⎬	7.69	11	110	2.2
10	13/8/37	1.5	3.4	7.69⎦		47	116	1.9
11	13/12/37	12.0	17.4	7.90	9.84	37	84	3.6
12	11/10/39	2.5	19.6	7.47⎤		9	128	1.5
13	26/3/40	14.0	27.6	11.60⎦	11.60	46	40	2.7
14	23/10/42	9.0	21.0	7.22⎤		12	124	2.4
15	5/12/42	3.5	11.0	7.40⎬	11.31	21	108	2.2
16	11/9/43	2.0	13.3	7.39⎦		13	132	1.8
17	2/9/44	5.0	17.4	9.43	9.43	13	140	2.0
18	16/10/44	3.0	11.2	9.00⎤ record		17	138	1.7
19	6/11/44	1.5	9.0	5.57⎦ incompl.		16	100	2.3

where D is duration of central burst of rainfall

P is depth of rainfall in central burst (mm)

Q_p is peak of observed hydrograph (m^3/s)

AMQ is annual maximum flood in that water year (m^3/s)

PR is percentage runoff

q_p is peak of triangular ½ hr unit hdrograph (m^3/s/100 Km²

T_p is time to peak of triangular ½ hr unit hydrograph (hr)

MODELLING BY FLOUT

Eight events were chosen for modelling by FLOUT, four events
corresponding to urban conditions at about 1930 (Nos 2, 3,
5, 6) and four events corresponding to urban conditions at abou
1940 (Nos 11, 13, 14, 17). In each group, two events represent
high percentage runoff events and two represent low percentage
runoff events. Percentage runoff for each event was set equal
to the observed value and was distributed between the subcatchments as follows. 80% runoff was assumed from impervious
surfaces (ie 32% of the URBAN area) as recommended by Packman
(1977b)and the residual runoff volume was distributed between
the subcatchments according to their percentage pervious area.
Subcatchment times to peak were estimated using the Flood
Studies Report equation (N.E.R.C., 1975). Reach travel times
were estimated from the difference in time to peak at the
upstream and downstream ends of the reach, but using the same
degree of urbanisation for the upstream as for the downstream
reach. (Some such assumption was necessary to avoid the

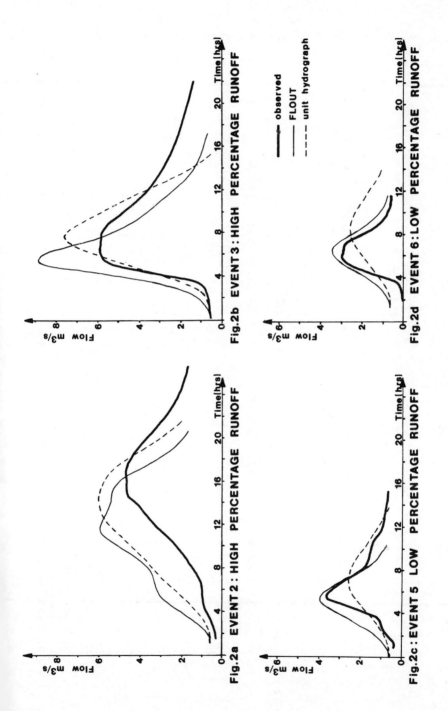

Fig.2a EVENT 2: HIGH PERCENTAGE RUNOFF

Fig.2b EVENT 3: HIGH PERCENTAGE RUNOFF

Fig.2c: EVENT 5 LOW PERCENTAGE RUNOFF

Fig.2d EVENT 6: LOW PERCENTAGE RUNOFF

observed
FLOUT
unit hydrograph

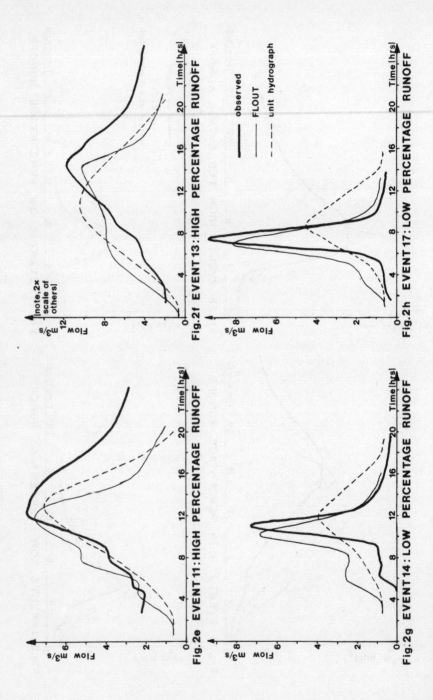

Fig.2e EVENT 11: HIGH PERCENTAGE RUNOFF

Fig.2f EVENT 13: HIGH PERCENTAGE RUNOFF

Fig.2g EVENT 14: LOW PERCENTAGE RUNOFF

Fig.2h EVENT 17: LOW PERCENTAGE RUNOFF

inconsistency of downstream time to peak being shorter than
upstream time to peak. This form of assumption seemed reasonable in that it assumed a uniformly developed catchment with
the degree of urbanisation of the downstream subcatchment determining travel time through the reach). Figures 2a-h show the
observed hydrographs and predicted hydrographs using FLOUT and
the lumped unit hydrograph procedure of the Flood Studies
Report (N.E.R.C., 1975). It must be emphasized that neither
model has been fitted to the event except in terms of direct
runoff volume; otherwise the models have been fitted using
only catchment characteristics and published regression equations. It is expected that base flow estimates would be
improved if values for antecedent Catchment Wetness Index were
available. Apart from events 2 and 3 the FLOUT simulation
represents an improvement over the lumped unit hydrograph
simulation, particularly for the low percentage runoff events
(nos 5, 6, 14, 17) when little runoff from the upstream subcatchments is predicted. FLOUT follows the form of the observed
response very well, but the tendency to overpredict the
beginning of the hydrograph and underpredict the recession
suggests some improvement in effective rainfall separation
could be made. In view of the somewhat disappointing simulation of events 2 and 3 FLOUT was run on these events again
assuming urban development was at its 1912 level. A considerable improvement in fit was obtained. This fact, together
with the sudden increase in annual maximum discharge around
1935 suggests that urban development in the catchment may not
have increased linearly from 1912 to 1935, but may have
remained at its 1912 level until the late 1920's. This however
is conjecture at this stage.

CONCLUSIONS - FLOUT AS A DESIGN FLOOD MODEL

FLOUT seems better able than a simple lumped unit hydrograph
model to simulate the flood response of a partly urbanised
catchment. There is however a significant increase in the
work required for fitting and running (greater by approximately
twice the number of subcatchments). However, since FLOUT
(in its fixed parameter Muskingham-Cunge form) is a linear
model, it could be used to derive a total catchment unit hydrograph for use with all storms. However, derived unit hydrograph shape is sensitive to the ratio of contributions of the
various subcatchments. Probability simulation and sensitivity
analysis (as used in the Flood Studies Report (N.E.R.C., 1975))
based on many catchments could be used to develop relationships
for a design choice of variables including the ratio of subcatchment contributions (giving unit hydrograph shape). Until
such work has been done FLOUT should be considered as a simulation model for use with a synthetic or historic record to
generate a series of annual maxima for subsequent flood
frequency analysis.

ACKNOWLEDGEMENTS

This paper is presented by kind permission of Dr J S G McCulloch, Director of the Institute of Hydrology. Ongoing work is supported by the Department of the Environment under Contract number DGR/480/38.

REFERENCES

Institute of Hydrology (1978) Modified Flood Prediction Methods for Urbanised Catchments. Flood Studies Supplementary Reports No 5 (in preparation).

Natural Environment Research Council (1975) Flood Studies Report. 5 volumes, London.

Packman, J.C. (1977a) The Effect of Urbanisation on Flood Magnitude and Frequency. Annual Meeting of Institute of British Geographers, Newcastle.

Packman, J.C. (1977b) The Effects of Urbanisation on Flood Discharges - discussion and recommended procedures. Conference of River Engineers, Cranfield.

Price, R.K.(1977) FLOUT - A River Catchment Flood Model. Hydraulics Research Station, Wallingford, Report No IT168.

THE GENERATION OF SUSPENDED SOLIDS LOADS IN URBAN STORMWATER

Tucker, C. G. J. & Mortimer G. H.

Severn-Trent Water Authority &
Loughborough University of Technology

INTRODUCTION

The causes of urban stormwater contamination may appear
obvious at first sight, since this contamination results from
the washing of dirt and other materials from catchment
surfaces under the action of falling rain and subsequent
runoff. However, a closer scrutiny reveals that several
factors interact in determining the nature and degree of
contamination arising at a particular location during a given
storm. These include the type and extent of sources of
contamination within the catchment, the physical characteristics of the catchment and the nature of land use, and also
weather, climatic and seasonal forces.

Very little literature on this subject existed before the
early 1970s, but significant among the published work are
the studies by Wilkinson (1956), Weibel et al (1964),
Pravoshinsky & Gatillo (1969), Soderlund & Lehtinen (1972)
and Sartor & Boyd (1972). These studies variously identified
such features as the 'first flush', the effects of dry
weather preceding rainfall, correlations between flow and
suspended solids concentrations, and the potential impact of
road gullies as sources of contamination.

CATCHMENT STUDIES IN NOTTINGHAM

This paper is derived from the results of a study carried out
in 1973-4, of runoff from two separately sewered suburban
catchments in Nottingham, England (Tucker 1975). The
objectives of this study were the measurement of solids loads
discharged from urban catchments over an extended period, and
assessment of the factors influencing these loads.

Suburban areas were chosen because it was considered that
these constitute the major proportion of the separately

sewered catchments in the UK. Although this type of
catchment is clearly not exposed to contamination on the same
scale as other catchment types such as inner city areas or
modern industrial estates, the similarity of land use in most
suburban areas is such that the results derived from experimental catchments in one location could be expected to be
applicable elsewhere, whereas this could not be assumed for
catchments in which land use is more variable.

The first area to be studied, Rise Park Estate, Nottingham,
is a recently developed housing estate, 62 ha in extent, of
which 19.3 ha is impermeable. The majority of the estate
consists of privately owned houses, in addition to which
there are blocks of flats, a shopping centre and a secondary
school. At the time of the study construction was complete
in all but one small section of the estate. The wholly
separate surface water system converges to a single outfall,
with a time of concentration (to the sampling chamber) of 11
minutes. The catchment forms a valley rising from south-
west to north-east with sewer gradients of up to 0.15
occurring. The underlying stratum is pervious sand and
gravel, and consequently base flow in the sewer is nil.

The second catchment, Welbeck estate, Hucknall, is situated
2.5 km from Rise Park. This is a council housing estate
dating from the early 1960s. Total area is 16 ha, of which
5.6 ha is impermeable. Housing consists mostly of small
terraces and there is also a small shopping centre and a
number of communal garage areas. As at Rise Park sewerage
is wholly separate and the surface water system converges to
a single outfall with a time of concentration in the region
of ten minutes. Sewer gradients up to 0.05 occur and since
the underlying stratum is primarily impervious red marl a
constant base flow of approximately 1 l/sec occurs.

The catchments are similar in most respects but two
differing factors influenced suspended solids in the runoff:
a) At the time of the study Welbeck was an established
catchment whereas at Rise Park construction was not yet
complete. Outcropping sand beds were used as a source of
building sand and large quantities of this were washed into
the sewer system during summer storms. Some of this sand is
likely to have been deposited in upstream sewers and
contributed to the solids loads in later storms.
b) Houses at Rise Park were centrally heated by oil or gas,
in contrast to Welbeck where coal fire heating was used.
Coal dust and soot therefore formed an additional source of
solids in the runoff.

Experimental details
Runoff from the catchments was monitored using an automatic
sampling and flow measurement system (Tucker 1974). This

system consisted of a flow actuated interval sampler coupled
with an automatic injection dilution gauging device using
lithium chloride as tracer. Samples were thus obtained at
5-10 minute intervals throughout each runoff period and
'labelled' with a lithium concentration inversely proportional
to flow. From analytical results instantaneous loads were
determined and also the total load discharged in each storm.

At Rise Park storms were monitored over the period August 1973
- August 1974, whereas at Welbeck the period covered was April
- August 1974. Over 100 storm events were monitored having
a total duration of 125 hours. The conditions producing these
storms ranged from prolonged low-intensity rainfall to short
intense thunderstorms.

Suspended solids loads discharged
The total measured quantity of solids discharged at Rise Park
was 5348 kg, of which 4138 kg (77%) was discharged in the
period April - August 1974. At Welbeck in the same period the
total measured quantity was 486 kg.

The large discrepancy between winter and summer totals is
the result of the massive loads discharged in summer thunderstorms. At Welbeck the thunderstorm of 16/6/74 produced 259
kg of solids, whereas at Rise Park on 13/7/74 3307 kg of
solids were discharged.

These quantities are far in excess of any other storm and
show the effect of solids scoured from unpaved surfaces under
the very high rainfall intensities encountered. Under more
normal rainfall conditions, runoff does not occur from unpaved
areas and solids are not directly derived from this source.
Therefore the thunderstorms may be excluded in comparing the
loads discharged from the two catchments. Expressing the
results in kg/imperv ha, the loads discharged were 44.8 and
40.5 kg/imperv ha for Rise Park and Welbeck respectively.
This close agreement demonstrates the importance of the
catchment impervious area in determining the loads discharged.

On the basis of similar quality for unmonitored runoff and a
total rainfall of 610 mm, the estimated annual loads
discharged (Sept 1973 - Aug 1974) are : Rise Park 13340 kg/yr
Welbeck 1920 kg/yr (692 and 342 kg/imperv ha/yr respectively).
These quantities may be compared with the annual load quoted
by Wilkinson (ibid) of 580 kg/imperv ha/yr for a suburban
catchment in a similar state of development to Rise Park.

The total mass of solids discharged in individual storms
varied between 0.013 and 1.21 kg/m³ of runoff, depending on
the nature of the storm and the preceding weather pattern.
However, despite this variability there is a strong overall
correlation between solids yield and discharge volume (see

Figure 1).

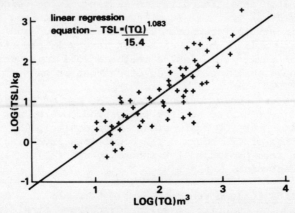

Figure 1   Relationship between Total measured storm flows
and Total measured storm Loads at Rise Park

Patterns of runoff quality
These results indicate that the major factors determining the
total solids loads discharged are the catchment impervious
area and the volume of runoff. However, the hydrographs and
solids concentration graphs for individual storms reveal that
other factors also influence the loads produced. The most
important of these are the duration of the preceding dry
spell and the rainfall intensities occurring in each storm.
Two phases or patterns of solids concentrations are normally
found, namely a first flush phase and a subsequent or
intensity related phase.

During the first flush, solids concentrations high in relation
to subsequent levels are normally found. Concentrations
decrease as runoff continues but the duration of the flush and
the concentrations observed depend on flow rate and the
duration of the preceding dry spell.

After the first flush, solids concentrations show an approximately linear correlation with flow, although a lag between
discharge and concentration peaks is normally apparent.

Storm analysis
In order to define these effects in more detail, the
individual storms were analysed according to the following
procedure.

The instantaneous storm load curve (ISL kg/min) ie the product
of flow and concentration curves, was integrated with respect
to time (T min), and plotted against a similar integration of
the square of instantaneous discharge (m^3/min). The slope of
the resulting curve is equal to the ratio

SS/Q $(\frac{kg}{m^3} \cdot \frac{min}{m^3}$ alternatively $\frac{mg}{l} \cdot \frac{hr}{mm})$ for the corresponding point in the storm.

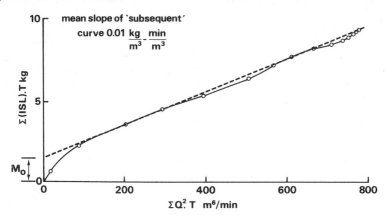

Figure 2   A plot of $\Sigma(ISL).T$ against $\Sigma Q^2 T$ for the storm at Rise Park on 3/5/74

A typical curve of this type is shown in Figure 2. The slope is initially high, corresponding to the first flush phase, and then falls to a nearly constant value, corresponding to the subsequent phase. This behaviour suggests that the mechanisms of generation and removal of solids are different in the two phases, and this is supported by the differences in the nature of the materials observed in each phase. First flush solids are primarily composed of fine low density material having a high organic content. This is considered to be derived from solids accumulating on catchment surfaces during the preceding dry spell, and also from solids deposited or produced by anaerobic degradation in road gullies. As can be seen from Figure 2 the rate of removal approximately follows an exponential decay or 'washing out' model, ie generation in runoff is largely independent of rainfall intensity. Solids obtained in subsequent runoff are primarily inert and inorganic, and are considered to be derived from scouring and erosion of surfaces under rainfall forces. From Figure 2 there is little evidence of 'washing out', and the generation of these solids is clearly dependant on rainfall intensity. Although separately distinguishable only in later runoff this intensity related effect must also apply during the first flush, and hence the slope of the 'subsequent' curve may be extrapolated to cut the vertical axis, defining the proportions of the total storm load (TL) due to each phase.

Comparison of all storms shows that the first flush phase is normally confined to initial runoff (from impervious area) of 0.4 - 0.9 mm. The mass of first flush solids (M_o)

produced in a storm is proportional to the duration of the preceding dry spell (see Figure 3). The value of the 'subsequent' slope varied from 13 - 200 $\frac{mg \cdot hr}{l \cdot mm}$, depending on the rainfall intensity in the first flush phase.

SOLIDS MODEL

The model is based on the principle of separate behaviour of 'first flush' and 'subsequent' solids. First flush solids removal is represented by Sartor and Boyds equation.

$$m_t = M_o \, \varepsilon^{-k_f \cdot r \cdot t} \tag{1}$$

where M_o, m_t = mass of first flush solids initially and at time $t (g/m^2)$

k_s = subsequent solids removal rate constant (mm^{-1})

k_f = first flush solids removal rate constant (mm^{-1})

r = rainfall intensity (mm/min)

t = time from start of storm (min)

Subsequent solids generation and removal is represented by the equation
$$M_t = b.r.e.^{-k_s \, r.t} \tag{2}$$

where M_t = mass of subsequent solids available for removal at time t, and b = constant (g.min/l)

Combining the first flush and subsequent solids equations using a step by step method over one minute intervals, we have

$$X_{t,t+1} = M_o \left[\varepsilon^{-k_f \left(\sum_{i=1}^{t} r_{i-1,i} \right)} \right] \left[1 - \varepsilon^{-k_f \cdot r_{t,t+1}} \right]$$

$$+ br_{t,t+1} \, \varepsilon^{-k_s \left[\left(\sum_{i=1}^{t} r_{i-1,i} \right) \right]} \left[1 - \varepsilon^{-k_s r_{t,t+1}} \right] \tag{3}$$

where $X_{t,t+1}$ = mass of solids generated on catchment surfaces (g/m^2) between time t and t+1 minutes.

FLOW MODEL

Viessman's (1966) synthetic one minute unit hydrograph is used to generate a hydrograph of runoff to the sewer. This method assumes linear reservoir characteristics for the ground surfaces. The resultant hydrograph from an area is then combined with other similar hydrographs from adjacent areas using the time-area method of Watkins (1962).

One Minute Hydrograph

Rising leg	ρ_t	$= r(1-\varepsilon^{-t/K})$	for t<1	(4)
Peak	ρ_{max}	$= Z = r(1-\varepsilon^{-1/K})$	t=1	(5)
Falling leg	ρ_t	$= Z\varepsilon^{-(t-1)/K}$	t>1	(6)

where ρ_t = ordinate of one minute hydrograph at time t (mm/min)

    k = reservoir storage constant (min)

The total runoff hydrograph is obtained by summing the one minute hydrograph

$$q_t = (1-\varepsilon^{-1/k}) \left[\sum_{i=1}^{t} r_{i-1,i} \cdot \varepsilon^{-(t-i)/K} \right] \text{(mm/min)} \tag{7}$$

This flow is combined with one minute incremental areas of the area-time diagram to produce the sewer discharge hydrograph

$$Q_t = \sum_{i=1}^{n} q_{t-i+1} \cdot a_i \tag{8}$$

where Q_t = sewer discharge of time t (1/min)

    a_i = one minute increments of area-time diagram (m²)

    n = total number of increments in area-time diagram

SOLIDS POLLUTION MODEL

The concentration of solids generated in each minute is combined with the corresponding one minute hydrograph and the resultant load is then routed through the system in the same manner as the flow.

The concentration generated by rainfall

$$C_{t-1,t} = X_{t-1,t}/r_{t-1,t} \text{ (g/1)} \tag{9}$$

The resultant load at time t is then

$$L_t = C_{t-1,t} \cdot Z_t = X_{t-1,t} (1-\varepsilon^{-1/K}) \text{ (g/m}^2 \text{.min)} \tag{10}$$

and the overall load in runoff is

$$Y_t = (1-\varepsilon^{-1/K}) \left[\sum_{i=1}^{t} X_{i-1,i} \cdot \varepsilon^{-(t-i)/K} \right] \text{(g/m}^2 \text{.min)} \tag{11}$$

The load in the sewer discharge

$$ISL_t = \sum_{i=1}^{n} Y_{t-i+1} a_i \quad \text{(g/min)} \tag{12}$$

and $SS_t = ISL_t/Q_t$ (g/1) $\tag{13}$

Features of the model and values of variables

In modelling of runoff alone rainfall 'losses' are normally extracted directly from the rainfall hyetograph, either using a percentage runoff factor or by subtracting losses at a constant rate. These methods can be used with the pollution model but each has disadvantages. In particular, since the generation of solids in the 'subsequent' phase is considered to depend on actual rainfall intensities it follows that losses should be accounted for at the point of entry to the sewer.

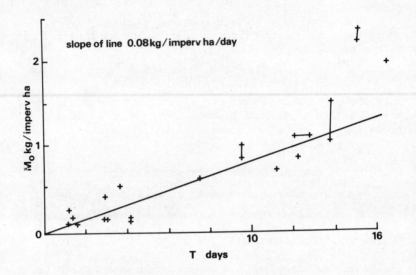

Figure 3   Relationship of first flush mass (M_O) to duration
            of preceding dry spell (T_D)

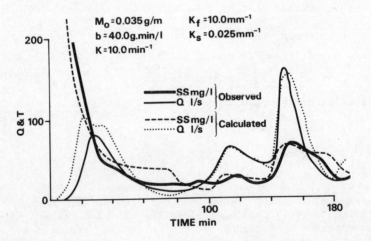

Figure 4   Comparison of recorded and calculated curves
            for the storm of 26/8/74 at Rise Park

The reservoir storage constant K was found by Viessman to lie
in the range 2-5 min for small impervious catchments. However,
it is found that a much higher value (K = 10 min) must be used
to obtain agreement between recorded and calculated hydrographs. This demonstrates the inadequacy of the simple time-
area method of flow routing through the sewer which does not
adequately represent the attenuation occurring in sewer
storage. Thus an artifically large attenuation must be
introduced on the ground surfaces.

The subsequent solids removal constant k_s is given the value
0.025 mm^{-1} as determined by Sartor and Boyd (ibid). The
constant k_f lies in the range 5-12 mm^{-1}, these limits
corresponding to 99% removal of the first flush in 0.9 and
0.4 mm of runoff (from impervious areas) respectively.

The constant b determines the subsequent load generated and
hence the slope of the graph Σ ISL.T vΣQ^2.T. For a given
storm the slope is proportional to b, and the relationship
may be found using a trial value. An example of the results
obtained with the model is shown in Figure 4.

CONCLUSION

A study of two suburban catchments in Nottingham, England
revealed that the major factors determining the loads of
suspended solids discharged in runoff are the impervious
catchment area and the total volume of runoff. However,
high intensity summer storms which cause direct runoff
from unpaved areas produce abnormally high solids loads.

The loads produced by individual storms are primarily
determined by the intensity of rainfall and the duration
of any preceding dry spell.

A mathematical model of solids generation based on the concept
of contributions to total suspended solids concentrations from
both 'first flush' and 'subsequent' solids produces reasonable
agreement with recorded data.

ACKNOWLEDGEMENT

This work was carried out under a Research Assistanship at
Trent Polytechnic, Nottingham, and the results were presented
to Loughborough University of Technology in a research thesis
for the degree of Master of Science.

REFERENCES

Pravoshinsky, N. A. and Gatillo, P. D. (1969) Determination of the pollutional effect of surface runoff. Proc. 4th Intern. Conf. on Wat. Pollut. Res., Prague, 187-203.

Sartor, J. D. and Boyd, G. B. Water pollution aspects of street surface contaminants. U.S.E.P.A. R2-72-081.

Soderlund, G. and Lehinen, H. Comparison of discharges from urban stormwater runoff, mixed storm overflow and treated sewage. Proc. 6th Intern. Conf. on Wat. Pollut. Res., Jerusalem, A/9/17/1-10.

Tucker, C. G. J. (1974) Stormwater pollution - sampling and measurement. J. Instn. Municipal Engrs., 101, 10: 269-273.

Tucker, C. G. J. (1975) Pollution loads in runoff from urban areas. M.Sc. Thesis, Loughborough University of Technology.

Viessman, W. (1966) The hydrology of small impervious areas. Wat. Resources Res., 2, 3: 405-412.

Watkins, L. H. (1962) The design of urban sewer systems. Road Res. Lab. Technical Paper No 55, H.M.S.O., Lond.

Weibel, S. R., Anderson, R. J. and Woodward, R. L.(1964) Urban land runoff as a factor in stream pollution. J. Wat. Pollut. Control Fed. 36 : 914-924.

Wilkinson, R. (1956) The quality of rainfall runoff water from a housing estate. J. Inst. Pub, Health Engrs. 55 , 2 : 70-84.

QUICK AND SLOW RESPONSE TO RAINFALL BY AN URBAN AREA

Jan A. van den Berg

IJsselmeerpolders Development Authority,
Scientific Division, Lelystad, The Netherlands

wer set in buoi dit lân to weak
en oft nou beammen om hwat droechte raze
gjin himelslûs giet earder op 'e heak
Syn eazjen sil har wieter blaze

Daniël Daen,   De iken fan Dodona

ABSTRACT

A rainstorm on an urban area causes a quick and a slow response.
The quick one consists of the surface runoff from the covered
area and can be simulated by a Nash cascade of linear reservoirs with a reservoir coefficient between some hundreds to some
thousands of seconds, according to the kind of the covered area.
Especially flat roofs decrease and delay the peak flow. The subsurface flow through the soil to the subsurface drainage system
is slow; the coefficient of the linear reservoir simulating the
depletion curve amounts to some tens of days.

The waterbalance of two different parts of an urban area shows
that quantitatively the discharge of the subsurface drains can
be of primary importance. Even for an almost completely paved
area, the subsurface drainage system may discharge 40 per cent
of the precipitation. It is an important factor for the waterma-
nagement because the water from the subsurface drains is of
good quality. However from the point of view of floods, the
discharge of the subsurface drains is slightly important.

The data are from four different urban catchment areas (flat
roof, parking lot, housing area and shopping centre) at the new
town of Lelystad, which is situated in a recently reclaimed pol-
der in the former Zuydersea.

1. INTRODUCTION AND BOUNDARY CONDITIONS

The research was set up at Lelystad, a new-town in one of the

four recently constructed polders in the Lake IJssel. In the
last two polders - Eastern and Southern Flevoland - town-building andnature preservation have become important factors next
to agricultural development. Two large new towns - Lelystad and
Almere - are under construction andplanned to house respectively 100.000 and 200.000 inhabitants in the year 2000.

The top-soil of the polders is a holocene deposit consisting
mainly of loam or clay. The little bearing power of this top-
soil, the originally high piezometric head of the aquifer below
the holocene deposits and the design of a town with separate
road systems were reason to raise the building sites in Lelystad and Almere with abouw 1 meter of sand. Groundwater control
in sand, overlaying an impermeable subsoil needs a subsurface
drainage system. Both during town-building and afterwards, the
drainage system is necessary to drain roads, streets and the
cellarage of the houses.

The urban areas in the polders are very flat. Hence there is
hardly any exchange of surface runoff from the unpaved area to
the covered area or conversily. It has been decided to construct
both in Lelystad and Almere separate sewer systems for waste
water and storm water. The storm water drains as well as the
subsurface drains discharge into open drains in the urban area.

2. CATCHMENT AREAS AND EQUIPMENT

A housing area,a parking lot and a shopping centre have been
fitted out as catchments (table 1). Moreover the discharge of
flat roofing of 350 m2 has been measuring recently. The storm
water sewer systems in the catchments are permanently filled
up for the greater part (static storage) in order to provide
a quick response when it starts to rain. It enables the calculation of the inflow into the storm drains (Van den Berg, 1976).

TABLE 1. Characteristics of the catchments

catchment	gross area (ha)	covered area (%)	nature of covered area		
			roofed (%)	asphalt (%)	bricks (%)
housing area	2.0	44	30	32	38
parking lot	0.7	99.6	--	45	55
shopping centre	2.5	100	42	6	52

A central data logging system measures continuously the following elements:
- rainfall (ground level raingauge);
- storm water discharge (rectangular Thomson V-notch);
- subsurface drainage discharge (electromagnetic flow meter);
- groundwater level.

Recording takes place only when there is a significant change in the value of an element.

3. THE WATERBALANCE

The ponderance of the different components of the urban discharge may be concluded partly from the water-balance. We distinguish the following terms expressed in mm depth related to the total area

$$P + K = Ea + Qd + Qs + Ec \qquad (1)$$

P is rainfall, K the seepage (positive if upwards), Ea the actual evapotranspiration from the unpaved area, Qd the discharge by the subsurface drains, Qs the discharge by the storm drains and Ec the evaporation from the covered area. Precaution has been employed to prevent the transfer of water across the boundaries of a catchment.

P, Qs and Qd are measured. Ec is incalculable and therefore disregarded. The seepage depends on the difference between the groundwater level and the head in the pleistocene aquifer, the top of which lays 4 à 6 m below the holocene deposits. The mean groundwater level amounts to about N.A.P. minus 4.75 m (N.A.P. = normal ordnance datum at Amsterdam, i.e. about mean sea level); the head lowered from N.A.P. minus 4.8 in 1971 to minus 5.0 m in 1975.

Though notice should be taken of little negative seepage it is also disregarded. The actual evapotranspiration from the unpaved area has been determined from the potential evapotranspiration with the method of Thornthwaite and Mather (1957). The potential evapotranspiration has been calculated multiplying the Penman evaporation by a reduction factor and is summed up to monthly values. If there is a monthly shortage of rainfall the potential evapotranspiration has been reduced proportionaly to the soil moisture deficit in the root zone. The yearly water balance has been made for the housing area and the parking lot for 5 resp. 4 years (1971-1975), table 2.

TABLE 2. Waterbalance (mm depth related to the total area)

year	P	Ea	housing area Qd	Qs	sum	parking lot Qd	Qs	sum
1971	531	221	263*	111	595	–	–	–
1972	726	254	250	152	656	332	304	636
1973	732	234	264	190	688	280	379	659
1974	733	233	325	186	744	289	387	676
1975	607	208	400*	143	751	215	327	542
mean	666	230	300	157	687			
	700					279	349	628
%	100	34	45	24	103	40	50	90

* drain-pipes have been cleaned in March 1971 and April 1975.

It can be seen that in a flat area the rainfall on an urban
area is drained for a considerable part by the subsurface drainage system; in the case of the almost completely paved parking
lot even forty per cent of rainfall is discharged by the subsurface drains, varying from 35 to 45 per cent in the individual years. Calculated over a long period as in a water balance,
amounts of discharge show only one of the different aspects of
urban water control; so the response of the covered area is
much quicker and more impetuous than that of the subsurface
drains as will be shown in the next sections.

4. STORM WATER DRAINS - INFLOW AND DISCHARGE

The response of the covered area will be considered separately
as well as in combination with the storm drains i.e. the inflow
and the discharge of the drains. In this section inflow and
discharge will be expressed in mm depth related to the covered
area.
For storms (N > 40) with a runoff volume of more than 5 mm the
runoff coefficient amounts to 0.66 for the parking lot and the
housing area and 0.56 for the shopping centre; the standard
deviation is 0.1 respectively 0.08.

The net rainfall has been estimated by distributing the loss of
rainfall exponentially during the rainstorm. The distribution
function doesn't depend on time but on a special variable r_j
which has been defined as the quotient of the part of rainfall
until time j and the total amount (Van den Berg, De Jong and
Schultz, 1977). The flat roof causes a initial loss of 3 à 5 mm.
In the case of the here considered rainstorms the roof surface
was "saturated" at the starting point; the remaining loss has
been distributed according to Φ-index.

How the catchments differ in response to a rainstorm will be
shown considering the values of the parameters of the Nash model. The optimalization of these parameters has been computeri-
zed together with other mathematical models like non-linear
reservoir, Laguerre functions and Volterra series (Van den Berg,
Van der Kloet, Ven and Van der Wal, 1978). Ten rainstorms have
been selected. For the shopping centre the discharge of the
storm drains had been measured of only five of these storms.
As for the roofs only some very recent measurements were available.

Discharge. The discharge measurements give the response of the
covered area inclusive the flow through the storm drains. The
results of the optimalization of the Nash parameters (n and k)
are summarized in table 3. It can be seen that the time lag n.k
increases with increasing percentage of (flat) roofs in the
catchment. This is caused mainly by the increase of the reservoir coefficient k. However, the values of the parameters are
not simply transferable because the storage of the storm drains

in the catchment is filled up for the greater part.

TABLE 3. Nash parameters for discharge of storm drains

	Parking lot (N = 10)			housing area (N = 10)			shopping centre (N = 5)		
	n (-)	k (sec)	n.k (sec)	n (-)	k (sec)	n.k (sec)	n (-)	k (sec)	n.k (sec)
mean	1.83	202	318	1.41	442	556	1.22	876	997
s	.89	77	64	.46	185	97	.42	273	260
v.c.	.49	.38	.20	.33	.41	.17	.34	.31	.26

s = standard deviation; v.c. = variation coefficient.

Inflow. The large static storage in the storm drains and an empirical relation between the stage at the V-notch and the dynamic storage enables us to calculate the inflow into the storm drains with the equation of continuity (Van den Berg, 1976). Only for the shopping centre it was not possible to calculate the inflow. The inflow from the flat roof has been measured separately.

TABLE 4. Nash parameters for the inflow into storm drains

	Parking lot (N = 10)			Housing area (N = 10)			flat roof (N = 3)		
	n (-)	k (sec)	n.k (sec)	n (-)	k (sec)	n.k (sec)	n (-)	k (sec)	n.k (sec)
mean	1.26	212	238	.69	423	273	1.27	3020	3865
s.	.57	81	70	.19	153	74	.15	193	717
v.c.	.45	.38	.29	.28	.38	.27	.12	.06	.18

s = standard deviation; v.c. = variation coefficient.

Again it can be seen that the flat roofs cause a relatively large time lag due to a large reservoir coefficient. Though comparable measurements from a catchment with sloping roofs are not yet disponible it can be concluded that the increase of the percentage of flat roofed area tends to reduce and to delay the peak flow. Moreover flat roofs have a large storage in comparison with streets, resulting in a greater loss at the beginning of rainfall.
For five mean return periods the peak inflow has been calculated from rainfall depth-duration data (Van den Herik and Kooistra, 1973). Figure 2 shows how the peak inflow depends on the parameters n and k of the Nash-model for three catchments.

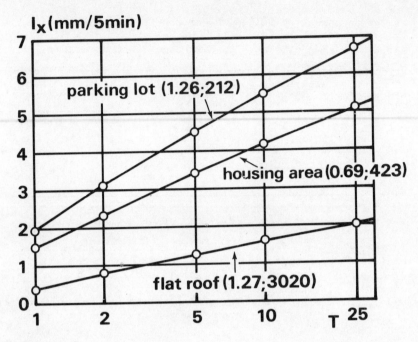

Figure 1.
Relation between peak inflow (I_x) and mean return period (T) for three catchment with different values of the parameters (n;k) of the Nash-model.

5. SUBSURFACE DRAINAGE DISCHARGE

As for the base flow, groundwater is considered often as a linear reservoir.

$$q_t = \alpha \cdot S = \frac{1}{k} \cdot S \qquad (2)$$

Then the depletion curve is given by

$$q_t = q_o \exp(-\alpha t)$$

Hence, if Δt = unit length of time (day)

$$\ln q_t = \ln q_{t+1} + \alpha \qquad (3)$$

However, not every part of a decreasing flow curve is exactly a depletion curve; it is difficult to know when the percolation has finished. In the graph of figure 2 all daily values (q_t, q_{t+1}) of 1972 are plotted from parts of the flow curve with decreasing subsurface drainage discharge from the housing area. The tangent touching the extreme points, represents the different depletion parts of the flow curve. From the graph α equals 0.06 day or k = 16.7 days.

The subsurface drainage flow is influenced by the precipitation of iron compounds in and around the drain pipes. This results into a higher groundwater level at the same discharge. Cleaning of the drain-pipes gives rise to a large discharge (see table 2, year 1975) and a quick lowering of the groundwater level with 0.3 à 0.8 m within three weeks. The influence of the drain cleaning on the value of α is still beeing studied.

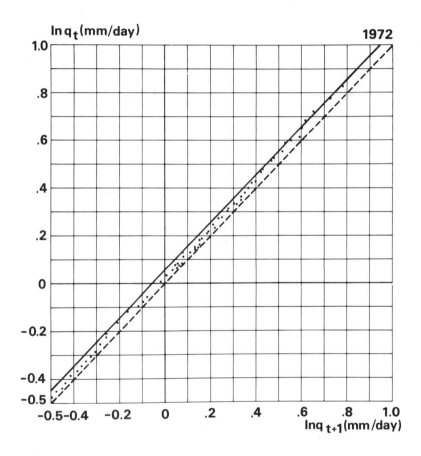

Figure 2. Plot of the logarithm of decreasing daily discharges from the subsurface drains of the housing area.

REFERENCES

Berg, J.A. van den (1976) Data analysis and system modelling in urban catchment areas (in the new town of Lelystad, the Netherlands). Hydrol.Sci.Bull. XXI, no. 1, 187-194.
Berg, J.A. van den, J. de Jong and E. Schultz (1977). Surface water in an urban area with separate sewer systems. In

"Effects of Urbanization and Industrialization on the
Hydrological Regime and on Water Quality".
(Proceedings of the Amsterdam Symposium, October 1977),
109-123: IAHS Publ. no. 123.

Berg, J.A. van den, P. van der Kloet, G.A. Ven and M. van der
Wal (1978). In preparation. Flevobericht. IJsselmeerpolders Development Authority, Lelystad.

Herik, A.G. van den and M.T. Kooistra (1973). Vijf-minuten re-
gens (in Dutch; "Rainfall data for time steps of five
minutes). Gront Mij, De Bilt.

Kidd, C.H.R. (1976). A non-linear urband runoff model.
Report 31. Institute of Hydrology, Wallingford, Oxon.

Thornthwaite, C.W. and J.R. Mather (1957). Instructions and
tables for computing the potential evapotranspiration
and the waterbalance. Publications in climatology, Vol.
X, nr. 2. Drexel Institute of Technology and Climatology
Centerton.

LIST OF PARTICIPANTS

Prof. P. Ackers,
Binnie & Partners,
Artillery House,
Artillery Row,
Westminster,
London, SW1P 1RX.

Mr. T. A. Anderson,
Department of Sewerage,
Strathclyde Regional Council,
Viceroy House,
India Street,
Glasgow.

Mr. D. Andrejevic,
Energoprojekt,
Kosovska 32,
Belgrade,
Yugoslavia.

Dr. S. Armanazi,
Babtie Shaw & Morton,
95 Bothwell Street,
Glasgow, G2 7HX.

Mr. V. Arnell,
Department of Hydraulics,
Chalmers University of
Technology,
Fack,
S-402 20 Göteborg,
Sweden.

Mr. R. E. G. Arnould,
Institut du Genié Civil
Université de Liège,
Quai Banning 6,
B-4000 Liège,
Belgium.

Mr. D. M. V. Aspinwall,
Yorkshire Water Authority,
Southern Division,
Castle Market Building,
Exchange Street,
Sheffield, S1 1GB.

Mr. T. C. Atkinson,
Lothian Regional Council,
Department of Drainage,
6 Cockburn Street,
Edinburgh, EH1 1NZ.

Mr. G. Austin,
Charnwood Borough Council,
Macaulay House,
Loughborough,
Leics.

Mr. J. M. Baird,
North West Water Authority,
Dawson House,
Great Sankey,
Warrington,
WA5 3LW.

Mr. J. Balek,
S.G. Consult. Comp.,
Prague 1,
M. Gorkeno 7,
Czechoslovakia.

Mr. D. J. Balmforth,
Sheffield City Polytechnic,
Department of Civil
Engineering,
Pond Street,
Sheffield, S1 1WB.

Mr. H. H. Barnes, Jr.,
U.S. Geological Survey,
National Center,
Mail Stop 415,
Reston,
Virginia 22092,
U.S.A.

Mr. R. E. Bartlett,
Consulting Engineer,
23 Lower Church Street,
Ashby-De-La-Zouch,
Leics.

Mr. A. Bell,
Kirk, McClure & Morton,
2 Elmwood Avenue,
Belfast,
BT9 6BA.

Mr. R. Bettess,
Hydraulics Research Station,
Howbery Park,
Wallingford,
Oxon, OX10 8BA.

Mr. R. Bishop,
James F. MacLaren Ltd.,
435 McNicoll Avenue,
Willowdale,
Ontario,
Canada, M2H 2R8.

Mr. D. Black,
Nicholas O'Dwyer & Partners,
Consulting Engineers,
Carrick House,
Dundrum,
Dublin, 14.

Mr. C. E. G. Bland,
Clay Pipe Development
Association,
Drayton House,
30 Gordon Street,
London,'WC1H OAN.

Dr. O. Bonnacci,
Gradevinski Institut,
Oour Fakultet Gradevinskih
Znanosti,
58000 Split - V. Maslese bb,
Yugoslavia.

Mr. J. F. Bonsall,
Department of the
Environment,
2 Marsham Street,
London, SW1P 3EB.

Mr. A. J. Boyd,
White Young & Partners,
Forum Chambers,
Town Centre,
Stevenage,
Herts, SG1 1EL.

Mr. B. Bozic,
Imperial College,
Yugoslav Association for
Hydrology,
Bulevar 23 Oktobra br 5/53,
21000 Novi Sad,
Yugoslavia.

Mr. R. S. Bray,
Severn Trent Water
Authority,
Avon House,
De Montfort Way,
Cannon Park,
Coventry, CV4 7EJ.

Mr. R. Buckingham,
Ministry of Agriculture
Fisheries and Food,
Great Westminster House,
Horseferry Road,
London, SW1P 2AE.

Mr. J. K. Budleigh,
Directorate of Operations,
Southern Water Authority,
Guildbourne House,
Chatsworth Road,
Worthing, BN11 1LD.

Mr. J. Bullen,
Southern Water Authority,
Hampshire Drainage Division,
Eastleigh House,
2 Market Street,
Eastleigh,
Hants.

Mr. J. Burke,
Sligo Regional Technical
College,
Sligo,
Eire.

Mr. R. Burrows,
Civil Engineering
Department,
University of Liverpool,
Brownlow Street,
P.O. Box 147,
Liverpool, L69 3BX.

Dr. H. Caspers,
Institüt für Hydrobiologin,
Palmaille 55,
2 Hambürg 50,
West Germany.

Dr. J. B. Chatterton,
Severn Trent Water Authority,
Abelson House,
Coventry Road,
Birmingham.

Mr. J. A. Clark,
P.O. Box 1073,
Cape Town,
Rep. of South Africa.

Mr. H. M. C. Cockbain,
Welsh National Water
Development Authority,
Cambrian Way,
Brecon,
Powys,
Wales.

Mr. J. A. Cole,
Water Research Centre,
P.O. Box 16,
Henley Road,
Medmenham,
Marlow, SL7 2HD.

Mr. R. H. F. Collar,
Department of Civil
Engineering,
John Anderson Building,
University of Strathclyde,
107 Rotten Row,
Glasgow.

Mr. P. J. Colyer,
Hydraulics Research Station,
Howbery Park,
Wallingford,
Oxon, OX10 8BA.

Mr. P. C. Cook,
Roadworks Ltd.,
Dobbs Lane,
Kesgrave,
Ipswich, IP5 7QQ.

Mr. L. C. Cordwell,
Trafford Borough Council,
Birch House,
Talbot Road,
Old Trafford,
Manchester, M16 OGH.

Mr. E. Creed,
Nicholas O'Dwyer & Partners,
Consulting Engineers,
Carrick House,
Dundrum,
Dublin 14.

Mr. R. C. Cresswell,
Severn Trent Water Authority,
Lower Severn Division,
Southwick Park,
Groucester Road,
Tewkesbury,
Glos, GL20 7DG.

Mr. S. Crook,
Liverpool City Council,
Steers House,
Canning Place,
Liverpool, L1 8JA.

Mr. B. Dahlström,
Swedish Meteorological
& Hydrological Institute,
Fack,
60101 Norköping,
Sweden.

Mr. E. P. D'Alton,
Cork Corporation,
City Hall,
Cork,
Ireland.

Mr. L. H. Davis,
Anglian Water Authority,
Five Acres,
Simpson,
Milton Keynes.

Mr. M. Desbordes,
Laboratoire d'Hydrologie
Mathematique,
Universite des Sciences et
Techniques du Languedoc,
Place Eugene Bataillon,
34060 Monpellier Cedex,
France.

Mr. E. Dinez,
Espay Muston & Assoc.,
3010 S. Laman,
Austin TX 78704,
U.S.A.

Mr. R. E. Dixon,
Oak Villa,
Townfield Lane,
Mollington,
Chester.

Mr. N. Djordjevic,
Faculty for Natural and
Mathematical Sciences,
P.O. Box 550,
Studentski trg 16,
11001 Belgrade,
Yugoslavia.

Mr. M. Dresdner,
Balasha Jalon Consultants
& Engineers Ltd.,
P.O. Box 1727,
Haifa,
Israel.

Mr. A. R. Eadon,
Severn Trent Water Authority,
Avon House,
De Montfort Way,
Cannon Park,
Coventry, CV4 7EJ.

Mr. H. Eagles,
Thames Water Authority,
Metropolitan Public Health
Division,
Broadway Building,
50-64 Broadway,
London, SW1H 0DB.

Mr. J. Eastwood,
Thames Water,
Rivers House,
Crossness Works,
Abbey Wood,
London, SE2 9AQ.

Dr. A. M. C. Edwards,
Yorkshire Water Authority,
Directorate of Resource
Planning,
21 Park Square South,
Leeds, LS1 2QG.

Mr. A. H. Elbeik,
Civil Engineer,
P.O. Box 3428,
Tripoli,
Libya.

Mr. J. B. Ellis,
Middlesex Polytechnic,
The Burroughs,
Hendon, NW4 4BT.

Mr. A. W. English,
Howard Humphries & Son,
Thorne Croft Manor,
Dorking Road,
Leatherhead,
Surrey, KT22 8JB.

Mr. G. P. Evans,
Water Research Centre,
P.O. Box 16,
Henley Road,
Medmenham,
Marlow, SL7 2HD.

Mr. M. Evans,
Stratford District Council,
The Council Offices,
Birmingham Road,
Stratford-on-Avon,
Warwickshire.

Mr. W. S. Eyre,
Geography Department,
University College London,
Gower Street,
London, WC1E 6BT.

Mr. J. Falk,
Department of Water Resources
Engineering,
Lund Institute of Technology,
University of Lund,
Fack 725, S-220 07,
Sweden.

Mr. W. R. Ferguson,
Lothian Regional Council,
Department of Drainage,
6 Cockburn Street,
Edinburgh, EH1 1NZ.

Mr. D. Fiddes,
Department of the Environment,
2 Marsham Street,
London, S.W.1.

Mr. R. Field,
Storm and Combined Sewer
Section,
Wastewater Research
Division,
Municipal Environmental
Research Lab.,
Woodbridge Avenue,
Building 10,
Edison,
New Jersey,
U.S.A.

Mr. I. J. Fletcher,
Department of Civil &
Structural Engineering,
Trent Polytechnic,
Burton Street,
Nottingham, NG1 4BU.

Mr. A. W. Fountain,
John H. Haiste & Partners,
Belmont House,
20 Wood Lane,
Leeds, LS 6 2AG.

Mr. W. J. Gardener,
Wavin Plastics Ltd.,
P.O. Box 12,
Rigby Lane,
Hayes,
Middlesex.

Mr. D. H. Garside,
Central Lancashire
Development Corporation,
Cuerden Hall,
Bamber Bridge,
Preston,
Lancashire, PR5 6AX.

Mr. H. Geiger,
Swiss Institute for
Forestry Research,
CM-8909 Birmensdorf,
Switzerland.

Dipl. Ing. W. F. Geiger,
8000 Munchen 71,
Strasslacher STR. 2,
FDR.

Mr. R. W. Gillett,
East Sussex Water &
Drainage Division,
Southern Water Authority,
28 Wellington Square,
Hastings,
Sussex.

Mr. E. P. Gouws,
Liebenberg & Stander,
P.O. Box 4733,
Cape Town 8000,
Republic of South Africa.

Mr.F. Gowdy,
Department of Environment
(NI),
Training Branch,
Room 372,
Stormont,
Belfast, BT4 355,
Northern Ireland.

Dr. A. J. Grass,
Department of Civil
Engineering,
University College London,
Gower Street,
London W.C.1.

Prof. N. S. Grigg,
Water Resources Research
Institute,
University of North
Carolina,
124 Riddick Building,
Raleigh,
North Carolina 27607,
U.S.A.

Mr. D. E. Grimshaw,
Wessex Water Authority,
Bristol Avon Division,
P.O. Box 95,
The Ambury,
Bath, BA1 2 YP

Mr. C. T. Haan,
Department of Agricultural
Engineering,
University of Kentucky,
Lexington,
Kentucky 40506,
U.S.A.

Mr. P. Hale,
Thames Water Authority,
Metropolitan Public Health
Division,
Broadway Building,
50-64 Broadway,
London, SW1H ODB.

Miss H. Hall,
Department of Geography,
University College,
London University,
Gower Street,
London, WC1E 6BT.

Mr. M. M. I. Harman,
Water Research Centre,
Stevenage Laboratory,
Elder Way,
Stevenage,
Herts, SG1 1TH.

Prof. P. Harremoës,
Department of Sanitary
Engineering,
Technical University of
Denmark,
2800 Lingby,
Denmark.

Mr. J. M. Harris,
Welsh Water Authority,
10 Kingsway,
Swansea,
Wales.

Mr. M. R. Hasan,
Sir William Halcrow &
Partners,
45 Notting Hill Gate,
London, W11 3JX.

Mr. D. J. Hay,
Environment Canada,
351 St. Joseph Boulevard,
13th Floor,
Place Vincent Massey,
Hull,
Quebec, KIA IC8,
Canada.

Mr. K. I. M. Henry,
Babtie Shaw & Morton,
95 Bothwell Street,
Glasgow, G2 7HX.

Mr. R. Hepworth,
Ward Ashcroft and Parkman,
Cunard Building,
Liverpool, L3 1ES.

Mr.F. Holly,
Sogreah,
B.P. 172,
Centre de TRI/38042,
Grenoble Cedex,
France.

Mr. J. Holmes,
Greater London Council,
Dept. of Public Health
Engineering,
10 Great George Street,
London, S.W.1.

Mr. J. E. V. Holmes,
Thames Water Authority,
Metropolitan Public Health
Division,
Broadway Building,
50-64 Broadway,
London, SW1H ODB.

Mr. C. D. D. Howard,
Charles Howard & Associates
Ltd.,
216-2025 Corydon Avenue,
Winnipeg,
Manitoba, R3P ON5,
Canada.

Dr. D. J. Howard,
General Engineering &
Computer Services Ltd.,
Cunard Building,
Liverpool, L3 1ES.

Mr. R. Howard,
John Taylor & Sons,
Artillery House,
Artillery Row,
London, SW1P 1RY.

Mr. A. C. Hoyle,
L. G. Mouchel & Partners,
19 Gay Street,
Bath, BA1 2PQ.

Dr. W. C. Huber,
University of Florida,
Department of Environmental
Engineering Sciences,
A. P. Black Hall,
Gainesville,
Florida 32611,
U.S.A.

Mr. T. Hugin,
Oslo Water and Sewage Works,
Trondheimsveien 5,
Oslo 1,
Norway.

Mr. C. F. Humphries,
Yorkshire Water Authority,
West Riding House,
67 Albion Street,
Leeds, LS1 5AA.

Mr. J. Hunter,
Severn Trent Water
Authority,
Lower Trent Division,
Mapperley Hall,
Lucknow Avenue,
Nottingham.

Mr. R. L. Hutchings,
South West Water Authority,
Manley House,
Kestrel Way,
Sowton Ind. Estate,
Exeter, EX2 7LQ.

Mr. P. Jacobsen,
Department of Sanitary
Engineering,
Technical University of
Denmark,
Bygning 115,
DK-2800 Lyngby,
Denmark.

Mr. P. L. Jacques,
Technical Services Dept.,
South Oxfordshire District
Council,
London Road,
Wheatley,
Oxford.

Mr. G. E. Jenkinson,
Watson Saudi Arabia,
P.O. Box 2023,
Jeddah,
Saudi Arabia.

Mr. R. K. Johnstone,
Chief Engineers Department,
North Tyneside Council,
6th Floor,
Amberley,
Killingworth,
Newcastle, NE12 0SB.

Mr. A. R. B. Jones,
Southern Water Authority,
West Kent Drainage Division,
54/58 College Road,
Maidstone,
Kent, ME15 6SJ.

Mr. K. R. Jones,
Mander, Raikes & Marshall,
Embassy House,
Queen's Avenue,
Bristol, BS8 1SB.

Dr. M. E. Jones,
Department of Geography,
Kings College,
Strand,
London, W.C.2.

Mr. L. Jordan,
North West Water,
Dawson House,
Great Sankey,
Warrington, WA5 3LW.

Prof. S. Jovanovic,
Gradjevinski Fakultet,
Bulevar Revolucije 73,
11000 Beograd,
Yugoslavia.

Mr. A. D. K. Kell,
Binnie & Partners,
Artillery House,
Artillery Row,
London, SW1P 1RX.

Mr. P. S. Kelway,
Northumbrian Water Authority,
Northumbria House,
Regent Centre,
Gosforth,
Newcastle-upon-Tyne,
NE3 3PX.

Dr. P. H. Kemp,
Department of Civil
Engineering,
University College,
Gower Street,
London, W.C.1.

Mr. J. Keser,
Institut für Wasserwirtschaft,
Technische Universität
Hannover,
Callinstr. 32,
D-3000 Hannover 1,
Germany.

Dr. C. H. R. Kidd,
Institute of Hydrology,
Wallingford,
Oxon.

Mr. D. King,
L. G. Mouchel & Partners,
19 Gay Street,
Bath,
Avon.

Mr. A. Knee,
Planning & Development Dept.,
Nuneaton Borough Council,
Council House,
Nuneaton, CV11 5AA.

Mr. D. A. Kraijenhoff,
Wageningen Agricultural
University,
Nieuwe Kanaal 11,
Wageningen,
The Netherlands.

Mr. T. Larsen,
University of Aalborg,
Post Box 159,
9100 Aalborg, Denmark.

Mr. J. Larsson,
Department of Hydraulics,
Royal Institute of
Technology,
S-100 44,
Stockholm, Sweden.

Mr. T. Lekane,
STE Traction et Electra SA,
Rue de la Science 31,
1040 Bruxelles, Belgium.

Mr. Lin,
Directorate of Operations,
Southern Water Authority,
Guildbourne House,
Chatsworth Road,
Worthing, BN11 1LD.

Mr. E. D. Lindesay,
Building Design Partnership,
Vernon Street,
Moor Lane,
Preston,
Lancashire.

Mr. O. Lindholm,
State Pollution Control
Authority (SFT),
P.O. Box 8100 Dep,
Oslo 1, Norway.

Mr. D. F. M. Lothian,
Department of Finance for
Northern Ireland,
Civil Engineering Branch,
Hydebank,
4 Hospital Road,
Belfast, BT8 8JL.

Dr. M. J. Lowing,
Institute of Hydrology,
Wallingford,
Oxon.

Mr. A. R. R. Lupton,
Binnie & Partners,
Artillery House,
Artillery Row,
London, SW1P 1RX.

Mr. E. Lygren,
Norwegian Institute for
Water Research,
P.O. Box 333 Blindern,
Oslo 3,
Norway.

Mr. J. T. Maddock,
294 Church Road,
St. Annes,
Lytham St. Annes,
Lancashire, FY8 3NR.

Mr. G. Mance,
Water Research Centre,
Stevenage Laboratory,
Elder Way,
Stevenage,
Herts, SG1 1TH.

Mr. C. R. Mann,
Rope, Kennard & Lapworth,
2/4 Sutton Court Road,
Sutton,
Surrey, SM1 4SS.

Mr. J. R. Marsden,
Fawcett & Partners,
21 Inner Park Road,
London, SW19 6ED.

Dr. C. T. Marshall,
North West Water,
Dawson House,
Great Sankey,
Warrington, WA5 3LW.

Mr. J. K. Marshall,
Mander Raikes & Marshall,
Lombard House,
145 Great Charles Street,
Birmingham,
B3 3JR.

Mr. C. Martin,
L. G. Mouchel & Partners,
19 Gay Street,
Bath,
Avon.

Dr. R. G. Mein,
Department of Civil
Engineering,
Monash University,
Clayton,
Vic. 3168,
Australia.

Mr. M. Melanen,
Helsinki University of
Technology,
Department of Civil
Engineering,
Otakaari 1,
02150 Espoo 15,
Finland.

Mr. A. Mezer,
Balasha Jalon Consultants
& Engineers Ltd.,
P.O. Box 1727,
Haifa,
Israel.

Mr. D. Mills,
Anglian Water Authority,
Peterborough Sewage Division,
Aqua House,
London Road,
Peterborough, PE2 8AG.

Mr. A. D. Moen,
Kloakkplankontoret,
Baerun Kommune,
1300 Sandvika,
Norway.

Mr. G. H. Mortimer,
Department of Civil
Engineering,
Loughborough University of
Technology,
Loughborough, LE11 3TU.

Mr. I. A. Morton,
Tayside Regional Council,
Water Services Department,
Bullion House,
Invergowrie,
Dundee, DD2 5BB.

Prof. J. S. McNown,
Department of Civil
Engineering,
Kansas University,
Lawrence KS,66045,
U.S.A.

Mr. R. W. J. Neill,
Kirk, McClure & Morton,
2 Elmwood Avenue,
Belfast, BT9 6BA.

Mr. P. Nequest,
Severn Trent Water Authority,
Lower Trent Division,
Mapperley Hall,
Lucknow Avenue,
Nottingham.

Dr. A. T. Newman,
Wessex Water Authority,
P.O. Box 9,
King Square,
Bridgwater,
Somerset.

Mr. J. Niemczynowicz,
Department of Water
Resources Engineering,
Lund Institute of
Technology,
University of Lund,
Fack 725,
S-220 07 Lund,
Sweden.

Mr. B. B. Nussey,
Sheffield City Polytechnic,
Pond Street,
Sheffield.

Mr. K. Oren,
Norwegian Institute for
Water Research,
P.O. Box 333 Blindern,
Oslo 3,
Norway.

Mr. B. L. W. Over,
London Borough Council
of Bromley,
Town Hall,
Church Avenue,
Beckenham, BR3 1EX.

Mr. J. C. Packman,
Institute of Hydrology,
Wallingford,
Oxon.

Mr. J. Payne,
City of Birmingham
Engineer's Department,
Baskerville House,
Civic Centre,
Birmingham, B1 2NF.

Mr. A. R. Perks,
The Proctor & Redfern
Group,
75 Eglinton Avenue East,
Toronto,
Ontario,
M4P 1H3,
Canada.

Mr. R. W. Pethick,
Hydraulics Research
Station,
Howbery Park,
Wallingford,
Oxon, OX10 8BA.

Mr. A. J. Price,
John Taylor & Sons,
Artillery House,
Artillery Row,
London, SW1 1RY.

Dr. R. K. Price,
Hydraulics Research Station,
Howbery Park,
Wallingford,
Oxon, OX10 8BA.

Mr. P. Pumphrey,
Anglian Water Authority,
(Lincs. Sewerage Division),
Waterside House,
Waterside North,
Lincoln.

Mr. M. A. Puttock,
Sir Frederick Snow
& Partners,
Ross House,
144 Southwark Street,
London, SW1 0SZ.

Mr. Z. Radic,
Gradjevinski Fakultet,
Bulevar Revolucije 73,
11000 Beograd,
Yugoslavia.

Mr. M. Ramon Dominguez,
Engineer Instituto,
Hydraulic Section,
Unam,
Mexico 20, D.F.

Mr. B. A. Randerson,
Ove Arup Partnership,
13 Fitzroy Street,
London, W1P 6BQ.

Mr. P. K. Read,
B. C. Tonkin & Associates,
408 King William Street,
Adelaide,
S. Australia 5000.

Mr. C. Reynolds,
Rochford District Council,
Council Offices,
Hockley Road,
Rayleigh,
Essex.

Mr. L. W. Richardson,
Clay Pipe Development
Association Ltd.,
Drayton House,
30 Gordon Street,
London,
WC1H 0AN.

Mr. K. J. Riddell,
C. H. Dobbie & Partners,
154 - 160 Croydon Road,
Beckenham,
Kent,
BR3 4BX.

Mr. P. Rouse,
Engineering Department,
Dublin Corporation,
28 Castle Street,
Dublin 2,
Ireland.

Mr.N. A. Saltveit,
Oslo Water and
Sewerage Works,
Trondheimsveien 6,
Oslo 1,
Norway.

Mr. J. T. Sansby,
Anglian Water Authority,
16 Nene View,
Oundle,
Peterborough.

Dr. E. J. Sarginson,
Department of Civil
Engineering,
University of Sheffield,
Mappin Street,
Sheffield,
S1 3JD.

Mr. N. G. Semple,
Scottish Development
Department,
Engineering Division,
Pentland House,
Robbs Loan,
Edinburgh,
EH11 1TY.

Mr. K. C. Shaw,
Trafford Borough Council,
Birch House,
Talbot Road,
Old Trafford,
Manchester, M16 0GH.

Mr. F. Sieker,
Technical University
Hannover,
Institut für Wasserwirtschaft,
Callinstr. 32,
3000 Hannover,
West Germany.

Mr. H. Sigurdsson,
HNIT Ltd.,
Sidumuli 34,
Reykjavik,
Iceland.

Mr. K. Sivaloganathan,
Department of Civil
Engineering & Building,
Lanchester Polytechnic,
Priory Street,
Coventry, CV1 5FB.

Mr. J. N. Skeet,
Trevor Crocker & Partners,
Drive House,
323-339 London Road,
Mitcham,
Surrey.

Dr. R. W. Slater,
Environment Canada,
135 St. Clair Av. W.,
Toronto,
Ontario M4V 1P5,
Canada.

Mr. B. Smisson,
186 Redland Road,
Bristol, BS6 6YH.

Mr. R. P. M. Smisson,
Parry Froud and Snook,
5 Elton Road,
Clevedon,
Avon, BS21 7RE.

Mr. J. B. Smith,
North West Water,
Dawson House,
Great Sankey,
Warrington, WA5 3LW.

Mr. J. A. Smyth,
Office of Public Works,
Marine House,
3 Clanwilliam Place,
Lower Mount Street,
Dublin 2.

Mr. P. R. Solway,
Department of Civil
Engineering,
Faculty of Engineering,
University of
Newcastle-Upon-Tyne,
Newcastle-upon-Tyne,
Tyne & Wear.

Mr. G. T. Spicer,
Principal Engineer (Agency
Liaison),
Avon and Dorset Division,
Wessex Water Authority,
2 Nuffield Road,
Poole, BH17 7RL.

Mr. G. J. Stanley,
Anglian Water Authority,
Diploma House,
Grammar School Walk,
Huntingdon.

Mr. K. D. Staples,
J. D. & D. M. Watson,
Terriers House,
Amersham Road,
High Wycombe,
Bucks, HP13 5AJ.

Mr. J. R. Stipp,
Dewberry, Nealon & Davis,
8411 Arlington Boulevard,
Fairfax, V.A. 22030,
U.S.A.

Mr. J. T. Stone,
L. G. Mouchel & Partners,
19 Gay Street,
Bath.

Mr. W. A. Strong,
Northern Ireland Polytechnic,
Newtownabbey,
Co. Antrim,
BT36 8AP.

Mr. T. H. Stuart,
Tayside Regional Council,
Water Services Department,
Bullion House,
Invergowrie,
Dundee, DD2 5BB.

Mr. P. W. Styles,
Severn-Trent Water Authority,
Abelson House,
2297 Coventry Road,
Sheldon,
Birmingham, B26 3PU.

Mr. G. Svensson,
Dept. of Water Supply
and Sewerage,
Chalmers University of
Technology,
Fack,
S-402 20 Göteborg,
Sweden.

Mr. G. P. Tan,
Head,
Drainage Department,
Ministry of the Environment,
Princess House,
Alexandra Road,
Singapore 3.

Mr. J. D. Taylor,
John Dossor & Partners,
West Huntington Hall,
York.

Mr. P. Telford,
Severn Trent Water Authority,
Abelson House,
2297 Coventry Road,
Sheldon,
Birmingham, B26 3PU.

Dr. A. B. Templeman,
University of Liverpool,
Department of Civil
Engineering,
P.O. Box 147,
Liverpool, L69 3BX.

Mr. J. L. Thompson,
Binnie & Partners,
Artillery House,
Artillery Row,
London, SW1P 1RX.

Mr. P. K. Thorpe,
Severn Trent Water Authority,
Abelson House,
2297 Coventry Road,
Sheldon,
Birmingham, B26 3PU.

Mr. W. Topping,
Department of Sewerage,
Strathclyde Regional
Council,
78 Queen Street,
Glasgow.

Mr. A. Townson,
Telford Development Corp.,
Priorslee Hall,
Telford,
Salop., TF2 9NT.

Mr. C. Tucker,
Seven Trent Water Authority,
No. 10 Throckmorton,
Pershore,
Worcs.

Mr. A. G. Tuckley,
Cheltenham Borough Council,
Borough Engineer's Dept.,
Municipal Offices,
Promenade,
Cheltenham.

Mr. D. A. Tupper,
Alberta Environment,
Edmonton,
Alberta,
Canada, T6A 2H1.

Mr. J. A. van den Berg,
p.a. Ryksdienst voor de
Ijsselmeer polders,
Post Box 600,
Lelystad,
The Netherlands.

Mr. D. J. Van den Heever,
Scott & De Waal Incorporated,
Private Bag 7,
Savonwold,
2132,
South Africa.

Prof. A. Van de Beken,
Vrije Universiteit Brussel,
Lab. of Hydrology, VUB,
Pleinlaan 2,
Brussels,
B-1050,
Belgium.

Mr. W. Verworn,
Technische Universität
Hannover,
Institut für Wasserwirtschaft,
Callinstr. 32,
D-3000 Hannover 1,
West Germany.

Mr. M. F. Vincent,
Thames Water Authority,
Chiltern Division,
Maple Lodge,
Denham Way,
Rickmansworth,
Herts, WD3 2SQ.

Mr. G. A. Walters,
Liverpool University,
Department of Civil
Engineering,
P.O. Box 147,
Liverpool,
L69 3BX.

Mr. M. P. Wanielista,
Florida Technical University,
P.O. Box 25000,
Orlando,
Florida 32016,
U.S.A.

Mr. L. H. Watkins,
Transport and Road
Research Laboratory,
Crowthorne,
Berkshire.

Mr. I. L. Watts,
Town Hall,
St. Aldates,
Oxford, OX1 1BX.

Mr.M. Whiteland,
Thames Water,
Rivers House,
Crossness Works,
Abbey Wood,
London, SW2 9AQ.

Mr. K. A. Whittaker,
Newport Borough Council,
Civic Centre,
Newport,
Gwent.

Mr. J. B. M. Wiggers,
DHV - P.O. Box 18,
Laan 1914,
Amenfort,
Holland.

Mr. W. B. Wilkinson,
Water Research Centre,
P.O. Box 16,
Henley Road,
Medmenham,
Marlow, SL7 2HD.

Mr. D. R. Woods,
Constantia,
Rectory Road,
Steppingley,
Bedford.

Dr. D. E. Wright,
Sir William Halcrow &
Partners,
Vineyard House,
44 Brook Green,
London, W.6.

Mr. M. L. Yates,
Severn Trent Water Authority,
Abelson House,
2297 Coventry Road,
Sheldon,
Birmingham,
B26 3PU.

Dr. B. C. Yen,
Hydrosystems Laboratory,
Department of Civil
Engineering,
University of Illinois,
Urbana,
Illinois 61801,
U.S.A.

Mr. D. Young,
North West Water,
Dawson House,
Great Sankey,
Warrington,
WA5 3LW.

ERRATA

p.49, Fig. 4,   upside down
p.66 & p.67,   reversed
p.541, Fig. 2,   for '0.001' read '0.01'
p.543, Fig. 3,   for 'D = 0.4 m' read 'D = 0.4 mm'
p.580, Table 4,   for 'kg/km^2 year' read 'kg/ha year'
p.582, Fig. 1,   for '(kg COD/km^2 year)'read'(kg COD/ha year)'
p.583, Fig. 3,   for '(kg N/km^2 year)' read '(kg N/ha year)'
p.587 & p.588,   reversed
p.592, Fig. 3,   for 'Nitrate' read 'Ammonium'